BE23858

W9-DHV-867

Intestinal Lipid Metabolism

Intestinal Lipid Metabolism

Edited by

Charles M. Mansbach II
University of Tennessee College of Medicine
Memphis, Tennessee

Patrick Tso
University of Cincinnati College of Medicine
Cincinnati, Ohio

and

Arnis Kuksis
University of Toronto
Toronto, Ontario, Canada

Kluwer Academic / Plenum Publishers
New York, Boston, Dordrecht, London, Moscow

Library of Congress Cataloging-in-Publication Data

Intestinal lipid metabolism/edited by Charles M. Mansbach II, Patrick Tso, and Arnis Kuksis.
   p.   cm.
   Includes bibliographical references and index.
   ISBN 0-306-46241-9
      1. Lipids—Metabolism.   2. Intestinal absorption.   I. Mansbach, Charles M.   II. Tso,
Patrick.   III. Kuksis, Arnis.

QP751 .I685 2000
612.3'97—dc21

00-028732

Cover illustration courtesy of David H. Alpers and the Computer Graphics Center, Washington University School of Medicine. The figure and descriptive caption are located in this volume on page 331.

ISBN 0-306-46241-9

©2001 Kluwer Academic/Plenum Publishers, New York
233 Spring Street, New York, New York 10013

http://www.wkap.nl/

10 9 8 7 6 5 4 3 2 1

A C.I.P. record for this book is available from the Library of Congress

To our wives
May Lynn Mansbach, Weiyl Li, and Inese Kuksis;
and our mentors
Malcolm P. Tyor, Wilfred J. Simmonds, and J. M. R. Beveridge

Contributors

David H. Alpers Division of Gastroenterology, Department of Medicine, Washington University School of Medicine, St. Louis, Missouri 63110-1010

Shrikant Anant Department of Internal Medicine, Washington University School of Medicine, St. Louis, Missouri 63110

Tak Yee Aw Department of Molecular and Cellular Physiology, Louisiana State University Medical Center, Shreveport, Louisiana 71130-3932

Bruce R. Bistrian Department of Medicine, Beth Israel Deaconess Medical Center, Boston, Massachusetts 02215

Dennis D. Black Department of Pediatrics, University of Tennessee School of Medicine, Crippled Children's Foundation Research Center at Le Bonheur Children's Medical Center, Memphis, Tennessee 38103

Howard Brockman The Hormel Institute, University of Minnesota, Austin, Minnesota 55912

S. Canaan Laboratoire de Lipolyse Enzymatique, 31 Chemin Joseph-Aiguier, 13402 Marseille Cédex 20, France

Nicholas O. Davidson Department of Internal Medicine and Molecular Biology and Pharmacology, Washington University School of Medicine, St. Louis, Missouri 63110

Stephen J. DeMichele Strategic Discovery Research and Development, Ross Products Division, Abbott Laboratories, Columbus, Ohio 43215-1724

Takashi Doi Department of Pathology, University of Cincinnati College of Medicine, Cincinnati, Ohio 45267-0529

Liliane Dupuis Laboratoire de Lipolyse Enzymatique, 31 Chemin Joseph-Aiguier, 13402 Marseille Cédex 20, France

Rami Eliakim Hadassah Medical School, Jerusalem, Israel

Michael J. Engle Division of Gastroenterology, Department of Medicine, Washington University School of Medicine, St. Louis, Missouri 63110-1010

Charlotte Erlanson-Albertsson Department of Cell and Molecular Biology, Section for Molecular Signalling, University of Lund, S-221 00 Lund, Sweden

F. Jeffrey Field Department of Internal Medicine, University of Iowa College of Medicine, Iowa City, Iowa 52242

Reynold Homan Cardiovascular Therapeutics, Parke-Davis Pharmaceutical Research Division, Ann Arbor, Michigan 48105

Philip N. Howles Department of Pathology and Laboratory Medicine, University of Cincinnati College of Medicine, Cincinnati, Ohio 45267-0529

Carl-Erik Høy Department of Biochemistry and Nutrition, Technical University of Denmark, DK-2800 Lyngby, Denmark

David Y. Hui Department of Pathology and Laboratory Medicine, University of Cincinnati College of Medicine, Cincinnati, Ohio 45267-0529

Mahendra Kumar Jain Department of Chemistry and Biochemistry, University of Delaware, Newark, Delaware 19716

Arnis Kuksis Banting and Best Department of Medical Research, University of Toronto, Toronto M5G 1L6 Canada

Terry S. LeGrand Department of Respiratory Care, The University of Texas Health Science Center, San Antonio, Texas 78245

Richard Lehner Banting and Best Department of Medical Research, University of Toronto, Toronto M5G 1L6 Canada

Mark E. Lowe Washington University School of Medicine, Departments of Pediatrics and of Molecular Biology and Pharmacology, St. Louis, Missouri 63110

Charles M. Mansbach II Department of Medicine, Division of Gastroenterology, The University of Tennessee, Memphis, Tennessee 38163 and The Office of Research and Development Medical Research Service, Department of Veterans Affairs, Memphis, Tennessee 38104

Keith R. Martin National Institute of Environmental Health Sciences, Research Triangle Park, North Carolina 27709-2233

Mohsen Meydani Vascular Biology Program, Jean Mayer USDA Human Nutrition Research Center on Aging at Tufts University, Boston, Massachusetts 02111

Huiling Mu Department of Biochemistry and Nutrition, Technical University of Denmark, DK-2800 Lyngby, Denmark

M. Rivière Laboratoire de Lipolyse Enzymatique, 31 Chemin Joseph-Aiguier, 13402 Marseille Cédex 20, France

John B. Rodgers Department of Medicine, Albany Medical College, Albany, New York 12208

Judith Storch Department of Nutritional Sciences, Rutgers University, New Brunswick, New Jersey 08901-8525

Alan B. R. Thomson Cell and Molecular Biology Collaborative Network in Gastrointestinal Physiology, Nutrition and Metabolism Research Group, Division of Gastroenterology, Department of Medicine, University of Alberta, Edmonton, Alberta T6G 2C2 Canada

Patrick Tso Department of Pathology, University of Cincinnati College of Medicine, Cincinnati, Ohio 45267-0529

R. Verger Laboratoire de Lipolyse Enzymatique, 31 Chemin Joseph-Aiguier, 13402 Marseille Cédex 20, France

Henkjan J. Verkade Groningen University Institute for Drug Exploration (GUIDE), Department of Pediatrics, Academic Hospital, 9700 RB Groningen, The Netherlands

John R. Wetterau Department of Metabolic Research, Bristol-Myers Squibb, Princeton, New Jersey 08543

C. Wicker-Planquart Laboratoire de Lipolyse Enzymatique, 31 Chemin Joseph-Aiguier, 13402 Marseille Cédex 20, France

Gary Wild Department of Medicine, Division of Gastroenterology, Department of Anatomy and Cell Biology, McGill University, Montreal H3A 2T5 Canada

Preface

This book was stimulated by the enthusiasm shown by attendees at the meetings in Saxon River, VT, sponsored by the Federation of American Societies for Experimental Biology (FASEB), on the subject of the intestinal processing of lipids. When these meetings were first started in 1990, the original organizers, two of whom are editors of this volume (CMM and PT), had two major goals. The first was to bring together a diverse group of investigators who had the common goal of gaining a better understanding of how the intestine absorbs lipids. The second was to stimulate the interest of younger individuals whom we wished to recruit into what we believed was an exciting and fruitful area of research. Since that time, the field has opened up considerably with new questions being asked and new answers obtained, suggesting that our original goals for the meetings were being met.

In the same spirit, it occurred to us that there has not been a recent book that draws together much of the information available concerning how the intestine processes lipids. This book is intended to reach investigators with an interest in this area and their pre- and postdoctoral students. The chapters are written by individuals who have a long-term interest in the areas about which they write, and many have been speakers at the subsequent FASEB conferences that have followed on the first. We are aware that many of the areas discussed are fast moving and that the material presented is but a snapshot of the subjects at the time the chapters were written. Nevertheless, we think the book will be a useful and up-to-date primer to initiate an understanding of this extremely complex and important area of biology.

It is well recognized that absorbed lipids contribute greatly to the development of atherosclerotic heart disease in those with a susceptible phenotype, especially those that eat lipid-laden diets. This fact makes this volume of interest to more than intestinal physiologists and biochemists, as is illustrated by the finding that chylomicrons, on their metabolism, contribute to HDL formation.

As in any area of biology, the work and questions of today are built upon the ideas of yesterday. This is particularly true of lipid absorption, where many of the ideas that were first promulgated in the late 1960s are just as true today as they were then. Indeed, the author of an important review of the subject in 1964 told one of the co-editors of this volume a few years ago that he had gotten out of the field of lipid absorption because he had said all that was ever going to be said about intestinal lipid absorption in his review! The editors think he was incorrect.

This book includes references to those "ancient texts" which are well beyond the reach of most computerized searches and thus do not show up when data banks are probed. At the other end of the spectrum, many entire areas that are discussed here were not even thought of in the late 1960s, many techniques that enable experiments to more closely answer important questions posed even then were simply not available to workers in those days but are in use now, and finally, many questions that seem so vital today could not be answered then because the conceptual framework on which the questions are based was not yet developed.

We hope that the readers of this book will be as stimulated as we have been by the wealth of questions that still remain to be answered in this ever-expanding, significant field of biology. Our ultimate hope is that the book will generate enough interest in this area of investigation to answer some of the many questions that are raised in the chapters that follow.

Contents

Chapter 4

Pancreatic Lipase: Physiological Studies 61
Howard Brockman

Chapter 5

**Biology, Pathology, and Interfacial Enzymology of Pancreatic
Phospholipase A_2** 81
Reynold Homan and Mahendra Kumar Jain

Chapter 11

Intestinal Synthesis of Triacylglycerols 185
Arnis Kuksis and Richard Lehner

Chapter 12

Triacylglycerol Movement in Enterocytes 215
Charles M. Mansbach II

Chapter 13

Regulation of Intestinal Cholesterol Metabolism 235

F. Jeffrey Field

Chapter 14

Regulation of Intestinal Apolipoprotein Gene Expression 263

Dennis D. Black

Chapter 15

**Recent Progress in the Study of Intestinal Apolipoprotein
B Gene Expression** 295
Nicholas O. Davidson and Shrikant Anant

Chapter 16

The Role of Apolipoprotein A-IV as a Satiety Factor 307
Patrick Tso and Takashi Doi

CHAPTER 1

Biophysics of Intestinal Luminal Lipids

HENKJAN J. VERKADE and PATRICK TSO

1. Introduction

By definition, lipids are more soluble in apolar, organic fluids than in aqueous environments. This feature underlies the compartmentalization in cells and in organisms in different metabolic processes. Their apolar nature engenders specific demands to enable vectorial transport of lipids through aqueous environments. Interestingly, the mechanisms by which lipids are transported from one compartment in the organism (organ, cell, subcellular organelle) to another are diverse and depend on both the compartments involved and the lipid species. In this chapter the physiological transport moieties for lipids inside the intestinal lumen are discussed. The influx of lipids into the intestinal lumen is accounted for by the diet, by bile secretion, and by sloughing of cells from the intestinal mucosa. Quantitatively the most important lipid species entering the intestine is triacylglycerol, originating from the diet. Other species include phospholipids and cholesterol (diet, bile, cells), plant sterols (diet), lipid-soluble vitamins, and other trace lipids (steroids, PCBs, organic pollutants).

The efficacy of absorption of dietary triacylglycerols under physiological conditions in human adults is above 95% (Carey *et al.*, 1983). As discussed elsewhere in this book, quantitative uptake of triacylglycerol lipids can only be achieved after at least partial hydrolysis of fatty acids from the glycerol molecule. Hydrolysis of triacylglycerols results in the formation of metabolites that have a higher tendency to interact with water than their parent compounds. The physicochemical characterization of the different lipids with respect to their spontaneous behaviour in aqueous environments is reviewed in section 2. The various physicochemical, intraluminal phases of lipid absorption are then discussed in sequential order: emulsification in the stomach and in the duodenum and the behavior of lipids at the oil–water interface, in the aqueous phase and in the unstirred water layer lining the intestinal mucosa (section 3).

HENKJAN J. VERKADE • Groningen University Institute for Drug Exploration (GUIDE), Department of Pediatrics, Academic Hospital, 9700 RB Groningen, The Netherlands.   PATRICK TSO • Department of Pathology, University of Cincinnati College of Medicine, Cincinnati, Ohio 45267-0529.
Intestinal Lipid Metabolism, edited by Charles M. Mansbach II *et al.,* Kluwer Academic/Plenum Publishers, 2001.

The available insights in the physicochemical events under physiological conditions have been challenged by studies under experimental or pathophysiological conditions in which the absorption efficacy of dietary lipids was, according to available paradigms of solubilization, unexpectedly high. These observations and the still unresolved issues on intestinal solubilization of lipids are highlighted, indicating potential directions of future research on these topics (section 4).

2. Physicochemistry of Different Lipids

Small and Carey classified the lipids on the basis of their tendency to interact with water (Carey and Small, 1970; Small, 1970). Figure 1 shows the classification in schematic fashion. The group of nonpolar lipids do not interact with water to an appreciable extent, but rather will be present in an aqueous environment either as a lens of oil or as a crystal (Carey and Small, 1972). Examples of nonpolar lipids are paraffin oil and benzpyrenes. Most nonpolar lipids are not biological. However, certain (biological) sterol esters, such as cholesteryl oleate and cholesteryl linoleate, behave like nonpolar lipids, despite the presence of polar groups. The group of polar lipids are further subdivided into three classes (Carey and Small, 1970; Small, 1970; Fig. 1). Class I polar lipids are also known as insoluble nonswelling amphiphiles and are exemplified by triacylglycerols, diacylglycerols, protonated long-chain fatty acids, cholesterol, and the fat-soluble vitamins A, D, E, and K. These lipids have a very limited solubility in water, but they do form a stable monolayer at the water surface. Class II polar lipids are insoluble swelling amphiphiles, such as phospholipids, monoacylglycerols, and fatty acid soaps. In addition to forming a monolayer on top of an aqueous solution, these lipids can penetrate into the bulk (aqueous) phase in the form of "liquid crystals." Liquid crystals are characterized by lipid bilayers, interdigitated by hydrated planes of polar head groups. Class III polar lipids are the so-called micelle-forming or soluble amphiphiles. Some of these lipids, called class III A lipids, such as sodium or potassium salts of long-chain fatty acids and lysophosphatidylcholine, can also form liquid crystals under certain conditions. Other soluble amphiphiles, class III B lipids, such as conjugated and unconjugated bile salts and sulfated bile alcohols, only form micelles and not liquid crystals. Monolayers of class III polar lipids are unstable in nature, due to their solubility in water.

The physicochemistry of the intraluminal phase of lipid absorption can be characterized by two main processes. First, lipolysis (discussed elsewhere in this volume) results in the shift of relatively apolar lipids toward increased polarity. For example, the sn-1- and sn-3-acyl chains of dietary triacylglycerol (a class I polar lipid) are hydrolyzed by pancreatic lipase, resulting in two free fatty acids (class II) and monoacylglycerol (class III). Second, the secretion of bile into the intestine results in the influx of significant amounts of bile salts (class III polar lipids) and phospholipids (class II polar lipids, swelling amphiphiles). Above a certain critical micellar concentration, class III polar lipids self-aggregate into micelles in an aqueous environment, as stated previously. Another feature of these class III micelles is equally important for the intraluminal phases of lipid absorption: micelles composed of class III lipids can incorporate class II and, to a limited extent, even class I polar lipids into their aggregates, leading to the formation of so-called mixed micelles. Through the formation of mixed micelles the concentration of class I and II polar lipids in the aqeous phase can be increased 100- to 1000-fold (Hofmann, 1976).

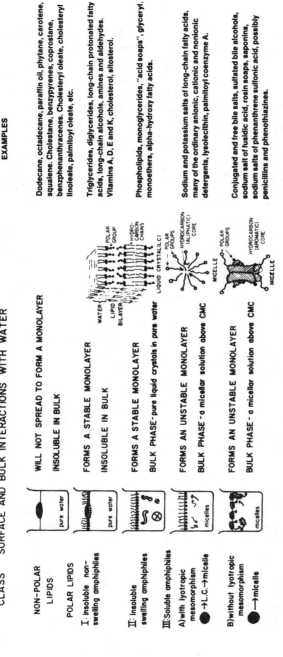

Figure 1. Classification of biologically active lipids. On the left the major lipid classes are listed. The center depicts the physiological state of the lipid in the bulk aqueous system (i.e., within water) and on the surface of water. The insets on the right represent the conception of the arrangement of the molecules in: a lamellar liquid crystal ("myelin figures") formed by class II polar lipids (insoluble swelling amphiphiles, e.g., lecithin) in an excess of water; a spherical micelle formed by class IIIA lipids (soluble amphiphiles, e.g., typical detergents); the micelle formed by class IIIB lipids (soluble amphiphiles, in this case the bile salts). Examples of lipids typical of each class are given in the column on the right. Figure and legend are reprinted from the American Journal of Medicine, Volume 49, Carey, M. C., and Small, D. M. The Characteristics of Mixed Micellar Solutions with Particular Reference to Bile, pp. 590–608, 1970, with the kind permission of the copyright owner, Excerpta Medica Inc.

3. Physicochemistry of Intraluminal Phases of Lipid Absorption

3.1. Emulsification in the Stomach

Triacylglycerols constitute the bulk (approx. 95%) of lipids entering the intestinal lumen (Carey and Hernell, 1992). Because triacylglycerols are class I polar lipids (insoluble, nonswelling), a two-phase system is present at the initial stages of lipid absorption, namely an oil phase and an aqueous (water) phase. As stated previously, conversion into more polar class II and III lipids is needed to allow significant solubilization of the dietary lipids into the aqueous phase of the small intestinal lumen. The enzymes responsible for the hydrolysis of dietary triacylglycerols (gastric lipase, colipase-dependent lipase) express the highest lipolytic activity at the oil–water interface (Carriere *et al.*, 1995). Indeed, their rate of lipolysis is strongly related to the surface area of the oil–water interface, as well as to its physicochemical properties (Schmit *et al.*, 1996; Dahim and Brockman, 1998). Under physiological conditions, the area of the oil–water interface of dietary lipids is greatly expanded by the process called emulsification. The oil droplet or lipid particles (the liquid or solid state depends on the transition temperature of the constituent lipids) are transferred into a fine, stable oil-in-water emulsion. Emulsification is partly obtained by shear forces in the stomach: the strong muscular contractions of the stomach cause its contents to be squirted back into the corpus as well as toward the duodenum (Carey and Hernell, 1992). The principle of forcing a crude emulsion under high pressure through a narrow opening is also used industrially to obtain fine emulsions. However, a fine emulsion of pure triacylglycerols in water would not be stable and readily coalesce, unless mechanical energy is delivered continuously to the emulsion. In physicochemical terms, the energy state of a fine oil-in-water emulsion of triacylglycerols is higher than that of a crude emulsion. To prevent a fine emulsion from (re-)coalescing, the emulsion needs stabilization. The physicochemistry of stabilization involves the inclusion of molecules with an amphiphilic character at the oil–water interface; in biophysical terms, relatively apolar lipids (nonpolar or class I polar lipids) are to be shed from the aqueous phase by more polar lipids such as class II and III polar lipids. In the stomach, candidate emulsifiers are class II polar lipids, such as phospholipids from the diet. Phospholipids orient themselves with their hydrophobic acyl chains toward the hydrophobic core of the triacylglycerol/emulsion particle, and their hydrophilic head group toward the aqueous phase (Borgström *et al.*, 1985). Phospholipids, however, are not the only substances that can function as stabilizers of the emulsions; denatured proteins and polysaccharides can also act to stabilize the emulsion in the acidic environment of the stomach (Borgström and Erlanson, 1978b; Carey and Hernell, 1992c). Apart from these mechanical forces, emulsification is aided by partial lipolysis of dietary triacylglycerols by gastric lipase in the stomach (see chapter 2). As detailed elsewhere in this volume, gastric lipase can hydrolyze up to 20% of dietary triacylglycerols in the stomach (Borgström *et al.*, 1957; Gargouri *et al.*, 1986a). Because gastric lipase is a "true" lipase, implying that it expresses its highest activity at the oil–water interface (Nury *et al.*, 1987), the hydrolytic activity of the enzyme is increased by the mechanical emulsification due to the muscular contractions of the stomach. The substrates for gastric lipase are dietary triacylglycerols, which are partially hydrolyzed into free fatty acids and sn-1,2-diacylglycerols (Gargouri *et al.*, 1986b; Gargouri *et al.*, 1989). Long-chain diacylglycerols, which belong to the class I polar lipids, will predominantly distribute themselves into the oil phase (emulsion cores) of the emulsion. In the low pH environment of the stomach, the fatty acids will be protonated and will also concentrate themselves in the lipid cores of the emulsion (class I polar lipids).

The emulsification by shear forces and the partial lipolysis by gastric lipase normally results in the presence of a fine, stable oil-in-water emulsion in the proximal part of the duodenum. The particle sizes of this emulsion have been reported to be smaller than 500 nm (Frazer, 1946). The next phase in lipid absorption is characterized by the secretion of pancreatic fluid and bile into the duodenum. It is beyond the scope of this review to discuss the physiological signal for pancreas secretion and gallbladder contraction, which is mediated by the uptake of free fatty acids.

3.2. Emulsification in the Duodenum

3.2.1. Pancreatic Bicarbonate Secretion

The secretion of the bicarbonate-rich pancreatic fluid induces a rapid neutralization of the pH in the duodenum, with profound effects on the emusified lipids present in the duodenum. The fatty acids that had been hydrolyzed from dietary triacylglycerols by gastric lipase had diffused to the core of the emulsion particles in protonated form under the acidic conditions in the stomach (class I polar lipids). Yet, upon the neutralization of the pH in the proximal part of the duodenum, the long-chain fatty acids become partially ionized, causing an increase in their polarity (fatty acid soap, class II polar lipids). The increased fatty acid polarity induces a shift of these molecules from the lipid core of the emulsion toward the oil–water interface (Shankland, 1970; Saleeb *et al.*, 1975; Carey and Hernell, 1992). The increased content of polar lipids at the oil–water interface acts not only to prevent coalescence of the oil-in-water emulsion; it also induces the transition of relatively large emulsion particles into smaller ones (Frazer *et al.*, 1964). At this stage, the surface of the emulsion particles is covered with dietary phospholipids, partially hydrogenated fatty acids ("soaps"), (partially digested) amphipathic proteins, and traces of diacylglycerols and to a lesser extent triacylglycerols. The adsorbance of pancreatic lipase and colipase to the oil–water interface (Borgström, 1977) and their interaction with biliary bile salts and lipids (Borgström, 1980; Borgström and Erlanson-Albertsson, 1982) is discussed elsewhere in this volume (chapters 3 and 4).

3.2.2. Bile Secretion

Finally, the process of emulsification is aided by the secretion of bile into the intestine. Apart from the class I polar lipid cholesterol, human gallbladder bile contains significant amounts of phosphatidylcholine (class II polar lipid, concentration 20–40 mM) and of bile salts (class III polar lipid, concentration 100–200 mM; Hofmann, 1990). The physicochemical state of these lipids in bile largely depends on their relative and total concentrations. The lipid concentration of gallbladder bile is approximately five times higher than that of hepatic bile. Duodenal bile salt and phospholipid concentrations during fat digestion in healthy adults were found to be around 15 mM and 5 mM, respectively (Hernell *et al.*, 1990). At the initial stages of bile formation, canalicular bile salts act on the external leaflet of the bile canalicular membrane, leading to the secretion into the bile of phospholipids (mainly phosphatidylcholine) and cholesterol (Verkade *et al.*, 1995; Verkade, Havinga, *et al.,* 1995). Although the actual molecular processes have not yet been elucidated, available data strongly indicate that the phospholipids and cholesterol are, at least partly, secreted in liquid crystalline form as unilamellar vesicles (Cohen *et al.,* 1989; Cohen and Carey, 1990; Cohen *et al.,* 1992; Crawford *et al.,* 1995; Mockel *et al.,* 1995; Luk *et al.,* 1997). After secretion into the bile of the individual components (bile salts, phospholipids, cholesterol), a transfer of the lipids toward a physicochemical equilibrium state starts. The equilibrium state can contain bile salt monomers, simple bile salt micelles, mixed micelles (bile salts, phospholipids, cholesterol), phospholipid–cholesterol vesicles, and cholesterol crystals, depending on the relative contribution and total concentration of the lipids. Fundamental studies by the groups of Carey, Donovan, and others (Donovan and Carey, 1990; Donovan *et al.,* 1991; Carey and Lamont, 1992) have characterized the relationships beween the concentrations of biliary bile salts and lipids on the one hand and their physicochemical states on the other hand. It is not likely that the physicochemical equilibrium has always been reached upon entry of bile in the duodenum, where the bile mixes with pancreatic juice and with the nutrients from the stomach.

As is discussed in the following section, biliary bile salts and phosphatidylcholine play significant roles in the micellarization of lipolysis breakdown products. Even in the absence of actual lipolysis, however, biliary compounds interfere with the oil-in-water emulsion present in the duodenum. In general, biliary phospholipids act to stabilize the emulsion, based on their physicochemical chararcteristics as discussed previously. Bile salts, however, tend to destabilize the emulsion, partially by the incorporation of class II polar lipids as phospholipids into mixed micelles (Borgström *et al.,* 1985). The bile salt-induced "destabilization" of the emulsion affects the physicochemical characteristics of the substrate interface, which leads to the increased availability of triacylglycerols and diacylglycerols for lipolysis by the pancreatic lipase–colipase complex (Borgström and Erlanson, 1978a; Borgström *et al.,* 1985).

3.3. Physicochemistry of Intestinal Lipolysis at the Oil–Water Interface

In 1979, Patton and Carey reported on the visualization of lipid digestion *in vitro.* Upon mixing olive oil droplets, bile salts, and human pancreatic lipase and colipase, light microscopic techniques showed the formation of a liquid crystalline product phase, which was demonstrated to be composed of calcium and fatty acids. Later, a clear "viscous isotropic" phase was observed, which appeared to contain monoglycerides and protonated fatty acids (Patton and Carey). The two phases could be solubilized by an excess of bile salts into a micellar phase. Other studies on the physicochemistry of lipids during intestinal lipolysis frequently applied either ultracentrifugation (Hofmann and Borgström, 1964) or ultrafiltration (Porter and Saunders, 1971; Mansbach, II, *et al.,* 1975) with or without inhibition of lipolysis by initial heating of duodenal aspirates at 70 °C. Since these reports, it has become appreciated that these techniques can greatly affect the distributions of individual bile salts and lipids over the various phases present (Donovan and Carey, 1990). It should also be noted that these techniques do not provide insights into the actual processes occurring at the oil–water interface.

Hernell, Staggers, and Carey investigated the physicochemical behavior of dietary lipids during fat digestion in model systems and in healthy human adults (Hernell *et al.,* 1990; Staggers *et al.,* 1990). A cocktail of lipase inhibitors was applied to inactivate lipases in duodenal aspirates obtained from healthy humans during fat digestion (Hernell *et al.*, 1990). Based on model studies, 2.5 to 4% of emulsified triacylglycerols is present at the oil–water interface (Miller and Small, 1982). These triacylglyerols act as the substrate for pancreatic lipase. The adsorption of pancreatic lipase to the emulsion surface, including its anchoring by pancreatic colipase, is beyond the scope of this chapter. For this topic, the reader is referred to some key publications (Borgström, 1977, 1980; Lairon *et al.*, 1980; Borgström and Erlanson-Albertsson, 1982; Schmit *et al.*, 1996; Dahim and Brockman, 1998). Upon lipolysis, the metabolized triacylglycerols at the surface are rapidly replenished by triacylglycerols from the emulsion core. At the conditions persent in the duodenum during lipid digestion, the lipolytic products, fatty acids and monoacylglyerols, behave as class II polar lipids, implying that they can form liquid crystals in the bulk phase of the aqueous phase. Under physiological conditions, the pancreatic lipase secretion is abundant: it has been calculated that a 100-fold excess of enzyme is available for the lipolysis of common portions of dietary lipids (Patton, 1981; Borgström, 1985). At least during the initial stages, the rate of lipolysis significantly exceeds the capacity to incorporate the resulting fatty acids and monoacylglycerols into mixed bile salt micelles (Carey and Hernell, 1992). Rather, these class II polar lipids accumulate at their site of origin, the oil–water interface, and orient themselves into a liquid crystalline phase in the form of lipid–water lamellae, reminiscent of phospholipid bilayers (Fig. 2). This liquid crystalline phase corresponds to the light microscopic observations of the initial events of lipid digestion of Patton and Carey (Patton and Carey, 1979). It is likely that this liquid crystalline phase also incorporates surface components from the original oil–water interface, such as unesterified cholesterol, phospholipids, and bile salts.

The next phase has long been considered to be characterized solely by micellar solubilization of the products of lipolysis, mainly by biliary bile salts (Porter and Saunders, 1971; Mansbach, II, *et al.*, 1975; Thomson *et al.*, 1989). The concept of an oil phase from which hydrolyzed lipids are transferred to a micellar phase as the transport vehicle toward the intestinal mucosa has become known as the Hofmann–Borgström hypothesis, based on their original 1964 paper (Hofmann and Borgström, 1964). In this study, duodenal aspirates were analyzed after ultracentrifugation. A more detailed insight into these processes was obtained using techniques whose interference with the distribution of lipids in the samples was considerably less than that of ultracentrifugation, such as light microscopy (see the previous paragraph), quasi-elastic light scattering, and freeze fracture electron microscopy, employing ultrajet freezing of the sample (Rigler *et al.*, 1986; Hernell *et al.*, 1990; Staggers *et al.*, 1990). Applying these techniques, it became apparent from studies in model bile and in human bile that, under physiological conditions, phospholipids and cholesterol could be present stably in the vesicular form (liquid crystals; Mazer *et al.*, 1980; Mazer and Carey, 1983; Somjen and Gilat, 1983; Gilat and Somjen, 1996). The conditions in which the vesicular form was stably present included relatively low bile salt concentrations and high total lipid concentrations. Postprandial bile salt concentration in the duodenum during fat absorption in humans does not exceed 10 to 15 mM (Hernell *et al.*, 1990). Staggers *et al.* theorized that the condensed phase diagram for typical physiological, duodenal conditions was similar to that of a dilute model bile (Staggers *et al.*, 1990). Given the high lipid concentrations after a fat meal, the presence of a vesicular phase at this stage of lipid absorption was likely. Using freeze fracture electron microscopy, Rigler *et al.* visualized digestion *in vitro* and *in vivo* in the killifish using freeze fracture techniques (Honkanen *et al.*, 1985;

Rigler *et al.*, 1986). At the initial stages of lipolysis of olive oil droplets *in vitro*, spherical vesicles emerged at the oil–water interface (their diameters mostly between 11 and 30 nm) in the presence of bile salts (4 mM; Rigler *et al.*, 1986). Upon further lipolysis, the ratio of the fatty acid and monoacylglycerol concentration to the bile salt concentration increased above 1, which was associated with the appearance of lamellar product phases at the surface of the oil droplets. At 30 min after the initiation of lipolysis, multilamellar vesicles could be observed with diameters between 50 and 250 nm (Rigler *et al.*, 1986). The lipid product phases in the killifish intestine after a 10% fat meal were very similar to those observed in the *in vitro* models (Honkanen *et al.*, 1985; Rigler *et al.*, 1986). Rigler *et al.* concluded that the multilayered lipid phase corresponds to the "viscous isotropic" phase observed by light microscopy (Rigler *et al.*, 1986), based on its ultrastructural morphology and on its characteristic that it could be dissolved by high concentrations of bile salt. Reversed micellar L2 phases at the vicinity of lamellar phases covering the oil droplet were not observed by the freeze fracture technique nor by later studies employing quasi-elastic light scattering techniques (Hernell *et al.*, 1990), implying that these phases are either not present or may be transitory during lipolysis.

Figure 2 (obtained from Rigler *et al.*, 1986) shows a schematic summary of the morphologies during lipolysis as derived from the studies described previously. The products of lipolysis (fatty acids, monoacylglycerols) are initially present in small unilamellar vesicles. When the concentration of fatty acids and of monoacylglycerols (class II and IIIA polar lipids) significantly exceeds the bile salt concentration, a liquid crystalline phase is formed, characterized by multiple lipid layers covering the lipolyzed emulsion. These multilayers can "bud off" into multilamellar vesicles, which thereafter can be dispersed into mixed micelles if unsaturated micelles are available.

3.4. Physicochemistry of Lipolytic End-Products in the Aqueous Phase

According to the Hofmann-Borgström hypothesis (Hofmann and Borgström, 1964; Hofmann, 1976), lipolytic products are present in the intestinal aqueous phase as monomers (in very low concentrations) or as mixed micelles, consisting of bile salts, phospholipids, cholesterol, fatty acids and monoacylglycerols (Mansbach, II, *et al.*, 1975). Since 1980, the presence of other lipid phases during lipid digestion was demonstrated by various techniques, as discussed previously. Yet, incorporation of lipolytic end products into mixed micelles still stands as a prominent and highly effective means of lipid transport through the aqueous phase of the intestinal lumen. Micellar solubilization requires the presence of at least one class III polar lipid, such as bile salt (Carey and Small, 1970). Bile salt micelles can incorporate into their micellar stucture class I and/or class II polar lipids to a limited extent. In model systems mimicking the duodenal environment with pH between 5.5 and 7.5, fatty acids are predominantly present in the form of fatty acid soaps (class II polar lipids), a 1:1 mixture of protonated fatty acid and fatty acid sodium salts (Staggers *et al.*, 1990). Carey and Small reviewed the physicochemistry and nomenclature of the various possible combinations of mixed micelles, depending on the inclusion of an apolar or class I polar lipid, a class II polar lipid, or another class III polar lipid (Carey and Small, 1970, 1972). Hofmann estimated the physiological effect of incorporating lipolytic end products into bile-salt-containing mixed micelles by calculating the diffusive flux for a fatty acid as a monomer as opposed to a fatty acid in a micellar solution (Hofmann, 1976). The quanti-

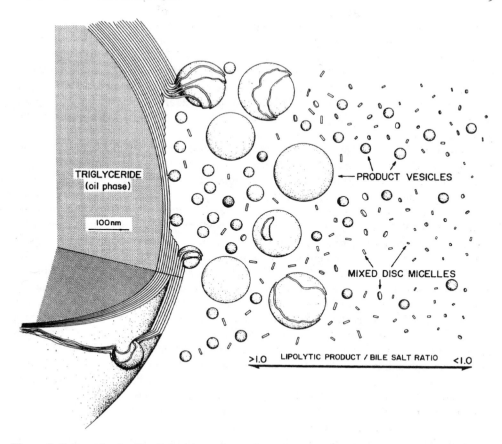

Figure 2. Schematic visualization of the vesiculation and micellar solubilization of lipolyzed triacylglyceols in the intestinal lumen during fat digestion. Upon lipolysis at the oil–water interface, a multilamellar phase is formed, which then transgresses from multilamellar vesicles to mixed disc micelles, under the influence of bile components (bile salts, phospholipids). Figure is reproduced from the Journal of Lipid Research, Volume 27, Rigler, M. W., Honkanen, R. E., and Patton, J. S. Visualization by freeze fracture, in vitro and in vivo, of the products of fat digestion. pp. 836–57, 1986, with the kind permission of the copyright owner, Lipid Research Inc.

ty of diffusive flux equals the product of the aqueous concentration of fatty acids and the diffusion coefficient. The diffusion coefficient of monomeric fatty acids is five- to sixfold higher than for the considerably larger mixed micelles. In a micellar solution, however, the aqueous concentrations of fatty acids is 100- to 1000-fold higher than in monomeric solution, resulting in an estimated 150-fold increase in quantitative diffusive flux for micelles versus monomers (Hofmann, 1976). Westergaard and Dietschy studied the effects of fatty acid incorporation into mixed micelles on lipid absorption in rabbit intestinal mucosa disks (Westergaard and Dietschy, 1974, 1976). The results of these studies were compatible with a model in which incorporation of fatty acids in mixed micelles serves to overcome efficiently the diffusional barrier of the aqueous phase between the site of lipolysis and the mucosal membrane. Yet, the actual uptake rate appeared proportional to the aqueous monomeric concentration of the fatty acid involved (Westergaard and Dietschy, 1976). These observations indicate that the mixed micelles function as a reservoir of lipolytic end products, particularly at the vicinity of the apical mucosal membrane. According to this model, upon the uptake of monomeric fatty acids from the solute phase, a rapid replenishment takes place through the transition of fatty acids (monoacylglycerols) from mixed micellar to the monomeric phase.

Although the role of micellarization of fatty acids and monoacylglycerols is still being appreciated along the lines of the Hofmann–Borgström hypothesis and the studies described previously, several observations have indicated that the actual processes may involve a higher degree of complexity. First, lipid absorption under bile-deficient conditions, such as in bile-fistula patients (Porter *et al.*, 1971) and in rats with chronic bile diversion (Minich *et al.*, 1997), is rather well maintained (to be discussed). Second, studies by Carey and colleagues indicated large individual variations in the relative lipid compositions of "subphases," obtained by ultracentrifugation of duodenal aspirates of healthy adults after lipid ingestion (Hernell *et al.*, 1990). Relative lipid compositions in different ultracentrifugation isolates indicated the presence of two lipid phases, as derived from plotting their relative lipid compositions in ternary phase diagrams. The two phases included mixed micelles of bile salts, hydrogenated fatty acids, fatty acid ions, monoacylglycerols, phospholipids and cholesterol (mean diameter <4 nm), and unilamellar ("liquid crystalline") vesicles composed of the same lipids (mean diameter 20–60 nm; Hernell *et al.*, 1990). Upon reviewing available data in the literature on "micellar" subphases of aqueous duodenal contents in healthy humans, Hernell *et al.* consistently plotted relative lipid compositions in the two-phase zone (micelles plus vesicles) of the phase diagram (Hernell *et al.*, 1990). Similar to the microscopic evaluations, the two-phase model with vesicles and lipids appeared frequently not to be in equilibrium condition; rather, the vesicles tended to dissolve spontaneously into mixed micelles (Hernell *et al.*, 1990). The dynamics of these processes *in vivo* have not been well characterized; in particular, it has not been elucidated whether the equilibrium conditions are ever being reached under the continuously changing conditions in the intestine (bile secretion, lipid and fluid absorption, lipolysis). It is also unclear what the potential role of vesicles is with respect to replenishment of the monomeric fatty acids after actual translocation of the latter across the apical mucosal membrane. The significantly larger diameters of the vesicles in comparison with the micelles certainly limits their diffusional capacity. However, in the intestine of the killifish, Rigler did report the presence of vesicles even up to the close vicinity of the villi of the intestinal mucosa (Honkanen *et al.*, 1985; Rigler *et al.*, 1986). Additionally, using small angle X ray (SAXS) and neutron scattering (SANS) techniques, Hjelm *et al.* described the presence of rodlike mixed micelles under specific conditions in model systems of intestinal contents (Hjelm *et al.*, 1995). In monoolein–cholylglycine mixtures, at relatively low total lipid concentrations, vesicles coexisted with rodlike forms which most likely were composed of stacked mixed micelles in which the bile salts and monoacylglycerol molecules were radially oriented ("radial-shell model"; Hjelm *et al.*, 1995). Similar results were reported on mixtures of bile salt and lecithin (Hjelm *et al.*, 1995). Although the concentration of the individual components as well as the conditions chosen in the model system are close to physiological levels, no firm proof of the (transient) existence of rodlike mixed micelles during intestinal lipid digestion has become available.

3.5. Physicochemistry of Solubilization in the Unstirred Water Layer

The molecular mechanism of translocation is still a matter of scientific debate, in which strong indications in support for its passive nature (Tso and Simmonds, 1977; Thomson *et al.,* 1989) have been challenged by reports from the groups of Stremmel *et al.* (Stremmel and Hofmann, 1990; Schoeller *et al.,* 1995; Fitscher *et al.,* 1996) and Hauser *et al.* (Compassi *et al.,* 1995; Schulthess *et al.,* 1996; Boffelli, Compassi, *et al.,* 1997; Boffelli, Weber, *et al.,* 1997), suggesting active transport involving specific proteins in the mucosal membrane. Yet, it is generally accepted that fatty acids translocate across the apical membrane of the mucosal cells in monomeric form (Wilson *et al.,* 1971; Tso and Simmonds, 1977). Wilson *et al.* provided strong indications that the actual translocation process (which, according to their studies, is purely passive) was not predominantly rate limited by the apical membrane of the mucosal cells (Wilson *et al.,* 1971). Instead, an "unstirred water layer" was proposed as the most important rate-limiting barrier for absorption of long chain fatty acids. The unstirred water layer has been described by Borgström as "an infinite number of water lamellae arranged in parallel with the enterocyte membrane. The closer the water lamellae are situated to the enterocyte membrane, the lower the relative rate of stirring is" (Borgström *et al.,* 1985, p. 412). The concept of an unstirred water layer as a transport barrier for especially long-chain fatty acids has been upheld since the original studies by Wilson *et al.* (1971). The actual size of the unstirred water layer was originally calculated to be 100 to 500 µm (Westergaard and Dietschy, 1974, 1976). More recent studies by Levitt and coworkers have demonstrated that *in vivo* the luminal contents are efficiently stirred due to the intestinal motility, leading them to estimate the actual size of the unstirred water layer between 35 to 50 µm (Levitt *et al.,* 1988; Levitt, Furne, *et al.,* 1992; Levitt, Strocchi, *et al.,* 1992; Strocchi *et al.,* 1996). An interplay between the unstirred water layer and the mucus layer overlying the mucosal cells has been suggested (Borgström *et al.,* 1985; Carey and Hernell, 1992), but a detailed insight has not yet been obtained. In analogy to the calculation described for the aqueous phase (see the previous text), the diffusional barrier of the unstirred water layer, in combination with components of the mucus layer (proteoglycans, glycocalyx), is inversely related to the radius of the diffusing particle. Carey and Hernell calculated the relative fluxes through an unstirred water layer of a fatty acid present in a mixed micelle or in a vesicle as compared to monomeric fatty acids (Carey and Hernell, 1992). Using the assumption that both vesicular fatty acid concentrations and vesicle diffusivity were one tenth of that of micelles, the relative flux of vesicular fatty acids was 100-fold lower than that of micellar fatty acids (Carey and Hernell, 1992). These calculations are compatible with the concept that, even though mixed micelles are no longer regarded as the exclusive transport moiety for lipolytic products as the original Hofmann–Borgström hypothesis postulated, micellar transport of lipolytic products still represents functionally the most important moiety under physiological conditions.

The actual translocation of fatty acids across the apical mucosal membrane in monomeric form implies the transfer of fatty acids from their aggregated states (micelles, vesicles) toward monomers dissolved as such in the aqueous unstirred water layer. Originally it was postulated that the translocation of monomeric fatty acids would pose a concentration gradient, leading to the flux of fatty acids from the mixed micelles into the aqueous phase, given their equilibrium relationship (Sallee, 1974; Westergaard and Dietschy, 1976). Shiau proposed that an acidic microclimate in the unstirred water layer facilitated fatty acid translocation across the mucosal membrane (Shiau and Levine, 1980; Shiau, 1981). The existence of an acidic microclimate adjacent to the intestinal cells had already been suggested by Hogben *et al.* (1959), and was confirmed by Lucas *et al.* by direct pH measurements in rat jejunum (Lucas *et al.*, 1975). The pH of the acidic microclimate has been estimated at between 5 and 6, or at least 1 pH-point lower than the bulk phase (Shiau, 1981). The physiological origin of the acidity has been attributed to intestinal hydrogen ion secretion (Shiau *et al.*, 1985). Manipulation of the pH of the microclimate in everted sacs of rat small intestine showed that the rate of fatty acid uptake was inversely related to the pH of the microclimate (Shiau, 1990). Shiau *et al.* provided indications that, apart from the pH itself, the intestinal mucus may play a role in facilitating diffusion of fatty acids (Shiau *et al.*, 1990). The postulated mechanism by which an acidic microclimate would favor translocation of fatty acids across the intestinal membrane involves at least three processes:

1. The fatty acids shift toward their protonated state at low pH, which implies in physicochemical terms a shift from a class II to a class I polar lipid. Because protonated fatty acids are incorporated in mixed micelles to a considerably lower extent than fatty acid soaps or salts, a flux of fatty acids out of the micelles into the unstirred water layer takes place. Under *in vitro* conditions, this event could even be visualized by the naked eye by the generation of an emulsified phase upon acidification of mixed micelles containing cholyltaurine and oleate (Shiau *et al.*, 1990).

2. The critical micellar concentration of a bile salt increases upon decreasing the pH (Thomson *et al.*, 1993). In the acidic microclimate, bile salts will shift from the micelles toward the aqueous phase. The flux of micellar fatty acids and bile salts to the aqueous phase functionally implies micellar dissociation due to the low pH environment adjacent to the mucosal cells.

3. The protonation of the fatty acids at the low pH conditions decreases their polarity, which enhances their capacity to partition into the apical membrane of the intestinal mucosal cell.

The concept of the acidic microclimate in aiding micelle dissociation and fatty acid translocation has also become popular because this concept would explain why fatty acid absorption is not solely dependent on fatty acid monomer concentration, according to the original concepts (Hofmann and Borgström, 1964; Westergaard and Dietschy, 1976; Westergaard *et al.*, 1986), but is also related to the concentration of the bile salts (Shiau *et al.*, 1990; Thomson *et al.*, 1993). No studies have yet been published in which the effects of an acidic microclimate on the physicochemistry of vesicular fatty acids has been determined.

4. Unresolved Issues on the Biophysics of Lipid Absorption

Although major progress in insight in the biophysics of intestinal lipids has been obtained since the mid-1970s, several aspects have remained unresolved. It has been demonstrated that lipid absorption can still occur reasonably efficiently in the bile-salt-deficient state. For example, Porter *et al.* reported on a patient with a complete bile fistula who still absorbed 75% of his dietary lipids, despite the considerable ingestion of 120 g of fat daily (Porter *et al.,* 1971). As expected in bile salt deficiency, the fatty acid concentration in the aqueous phase of the small intestine was decreased by a factor of 100 (Porter *et al.,* 1971). A similar degree of decrease in fatty acid concentration in the aqueous phase was described by Mansbach *et al.* in patients with bile acid malabsorption, but again with a minimal degree of lipid malabsorption (Mansbach *et al.,* 1980). Our studies in rats with chronic interruption of enterohepatic circulation indicated that the extent of lipid malabsorption is related to the amount of dietary lipids ingested; on a regular chow diet (6 wt% equals 14% of energetic value), absorption of dietary lipids was highly efficient (87%), but this value declined to only 54% when rats were fed a high-fat chow (16 wt% equals 35% of energetic value; Minich *et al.,* 1997). The relationship between the dietary lipid load and the efficacy of lipid absorption suggests that the compensatory mechanisms to overcome the absence of bile in the intestine have a limited capacity. Based on studies by Porter *et al.* and by others, one compensatory mechanism can be identified as being the absorptive reserve of the small intestine (Porter *et al.,* 1971; Knoebel, 1972). Whereas under physiological conditions only the proximal part of the small intestine is actually involved in lipid absorption, the more distal parts become involved whenever the absorptive process in the proximal part has not been complete (Knoebel, 1972). Yet, the actual biophysical state of the lipids under low bile salt conditions is still intriguing. Hernell *et al.* reviewed data from literature on patients with various low bile salt states, such as cystic fibrosis, ileectomy, or cirrhosis (Hernell *et al.,* 1990). Although the absolute lipid concentrations in the aqueous phase were invariably significantly lower than as observed in healthy individuals, the lipid compositions, plotted in appropriate phase diagrams, generally fell within the two-phase zone, indicating the coexistence of mixed micelles and vesicles, similar to the physiological situation. It has been speculated that, at low bile salt conditions, the multilamellar or unilamellar vesicles may serve to guarantee the replenishment of translocated monomeric fatty acids at the acidic microclimate of the unstirred water layer (Hernell *et al.,* 1990; Carey and Hernell, 1992). The actual importance of vesicles for lipid absorption, however, has not been proven. In our rat model of chronic interruption of enterohepatic circulation we demonstrated that the plasma concentrations of ^{13}C-labeled palmitic acid or linoleic acid after their intraduodenal administration were three- to sixfold lower when compared to rats with an intact enterohepatic circulation (Minich *et al.,* 1997). Yet, intraduodenal administration of lecithin–cholesterol unilamellar vesicles in a dose similar to the biliary secretion of these lipids failed to increase plasma ^{13}C-linoleic acid concentrations compared to bile-deficient rats, indicating that either the liquid crystalline state of the lipolytic products itself is sufficient to guarantee the observed absorption efficiency or that the rate-limiting step for lipid absorption under bile deficient conditions is not at the level of providing transport moieties for lipolytic products (Nishioka *et al.,* 2000).

These observations highlight another unresolved issue on the biophysics of lipid absorption, namely the *quantitative* importance of the various transport moieties that have been identified in the intestinal lumen. Given the qualitative identification of the various biophysical states, the next challenge is to address their functional importance under physiological as well as under pathophysiological conditions, such as during low bile salt states, impaired lipolysis, inefficient neutralization of duodenal contents, altered mucus formation, and so forth. The compensatory potency for nonmicellar transport moieties needs further quantification, as illustrated by efficient lipid absorption during the (virtual) absence of bile salts in the duodenum (see the previous text). Insights into these processes may evolve into the development of novel therapeutic strategies to guarantee the absorption of lipids in general or of specific lipids (e.g., essential fatty acids, fat-soluble vitamins) under various pathophysiological states with lipid malabsorption.

Finally, whereas this chapter has concentrated on the absorption of fatty acids, insight into the biophysics of non-fatty acid lipids in the intestine is still comparatively minimal. Although numerous quantitative data are available on the intestinal absorption of dietary or biliary cholesterol and phospholipid, it remains to be established to what extent nonmicellar transport moieties could functionally be important under low bile salt conditions. Additionally, the actual mechanism by which nonpolar lipids, including pollutants such as polychlorobiphenyls (PCBs), are intestinally absorbed is still unclear. Whereas a limited partitioning in liquid crystalline vesicles can be envisioned for class I polar lipids, such as the fat-soluble vitamins, it has not been demonstrated that similar processes take place under physiogical circumstances for nonpolar lipids.

5. Conclusion

In conclusion, the biophysics of lipids during their absorption from the intestine has become reasonably well described in qualitative terms. Emulgation in the upper gastrointestinal tract, involving both shear forces and biochemical processes (lipolysis, neutralization), results in the generation of a fine stable emulsion in the proximal part of the duodenum. It has become appreciated that, upon lipolysis, various lipid phases are physiologically present in the intestinal lumen, including mixed micelles, multilayer lamellae, unilamellar vesicles, and, possibly, rodlike mixed micelles. The role of an acidic microclimate adjacent to the mucosal cells in aiding translocation of fatty acids across the apical membrane has become generally accepted. Unresolved issues include the quantitative contribution of the various lipid phases under physiological as well as pathophysiological conditions, as well as the biophysical states of non-fatty acid lipids.

References

Boffelli, D., Compassi, S., Werder, M., Weber, F. E., Phillips, M. C., Schulthess, G., and Hauser, H., 1997, The uptake of cholesterol at the small-intestinal brush border membrane is inhibited by apolipoproteins, *FEBS Lett.* **411**:7–11.

Boffelli, D., Weber, F. E., Compassi, S., Werder, M., Schulthess, G., and Hauser, H., 1997, Reconstitution and further characterization of the cholesterol transport activity of the small-intestinal brush border membrane, *Biochemistry* **36**:10784–10792.

Borgström, B., 1977, The action of bile salts and other detergents on pancreatic lipase and the interaction with colipase, *Biochim. Biophys. Acta* **488**:381–391.

Borgström, B., 1980, Importance of phospholipids, pancreatic phospholipase A2, and fatty acid for the digestion of dietary fat: *In vitro* experiments with the porcine enzymes, *Gastroenterology* **78**:954–962.

Borgström, B., 1985, Fat Assimilation, in: *Bockus Gastroenterology,* (J. E. Berk, ed.), WB Saunders Company, Philadelphia, pp. 1510–1519.

Borgström, B., and Erlanson, C., 1978a, Interactions of serum albumin and other proteins with porcine pancreatic lipase, *Gastroenterology* **75**:382–386.

Borgström, B., and Erlanson, C., 1978b, Lipase, colipase, amphipathic dietary proteins, and bile acids [letter], *Gastroenterology* **75**:766.

Borgström, B., and Erlanson-Albertsson, C., 1982, Hydrolysis of milk fat globules by pancreatic lipase. Role of colipase, phospholipase A2, and bile salts, *J. Clin. Invest.* **70**:30–32.

Borgström, B., Dahlqvist, A., Lundh, G., and Sjovall, J., 1957, Studies on intestinal digestion and absorption in the human, *J. Clin. Invest.* **36**:1521–1536.

Borgström, B., Barrowman, J. A., and Lindstrom, M., 1985, Roles of bile acids in intestinal lipid digestion and absorption, in: *Sterols and Bile Acids.* (H. Danielsson and J. Sjovall, eds.), Elsevier Science Publishers BV, Amsterdam, pp. 405–425.

Carey, M. C., and Hernell, O., 1992, Digestion and absorption of fat, *Semin. Gastroint. Dis.* **3**:189–208.

Carey, M. C., and Lamont, J. T., 1992, Cholesterol gallstone formation. 1. Physical-chemistry of bile and biliary lipid secretion, *Prog. Liver. Dis.* **10**:139–163.

Carey, M. C., and Small, D. M., 1970, The characteristics of mixed micellar solutions with particular reference to bile, *Am. J. Med.* **49**:590–608.

Carey, M. C., and Small, D. M., 1972, Micelle formation by bile salts. Physical-chemical and thermodynamic considerations, *Arch. Intern. Med.* **130**:506–527.

Carey, M. C., Small, D. M., and Bliss, C. M., 1983, Lipid digestion and absorption, *Annu. Rev. Physiol.* **45**:651–677.

Carriere, F., Verger, R., Lookene, A., and Olivecrona, G., 1995, Lipase structures at the interface between chemistry and biochemistry, *Interface Between Chemistry and Biochemistry* **73**:3–26.

Cohen, D. E., and Carey, M. C., 1990, Physical chemistry of biliary lipids during bile formation, *Hepatology* **12**:143S–147S.

Cohen, D. E., Angelico, M., and Carey, M. C., 1989, Quasielastic light scattering evidence for vesicular secretion of biliary lipids, *Am. J. Physiol.* **257**:G1–8.

Cohen, D. E., Leighton, L. S., and Carey, M. C., 1992, Bile salt hydrophobicity controls vesicle secretion rates and transformations in native bile, *Am. J. Physiol.* **263**:G386–395.

Compassi, S., Werder, M., Boffelli, D., Weber, F. E., Hauser, H., and Schulthess, G., 1995, Cholesteryl ester absorption by small intestinal brush border membrane is protein-mediated, *Biochemistry* **34**:16473–16482.

Crawford, J. M., Mockel, G. M., Crawford, A. R., Hagen, S. J., Hatch, V. C., Barnes, S., Godleski, J. J., and Carey, M. C., 1995, Imaging biliary lipid secretion in the rat: Ultrastructural evidence for vesiculation of the hepatocyte canalicular membrane, *J. Lipid. Res.* **36**:2147–2163.

Dahim, M., and Brockman, H., 1998, How colipase–fatty acid interactions mediate adsorption of pancreatic lipase to interfaces, *Biochemistry* **37**:8369–8377.

Donovan, J. M., and Carey, M. C., 1990, Separation and quantitation of cholesterol "carriers" in bile, *Hepatology* **12**:94S–104S.

Donovan, J. M., Timofeyeva, N., and Carey, M. C., 1991, Influence of total lipid concentration, bile salt:lecithin ratio, and cholesterol content on inter-mixed micellar/vesicular (non-lecithin-associated) bile salt concentrations in model bile, *J. Lipid. Res.* **32**:1501–1512.

Fitscher, B. A., Elsing, C., Riedel, H. D., Gorski, J., and Stremmel, W., 1996, Protein-mediated facilitated uptake processes for fatty acids, bilirubin, and other amphipathic compounds, *Proc. Soc. Exp. Biol. Med.* **212**:15–23.

Frazer, A. C., 1946, The absorption of triglyceride fat from the intestine, *Physiol. Rev.* **26**:103–119.

Frazer, A. C., Schulman, J. H., and Stewart, H. C., 1964, Emulsification of fat in the intestine of the rat and and its relationship to absorption, *J. Physiol.* **103**:306–316.

Gargouri, Y., Pieroni, G., Riviere, C., Lowe, P. A., Sauniere, J. F., Sarda, L., and Verger, R., 1986a, Importance of human gastric lipase for intestinal lipolysis: An *in vitro* study, *Biochim. Biophys. Acta* **879**:419–423.

Gargouri, Y., Pieroni, G., Riviere, C., Sauniere, J. F., Lowe, P. A., Sarda, L., and Verger, R., 1986b, Kinetic assay of human gastric lipase on short- and long-chain triacylglycerol emulsions, *Gastroenterology* **91**:919–925.

Gargouri, Y., Moreau, H., and Verger, R., 1989, Gastric lipases: Biochemical and physiological studies [published erratum appears in Biochim Biophys Acta 1990 Feb 23; 1042(3):421], *Biochim. Biophys. Acta* **1006**:255–271.

Gilat, T., and Somjen, G. J., 1996, Phospholipid vesicles and other cholesterol carriers in bile, *Bba-Rev. Biomembranes* **1286**:95–115.

Hernell, O., Staggers, J. E., and Carey, M. C., 1990, Physical–chemical behavior of dietary and biliary lipids during intestinal digestion and absorption. 2. Phase analysis and aggregation states of luminal lipids during duodenal fat digestion in healthy adult human beings, *Biochemistry* **29**:2041–2056.

Hjelm, R. P., Thiyagarajan, P., Schteingart, C. D., Hofmann, A. F., Alkan-Onyuksel, H., and Ton-Nu, H. T., 1995,

Structure of mixed micelles present in bile and intestinal contents based on studies in model systems, in: *Bile Acids in Gastroenterology. Basic and Clinical Advances.* (A. F. Hofmann, G. Paumgartner, and A. Stiehl, eds.), Kluwer Academic Publishers, Dordrecht, pp. 41–58.

Hofmann, A. F., 1976, Fat digestion: The interaction of lipid digestion products with micellar bile acid solutions, in: *Lipid Absorption: Biochemical and Clinical Aspects* (K. Rommel, H. Goebell, and R. Boehmer, eds.), MTP Press Ltd. Lancaster, England, pp. 3–22.

Hofmann, A. F., 1990, Bile acid secretion, bile flow and biliary lipid secretion in humans, *Hepatology* 12:17S–22S.

Hofmann, A. F., and Borgström, B., 1964, The intraluminal phase of fat lipid digestion in man: The lipid content of the micellar and the oil phases of intestinal contents obtained during fat digestion and absorption, *J. Clin. Invest.* 43:247–257.

Hogben, C. A. M., Tocco, D. J., Brodie, B. B., and Schanker, L. S., 1959, On the mechanism of intestinal absorption of drugs, *J. Pharmacol. Exp. Ther.* 125:275–282.

Honkanen, R. E., Rigler, M. W., and Patton, J. S., 1985, Dietary fat assimilation and bile salt absorption in the killifish intestine, *Am. J. Physiol.* 249:G399–407.

Knoebel, L. K., 1972, Intestinal absorption *in vivo* of micellar and nonmicellar lipid, *Am. J. Physiol.* 223:255–261.

Lairon, D., Nalbone, G., Lafont, H., Leonardi, J., Vigne, J. L., Chabert, C., Hauton, J. C., and Verger, R., 1980, Effect of bile lipids on the adsorption and activity of pancreatic lipase on triacylglycerol emulsions, *Biochim. Biophys. Acta* 618:119–128.

Levitt, M. D., Kneip, J. M., and Levitt, D. G., 1988, Use of laminar flow and unstirred layer models to predict intestinal absorption in the rat, *J. Clin. Invest.* 81:1365–1369.

Levitt, M. D., Strocchi, A., and Levitt, D. G., 1992, Human jejunal unstirred layer: Evidence for extremely efficient luminal stirring, *Am. J. Physiol.* 262:G593–596.

Levitt, M. D., Furne, J. K., and Levitt, D. G., 1992, Shaking of the intact rat and intestinal angulation diminish the jejunal unstirred layer, *Gastroenterology* 103:1460–1466.

Lucas, M. L., Schneider, W., Haberich, F. J., and Blair, J. A., 1975, Direct measurement by pH-microelectrode of the pH microclimate in rat proximal jejunum, *Proc. R. Soc. Lond. B. Biol. Sci.* 192:39–48.

Luk, A. S., Kaler, E. W., and Lee, S. P., 1997, Structural mechanisms of bile salt-induced growth of small unilamellar cholesterol-lecithin vesicles, *Biochemistry* 36:5633–5644.

Mansbach, C. M.,II, Cohen, R. S., and Leff, P. B., 1975, Isolation and properties of the mixed lipid micelles present in intestinal content during fat digestion in man, *J. Clin. Invest.* 56:781–791.

Mansbach, C. M.,II, Newton, D., and Stevens, R. D., 1980, Fat digestion in patients with bile acid malabsorption but minimal steatorrhea, *Dig. Dis. Sci.* 25:353–362.

Mazer, N. A., and Carey, M. C., 1983, Quasi-elastic light-scattering studies of aqueous biliary lipid systems. Cholesterol solubilization and precipitation in model bile solutions, *Biochemistry* 22:426–442.

Mazer, N. A., Benedek, G. B., and Carey, M. C., 1980, Quasi-elastic light-scattering studies of aqueous biliary lipid systems. Mixed micelle formation in bile salt-lecithin solutions, *Biochemistry* 19:601–615.

Miller, K. W., and Small, D. M., 1982, The phase behavior of triolein, cholesterol, and lecithin emulsions, *J. Coll. Interface. Sci.* 89:466–478.

Minich, D. M., Kalivianakis, M., Havinga, R., Stellaard, F., Kuipers, F., Vonk, R. J., and Verkade, H. J., 1997, Absorption and metabolism of ^{13}C-linoleic acid in rats with an intact or permanently interrupted enterohepatic circulation, *Gastroenterology* 112:A894(Abstract).

Mockel, G. M., Gorti, S., Tandon, R. K., Tanaka, T., and Carey, M. C., 1995, Microscope laser light-scattering spectroscopy of vesicles within canaliculi of rat hepatocyte couplets, *Amer. J. Physiol-Gastrointest. L.* 32:G73–G84.

Nishioka, T., Havinga, H., Tazuma, S., Stellaard, F., Kuipers, F., and Verkade, H. J., 2000, Enteral administration of phosphatidylcholine-cholesterol liposomes partially overcomes intestinal fat malabsorption in bile-deficient rats (abstract), *Gastroenterology* 118:A572.

Nury, S., Pieroni, G., Riviere, C., Gargouri, Y., Bois, A., and Verger, R., 1987, Lipase kinetics at the triacylglycerol–water interface using surface tension measurements, *Chem. Phys. Lipids* 45:27–37.

Patton, J. S., 1981, Gastrointestinal lipid digestion, in: *Physiology of the Gastrointestinal Tract* (L. R. Johnson, ed.), Raven Press, New York, pp. 1123–1146.

Patton, J. S., and Carey, M. C., 1979, Watching fat digestion, *Science* 204:145–148.

Porter, H. P., and Saunders, D. R., 1971, Isolation of the aqueous phase of human intestinal contents during the digestion of a fatty meal, *Gastroenterology* **60:**997–1007.

Porter, H. P., Saunders, D. R., Tytgat, G., Brunser, O., and Rubin, C. E., 1971, Fat absorption in bile fistula man. A morphological and biochemical study, *Gastroenterology* **60:**1008–1019.

Rigler, M. W., Honkanen, R. E., and Patton, J. S., 1986, Visualization by freeze fracture, *in vitro* and *in vivo,* of the products of fat digestion, *J. Lipid Res.* **27:**836–857.

Saleeb, F. Z., Cante, C. J., Streckfus, T. K., Frost, J. R., and Rosano, H. L., 1975, Surface pH and stability of oil–water emulsions derived from laurate solutions, *J. Am. Oil Chem. Soc.* **52:**208–212.

Sallee, V. L., 1974, Apparent monomer activity of saturated fatty acids im micellar bile salt solutions measured by a polyethylene partitioning system, *J. Lipid Res.* **15:**56–64.

Schmit, G. D., Momsen, M. M., Owen, W. G., Naylor, S., Tomlinson, A., Wu, G., Stark, R. E., and Brockman, H. L., 1996, The affinities of procolipase and colipase for interfaces are regulated by lipids, *Biophys. J.* **71:**3421–3429.

Schoeller, C., Keelan, M., Mulvey, G., Stremmel, W., and Thomson, A. B. R., 1995, Role of a brush border membrane fatty acid binding protein in oleic acid uptake into rat and rabbit jejunal brush border membrane, *Clin. Invest. Med.* **18:**380–388.

Schulthess, G., Compassi, S., Boffelli, D., Werder, M., Weber, F. E., and Hauser, H., 1996, A comparative study of sterol absorption in different small-intestinal brush border membrane models, *J. Lipid Res.* **37:**2405–2419.

Shankland, W., 1970, The ionic behavior of fatty acids solublized by bile salts. *J. Colloid Interface Sci.* **34:**9–25.

Shiau, Y. F., 1981, Mechanisms of intestinal fat absorption, *Am. J. Physiol.* **240:**G1–9.

Shiau, Y. F., 1990, Mechanism of intestinal fatty acid uptake in the rat: The role of an acidic microclimate, *J. Physiol. (Lond)* **421:**463–474.

Shiau, Y. F., and Levine, G. M., 1980, pH dependence of micellar diffusion and dissociation, *Am. J. Physiol.* **239:**G177–182.

Shiau, Y. F., Fernandez, P., Jackson, M. J., and McMonagle, S., 1985, Mechanisms maintaining a low-pH microclimate in the intestine, *Am. J. Physiol.* **248:**G608–617.

Shiau, Y. F., Kelemen, R. J., and Reed, M. A., 1990, Acidic mucin layer facilitates micelle dissociation and fatty acid diffusion, *Am. J. Physiol.* **259:**G671–675.

Small, D. M., 1970, Surface and bulk interactions of lipids and water with a classification of biologically active lipids based on these interactions, *Fed. Proc.* **29:**1320–1326.

Somjen, G. J., and Gilat, T., 1983, A non-micellar mode of cholesterol transport in human bile, *FEBS. Lett.* **156:**265–268.

Staggers, J. E., Hernell, O., Stafford, R. J., and Carey, M. C., 1990, Physical–chemical behavior of dietary and biliary lipids during intestinal digestion and absorption. 1. Phase behavior and aggregation states of model lipid systems patterned after aqueous duodenal contents of healthy adult human beings, *Biochemistry* **29:**2028–2040.

Stremmel, W., and Hofmann, A. F., 1990, Intestinal absorption of unconjugated dihydroxy bile acids: Non-mediation by the carrier system involved in long chain fatty acid absorption, *Lipids* **25:**11–16.

Strocchi, A., Corazza, G., Furne, J., Fine, C., Di Sario, A., Gasbarrini, G., and Levitt, M. D., 1996, Measurements of the jejunal unstirred layer in normal subjects and patients with celiac disease, *Am. J. Physiol.* **270:**G487–491.

Thomson, A. B., Keelan, M., Garg, M. L., and Clandinin, M. T., 1989, Intestinal aspects of lipid absorption: In review, *Can. J. Physiol. Pharmacol.* **67:**179–191.

Thomson, A. B., Schoeller, C., Keelan, M., Smith, L., and Clandinin, M. T., 1993, Lipid absorption: Passing through the unstirred layers, brush-border membrane, and beyond, *Can. J. Physiol. Pharmacol.* **71:**531–555.

Tso, P., and Simmonds, W. J., 1977, Importance of luminal lecithin in intestinal absorption and transport of lipid in the rat, *Aust. J. Exp. Biol. Med. Sci.* **55:**355–357.

Verkade, H. J., Havinga, R., Kuipers, F., and Vonk, R. J., 1995, Mechanism of biliary lipid secretion, in: *Bile Acids in Gastroenterology: Basic and Clinical Advances.* (A. F. Hofmann, G. Paumgartner, and A. Stiehl, eds.), Kluwer Academic Publishers, Lancaster, pp. 230–246.

Verkade, H. J., Vonk, R. J., and Kuipers, F., 1995, New insights into the mechanism of bile acid-induced biliary lipid secretion, *Hepatology* **21:**1174–1189.

Westergaard, H., and Dietschy, J. M., 1974, Delineation of the dimensions and permeability characteristics of the two major diffusion barriers to passive mucosal uptake in the rabbit intestine, *J. Clin. Invest.* **54:**718–732.

Westergaard, H., and Dietschy, J. M., 1976, The mechanism whereby bile acid micelles increase the rate of fatty acid and cholesterol uptake into the intestinal mucosal cell, *J. Clin. Invest.* **58:**97–108.

Westergaard, H., Holtermuller, K. H., and Dietschy, J. M., 1986, Measurement of resistance of barriers to solute transport in vivo in rat jejunum, *Am. J. Physiol.* **250:**G727–735.

Wilson, F. A., Sallee, V. L., and Dietschy, J. M., 1971, Unstirred water layer in intestine: Rate determinant of fatty acid absorption from micellar solutions, *Science* **174:**1031–1033.

CHAPTER 2

Preduodenal Lipases and their Role in Lipid Digestion

L. DUPUIS, S. CANAAN, M. RIVIÈRE, R. VERGER, and C. WICKER-PLANQUART

1. Introduction

Under normal physiological conditions the digestion and absorption of dietary lipids are highly efficiently processed. In humans, the diet generally contains 90 to 120 g of lipids (mostly triacylglycerols), more than 95% of which are absorbed, due to the interplay between the stomach, the small intestine, the liver, and the pancreas (Carey *et al.,* 1983). Several steps can be distinguished in the processing of dietary lipids, including their emulsification, hydrolysis and solubilization, and, last, their uptake into the enterocyte. The emulsification of lipids starts in the stomach and is mediated by physical forces and is facilitated by the partial lipolysis of the dietary lipids (Carey *et al.,* 1983). For a long time, the hydrolysis of dietary triglycerides was thought to begin in the intestinal lumen and to be catalyzed entirely by pancreatic lipase. The stomach was thought to be a transient storage organ, the role of which was limited to mixing lipids with the other nutriments and dispersing them as required. Although many authors observed the occurrence of lipolysis at the preduodenal level in humans and in several other species, the gastric phase of lipolysis was assumed to be negligible and to be of little or no significance in comparison with the intestinal step. Gastric lipolysis was even attributed to pancreatic contamination resulting from a duodeno–gastric reflux. At the beginning of the twentieth century, however, it was observed that gastric juice could hydrolyze fat. In 1901, Volhard stated that gastric lipase was the "ferment" present in gastric juice that was responsible for fat hydrolysis. Finally, the gastric origin of the lipase present in dog gastric juice was established by Hull and Keaton (1917) in dogs with Pavlov stomach under conditions precluding the possibility of any pancreatic contamination.

The idea that gastric lipase might be of some physiological importance was suggest-

L. DUPUIS, S. CANAAN, M. RIVIÈRE, R. VERGER, and C. WICKER-PLANQUART • Laboratoire de Lipolyse Enzymatique, 31 Chemin Joseph-Aiguier, 13402 Marseille Cédex 20, France.
Intestinal Lipid Metabolism, edited by Charles M. Mansbach II *et al.,* Kluwer Academic/Plenum Publishers, 2001.

ed by studies on pathological situations involving exocrine pancreatic insufficiency, such as the late stage of chronic pancreatitis or cystic fibrosis. In these cases, even in the complete absence of pancreatic lipase, patients still absorb a high percentage of the dietary fat ingested (Ross, 1955; Ross & Sammons, 1955; Lapey *et al.*, 1974; Muller *et al.*, 1975). In newborn infants, the partial prehydrolysis of milk fat globules or of infant formulas in the stomach is necessary to the digestion of fat (Hernell & Blackberg, 1994). The fat in milk takes the form of milk fat globules, the core of which consists of triglycerides and the membrane of which consists of more polar lipids (phospholipids and cholesterol) as well as proteins. Newborn infants indeed secrete only low amounts of pancreatic lipase and bile salts. Furthermore, it has been demonstrated that pancreatic lipase alone does not readily hydrolyze native milk fat globules. After preincubating milk with preduodenal lipase, however, pancreatic lipase was found to hydrolyze milk fat droplets very efficiently (Bernbäck *et al.*, 1989). Gastric and pancreatic lipase between them hydrolyze two thirds of the ester bonds present in these droplets. Total digestion requires the action of bile-salt-activated lipase, which is triggered by the gastric lipolysis occurring in the previous steps (Bernbäck *et al.*, 1990). Among the gastrointestinal lipases, preduodenal lipase has the unique ability to degrade maternal milk fat globules. In addition, the gastric digestion of milk fat may play an important protective role, and may thus serve to prevent gastrointestinal infections. It has been reported that after the digestion of human milk or infant formulas, an extract of the lipid fraction present in the stomach displayed antimicrobial activity (Isaacs *et al.*, 1990). More specifically, medium-chain saturated and long-chain unsaturated fatty acids or monoacylglycerol components of these fatty acids showed high levels of antiviral and antibacterial activity *in vitro* (Thormar *et al.*, 1987; Isaacs *et al.*, 1992).

For many years, the exact physiological contribution of preduodenal lipase to the overall process of lipolysis was not known. A few *in vitro* data suggested that gastric lipase might act preferentially under duodenal conditions (Gargouri *et al.*, 1986a) in synergy with pancreatic lipase (Gargouri *et al.*, 1986b; Bernbäck *et al.*, 1989). On the other hand, purified preduodenal lipases are known to be inactivated by pancreatic proteases (Bernbäck *et al.*, 1987). Roberts and Hand (1988) reported that rat lingual lipase was inactivated in the intestine. In humans, Abrams *et al.* (1984, 1987) demonstrated that preduodenal lipase amounted to 7% of the total lipolytic activity present at the angle of Treitz during the first two hours of digestion. In the latter studies, the preduodenal lipase activity was titrated at pH 5.0 using triolein as the substrate, that is, under conditions where the possibility that pancreatic lipase may have interfered could not be ruled out. Carrière *et al.* (1993) have clearly distinguished and measured the gastric and pancreatic lipase outputs and monitored the amounts of lipolytic products produced during the digestion of a liquid test meal. The subjects (healthy volunteers), after fasting overnight, were intubated with a triple-lumen duodenal tube and a separate single lumen naso-gastric tube. These authors established for the first time that most of the human gastric lipase (HGL) secreted in the stomach was still active in the duodenum. They estimated the relative contributions of HGL and human pancreatic lipase (HPL) to the overall digestion of dietary triglycerides (see Fig. 1). During the whole digestion period, gastric lipase hydrolyzes 17.5% of all the triglyceride acyl chains. Considering that only two acyl chains out of the three from a triglyceride molecule need to be hydrolyzed to ensure complete intestinal absorption, gastric lipase contribution in the hydrolysis of triglyceride is about 25%.

The secretion of HGL and its contribution to digestion in patients suffering from se-

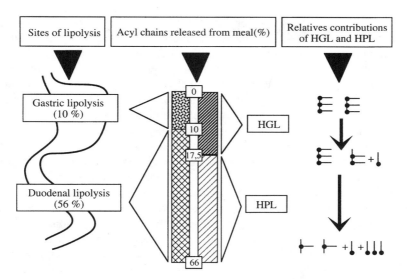

Figure 1. Diagram of the respective contributions of HGL and HPL to the overall digestion of dietary triacylglycerols (Reprinted with permission from Carriere *et al.,* 1993).

vere pancreatic deficiencies have also been studied (Mossi *et al.,* 1994). The HPL concentration recorded in the duodenum was found to be 20% below the lower limit observed in normal subjects. No compensatory hypersecretion of HGL was observed in a group of patients with severe pancreatic deficiencies. The preservation of normal HGL secretion is essential to the overall lypolysis, however.

The main products of gastric lipolysis present in the stomach are diacylglycerols and free fatty acids. The main function of gastric lipolysis seems to be to generate the fatty acids that are necessary for several reasons: they serve as emulsifiers when they become partially ionized in the small intestinal contents and they release cholecystokinin (CCK) and gastric inhibitory peptide. It is also worth mentioning that HGL contributes significantly to fat digestion. Little is known so far, however, about its possible neurohormonal regulatory role in humans. Borovicka *et al.* (1997) have studied the role of the CCK and cholinergic mechanisms in the postprandial regulation of HGL and HPL secretion in six healthy subjects. These authors demonstrated that the postprandial gastric secretion of lipase, pepsin, and acid occurs in parallel and is stimulated by cholinergic mechanisms but is inhibited by CCK. HPL secretion is enhanced by cholinergic mechanisms and by CCK. The regulation of the postprandial digestive secretions by CCK therefore consists of a positive pancreatic stimulation and an inhibitory feedback mechanism controlling HGL secretion.

In humans, several conditions, including cystic fibrosis, pancreatitis, premature birth, and alcoholism, are associated with pancreatic lipase insufficiency. In order to palliate these insufficiencies, enzyme replacement therapy with porcine pancreatic lipase has been widely used for a long time (Bénicourt *et al.,* 1993). To reduce the malabsorption of fat, the enzymes delivered into the duodenum must amount to 5 to 10% of the quantities usually present after maximum stimulation of the pancreas (DiMagno *et al.,* 1973). Under optimum conditions, if no inactivation of the supplemental enzymes occurs in the stomach or the duodenum, approximately 30,000 IU of lipase must be taken with each meal (DiMagno, 1982), which corresponds to about 5 to 10 g of lyophilized pancreatic powder per day. In the past, preparations of this kind were far from satisfactory, because a large proportion of the enzymes administered were denatured in the stomach due to the extremely high acidity of the gastric juice. Pancreatic lipase is irreversibly inactivated from pH levels of 4.0 downward. This problem has been partly solved by using enteric-coated microspheric pancreatic preparations. Lipase activity is released from the microspheres at pH levels of around 5.5. It is necessary to bear in mind, however, that a considerable loss of activity occurs along the gastrointestinal tract, because only 8% of the exogenous lipase was recovered in the previously discussed study. The lipase was degraded at the duodenal level by proteases present in the pancreatic extracts. The use of acid-resistant lipases should in principle yield more satisfactory results than the pancreatic preparations currently in use. Concomitantly administering acid lipases, which hydrolyze dietary lipids under acidic conditions, should help to answer some of the problems associated with lipid malabsorption and steatorrhea. Physiological studies have shown that preduodenal lipases are capable of acting not only in the stomach but also in the duodenum in synergy with pancreatic lipase (Carrière *et al.*, 1993). Various clinical studies have been conducted on both animals and humans to assess the efficiency of enzyme replacement therapies based on the use of acid-resistant lipases of bacterial origin to treat exocrine pancreatic insufficiency (Suzuki *et al.*, 1997). Although this treatment significantly increased the body weight and reduced the steatorrhea in dogs, these enzymes are difficult to use because of their sensitivity to the proteolytic action of gastric pepsin, in addition to the legal obstacles relating to their use.

2. Purification and Characterization of Acid Lipases

2.1. Preduodenal Lipases

Acid lipases have been screened in various species from the tongue to the pylorus (DeNigris *et al.*, 1988; Moreau *et al.*, 1988b; Figure 2). Six preduodenal lipases have been purified and biochemically characterized so far: rat lingual lipase (RLL), human gastric lipase (HGL), calf pharyngeal lipase (CPL), lamb pharyngeal lipase (LPL), dog gastric lipase (DGL), and rabbit gastric lipase (RGL). All these preduodenal lipases have molecular masses ranging between 45 and 51 kDa.

RLL was isolated from the serous glands of the rat tongue by ammonium sulphate precipitation followed by acetone precipitation and hydrophobic interaction chromatography on octyl agarose (Hamosh *et al.*, 1979; Field and Scow, 1983; Roberts *et al.*, 1984).

Gastric lipases have been purified from gastric juice and from fundic gastric mucosae and have been characterized *in vitro*. HGL was first purified by Tiruppathi and Balasubramanian (1982), and its secretion was stimulated by injecting pentagastrin (Szafran *et al.*, 1978; Moreau *et al.*, 1988d). Nowadays, the enzyme is purified using cation-exchange (Mono S) chromatography followed by an immunoaffinity column procedure using monoclonal antibodies (Aoubala *et al.*, 1993). Lipase activity can be measured potentiometrically on short-chain (tributyroylglycerol, TC4), medium-chain (trioctanoylglycerol, TC8) and long-chain (Intralipid) triacylglycerols at 37°C using a pH-stat (Gargouri *et al.*, 1986c). The

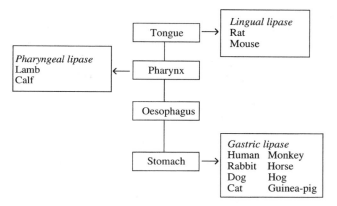

Figure 2. Anatomical sites of preduodenal lipases in several mammals.

maximum specific activities obtained have been 1160 units/mg at pH 6, 1110 units/mg at pH 6, and 600 units/mg at pH 5 on TC4, TC8, and Intralipid, respectively. HGL cDNA was fused with the maltose-binding protein cDNA and expressed in *E. coli*. A soluble fusion protein of about 80 kDa was produced, which was recognized by rabbit polyclonal antibodies but showed no activity toward triglycerides (Wicker-Planquart *et al.*, 1996). Other authors (Bodmer *et al.*, 1987; Smerdon *et al.*, 1995; Crabbe *et al.*, 1996) have expressed the protein in yeast. Smerdon *et al.* (1995) and Crabbe *et al.* (1996) succeeded in obtaining catalytically active material and noted that HGL was mostly associated with the cell wall. Most of the HGL secreted in *Schizosaccharomyces pombe* had a higher molecular mass than native HGL because of the hyperglycosylation of the protein. To overcome the hyperglycosylation problems encountered in yeast, HGL was expressed in the baculovirus/insect cell system (Wicker-Planquart *et al.*, 1996). The maximum activities of the purified recombinant HGL were 434, 730 and 562 units/mg on Intralipid, TC8, and TC4, respectively.

CPL, which is also called pregastric esterase (PGE), was purified from calf pharyngeal tissue. The purification procedure consisted of performing chromatography on octyl-Sepharose and lentil-lectin-Sepharose followed by gel filtration (Bernbäck *et al.*, 1985). Timmermans (1996) has developed a new procedure including anion exchange chromatography, ammonium sulphate precipitation, and gel filtration but no delipidation step. The enzyme has also been expressed in *E. coli* in the form of a fusion protein with glutathione S-transferase. The largely inactive part of the fusion protein was found to consist of insoluble inclusion bodies. Eukaryotic expression has been performed in the yeast *Pichia pastoris* and in insect cells. In the baculovirus insect cell system, CPL was produced in the form of a 45 kDa glycoprotein, showing a clearly visible lipase activity of 0.25 U/ml (Timmermans, 1996).

LPL has been purified from lamb pharyngeal tissue, using an aqueous extract containing 0.7% Tween 80, which was chromatographed on a DEAE-cellulose anion exchanger and absorbed on HA-Ultrogel before undergoing gel filtration on Ultrogel AcA-54 (De Caro *et al.*, 1995). The pure enzyme displayed specific activities of 950 units/mg, 300 units/mg, and 30 units/mg on TC4, TC8, and TC18, respectively.

DGL (Carrière *et al.,* 1991) was purified from the gastric juice of dogs with chronic gastric fistulae. Little if any DGL activity was observed in the basal or stimulated gastric secretion, whereas the gastric mucosa was found to contain substantial amounts of lipolytic activity. This paradox can be explained by the occurrence of an irreversible inactivation of DGL at the low pH levels present in the stomach under fasting conditions. By buffering the acidic secretion *in vivo,* or by using an antiacid secretion drug such as omeprazol during stimulation, the effects of several gastric secretagogues on the secretion of DGL were determined (Carrière *et al.,* 1992). Unlike HGL secretion, DGL secretion was poorly stimulated by pentagastrine, whereas urecholine, 16,16-dimethyl prostaglandine E2, and secretin were potent secretagogues. DGL was extracted from soaked dog gastric mucosa and was purified after cation exchange, anion exchange, and gel filtration chromatography. The maximum specific activities were 550 units/mg on TC4 at pH 5.5, 750 units/mg on TC8 at pH 6.5, and 950 units/mg on Intralipid at pH 4 (Carrière *et al.,* 1991). DGL is nearly twice as active on long-chain triacylglycerol as on short-chain triacylglycerol, and in this respect, it is unique among all the acid lipases investigated so far.

RGL was purified from an acetonic powder prepared from rabbit stomach fundus by performing ammonium sulphate fractionation, Sephadex G-100 gel filtration, and mono S column cation exchange (Moreau *et al.,* 1988a). The pure enzyme had specific activities of 1200 units/mg, 850 units/mg, and 280 units/mg, using TC4, TC8, and soybean oil as substrates, respectively.

All the preduodenal lipases specifically hydrolyze triacylglycerols but not the ester bonds of phospholipids or cholesteryl esters. In milk fat, the short- and medium-chain fatty acids naturally occurring at the *sn*-3 position are preferentially removed due to the *sn*-3 specificity of acid lipases. The apparent *sn*-3 selectivity of these lipases toward short- and medium-chain fatty acids resulted from the fact that these fatty acids are almost entirely located at the *sn*-3 position. Long-chain fatty acids are also hydrolyzed by these lipases, but to a much lesser extent. DGL is the only exception: it hydrolyzes long-chain triacylglycerols with a high degree of specificity (Carrière *et al.,* 1991). The stereoselectivity of DGL and DPL was investigated both *in vitro* under simulated physiological conditions and *in vivo* during the digestion of a liquid test meal (Carrière *et al.,* 1997). It was observed that although both lipases have a stereospecificity for the *sn*-3 position in triglycerides, the stereopreference of DGL was about three times higher *in vitro* than that of DPL. On the other hand, both lipases clearly showed a comparable degree of enantioselectivity for the *sn*-1 position when a racemic diolein mixture was used as the substrate. The same authors clearly showed that a high enantiomeric excess of 1,2-*sn*-diglyceride was due to the predominant contribution of gastric lipase to the hydrolysis of triglycerides.

Purified gastric lipases are highly sensitive to interfacial denaturation (Gargouri *et al.,* 1986a). When amphiphilic proteins such as BSA or β-lactoglobulin were not preincluded in the reaction mixture before the purified enzyme was added, no activity was detected. This interfacial activation is irreversible because supplementing the reaction mixture with amphiphiles after adding the enzyme failed to restore the activity.

2.2. Lysosomal Acid Lipase

Human lysosomal acid lipase (HLAL) is an enzyme that catalyzes the deacylation of both cholesteryl esters and triacylglycerols after they have been internalized via the receptor-mediated endocytosis of lipoprotein particles (Assmann and Seedorf, 1995). In this process, the free cholesterol released regulates its own endogenous synthesis and esterification, as well as the uptake of low-density lipoprotein (Brown and Goldstein, 1986). HLAL plays an important role in the regulation of the intracellular cholesterol flux, and functional impairment of this lipase is thought to be responsible for the development of atherosclerosis (Coates *et al.*, 1986). HLAL is a member of the acid lipase family, along with the enteric gastric and lingual lipases (Anderson & Sando, 1991). This lipase family is completely separate from the well-documented pancreatic, lipoprotein, and hepatic neutral lipase families (Hide *et al.*, 1992).

In the early 1980s, many attempts were made to purify HLAL from human liver (Warner *et al.*, 1981), placenta (Burton and Mueller, 1980), aorta (Sakurada *et al.*, 1976) and leukocytes (Rindler-Ludwig *et al.*, 1977). Purified LAL was also obtained from rat and rabbit liver homogenates (Brown and Sgoutas, 1980; Imanaka *et al.*, 1984; Klemets and Lundberg, 1986). With the procedures used, purified protein yield was generally low, however, and the enzyme was found to be unstable after being isolated. In 1985, Sando and Rosenbaum developed a new method for purifying HLAL from fibroblast secretions grown on microcarriers. This procedure resulted in a 1500-fold purification of the enzyme and a 25 to 30% yield. Studies in which HLAL was characterized in details showed that the protein had an apparent molecular mass of 49 kDa on SDS-polyacrylamide gels and preferentially hydrolyzed fatty acyl esters with intermediate chain lengths and cis-unsaturated long-chain fatty acyl esters. Ameis *et al.* (1994) have also purified liver HLAL to apparent homogeneity, and these authors reported that the mature enzyme was a glycosylated 42-kDa monomer, with a high mannose sugar residue content. HLAL has also been expressed in low amounts using heterologous systems such as Cos-1 cells (Anderson and Sando, 1991) and baculovirus/insect cell systems (Sheriff *et al.*, 1995; Dupuis *et al.*, 1997).

3. Gene Sequence of Acid Lipases

The cDNAs coding for RLL (Docherty *et al.*, 1985), CPL (Timmermans *et al.*, 1994), HGL (Bodmer *et al.*, 1987) and DGL and RGL (Bénicourt *et al.*, 1993) have been cloned. A high degree of sequence homology (up to 80%) was found to exist between HGL and the other preduodenal lipases (Fig. 3). Upon cloning the HLAL cDNA (Anderson & Sando, 1991), it was subsequently observed that the protein is a member of mammalian acid lipase gene family, and the 378 amino acid sequence was deduced and was found to be 58% and 57% identical to the deduced sequences of HGL and RLL, respectively.

The HLAL gene spans a 36 kb stretch including 10 exons separated by 9 introns (Anderson *et al.*, 1994; Aslanidis *et al.*, 1994). Lohse *et al.* (1997c) have shown that the gene locus was slightly larger (38.8 kb) and attributed the difference to DNA polymorphism. The last exon is the largest and encodes the C-terminal domain of the protein as well as the 3'-untranslated region of the gene. The structures of the HGL and RLL genes were determined by intron amplification (Lohse *et al.*, 1997c) using cDNA primers designed to flank the putative exon/intron borders, based on the expected homology with the previously determined HLAL gene structure. The genes encoding HGL and RLL are also composed of 10 exons separated by 9 introns and span roughly 14 kb and 18.7 kb stretches of genomic DNA, respectively. The differences recorded in size are mainly due to variations in the intron length,

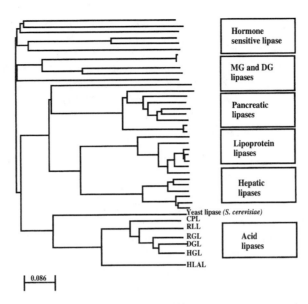

Figure 3. Dendrogram of sequence alignment within the lipase family. This dendrogram is based on the pairwise amino acid identities. The multisequence alignment was carried out by using the CLUSTAL software program from the PC-GENE package.

although the structure of the HGL and RLL genes is identical. The positions of the intervening sequences in HLAL are also well conserved, except for the location of intron 1. These results strongly suggest that the HGL/RLL and HLAL were generated by a gene duplication mechanism and adopted different roles in the neutral lipid metabolism, depending on their protein sequences and on their tissue-specific patterns of expression.

4. Preduodenal Lipases are Glycoproteins

The amino acid sequence of HGL was determined by cloning the cDNA (Bodmer *et al.*, 1987), and HGL was found to consist of a single 379 amino acid polypeptide with a molecular mass of 43 kDa. The experimentally determined molecular mass (50 kDa) of the native enzyme purified from gastric juice indicated that HGL, RGL, and DGL are all glycoproteins containing around 15% carbohydrate. Endoglycosidase F (Moreau *et al.*, 1992) treatment of HGL yielded three major bands, which could be separated. Based on the known specificity of endoglycosidase F and its ability to remove all the carbohydrates bound to HGL, the glycan moiety of HGL can be said to consist of asparagine-linked carbohydrates. HGL possesses four potential N-glycosylation sites (asparagine 15, 80, 252, and 308) having the consensus sequence Asn-X-Ser or Asn-X-Thr. CPL is also a glycoprotein with an N-linked carbohydrate content amounting to 20% of the total molecular mass of the native protein. After endo F treatment, CPL conserved only 20% of its original activity (Timmermans, 1996), unlike HGL, in which 100% of the activity was recovered after endo F treatment (Moreau *et al.,* 1992). After 3 hours of PNGase F treatment, the molecular mass of the deglycosylated LPL was reduced to 43 kDa and the remaining enzyme activity amounted to 30% of its initial value (De Caro *et al.,* 1995). The high level of glycosylation of preduodenal lipases results in the charge heterogeneity of these lipases. Although HGL, DGL, and RGL display a single protein band on SDS-PAGE, they showed multiple bands when migrating in an isoelectrofocusing gel. The isoelectric points observed ranged between 5.8 and 7.2 in the case of RGL (more than 7 isoforms), 6.8 and 7.8 in that of HGL (4 isoforms), and 6.3 and 6.5 in that of DGL (at least two isoforms; Moreau *et al.,* 1992). Recombinant HGL migrated with an apparent molecular mass of 45 kDa, under SDS-PAGE analysis (as compared with 50 kDa in the case of native HGL), due to the limited capacity of insect cells to glycosylate proteins (Wicker-Planquart *et al.,* 1996). Some heterogeneity therefore exists in recombinant HGL, although it is less marked than in the natural form, because three isoforms were observed in recombinant HGL with estimated isoelectric points of 7.0 (minor band), 7.5, and 8.1, respectively.

HLAL contains six glycosylation consensus sequences (Asn-X-Ser/Thr), three of which (Asn^{15}, Asn^{80} and Asn^{252}) are conserved among all the members of the acid lipase gene family (Anderson and Sando, 1991). The enzyme is transported to the lysosome via the mannose 6-phosphate receptor system (Sando & Henke, 1982). Sheriff *et al.*(1995) observed that the HLAL enzymatic activity was conserved after deglycosylation, which indicates that HLAL glycosylation is not necessary for the expression of catalytic activity. On the other hand, Tunicamycin treatment of Sf9 insect cells resulted in the production of an inactive HLAL, which could be detected intracellularly. The authors concluded that protein folding and production of a catalytically active enzyme require cotranslational glycosylation of HLAL. Once the protein has been suitably folded, the occupancy of the glycosylation sites is no longer necessary for a catalytically active conformer to be conserved.

5. Preduodenal Lipases are Serine Enzymes

Biochemical and crystallographic studies have shown that all lipases have a common feature: they all include a catalytic triad composed of serine, histidine, and aspartate or glutamate (Brady *et al.,* 1990; Winkler *et al.,* 1990; Brzozowski *et al.,* 1991; Schrag *et al.,* 1991). Nevertheless, the residues involved in the active site of HLAL as well as that of HGL have not yet been clearly defined. An essential serine seems to contribute importantly to acid lipase activity, because HLAL (Anderson & Sando, 1991) and CPL (Timmermans, 1996), like HGL and RLL, are specifically inhibited by micellar diethyl p-nitrophenyl phosphate E600 (Moreau *et al.,* 1991), which is a well-known serine esterase inhibitor (Fig. 4). RGL, CPL, and HLAL showed the same inhibition profile, that is, the lipolytic activity was completely abolished during the first hour of incubation; whereas in the case of HGL, the inhibition proceeded at a lower rate, reaching a plateau at 20% remaining activity after 5 hr of incubation. This difference in the rate of inhibition might reflect the fact that the accessibility differs between the active sites. Futhermore, acidic lipases show the consensus sequence characteristic of serine enzymes "(LIV)-X-(LIVFY)-(LIVST)-G-(HYWV)-S-X-G-(GSTAC)" (Ollis *et al.,* 1992), which is located around Ser 153. In pancreatic lipases, the active serine residue (Ser 152) is located at the hinge of the nucleophilic elbow and has been

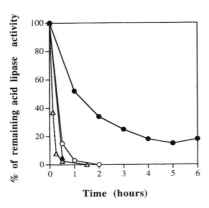

Y-axis: % of remaining acid lipase activity (0, 20, 40, 60, 80, 100)
X-axis: Time (hours) (0, 1, 2, 3, 4, 5, 6)

Figure 4. Time course of the inhibition of HGL, RGL, CPL, and HLAL during incubation with E_{600}. In studies by Moreau *et al.* (1991) on HGL (●) and RGL (○) inhibition, both enzymes (25 nmol) were incubated with 2.1 μmol of E_{600} (200 μCi) at 25°C in 50 mM acetate buffer (pH 6.0), 50 mM NaCl, 25 mM $CaCl_2$, and 3 mM NaTDC. Residual lipase activity was measured on a tributyrin substrate. In studies by Timmermans (1996) on CPL (△) inhibition, the enzyme was incubated with 0.48 mM E_{600} at room temperature in a pH-stat at pH 5.5 using a 2.5 ml tributyrin emulsion substrate. The emulsion contained 2.5ml of gum arabic, 0.9% NaCl, and 2 μM BSA. In studies by Sheriff *et al.* (1995) on HLAL (▲) inhibition, the enzyme (5 μl of medium from Sf9 cells) was incubated with 20 μM E_{600}. The HLAL activity was estimated with the fluorogenic substrate 4-methylumbelliferyl oleate at a concentration of 250 μM.

found, as in other lipases, to participate in the charge relay system, together with a histidine and an aspartic (or glutamic) residue.

In order to further identify the amino acids of HLAL and HGL involved in the active site during the catalytic breakdown of ester bonds, Lohse *et al.* (1997a) have replaced the conserved serine, histidine, and aspartic acid residues within the acid lipase family by threonine, glycine, and glutamine, respectively. Site mutagenesis of Ser 153 abolished the activity of HLAL toward both glycerol trioleate and cholesteryl oleate, whereas replacing Ser 99 was not detrimental to the enzymatic activity of HLAL. The transiently expressed mutant enzyme conserved about 39 to 41% (as expressed in Cos-7 cells) and 63 to 66% (as expressed in Ltk- cells) of the catalytic activity measured on the two substrates. Based on the expression of S153A HLAL and S99A HLAL in insect cells, it was previously suggested by Sheriff *et al.* (1995) that Ser 153 might play an essential role in HLAL. Likewise, transient expression studies of S153T HGL resulted in an almost inactive enzyme, because less than 4% and 6% of the wild-type activity was conserved in the extracts and media, respectively. The mutation of Ser 99, on the contrary, yielded a partially functional enzyme that was still able to hydrolyze 41% and 45% of the triolein present in the Cos-7 cell extracts and media, respectively. Although His 262, 298, and 345 are completely conserved in HLAL, HGL, and RLL, their replacement in HLAL did not result in any significant decrease in the lipolytic activity drop in either cell line, whereas replacing the His 65 and 353 in HLAL and in HGL induced an almost complete loss of the esterolytic properties of both enzymes (Lohse *et al.*, 1997a). These authors attributed the essential functional role there to the His 353 residue, based on the assumption that catalytic His residue is usually located further along the amino acid sequence, on Ser 153. Replacing His 274 led to an inactive HLAL mutant and to a partially active HGL. A large excess of diethylpyrocarbonate, which reacts with accessible His residues, partially inhibited RGL, however (50% activity residual, J. De Caro, personal communication).

Based on a similar assumption to that discussed previously, the third partner in the putative catalytic triad (Asp or Glu) may be located between Ser 153 and His 353. In this context, only 4 Asp residues (257, 324, 328, and 331) were thought to be plausible candidates for carboxylic acid. Consecutive mutations performed on these 4 Asp residues led these authors to conclude that Asp 324 was the third partner in the triad (Lohse *et al.* 1997a).

No crystallographic data on any acid lipases are available so far, although several attempts have been made along these lines (Moreau *et al.*, 1992) trying to definitely confirm the previous hypotheses.

6. Sulfhydryl Group in Acid Lipases

Three cysteine residues (Cys 227, Cys 236, and Cys 244) are present in both HGL and DGL. Titration of HGL (Gargouri *et al.*, 1988), RGL (Moreau *et al.*, 1988c) and DGL (Carrière *et al.*, 1991) was performed by incubating the lipase with sulfhydryl reagents such as dithio-nitrobenzoic acid (DTNB), lauroyl thio-nitrobenzoic acid (C12-TNB), or di(thiopyridine) (4-PDS). One titrable sulfhydryl group was found to exist per molecule of enzyme. Further adding a denaturing agent such as SDS or urea did not increase the number of modified SH groups. As only one free sulfhydryl group was titrated, it can be deduced that gastric lipases contain a single disulfide bridge. Futhermore, the modification of a single sulfhydryl group was accompanied by a concomitant loss of gastric lipase activity (Fig. 5). The authors concluded that HGL, RGL, and DGL possess a cysteine residue that seems to be either directly or indirectly involved in expressing the catalytic activity of the enzyme. Using the monomolecular film technique, it was established that after the modification of one sulfhydryl group, gastric lipase still binds to phosphatidylcholine and dicaprin films.

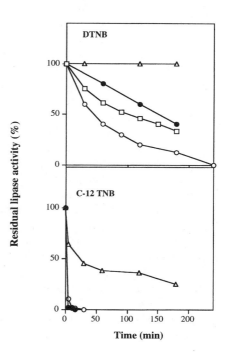

Figure 5. Variations in HGL, DGL, RGL and CPL activity during incubation with DTNB and C12:0-TNB. HGL (●) and RGL (○) (26μM) were incubated at 37°C, in a final volume of 150 μl of 50 mM acetate buffer (pH6.0), 150 mM NaCl and 26 mM C12:0-TNB (from Gargouri *et al.*, 1991). DGL (□) (13 nmol), HGL (24 nmol), and RGL (20 nmol) were incubated at pH 8.0 and 25°C with DTNB (1200 nmol; Gargouri *et al.*, 1988; Moreau *et al.*, 1988c; Carrière *et al.*, 1994). CPL (△) was incubated with a 100 molar excess of DTNB and C12:0 TNB (Timmermans, 1996).

Table 1. Cysteine Alignment in Acid Lipases. To Optimize the Amino Acid Sequence Alignment
of CPL with the Other Preduodenal Lipases, A Gap of 1 Amino Acid was Introduced
After Amino Acid Residue 4. Cysteine Residues Are Given in Bold Characters
in the Sequence and Are Underlined in the Signal Peptide.

	-11	-1	27	41	188	227	236	240	244
HGL	S	G	N	E	T	C	C	L	C
HLAL	C	G	S	C	C	C	C	C	C
RGL	A	G	S	E	T	C	C	L	C
RLL	S	G	C	G	T	C	C	L	C
DGL	ND	ND	A	G	T	C	C	L	C
CPL	C	C	S	Q	T	C	C	L	T

The modification of the sulfhydryl group therefore seems to prevent the access of the substrate to the catalytic site and not to interfere with the interfacial binding step (Gargouri *et al.*, 1988).

Three of these cysteines are well conserved at positions 227, 236, and 244 in all the members of the lipase family except for CPL, which possesses only two cysteines (Table 1) and RLL, which has an additional fourth cysteine residue at position 27. HLAL, on the contrary, contains six cysteines at positions 41, 188, 227, 236, 240, and 244 and an additional cysteine within the 27 amino acid signal peptide (Anderson and Sando, 1991). Pagani *et al.* (1997) have reported that treating HLAL with the reducing agent dithiothreitol affected the triacylglycerol and cholesterol esterase activity differently, which suggested that cysteine residues may contribute to determining the substrate specificity. These authors carried out an extensive functional study using site-directed HLAL mutants in order to determine the role of cysteine residues in the substrate specificity and catalysis. They observed that Cys 227, Cys 236, Cys 240, and Cys 244 play a crucial role in determining the HLAL substrate specificity, and they suggested that these cysteine residues may be involved in the hydrolysis of cholesteryl ester by selectively blocking the access of this substrate to the active catalytic site. In addition, the fact that the catalytic activity is never completely abolished in these mutants shows that HLAL is not a thiol enzyme. Using a site-directed mutagenesis approach, Lohse *et al.* (1997b) have shown that Cys 227 and Cys 236 are essential to the hydrolysis of cholesteryl esters and to a lesser extent, to the catalysis of triolein and tributyrin. The well conserved Cys 244 and the three additional HLAL cysteines appear however not to play an important role in the catalysis.

CPL, a preduodenal lipase showing 75% and 71% amino acid sequence homology with HGL and RLL, respectively, contains only two cysteines, at positions 226 and 235, whereas Cys 244 is replaced by a threonine (Timmermans *et al.*, 1994). In addition, Timmermans *et al.* (1996) have reported that CPL was not inhibited by DTNB even after more than 3 hours of incubation (Fig. 5A). By contrast, C12:0 TNB partially inhibits the enzyme, although this effect is less pronounced and much slower than in the case of HGL and DGL (Fig. 5B). These authors concluded that CPL has no functionnal free thiol group, which suggests that the conserved cysteines may form the single disulfide bridge in CPL. Based on the data on CPL and on their own results, Lohse *et al.* (1997b) assigned the disulfide bridge in HGL and HLAL to a location between Cys 227 and Cys 236, which meant that Cys 244 was the single free sulfhydryl. These conclusions are contradictory to those obtained by Aoubala *et al.* (1994), based on HGL and RGL peptide analyses. The latter authors took the disulfide bridge in HGL to lie between Cys 236 and Cys 244, while Cys 227 was taken to be the free thiol group. De Caro *et al.* (1998) recently observed that RGL appears to have a heterogenous pattern of cysteine residues. The 30% enzymatic activity of RGL that persisted after trypsin treatment may be attribuable to the 45 kDa molecular form (with the Cys227–Cys236 or Cys227–Cys244 disulfide bridge). In the absence of any information on the three-dimensional structure of acid lipases, which constitute a separate family of lipases, it seems to be premature to draw any definite conclusions about the possible role of sulfhydryl groups in acid lipases.

7. Conclusion

Preduodenal lipases, along with pancreatic lipases, play an essential role in lipid digestion. Conventional treatments for obesity have focused largely on strategies for controlling the energy intake. These approaches are not very efficacious in the long term, however (Bennet, 1987). Reducing dietary fat absorption by inhibiting the digestive lipases constitutes a new promising approach to the treatment of obesity. It has been established that tetrahydrolipstatin (THL), which is derived from lipstatin and is produced by *Streptomyces toxytricini,* strongly inhibits pancreatic and gastric lipases as well as cholesteryl ester hydrolase *in vitro* (Borgström, 1988; Hadvàry *et al.,* 1988; Gargouri *et al.,* 1991; Ransac *et al.,* 1991). Several studies are now avalaible on the potential clinical applications of THL in the treatment of human obesity (Drent *et al.,* 1995; Zhi et al., 1995; Schwizer *et al.,* 1997; Hildebrand *et al.,* 1998).

References

Abrams, C. K., Hamosh, M., Hubbard, V. S., Dutta, S. K., and Hamosh, P., 1984, Lingual lipase in cystic fibrosis. Quantification of enzyme activity in the upper small intestine of patients with exocrine pancreatic insufficiency, *J. Clin. Invest.* **73**:374–382.

Abrams, C. K., Hamosh, M., Dutta, S. K., Hubbard, V. S., and Hamosh, P., 1987, Role of nonpancreatic lipolytic activity in exocrine pancreatic insuffciency, *Gastroenterology* **92**:125–129.

Ameis, D., Merkel, M., Eckerskorn, C., and Greten, H., 1994, Purification, characterization and molecular cloning of human hepatic lysosomal acid lipase, *Eur. J. Biochem.* **219**:905–914.

Anderson, R. A., and Sando, G. N., 1991, Cloning and expression of cDNA encoding human lysosomal acid lipase/cholesteryl ester hydrolase. Similitaries to gastric and lingual lipases, *J. Biol. Chem.* **266**:22479–22484.

Anderson, R. A., Byrum, R. S., Coates, P. M., and Sando, G. N., 1994, Mutations at the lysosomal acid cholesteryl ester hydrolase gene locus in Wolman disease, *Proc. Natl. Acad. Sci. USA* **91**:2718–2722.

Aoubala, M., Douchet, I., Laugier, R., Hirn, M., Verger, R., and De Caro, A., 1993, Purification of human gastric lipase by immunoaffinity and quantification of this enzyme in the duodenal contents using a new ELISA procedure, *Biochim. Biophys. Acta* **1169**:183–188.

Aoubala, M., Bonicel, J., Bénicourt, C., Verger, R., and De Caro, A., 1994, Tryptic cleavage of gastric lipases: Location of the single disulfide bridge, *Biochim. Biophys. Acta* **1213**:319–324.

Aslanidis, C., Klima, H., Lackner, K. J., and Schmitz, G., 1994, Genomic organization of the human lysosomal acid lipase gene (LIPA), *Genomics* **20**:329–331.

Assmann, G., and Seedorf, U., 1995, Acid lipase deficiency: Wolman disease and cholesteryl ester storage disease, in: *The Metabolic Bases of Inherited Diseases* (C. R. Scriver, A. L. Beaudet, W. S. Sly, and D. Valle, eds.), McGraw-Hill, New York, pp. 2563–2587.

Bénicourt, C., Blanchard, C., Carrière, F., Verger, R., and Junien, J.-L., 1993, Potential use of a recombinant dog gastric lipase as an enzymatic supplement to pancreatic extracts in cystic fibrosis, in: *Clinical Ecology of Cystic Fibrosis* (H. Escobar, C. F. Baquero, and L. Suárez, eds.), Elsevier Science Publishers B.V, Amsterdam, pp. 291–295.

Bennet, W., 1987, Dietary treatments of obesity, *Ann. N. Y. Acad. Sci.* **449:**250–263.

Bernbäck, S., Hernell, O., and Blackberg, L., 1985, Purification and molecular characterization of bovine pregastric lipase, *Eur. J. Biochem.* **148:**233–238.

Bernbäck, S., Hernell, O., and Bläckberg, L., 1987, Bovine pregastric lipase: A model for the human enzyme with respect to properties relevant to its site of action, *Biochim. Biophys. Acta* **922:**206–213.

Bernbäck, S., Bläckberg, L., and Hernell, O., 1989, Fatty acids generated by gastric lipase promote human milk triacylglycerol digestion by pancreatic colipase-dependent lipase, *Biochim. Biophys. Acta* **1001:**286–291.

Bernbäck, S., Blackberg, L., and Hernell, O., 1990, The complete digestion of milk triacylglycerol *in vitro* requieres gastric lipase, pancreatic colipase-dependant lipase and bile salt-stimulated lipase, *J. Clin. Invest.* **85:**1221–1226.

Bodmer, M. W., Angal, S., Yarranton, G. T., Harris, T. J. R., Lyons, A., King, D. J., Piéroni, G., Rivière, C., Verger, R., and Lowe, P. A., 1987, Molecular cloning of a human gastric lipase and expression of the enzyme in yeast, *Biochim. Biophys. Acta* **909:**237–244.

Borgström, B., 1988, Mode of action of tetrahydrolipstatin: A derivative of the naturally occuring lipase inhibitor lipstatin, *Biochim. Biophys. Acta* **962:**308–316.

Borovicka, J., Schwizer, W., Mettraux, C., Kreiss, C., Remy, B., Asal, K., Jansen, J. B. M. J., Douchet, I., Verger, R., and Fried, M., 1997, Regulation of gastric and pancreatic lipase secretion by CCK and cholinergic mechanisms in humans, *Am. J. Physiol.* **273:**G374–G380.

Brady, L., Brzozowski, A. M., Derewenda, Z. S., Dodson, E., Dodson, G., Tolley, S., Turkenburg, J. P., Christiansen, L., Huge-Jensen, B., Norskov, L., Thim, L., and Menge, U., 1990, A serine protease triad forms the catalytic centre of a triacylglycerol lipase, *Nature* **343:**767–770.

Brown, M. S., and Goldstein, J. L., 1986, A receptor-mediated pathway for cholesterol homeostasis, *Science* **232:**34–47.

Brown, W. J., and Sgoutas, D. S., 1980, Purification of rat liver lysosomal cholesteryl ester hydrolase, *Biochim. Biophys. Acta* **617:**305–317.

Brzozowski, A. M., Derewenda, U., Derewenda, Z. S., Dodson, G. G., Lawson, D. M., Turkenburg, J. P., Bjorkling, F., Huge-Jensen, B., Patkar, S. A., and Thim, L., 1991, A model for interfacial activation in lipases from the structure of a fungal lipase-inhibitor complex, *Nature* **351:**491–494.

Burton, B. K., and Mueller, H. W., 1980, Purification and properties of human placental acid lipase, *Biochim. Biophys. Acta* **618:**449–460.

Carey, M. C., Small, D. M., and Bliss, C. M., 1983, Lipid digestion and absorption of fat, *Ann. Rev. Physiol.* **45:**651–677.

Carrière, F., Moreau, H., Raphel, V., Laugier, R., Bénicourt, C., Junien, J.-L., and Verger, R., 1991, Purification and biochemical characterization of dog gastric lipase, *Eur. J. Biochem.* **202:**75–83.

Carrière, F., Raphel, V., Moreau, H., Bernadac, A., Devaux, M.-A., Grimaud, R., Barrowman, J. A., Bénicourt, C., Junien, J.-L., Laugier, R., and Verger, R., 1992, Dog gastric lipase: Stimulation of its secretion *in vivo* and cytolocalization in mucous pit cells, *Gastroenterology* **102:**1535–1545.

Carrière, F., Barrowman, J. A., Verger, R., and Laugier, R., 1993, Secretion and contribution to lipolysis of gastric and pancreatic lipases during a test meal in humans, *Gastroenterology* **105:**876–888.

Carrière, F., Gargouri, Y., Moreau, H., Ransac, S., Rogalska, E., and Verger, R., 1994, Gastric lipases: Cellular, biochemical and kinetic aspects, in: *Lipases: Their Structure, Biochemistry and Application* (P. Wooley and S. B. Petersen, eds.) Cambridge University Press, Cambridge, England, pp. 181–205.

Carrière, F., Rogalska, E., Cudrey, C., Ferrato, F., Laugier, R., and Verger, R., 1997, *In vivo* and *in vitro* studies on the stereoselective hydrolysis of tri- and diglycerides by gastric and pancreatic lipases, *Bioorganic and Medicinal Chemistry* **5:**429–435.

Coates, P. M., Langer, T., and Cortner, J. A., 1986, Genetic variation of human mononuclear leukocyte lysosomal lipase activity. Relationship to atherosclerosis, *Atherosclerosis* **62:**11–20.

Crabbe, T., Weir, A. N., Walton, E. F., Brown, M. E., Sutton, C. W., Tretou, N., Bonnerjea, J., Lowe, P. A., and Yarranton, G. T., 1996, The secretion of active recombinant human gastric lipase by Saccharomyces cerevisiae, *Protein Expr. Purif.* **7:**229–236.

De Caro, J., Ferrato, F., Verger, R., and De Caro, A., 1995, Purification and molecular characterization of lamb pregastric lipase, *Biochim. Biophys. Acta* **1252:**321–329.

De Caro, J., Verger, R., and De Caro, A., 1998, An enzymatically active truncated form (-55N-terminal residues) of rabbit gastric lipase. Correlation between the enzymatic activity and disulfide bond oxydo-reduction state, *Biochim. Biophys. Acta* **386**:39–49.

DeNigris, S. J., Hamosh, M., Kasbekar, D. K., Lee, T. C., and Hamosh, P., 1988, Lingual and gastric lipases: Species differences in the origin of prepancreatic digestive lipases and in the localization of gastric lipase, *Biochim. Biophys. Acta* **959**:38–45.

DiMagno, E. P., 1982, Controversies in the treatment of exocrine pancreatic insufficiency, *Dig. Dis. Sci.* **27**:481–484.

DiMagno, E. P., Go, V. L. W., and Summerskill, W. H. J., 1973, Relations between pancreatic enzyme outputs and malabsorption in severe pancreatic insufficiency, *N. Engl. J. Med.* **288**:813–815.

Docherty, A. J. P., Bodmer, M. W., Angal, S., Verger, R., Rivière, C., Lowe, P. A., Lyons, A., Emtage, J. S., and Harris, T. J. R., 1985, Molecular cloning and nucleotide sequence of rat lingual lipase cDNA, *Nucleic Acid Res.* **13**:1891–1903.

Drent, M. L., Larsson, I., Williamolsson, T., Quaade, F., Czubayko, F., Vonbergmann, K., Strobel, W., Sjostrom, L., and Vanderveen, E. A., 1995, Orlistat (RO 18–0647), a lipase inhibitor, in the treatment of human obesity: A multiple dose study, *Int. J. Obes.* **19**:221–226.

Dupuis, L., Canaan, S., Rivière, M., and Wicker-Planquart, C., 1997, Influence of various signal peptides on secretion of mammalian acidic lipases in baculovirus-insect cell system, in: *Methods in Enzymology* (A. Rubin and E. A. Dennis, eds.) Academic Press, INC, San Diego, pp. 261–272.

Field, R., and Scow, R. O., 1983, Purification and characterization of rat lingual lipase, *J. Biol. Chem.* **258**:14563–14569.

Gargouri, Y., Piéroni, G., Lowe, P. A., Sarda, L., and Verger, R., 1986a, Human gastric lipase. The effect of amphiphiles, *Eur. J. Biochem.* **156**:305–310.

Gargouri, Y., Piéroni, G., Rivière, C., Lowe, P. A., Saunière, J.-F., Sarda, L., and Verger, R., 1986b, Importance of human gastric lipase for intestinal lipolysis: An *in vitro* study, *Biochim. Biophys. Acta* **879**:419–423.

Gargouri, Y., Piéroni, G., Rivière, C., Saunière, J.-F., Lowe, P. A., Sarda, L., and Verger, R., 1986c, Kinetic assay of human gastric lipase on short- and long-chain triacylglycerol emulsions, *Gastroenterology* **91**:919–925.

Gargouri, Y., Moreau, H., Piéroni, G., and Verger, R., 1988, Human gastric lipase: A sulfhydryl enzyme, *J. Biol. Chem.* **263**:2159–2162.

Gargouri, Y., Chahinian, H., Moreau, H., Ransac, S., and Verger, R., 1991, Inactivation of pancreatic and gastric lipases by THL and $C_{12:0}$-TNB: A kinetic study with emulsified tributyrin, *Biochim. Biophys. Acta* **1085**:322–328.

Hadvàry, P., Lengsfeld, H., and Wolfer, H., 1988, Inhibition of pancreatic lipase *in vitro* by the covalent inhibitor tetrahydrolipstatin, *Biochem. J.* **256**:357–361.

Hamosh, M., Ganot, D., and Hamosh, P., 1979, Rat lingual lipase: Characteristics of enzyme activity, *J. Biol. Chem.* **254**:12121–12125.

Hernell, O., and Blackberg, L., 1994, Molecular aspects of fat digestion in the newborn, *Acta Paediat.* **83**:65–69.

Hide, W. A., Chan, L., and Li, W. H., 1992, Structure and evolution of the lipase superfamily, *J. Lipid Res.* **33**:167–178.

Hildebrand, P., Petrig, C., Burckhardt, B., Ketterer, S., Lengsfeld, H., Fleury, A., Hadvary, P., and Beglinger, C., 1998, Hydrolysis of dietary fat by pancreatic lipase stimulates cholecystokinin release, *Gastroenterology* **114**:123–129.

Hull, M., and Keaton, R. W., 1917, The existence of a gastric lipase, *J. Biol. Chem.* **32**:127–140.

Imanaka, T., Amanuma-Muto, K., Ohkuma, S., and Takano, T., 1984, Characterization of lysosomal acid lipase purified from rabbit liver, *Biochim. Biophys. Acta* **665**:322–330.

Isaacs, C. E., Kashyap, S., Heird, W. C., and Thormar, H., 1990, Antiviral and antibacterial lipids in human milk and infant formula feed, *Arch. Dis. Child.* **65**:861–865.

Isaacs, C. E., Litov, R. E., Marie, P., and Thormar, H., 1992, Addition of lipases to infant formulas produces antiviral and antibacterial activity, *J. Nutr. Biochem.* **3**:304–308.

Klemets, R., and Lundberg, B., 1986, Substrate specificity of lysosomal cholesteryl ester hydrolase isolated from rat liver, *Lipids* **21**:481–485.

Lapey, A., Kattwinkel, J., di Sant Agnese, P. A., and Laster, L., 1974, Steatorrhea and azotorrhea and thier relation to growth and nutrition in adolescents and young adults with cystic fibrosis, *J. Pediatr.* **84**:328–334.

Lohse, P., Chahrokh-Zadeh, S., Lohse, P., and Seidel, D., 1997a, Human lysosomal acid lipase/cholesteryl ester hydrolase and human gastric lipase: Identification of the catalytically active serine, aspartic acid and histidine residues, *J. Lipid Res.* **38**:892–903.

Lohse, P., Lohse, P., Chahrokh-Zadeh, S., and Seidel, D., 1997b, Human lysosomal acid lipase/cholesteryl ester hydrolase and human gastric lipase: Site-directed mutagenesis of Cys227 and Cys236 results in substrate-dependent reduction of enzymatic activity, *J. Lipid Res.* **38:**212–221.

Lohse, P., Lohse, P., Chahrokh-Zadeh, S., and Seidel, D., 1997c, The acid lipase gene family: Three enzymes, one highly conserved gene structure, *J. Lipid Res.* **38:**880–891.

Moreau, H., Gargouri, Y., Lecat, D., Junien, J.-L., and Verger, R., 1988a, Purification, characterization and kinetic properties of the rabbit gastric lipase, *Biochim. Biophys. Acta* **960:**286–293.

Moreau, H., Gargouri, Y., Lecat, D., Junien, J.-L., and Verger, R., 1988b, Screening of preduodenal lipases in several mammals, *Biochim. Biophys. Acta* **959:**247–252.

Moreau, H., Gargouri, Y., Piéroni, G., and Verger, R., 1988c, Importance of sulfhydryl group for rabbit gastric lipase activity, *FEBS Lett.* **236:**383–387.

Moreau, H., Saunière, J.-F., Gargouri, Y., Piéroni, G., Verger, R., and Sarles, H., 1988d, Human gastric lipase: Variations induced by gastrointestinal hormones and by pathology, *Scand. J. Gastroenterol.* **23:**1044–1048.

Moreau, H., Moulin, A., Gargouri, Y., Noël, J.-P., and Verger, R., 1991, Inactivation of gastric and pancreatic lipases by diethyl *p*-nitrophenyl phosphate, *Biochemistry* **30:**1037–1041.

Moreau, H., Abergel, C., Carrière, F., Ferrato, F., Fontecilla-Camps, J. C., Cambillau, C., and Verger, R., 1992, Isoform purification of gastric lipases: Towards crystallization, *J. Mol. Biol.* **225:**147–153.

Mossi, S., Aoubala, M., de Caro, A., Verger, R., and Laugier, R., 1994, Gastric lipase (GL) becomes essential for the lipid digestion during severe pancreatic exocrine insuffiency (PEI), *Digestion* **55:**328.

Muller, D. P. R., McCollum, J. P. K., Tompeter, R. S., and Harries, J. T., 1975, Studies on the mechanism of fat absorption in congenital isolated lipase deficiency, *Gut.* **16:**838.

Ollis, D. L., Cheah, E., Cygler, M., Dijkstra, B., Frolow, F., Franken, S. M., Harel, M., Remington, S. J., Silman, I., Schrag, J., Sussman, J. L., Verschueren, K. H. G., and Goldman, A., 1992, The α/β hydrolase fold, *Protein Eng.* **5:**197–211.

Pagani, F., Pariyarath, R., Stuani, C., Garcia, R., and Baralle, F. E., 1997, Cysteine residues in human lysosomal acid lipase are involved in selective cholesteryl esterase activity, *Biochem J.* **326:**265–269.

Ransac, S., Gargouri, Y., Moreau, H., and Verger, R., 1991, Inactivation of pancreatic and gastric lipases by tetrahydrolipstatin and alkyl-dithio-5-(2-nitrobenzoic acid). A kinetic study with 1,2-didecanoyl-sn-glycerol monolayers, *Eur. J. Biochem.* **202:**395–400.

Rindler-Ludwig, R., Patsch, W., Saimler, S., and Braunsteiner, H., 1977, Characterization and partial purification of acid lipase from human leucocytes, *Biochim. Biophys. Acta* **488:**294–304.

Roberts, I. M., and Hand, S. I., 1988, Stability of lingual lipase *in vivo:* Studies of the iodinated enzyme in the rat stomach and duodenum, *Biochim. Biophys. Acta* **960:**107–110.

Roberts, I. M., Montgomery, R. K., and Carey, M. C., 1984, Lingual lipase: Partial purification, hydrolytic properties and comparison with pancreatic lipase, *Am. J. Physiol.* **247G:**385–393.

Ross, C. A. C., 1955, Fat absorption studies in the diagnosis and treatment of pancreatic fibrosis, *Arch. Dis. Child* **30:**316–320.

Ross, C. A. C., and Sammons, H. C., 1955, Non-pancreatic lipase in children with pancreatic fibrosis, *Arch. Dis. Child* **30:**428–431.

Sakurada, T., Orimo, H., Okabe, H., Noma, A., and Murakami, M., 1976, Purification and properties of cholesterol ester hydrolase from human aortic intima and media, *Biochim. Biophys. Acta* **424:**204–212.

Sando, G. N., and Henke, V. L., 1982, Recognition and receptor-mediated endocytosis of the lysosomal acid lipase secreted by cultured human fibroblasts, *J. Lipid Res.* **23:**114–123.

Sando, G. N., and Rosenbaum, L. M., 1985, Human lysosomal acid lysase/cholesteryl ester hydrolase: purification and properties of the form secreted by fibroblasts in microcarrier cultures, *J. Biol. Chem.* **260:**15186–15193.

Schrag, J. D., Li, Y., Wu, S., and Cygler, M., 1991, Ser-His-Glu triad forms the catalytic site of the lipase from *Geotrichum candidum, Nature* **351:**761–764.

Schwizer, W., Asal, K., Kreiss, C., Mettraux, C., Borovicka, J., Remy, B., Guzelhan, C., Hartmann, D., and Fried, M., 1997, Role of lipase in the regulation of upper gastrointestinal function in humans, *Am. J. Physiol.* **273:**G612–G620.

Sheriff, S., Du, H., and Grabowski, G. A., 1995, Characterization of lysosomal acid lipase by site-directed mutagenesis and heterologous expression, *J. Biol. Chem.* **270:**27766–27772.

Smerdon, G. R., Aves, S. J., and Walton, E. F., 1995, Production of human gastric lipase in the fission yeast Schizosaccharomyces pombe, *Gene* **165:**313–318.

Suzuki, A., Mizumoto, A., Sarr, M. G., and Dimagno, E. P., 1997, Bacterial lipase and high-fat diets in canine exocrine pancreatic insufficiency: A new therapy of steatorrhea? *Gastroenterology* **112:**2048–2055.

Szafran, Z., Szafran, H., Popiela, T., and Trompeter, G., 1978, Coupled secretion of gastric lipase and pepsin in man following pentagastrin stimulation, *Digestion* **18:**310–318.

Thormar, H., Isaacs, C. E., Brown, H. R., Barshatzky, M. R., and Pessolano, T., 1987, Inactivation of enveloped viruses and killing of cells by fatty acids and monoglycerides, *Antimicrob. Agents Ch.* **31:**27–31.

Timmermans, M., 1996, Purification, molecular cloning and expression of the cDNA of bovine pregastric esterase, *Ph.D. Thesis,* Limburgs Universitair Centrum, Diepenbeek, pp. 1–132.

Timmermans, M. Y. J., Teuchy, H., and Kupers, L. P. M., 1994, The cDNA sequence encoding bovine pregastric esterase, *Gene* **147:**259–262.

Timmermans, M. Y., Reekmans, G., Teuchy, H., and Kupers, L., 1996, Inhibition studies on calf pregastric esterase: The enzyme has no functional thiol group, *Biochem J* **314:**931–936.

Tiruppathi, C., and Balasubramanian, K. A., 1982, Purification and properties of an acid lipase from human gastric juice, *Biochim. Biophys. Acta* **712:**692–697.

Volhard, F., 1901, Über das fettspaltende Ferment des Magens, *Z. Klind. Med.* **42:**414–429.

Warner, T. G., Dambach, L. M., Shin, J. J., and O'Brien, J. S., 1981, Purification of the lysosomal acid lipase from human liver and its role in lysosomal lipid hydrolysis, *J. Biol. Chem.* **256:**2952–2957.

Wicker-Planquart, C., Canaan, S., Rivière, M., Dupuis, L., and Verger, R., 1996, Expression in insect cells and purification of a catalytically active recombinant human gastric lipase, *Protein Engineering* **9:**1225–1232.

Winkler, F. K., d'Arcy, A., and Hunziker, W., 1990, Structure of human pancreatic lipase, *Nature* **343:**771–774.

Zhi, J. G., Melia, A. T., Eggers, H., Joly, R., and Patel, I. H., 1995, Review of limited systemic absorption of orlistat, a lipase inhibitor, in healthy human volunteers, *J. Clin. Pharmacol.* **35:**1103–1108.

CHAPTER 3

Molecular Mechanisms of Pancreatic Lipase and Colipase

MARK E. LOWE

1. Introduction

A healthy diet includes dietary fats (Carey and Hernell, 1992). Fats provide an important energy source, a vehicle for fat soluble vitamins, and acyl chain precursors for hormones, inflammatory mediators, and cellular membrane components. Fats also improve the palatability of foods and largely govern postprandial satiety, whereas an inadequate supply of dietary fat affects growth and leads to deficiencies of vitamins A, D, E, and K. Long-chain triglycerides contribute 95% of the dietary fats in the average Western diet.

The utilization of dietary triglycerides requires their degradation in the gut lumen because intestinal enterocytes can not absorb intact long-chain triglycerides. Instead, dietary fats enter enterocytes as sn-2 monoacylglycerols and as free fatty acids (Hofmann and Borgström, 1962; 1964). The concerted action of preduodenal and duodenal lipases releases these products from triglycerides (Carey and Hernell, 1992). The monoacylglycerols and fatty acids form mixed micelles with bile salts facilitating their entry into enterocytes. In humans, fat digestion begins in the stomach where gastric lipase releases about 15% of the fatty acids from dietary triglycerides (Carriere et $al.$, 1993). Lipases, secreted by pancreatic acinar cells into the small intestine, complete dietary fat digestion.

The exocrine pancreas secretes a variety of lipases that contribute to dietary fat digestion (Rinderknecht, 1993). One, pancreatic triglyceride lipase, (PTL) digests the majority of dietary triglycerides (Figarella et $al.$, 1980; Ghishan et $al.$, 1984). Two other exocrine proteins have marked homology to PTL and have been named pancreatic lipase related proteins 1 and 2 (Giller et $al.$, 1992). Although both related proteins resemble PTL and appear in pancreatic secretions, their role in dietary fat digestion has not been defined. Even so, the identification and characterization of these related proteins provided much information about the evolution and function of PTL and have stimulated hypotheses about the role of

MARK E. LOWE • Washington University School of Medicine, Departments of Pediatrics and of Molecular Biology and Pharmacology, St. Louis, Missouri 63110.

Intestinal Lipid Metabolism, edited by Charles M. Mansbach II et $al.$, Kluwer Academic/Plenum Publishers, 2001.

these proteins in dietary fat digestion (Lowe, 1997b). These findings and a number of recent studies have provided new insights into the biochemistry of lipolysis and the physiology of fat digestion.

2. The Lipase Gene Family

The similarities among the predicted amino acid sequences of three lipases—PTL, lipoprotein lipase, and hepatic lipase—led to the original hypothesis that a family of lipases evolved from a gene encoding an ancestral hydrolase (Kirchgessner *et al.*, 1987). Subsequent studies provided additional evidence for this hypothesis and the concept of a lipase gene family is now well accepted. The isolation of the genes encoding PTL, lipoprotein lipase, and hepatic lipase allowed comparison of their exon–intron organization (Kirchgessner *et al.*, 1989; Sims *et al.*, 1993; Deeb and Peng, 1989; Cai *et al.*, 1989; Ameis *et al.*, 1990). The comparison demonstrated conservation of the gene organization among these three lipases. Additionally, the genes encoding PTL and lipoprotein lipase contained a repetitive element, an Alu element, in corresponding introns—an observation that supported the relationship of the two genes (Sims *et al.*, 1993; Chuat *et al.*, 1992). Classification of the Alu sequences suggested that the two genes diverged within the last 40 to 50 million years.

The description of pancreatic lipase-related proteins 1 and 2 (PLRP1 and PLRP2) provided additional support for a lipase gene family (Giller *et al.*, 1992). Comparisons of the predicted amino acid sequences of these proteins with the sequence of PTL revealed 63 to 68% identity and 77 to 81% homology depending on the species (Lowe, 1997b). The conservation extended to determinants of secondary structure and of activity. All of the cysteine residues are conserved in the identical positions. Both related proteins contain the active site residues identified in PTL and they reside in the same relative positions in the primary sequence. The differences among these sequences indicate that separate genes encode each protein, whereas the similarities imply the genes evolved from a common ancestor of the lipase gene family.

Isolation and characterization of the genes encoding PLRP1 and PLRP2 increased the likelihood that the genes form a family. The three genes have similar exon–intron organization (Fig. 1). PTL and PLRP1 have 13 exons of nearly identical size (Sims *et al.*, 1993; Mickel *et al.*, 1989), and PLRP2 has only 12 exons, but exons II–XII coincide almost exactly with exons III–XIII in PTL and in PLRP1 (Kaplan *et al.*, 1996). The only difference lies in the first exons. The gene encoding PLRP2 has a large exon I corresponding to the smaller exons I and II in PTL and in PLRP1.

The chromosomal location of the genes encoding PTL and the related proteins also support the related lineage of these genes. They are clustered on chromosome 10 in humans and on chromosome 19 in mice (Warden *et al.*, 1993; Sims *et al.*, 1993; Davis *et al.*, 1991). The tight linkage of these genes and the identification of a PTL pseudogene near the human PTL gene locus suggest the genes arose through gene duplication events (Sims *et al.*, 1993). It is not possible, however, to determine which gene was the progenitor.

The lipase gene extends beyond lipases to other enzymes seemingly unrelated to PTL (Hide *et al.*, 1992; Cygler *et al.*, 1993). Despite poor primary sequence homology, a number of other esterases contain a unique, central three-dimensional structure also found in lipases (Fig. 2). The common folding pattern, termed the α/β hydrolase fold, conserves the

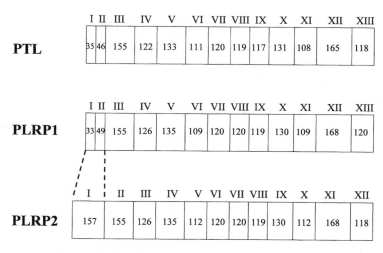

Figure 1. The exon–intron organization of the genes encoding PTL, PLRP1, and PLRP2. The exons are indicated by the boxes. The number of bases in each exon is given inside the box. The exons are numbered by Roman numerals. The figure is constructed from data in references (Sims *et al.,* 1993; Mickel *et al.,* 1989; Kaplan *et al.,* 1996).

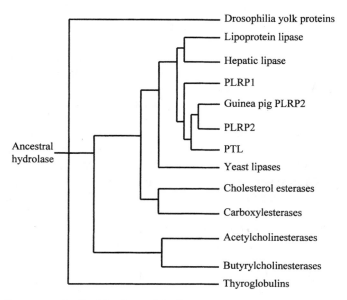

Figure 2. The lipase gene family. Members of the lipase gene family are mapped. Only approximate phylogenetic relationships are indicated.

topological position of the catalytic residues (Ollis *et al.,* 1992). The conserved folding pattern supports the evolution of these enzymes from a common gene encoding a primordial hydrolase.

3. Physiology

Expression of PTL is practically limited to the pancreas. There may be low levels of PTL mRNA expressed in other tissues under certain conditions, but the levels do not compare to the expression in pancreas (Terada *et al.,* 1993; Terada and Nakanuma, 1995). Pancreatic acinar cells synthesize and secrete PTL through both regulated and basal pathways (Von Zastrow and Castle, 1987; Arvan and Chang, 1987; Arvan and Castle, 1987). The secreted PTL enters the pancreatic duct and mixes with biliary lipids and bile salts before entering the duodenum to complete fat digestion. Although the expression patterns and physiology of PTL are widely accepted, less is known about the two related proteins. The pancreas expresses mRNA encoding PLRP1, but expression in other tissues has not been investigated (Giller *et al.,* 1992; Kerfelec *et al.,* 1986; Wicker-Planquart and Puigserver, 1992). The expression of PLRP2 is primarily in the pancreas. mRNA encoding PLRP2 was not detected in a variety of other tissues by RNA blot (Wishart *et al.,* 1993). PLRP2 has been found in cytotoxic T-cells after subculture in medium containing interleukin 4, but its presence in circulating lymphocytes has not been established (Grusby *et al.,* 1990). It remains plausible that tissues other than the pancreas express the related proteins.

The presence of mRNA does not ensure mRNA translation and protein expression. Repeated observations have demonstrated the presence of PTL in pancreatic juice, but evidence for the secretion of the related proteins is only beginning to accumulate. Isolated rat pancreatic acinar cells and a cell line with acinar properties, AR42-J, each secreted PLRP1 and PLRP2 into the medium, suggesting that the proteins should appear in pancreatic juice (Kullman *et al.,* 1996). In fact, both have now been identified in pancreatic juice. PLRP2 was found in secretions obtained from a cannulated pancreatic duct in the rat and PLRP1 was isolated from dog pancreatic juice (Wagner *et al.,* 1994; Roussel *et al.,* 1998a). Although the synthesis and secretion of the two related proteins by the pancreas suggests that they participate in fat digestion, the appearance in pancreatic juice does not prove that they function in this capacity.

The temporal pattern of mRNA expression for the three homologues during development in the rat suggested a potential role for the two related proteins in fat digestion (Payne *et al.,* 1994). The adult pancreas expresses all three mRNA species. The predominant species is the mRNA encoding PTL, however, the levels of mRNA encoding the related proteins are four fold lower for PLRP1 and 24-fold lower for PLRP2. A much different expression pattern occurs in the pancreas of newborns—mRNA encoding PLRP1 predominated followed by mRNA encoding PLRP2, and mRNA encoding PTL could not be detected. This initial observation suggested the discoordinate regulation of these mRNA species during development.

A more complete profile of the temporal expression confirmed that suckling animals express the two related proteins and have low levels of PTL expression. All three mRNA species were low prior to birth (Payne *et al.,* 1994). Only mRNA encoding PLRP1 and PLRP2 increased after birth with expression continuing into adulthood. In contrast, mRNA encoding PTL remained low until the suckling–weanling transition when the mRNA rose quickly to adult levels. Examination of the proteins secreted from acinar cells isolated from rats at different ages confirmed that protein synthesis mirrored the mRNA profile (Lowe, 1997c). The related proteins, but not PTL, were secreted by acinar cells taken from suckling animals. PTL did not appear in the secretions until near weaning. The temporal expression pattern of PTL and the two related proteins suggests that the related proteins may compensate for the low PTL activity in suckling animals to ensure efficient fat digestion in newborns who consume more fat per kilogram than at any other age.

Evidence for the participation of PLRP2 in newborn fat digestion came from studies of PLRP2-deficient mice (Lowe *et al.*, 1998). PLRP2-deficient mice were created by homologous recombination to insert a null allele at the locus encoding PLRP2. The mice appeared grossly normal at all ages, but closer examination revealed important differences. The amount of fecal fat in PLRP2-deficient suckling animals was much higher than seen in wild-type littermates (55% vs. 9%). As predicted from the mRNA data, this difference disappeared with age. Adult mice had no difference in fecal fat content even when placed on high-fat diets. Although the fat malabsorption did not runt the PLRP2-deficient suckling mice, their rate of weight gain was significantly decreased compared to wild-type and heterozygote pups. These two findings implicate PLRP2 in the digestion of dietary fats in newborn animals, providing an explanation for the high coefficient of fat absorption in suckling animals with relative PTL deficiency.

4. Protein Structure

A full understanding of lipase function requires detailed knowledge about the molecular details governing the enzymatic mechanism of these lipases. An essential first step to understanding the function of any protein is the delineation of the primary and the tertiary structure of the protein. Primary structures of PTL and the related proteins have been predicted from the cDNAs encoding these proteins in various species. As mentioned previously, the primary sequences of PTL and the related proteins are remarkably conserved. The PTL mRNA encodes a 465 amino acid protein (Winkler *et al.*, 1990; Lowe *et al.*, 1989). The sequence includes a short, 16 amino acid signal peptide that is cleaved to produce the secreted 449 amino acid PTL. Similarly, the PLRP1 mRNA encodes a protein with a short, 17 amino acid peptide that is cleaved to leave a 451 amino acid mature form in humans and a 456 residue protein in rats (Kerfelec *et al.*, 1986; Giller *et al.*, 1992; Wicker-Planquart and Puigserver, 1992). Additional amino acids at the C-terminus of the rat PLRP1 account for the difference in length. The size of unprocessed PLRP2 is not known because there are two potential transcription start sites (Wishart *et al.*, 1993; Payne *et al.*, 1994; Giller *et al.*, 1992; Grusby *et al.*, 1990; Thirstrup *et al.*, 1994). One predicts a 30 amino acid signal peptide and the other a 17 amino acid peptide. In either case, the length of the mature protein is 452 residues.

5. Tertiary Structure

One of the most important advances in understanding of lipase structure and function occurred when Winkler *et al.* solved the tertiary structure of human PTL (Winkler *et al.*, 1990). The 2.8 nm crystal structure showed that PTL consists of two distinct domains, a

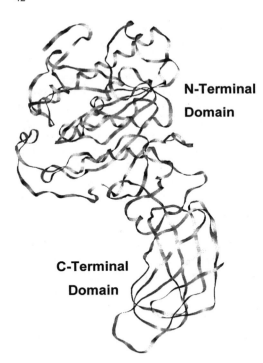

N-Terminal Domain

C-Terminal Domain

Figure 3. The three-dimensional structure of pancreatic triglyceride lipase. The tertiary structure of human PTL is shown as the α-carbon backbone. The coordinates of the colipase-PTL complex in the lid closed position were provided by Drs. Von Tilbeurgh and Cambillau (van Tilbeurgh *et al.,* 1992). The figure was constructed with Rasmol.

globular N-terminal domain formed by a central β-sheet core extending from amino acids 1 to 335 and a C-terminal domain of a β-sheet sandwich structure (Fig. 3). A short, unstructured stretch of amino acids joined the two domains, and seven disulfide bonds stabilized the overall structure. The folding pattern of the N-terminal domain resembles the α/β hydrolase fold later described in a variety of esterases and enolases (Ollis *et al.,* 1992). The three-dimensional structure of horse PTL has also been reported and it agrees with the human PTL structure completely (Bourne *et al.,* 1994).

Several other observations came from these crystal structures (Fig. 4). First, the topology and hydrogen bonding of Ser153, Asp177, and His264 duplicated the topology of the catalytic triad in trypsin (Winkler *et al.,* 1990). This finding supported earlier suspicions that the enzymatic mechanism of PTL resembles the mechanism of serine proteases and implicated these residues in the catalytic mechanism of PTL (Verger, 1984). Second, several surface loops sat over the catalytic site in a position to sterically hinder diffusion of substrate into the catalytic site. A disulfide bridge between Cys238 and Cys262 defines the largest of these loops, the lid domain. Residues 76 to 80, the β-5 fold, and residues 213 to 217 form two shorter loops that also block the catalytic site. Clearly, another conformation, one with the loops in a different position, must exist or PTL would be inactive.

The crystal structures of the two related proteins have also been solved. Dog PLRP1 has been crystallized and its structure has been determined and refined at a resolution of 2.1 nm (Roussel *et al.,* 1998a). The same group determined the rat PLRP2 structure at 1.9 nm (Roussel *et al.,* 1998b). As predicted from the homology of the primary sequences, both

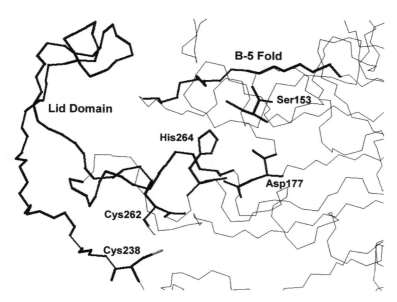

Figure 4. The active site of pancreatic triglyceride lipase. The catalytic residues, Ser153, Asp177, and His264 and the cysteines forming the disulfide bridge at the base of the lid are shown. The lid domain, the β-5 fold, and the 213–217 loop are shown in bold lines. The figure was prepared with Alchemy 2000 (Tripos, Inc., St. Louis, MO).

structures closely resemble the structure of human PTL and other reported lipase structures. The N-terminal residues superimpose within 1.0 nm. Rotations of a few degrees around the hinge region caused the C-terminal orientations to differ slightly. The catalytic triad superimposed in all three structures and the lid domain was present in a closed position in both dog PLRP1 and rat PLRP2. These results further emphasize the evolutionary relationship among these proteins.

6. Molecular Mechanism of Lipolysis

6.1. The Catalytic Triad

The crystal structure of human PTL provided the first direct evidence for a nucleophile–histidine–acidic amino acid triad in the catalytic site of PTL (Winkler *et al.,* 1990). Cocrystallization of an active site inhibitor with PTL supported the location of the proposed active site (Egloff *et al.,* 1995a). The inhibitor was located at the catalytic site. Site-specific mutagenesis of Ser153, Asp177, and His264 confirmed the importance of these residues in lipolysis (Lowe, 1992). Recombinant proteins with mutations at each position of the putative triad were created and expressed, and it was found that replacing Ser153 with any of eight amino acids or changing His264 to leucine completely inactivated PTL. In contrast, an Asp177 to glutamate mutant PTL retained 80% of its activity. This result differed from those reported for other enzymes with a Ser–His–acidic amino acid catalytic triad. In those studies, mutating the acidic group from the native residue to another acidic amino acid abolished activity in the mutant enzyme. The preserved activity of the D177E mutant suggested the assignment of Asp177 to the catalytic triad might be in error. Another observation supported this possibility. Asp206 could be rotated into a position suitable for an interaction with His264 (Derewenda *et al.*, 1994). Thus, Asp206 could function in the catalytic triad.

Additional mutagenesis studies resolved this point. Asp177, Asp206 or both residues were changed to alanine and the activity of the mutants was tested (Lowe, 1996). The D177A mutant had 20% of wild-type activity. It was found that the D206A retained full activity and the double mutation completely inactivated PTL. The combination of these mutagenesis studies and the crystal structure provides solid evidence for a Ser153–Asp177–His264 catalytic triad in PTL. Given that PLRP1 and PLRP2 contain the same residues in the same topology, they almost certainly utilize the same catalytic mechanism.

6.2. Substrate Specificity

PTL is a carboxyl esterase that attacks many types of esters (Verger, 1984). Of the dietary fats, PTL prefers acylglycerides over other dietary lipids, such as phospholipids, cholesterol esters, or galactolipids (Andersson *et al.*, 1996; Verger, 1984). The release of acyl chains by PTL from triglycerides proceeds in steps through diglycerides to monoglycerides to glycerol and free fatty acids (Brockerhoff and Jensen, 1974; Desnuelle, 1972). During the *in vivo* or *in vitro* hydrolysis of triglycerides by PTL, diglycerides and monoglycerides appear quickly, but glycerol appears much later after most of the triglyceride has been depleted (Fig. 5A). These studies suggested that PTL cleaves monoglycerides at a slow rate. In fact, PTL probably does not hydrolyze 2-monoacylglycerols at all. Studies have shown that PTL preferentially cleaves the acyl chains in the 1 and 3 positions, accounting for the rapid appearance of monoacylglycerides (Brockerhoff and Jensen, 1974; Desnuelle, 1972). The slow production of glycerol occurs because PTL does not hydrolyze 2-monoacylglycerides. Only after a chemical migration of the acyl chain in the 2 position to the 1 or 3 position of the monoglyceride does the acyl chain get released (Fig. 5B). It is the chemical step that accounts for the slow appearance of glycerol.

PTL efficiently hydrolyzes a broad range of acyl chain lengths in the α-position of triglycerides (Yang *et al.*, 1990; Savary, 1971; Brockerhoff, 1970). *In vitro* studies suggested that PTL hydrolyzed polyunsaturated long-chain fatty acid esters poorly, but *in vivo* studies showed the recovery of long-chain polyunsaturated fatty acids of marine oils in the fat tissues of rats and pigs (Storlein *et al.*, 1987; Garton *et al.*, 1952). Recent studies on menhaden oil triglycerides confirmed the ability of PTL to hydrolyze a broad range of acyl chains *in vitro*. PTL cleaved acyl chains from C_{14} to C_{22} with only a sixfold difference in rates between the best, 18:1(n−9), and the worst, 22:6(n−3), substrates (Yang *et al.*, 1990). Because of the large PTL excess in the duodenum, the sixfold difference is unlikely to be significant and dietary polyunsaturated long-chain fatty acids should be effectively released by PTL.

PLRP1 does not have a known substrate. It does not cleave triglycerides, phospholipids, cholesterol esters, or galactolipids under a variety of conditions (Payne *et al.*, 1994;

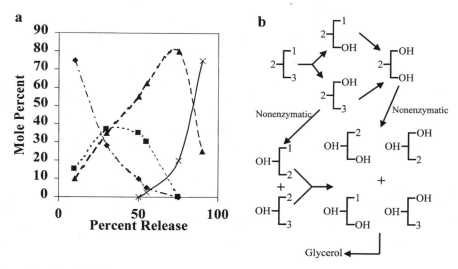

Figure 5. The hydrolysis of triglycerides by pancreatic triglyceride lipase. Panel A. The formation of products from triglyceride hydrolysis by PTL is shown. Diamonds, triglycerides; squares, diglycerides; triangles, monoglycerides; crosses, glycerol. The data is adapted from Constantin *et al.,* 1960. Panel B. Hydrolysis of triglycerides by PTL. The slow, nonenzymatic migration of acyl chains is indicated. The other reactions are catalyzed by PTL. The numbers 1 to 3 represent the stereospecific location of esterified acyl groups.

Giller *et al.,* 1992; Andersson *et al.,* 1996). The crystal structure of dog PLRP1 provided an explanation for the poor reactivity against triglycerides (Roussel *et al.,* 1998a). Very few amino acid substitutions occurred around the putative active site of PLRP1 when compared to the sequence of all known active lipases. Two of these, Val178 and Ala180, were always Ala178 and Pro180 in the active lipases. Furthermore, these residues were positioned to interfere with binding of a substrate acyl chain in the correct conformation for catalysis. The investigators tested this hypothesis directly by expressing a PLRP1 double mutant, V178A and A180P. This mutant had clear activity against tributyrin and trioctanoin, providing that the inability of PLRP1 to hydrolyze triglycerides could be explained by the inability of PLRP1 to correctly bind acyl chains.

Although this study provides one explanation for why PLRP1 does not hydrolyze triglycerides, the results do not identify the physiological function of PLRP1. Several potential explanations exist for the inability to demonstrate lipase activity in PLRP1. PLRP1 may act on substrates not yet tested, it may require a protein cofactor or other reaction conditions that have not been discovered, or it may be an inactive homologue of PTL, a vestigial protein. This last explanation seems unlikely for a number of reasons. In general, nonfunctional genes mutate at higher rates than their functional counterparts. This has not occurred with PLRP1. The conservation of PLRP1 among various species is as high as for PTL and for PLRP2. The changes that have occurred have not involved functional residues, such as the catalytic residues, and the major determinants of structure have been preserved, leaving the majority of divergent residues concentrated in other regions of the protein. Normally, random mutations would be expected in contrast to the pattern of nonrandom mutations observed in PLRP1. Certainly, mutations should have occurred in biologically important regions if the gene is nonfunctional. These findings argue strongly for a biological role of PLRP1, and additional substrates and reaction conditions must be tested and alternate functions should be considered before dismissing PLRP1 as an evolutionary remnant.

PLRP2, on the other hand, has easily detectable activity against a variety of substrates. PTL prefers triglycerides and can hydrolyze short-, medium-, and long-chain triglycerides (Jennens and Lowe, 1995a; Thirstrup *et al.,* 1994; Giller *et al.,* 1992). In contrast to PTL, PLRP2 has significant activity against phospholipids and galactolipids (Andersson *et al.,* 1996; Jennens and Lowe, 1995a; Roussel *et al.,* 1998b). The activity against phospholipids may be advantageous in digesting breast milk fat globules. Alternate reasons for the phospholipase have been suggested by the observations that PLRP2 binds tightly to the zymogen granule membrane and that cytotoxic T-cells express PLRP2 (Grusby *et al.,* 1990; Wishart *et al.,* 1993). PLRP2 could participate in the release of zymogen granule contents or in cytotoxicity by altering the membrane properties of acinar cells or of foreign cells through the hydrolysis of membrane phospholipids.

6.3. Interfacial Activation

The ability to hydrolyze water-insoluble, long-chain triglycerides distinguishes lipases from enzymes acting on water-soluble substrates (Verger, 1997). PTL has activity against water-soluble, monomeric substrates, but its specific activity increases greatly when the substrate forms water-insoluble particles such as micelles or emulsions. This property of PTL and other lipases has been termed interfacial activation. Speculations about the mechanism underlying the increased activity of PTL at an oil–water interface focus on two models—an enzyme model and a substrate model (Derewenda, 1995). In the substrate model, substrate particles either increase the local concentration of substrate or alter the conformation of the substrate to permit hydrolysis. The enzyme models center on conformational changes in the enzyme. Support for both models can be found.

The first direct evidence that a conformational change must happen before PTL can hydrolyze substrate came from the crystal structure of PTL that identified several surface loops obscuring the active site (Winkler *et al.,* 1990). Two additional studies describing the three-dimensional structure of PTL with its protein cofactor, colipase, provided additional support for the enzyme model. In these studies, crystals were obtained in the presence and in the absence of octylglucoside and phospholipid mixed micelles (van Tilbeurgh *et al.,* 1992, 1993). Without the interface, the lid domain of PTL remained in the closed position (Fig. 6). With the micelles providing an interface, several conformational changes occurred. Most important, a hinge movement of the lid domain brought this loop away from the catalytic site. The lid domain change and the accompanying movement of the β-5 loop opened and configured the active site to accept substrate. Separate studies on fungal lipases have also shown similar conformational changes in a surface loop homologous to the PTL lid domain (Brzozowski *et al.,* 1991; Grochulski *et al.,* 1993). These studies demonstrated the conformational changes predicted from the original structure of PTL and provided strong support for the enzyme model of interfacial activation.

The description of two PTL lid domain mutants provided additional evidence that the opening of the lid domain equates with interfacial activation (Jennens and Lowe, 1994). One mutant removed the entire lid domain and the other removed the α-helix covering the

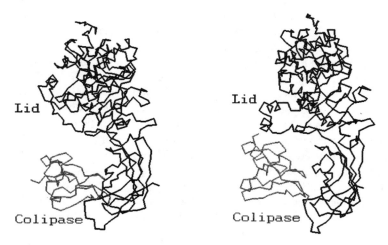

Figure 6. Tertiary structure of the colipase–pancreatic triglyceride lipase complex. Panel A. The complex structure with the PTL lid in the closed position. Panel B. The complex structure with the PTL lid in the open position. The coordinates were obtained from the Brookhaven Protein Data Bank and the figures were made with RasMol.

active site. Both mutants had decreased activity, but activity against emulsions of triglycerides and taurodeoxycholate was easily detectable. These mutants could also hydrolyze water-soluble, monomeric substrates at a high rate and did not display interfacial activation. These results and the description of a guinea pig lipase that does not have a lid domain and does not display interfacial activation support the hypothesis that the lid domain movement is the physical correlate of interfacial activation (Hjorth *et al.*, 1993).

The trigger for the lid domain movement remains speculative. Most have assumed that an oil–water interface triggers the conformational change in the lid. The kinetic demonstration of increased activity when tripropionin or *p*-nitrophenylbutyrate form interfaces supports this hypothesis for interfacial activation, but it does not prove a cause and effect relationship (Verger, 1997; Martinelle *et al.*, 1995). Other milieus may also induce opening of the lid. For instance, the presence of detergent micelles and colipase was sufficient to produce crystals of porcine PTL in the open conformation, suggesting that micelles and colipase could trigger lid domain movement (Hermoso *et al.*, 1996). The same group provided kinetic data for this speculation. They showed that PTL inhibition by di-isopropyl p-nitrophenylphosphate depended on having both detergent micelles and colipase in the solution. The implication of this result is that detergent micelles and colipase caused the lid to open, allowing the inhibitor access to the active site. In a follow-up study, this group examined the PTL-colipase crystals by neutron diffraction coupled to D_2O/H_2O contrast variation to identify a detergent micelle bound to the concave surface of colipase and the C-terminal end of PTL (Hermoso *et al.*, 1997). They speculated that the micelle bound in this position, a distance from the lid, triggered the lid domain to open.

Another model can explain all of the current experimental results. Rather than causing the lid domain to open, the presence of an oil–water interface or of colipase and detergent micelles may trap the lid in the open conformation. In this model, the lid is constantly opening and closing. The lid only stays open when stabilized in this conformation by an oil–water interface or by colipase and an interface of detergent micelles. Studies on the crystal structures of fungal lipases obtained in aqueous media show the lid in several different conformations and suggest that the lid is mobile (Derewenda *et al.,* 1995; Grochulski *et al.,* 1994). These studies lend support to the notion that interfacial activation results from lid domain movement and stabilization of the open conformation by hydrophobic forces.

Changes in the substrate may also influence interfacial activation. Recent studies on lipid monolayers demonstrated that lipids in monolayers undergo structural changes during phase transitions induced by increasing the surface pressures (Hwang *et al.,* 1995; Peters *et al.,* 1995). Lipid domain structures also vary with temperature and with lipid composition. Each of these physical changes in the monolayer could potentially alter lipase activity. Additionally, proteins added to the lipid surface can have large effects on the monolayer structure. A lung surfactant protein, SP-B, had large effects on the formation of condensed phases in a monolayer and also affected other physical properties of the monolayer (Lipp *et al.,* 1996). Clearly, the physical state of the substrate can affect the quality of the interface which, in turn, may influence lipolysis (Verger, 1980). Properties of the substrate and the enzyme can contribute to the ability of lipases to hydrolyze water-insoluble substrates. These properties may not be exclusive. Just as the interface may induce changes in PTL, PTL may alter the substrate, particularly when complexed with colipase, a small, surface-active protein that may alter the physical properties of the substrate.

7. Colipase and Lipolysis

A discovery in 1910 first suggested that PTL required another factor for activity (Rosenheim, 1910). Investigators extracted pancreas with glycerol to produce two inactive fractions, a precipitate and a filtrate, that, when mixed, restored full lipase activity. Another 50 years passed before investigators determined the nature of the cofactor. Ion exchange chromatography separated PTL from a small molecular weight, heat-stable protein (Baskys *et al.,* 1963). Eventually, two groups described the ability of the cofactor to restore activity to bile-salt inhibited PTL (Borgström and Erlanson, 1971; Morgan and Hoffman, 1971). This finding provided an explanation for the paradox that bile salts inhibited PTL *in vitro* but duodenal juice containing high concentrations of bile salts had high PTL activity.

Sternby and Borgström purified human colipase in 1979. The amino acid sequence gave a primary structure of 86 amino acids. Other studies of colipase from humans and from other species found additional forms of colipase (Borgström and Erlanson-Albertsson, 1984). Colipase isolated from porcine pancreas contained an N-terminal pentapeptide not present in colipase isolated from pancreatic juice (Figure 7; Erlanson-Albertsson and Larsson, 1981). The tissue-purified colipase had decreased activity compared to the form missing the N-terminal pentapeptide. Limited trypsin digestion removed the N-terminal pentapeptide producing a colipase with the same N-terminus and activity as colipase isolated from pancreatic juice. Based on these studies, the hypothesis that the pancreas secretes a less active form of colipase was postulated. In this model, the secreted colipase is converted to an active form in the duodenum by proteolytic cleavage of the pentapeptide. This hy-

Human	A P G P R	G I I I N
Rabbit	A P G P R	G I V I N
Rat	A P G P R	G L F I N
Chicken	A P G P R	G L I F N
Dog	V P D P R	G I I I H
Horse	V P D P R	G V I I N
Pig	V P D P R	G I I I N

Figure 7. The amino terminal sequence of colipase from various species. The first 10 amino acids of the amino terminus of procolipase are given in the single letter amino acid code. The propiece is indicated by the box.

pothesis accounts for the absence of a proform of PTL as found for many other exocrine enzymes. The secreted, tissue colipase was named procolipase.

Human procolipase was first identified in human pancreatic juice by amino acid sequence analysis of the N-terminus (Sternby and Borgström, 1984). The cDNA for human colipase confirmed the presence of a propiece in human colipase and suggested that additional processing occurs at the C-terminus (Lowe *et al.*, 1990). The cDNA encodes a protein with a predicted sequence of 112 amino acids with a molecular mass of 11.6 kD. The first 17 amino acids comprise a signal peptide followed by the N-terminal pentapeptide. The sequence of human colipase determined by chemical methods begins at amino acid 23 of the sequence predicted from the cDNA (Sternby *et al.*, 1984). From that point, the two sequences are identical until amino acid 91, where the sequence of the isolated protein ends, and the sequence predicted from the cDNA continues for 4 additional amino acids. Species of porcine colipase with differing C-terminal sequences have also been described (Borgström and Erlanson-Albertsson, 1984). Apparently, processing of procolipase occurs at both the N- and the C-termini.

The physiological role of procolipase processing remains speculative. No data suggests a role for C-terminal processing. Cleavage of amino acids from the C-terminus may be a consequence of the high protease concentrations in the duodenum. Alternately, processing of the C-terminus may regulate the activity or concentration of colipase in the duodenum. In contrast, several roles for the conversion of procolipase to colipase have been supported by experimental data. The decreased activity of procolipase led to the hypothesis that secretion of procolipase limited PTL activity in the pancreas (Wieloch *et al.*, 1981). More recent data obtained by assaying procolipase and colipase at neutral pH rather than the usual pH 8.0 demonstrated equal activity for the two forms (Larsson and Erlanson-Albertsson, 1991). Because the pH in these experiments resembled the pH in the duodenum, the authors felt that the results more accurately reflected the physiological conditions than did assays at alkaline pH. Even though the decreased activity of procolipase at alkaline pH may not be a factor in the duodenum, the differences may be important in the more alkaline pancreatic duct.

Recent studies have suggested another possible function for the cleavage of the pentapeptide from colipase (Erlanson-Albertsson, 1994). In these studies, rats decreased their voluntary fat intake when given the pentapeptide. The route of administration did not seem to matter and the peptides worked if given into the peritoneum, the lateral ventricle, or in the animal's feed. After feeding, both duodenal contents and serum contain the endogenous pentapeptide. These studies led to the hypothesis that the pentapeptide is a hormone, now called enterostatin, that regulates satiety. In this model, the release of enterostatin from procolipase provides a feedback mechanism to the central nervous system, which limits dietary fat intake through the sensation of satiety.

7.1. Tertiary Structure of Colipase

The three dimensional structure of colipase has been solved by both nuclear magnetic resonance (NMR) and X ray crystallography (Egloff *et al.,* 1995b; Breg *et al.,* 1995; van Tilbeurgh *et al.,* 1992). The crystal structure of colipase proved elusive until it was crystallized as a complex with PTL. The solution structure determined by NMR agrees well with the crystal structure, indicating that the conformation of colipase was not affected by the presence of PTL. Colipase is a flattened molecule with four loops protruding from a rigid core held together by five disulfide bonds and two short regions of β-sheet (Fig. 8). The disulfide bonds are the main stabilizing force of the structure. The bulk of the molecule lacks regular secondary structure, allowing great mobility in the protruding loops. Based on these studies, colipase belongs to the family of small cysteine-rich proteins lacking defined structure (Saudek *et al.,* 1991; Alder *et al.,* 1991).

7.2. Colipase Interacts with PTL

Although multiple investigations describe the reactivation of bile-salt inhibited PTL by colipase, the mechanism has been speculative until recently. Most early models centered on colipase anchoring PTL to the bile-salt covered substrate surface (Borgström and Er-

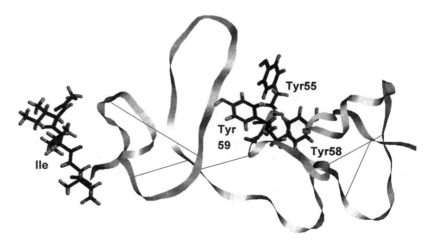

Figure 8. The tertiary structure of colipase. The α-carbon backbone of colipase is given. The locations of the disulfide bonds are indicated by the lines. The side chains of the amino terminal isoleucine residues and the neighboring tyrosines are shown.

lanson-Albertsson, 1984; Erlanson-Albertsson, 1992). Other roles for colipase were seldom proposed. Multiple studies utilizing a vast array of techniques, including gel filtration, microcalorimetry, ultracentrifugation, binding of colipase to immobilized PTL, and phase partition, support binding of colipase to PTL (Chaillan *et al.,* 1992; Mahe-Gouhier and Leger, 1988; Patton *et al.,* 1978a, 1978b; Sternby and Erlanson-Albertsson, 1982). Most evidence suggests that colipase binds to the C-terminal domain of PTL (Abousalham *et al.,* 1992; Chaillan *et al.,* 1992). The crystal structure of the colipase–PTL complex provides the most impelling evidence for the interaction between colipase and the C-terminal domain of PTL (van Tilbeurgh *et al.,* 1992, 1993; Egloff *et al.,* 1995b). Colipase forms 8 hydrogen bonds and about 80 van der Waals contacts with residues in the C-terminal domain of PTL. All of the interactions take place over a small surface area, about 600 nm^2. The small surface area and small number of hydrogen bonds predict a weak interaction between colipase and PTL in solution, as previously found experimentally (Patton *et al.,* 1978b).

A study of PTL mutants questioned the importance of the C-terminal interaction between colipase and PTL (Jennens and Lowe, 1995b). In this study, a stop codon placed in the region between the N-terminal and the C-terminal domain resulted in expression of just the N-terminal domain. This truncated lipase had decreased activity against triolein, indicating that the C-terminal domain contributes in some way to full activity. It is important to note that bile salts still inhibited the truncated PTL and colipase restored the activity. An interaction between colipase and the N-terminal domain provides the most plausible explanation for these results.

A second crystal structure of colipase and PTL cocrystallized with mixed micelles demonstrated interactions of colipase with the N-terminal domain (van Tilbeurgh *et al.,* 1993). As described previously, a large conformational change occurs in the structure of PTL when phospholipid binds in the active site. The change results in additional hydrogen bonds and van der Waals forces between colipase and the N-terminal domain of PTL, which adds 300 nm^2 to the interaction surface between colipase and PTL (Egloff *et al.,* 1995b). Now, the total interaction surface and number of hydrogen bonds lies within the range reported for other specific protein–protein interactions (Janin and Chothia, 1990). The increased bond force explains why the apparent association constant of PTL and colipase increases in the presence of an oil–water interface (Patton *et al.,* 1978a, 1978b; Sternby and Erlanson-Albertsson, 1982).

In the open lid conformation induced by the mixed micelles, important hydrogen bonds form between Glu15 of colipase and Asn241 of PTL and between Arg38 and Val246 of PTL (Egloff *et al.,* 1995b). In addition, a water molecule bridges Arg38 with Ser243 and Val246 (Figure 9). Each of the PTL residues is located in the lid domain of PTL, giving credence to the theory that colipase stabilizes the lid domain in the open position. Site-specific mutagenesis of colipase provided direct evidence for the role of Glu15 in colipase reactivation of PTL (Lowe, 1997a). Mutation of Glu15 to any one of several different amino acids greatly decreased activity. One mutant, E15R, was studied in more detail and was found to be 175-fold less active than wild-type colipase. Although the mutant colipase had decreased ability to restore activity to bile-salt inhibited PTL, the E15R mutant was still able to facilitate PTL binding to substrate. These findings suggest another function for colipase in addition to anchoring PTL at the substrate interface. Colipase can act to stabilize the lid in the open position and thereby maintain PTL in an active conformation.

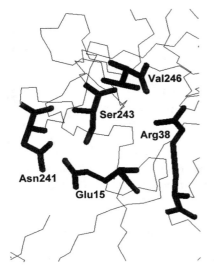

Figure 9. Contacts between colipase and the lid domain of pancreatic triglyceride lipase. A close-up of the interaction between colipase and the lid is shown. The α-carbon backbone is given by the thin lines. The side-chains of the indicated residues are indicated by tubes. Glu15 and Arg 38 are colipase residues. The others are PTL residues.

7.3. Colipase and Substrate

If colipase anchors PTL to the substrate surface and one surface of colipase binds to PTL, then another surface should bind to the lipid interface. The side opposite the lipase binding site contains the tips of the protruding loops (Egloff *et al.*, 1995b). These tips form a large surface containing a number of exposed hydrophobic residues. Together with the hydrophobic surface of the lid domain, which is exposed in the open conformation, this colipase surface forms a 50 nm long region of hydrophobic residues (van Tilbeurgh *et al.*, 1993). Although no detergent or phospholipid residues were observed next to this region in the crystal structure of the colipase–PTL complex, it seems reasonable to postulate that the surface represents the lipid binding domain of colipase.

Other studies support the hypothesis that the hydrophobic plateau mediates colipase binding to lipid surfaces. Proteolytic cleavage of the N-terminal isoleucines from pig colipase decreased the ability of the shortened colipase to bind a lipid substrate, Intralipid (Erlanson-Albertsson and Larsson, 1981). The isoleucines lie next to the hydrophibic face of the PTL lid domain and contribute to the hydrophobic surface of the complex, suggesting that these residues can contribute to the lipid binding surface of colipase.

The tip of the third loop contains three absolutely conserved neighboring tyrosine residues at positions 55, 58, and 59 (Egloff *et al.*, 1995b). Chemical modification of these residues and spectral analysis suggested that two of the three tyrosines interact with the lipid substrate. One of the earliest studies showed changes in the ultraviolet spectrum of two tyrosines in the presence of taurodeoxycholate micelles, a result suggesting the tyrosines participate in micelle binding (Sari *et al.*, 1978). A subsequent study found that acetylation of two tyrosines created a modified colipase with activity against tributyrin and olive oil, but not against Intralipid, indicating that the tyrosines mediate colipase binding to phospholipid (Erlanson-Albertsson, 1980). Neither of these studies identified the involved tyrosines. NMR evaluation of colipase suggested that either Tyr58 or Tyr59 reside on the surface and that Tyr55 is buried (Wieloch *et al.*, 1979). More recent studies of dansyltyrosine derivatives of porcine colipase contradicted the NMR findings (McIntyre *et al.*, 1987, 1990a, 1990b). These studies placed Tyr55 on the surface where it could interact with bile salt micelles and showed that dansyl-modified Tyr59 did not interact with bile salt micelles. The authors did not determine whether either tyrosine interacted with triglyceride substrates. Tyr58 was not modified and its position or contribution to bile salt micelle interactions could not be assessed.

The three-dimensional structure of colipase determined by NMR and by X ray crystallography resolved the controversy of which tyrosines reside on the exposed hydrophobic surface (van Tilbeurgh *et al.*, 1992, 1993; Egloff *et al.*, 1995b; Breg *et al.*, 1995). Examination of the crystal structure revealed that the side chains of tyrosines 55 and 59 extend away from colipase, toward the substrate, whereas the side chain of Tyr58 pointed inward. NMR analysis of colipase agrees with the location of the three tyrosines that was determined by crystallography. These structures support the hypothesis that Tyr55 interacts with bile salt micelles, and, by analogy, with substrate interfaces. The structures also place Tyr59 on the surface where it could interact with the lipid substrate, in contrast to the findings with dansyl-modified Tyr55.

A recent study investigated the role of the neighboring tyrosines by site-directed mutagenesis (Cordle and Lowe, 1998). A charged residue, aspartic acid, replaced each tyrosine, creating three single mutants, Y55D, Y58D, and Y59D, and one double mutant, Y55D/Y59D. The Y58D mutant had wild-type activity, indicating that the contribution of Y58 to colipase function is minimal. The Y55D and Y59D mutants had decreased activity against long-chain triglycerides and had impaired ability to anchor PTL to long-chain triglycerides. The double mutant had no activity against short- or long-chain triglycerides and also did not mediate PTL binding to substrate. This study demonstrated the contribution of Tyr55 and Tyr59 to the hydrophobic surface of colipase and implicated that surface in lipid binding.

The fourth loop also contains a number of hydrophobic residues. The role of these residues in lipid binding has not been directly studied, but one study suggests that the loop contributes to the lipid binding surface. In this paper, the authors compared two different preparations of colipase (Rugani *et al.*, 1995). One was prepared from fresh pancreas and the other from a side fraction separated during the production of porcine insulin. Electrospray mass spectra of the two fractions demonstrated that the mass of the colipase prepared from the side fraction was smaller than the mass of colipase prepared from fresh tissue. A specific cleavage between Ile79 and Thr80 accounted for the difference. The two resultant peptides remained associated because of disulfide bonds between them. Evaluation of their activity revealed that lipase assays with the cleaved colipase had a longer period before hydrolysis started, a lag time, than did assays with the intact colipase. The increased lag time reflects the time for colipase to anchor PTL and may be a surrogate for binding to substrate. These findings are consistent with the proposed role for the fourth loop in the lipid-binding properties of colipase.

Colipase may do more than simply bind to lipid interfaces. There is now evidence that colipase may influence the properties of the lipid substrate. Studies on monolayers show that colipase preferentially binds to substrates as opposed to non-PTL substrates, such as phospholipids (M. M. Momsen *et al.*, 1997; W. E. Momsen *et al.*, 1995). The preference of colipase is to bind substrate direct PTL binding to regions rich in substrates. Additional monolayer data shows that colipase may change the lateral distribution of lipids. Analysis of the cross-sectional areas for fatty acids and diacylglycerol in the monolayer suggests that substrate and product multilayers form around bound colipase. Colipase binding to a lipid–water interface may recruit substrate to its vicinity, creating patches of substrate colipase where PTL can bind. Because colipase shows differences in binding affinity for various substrates and can influence the lateral distribution of substrate, colipase may indirectly influence PTL substrate selectivity.

8. The Colipase Gene

The gene encoding colipase is not a member of the lipase gene family. The gene is located on human chromosome 6 apart from the gene encoding PTL on chromosome 10 (Sims and Lowe, 1992). The entire structural region of the gene spans about 2.3 kb and contains only three exons. The first exon contains the 5′-untranslated region of the mRNA, the sequence encoding the signal peptide, enterostatin, and the first six amino acids of the mature protein. This pattern differs from the organization of other genes encoding pancreatic exocrine proteins in which a separate exon encodes the signal peptide. The remaining two exons encode the rest of the protein and the 3′-untranslated region.

8.1. Regulatory Elements in the Colipase Gene

Analysis of the 5′-upstream sequence of the gene encoding human procolipase revealed several regulatory elements. A region from -122 and -150 bp was identified as a tissue specific enhancer element (Sims and Lowe, 1992). This region was associated with pancreatic cell specific expression in tissue culture cells. The sequence between -122 and -150 bp shows homology to the rat-pancreatic-specific enhancer element found in the genes encoding many pancreatic exocrine proteins. Additional studies showed that the region just upstream from the pancreatic-specific enhancer element mediated up-regulation of colipase expression by dexamethasone (Lowe *et al.,* 1994; Li and Lowe, 1993). A complex element lies between bp -150 and -215. The region does not contain a glucocorticoid-regulatory element and the entire region appears to be critical for the response. Band shift and DNAse I footprint analysis demonstrated binding of pancreatic acinar cell nuclear proteins to the region, but these proteins have not been identified.

9. Conclusion

Although the importance of PTL in fat digestion was recognized more than a century ago, the molecular details of lipolysis are only now beginning to be understood. Advances in molecular biology and in X ray crystallography have identified the active site of PTL, the catalytic triad, and clarified the interaction between colipase and PTL. The mechanism of colipase and PTL function still includes binding of the two proteins. The interaction may occur in bulk phase, but, more likely, occurs on the oil–water interface of the substrate, perhaps, after colipase has altered the lateral distribution of the lipids to gather substrates in its vicinity. Contact with the interface and colipase causes the lid domain to open and to configure the active site. Binding of the lid domain to colipase stabilizes the open, active PTL conformation and lipolysis proceeds at a high rate.

Many questions about the mechanism of lipolysis remain. The details of bile-salt inhibition and the mechanism of colipase reactivation are not known. The differences in substrate specificity between PTL and PLRP2 must lie in the detailed conformation of the active sites. What determines stability at acidic or at alkaline pH? The physiological function of PLRP1, if one exists, must be discovered. The information from these studies will lead to new approaches for nutritional therapy in patients with pancreatic insufficiency, or with obesity, or for premature infants who have relative PTL deficiency.

References

Abousalham, A., Chaillan, C., Kerfelec, B., Foglizzo, E., and Chapus, C., 1992, Uncoupling of catalysis and colipase binding in pancreatic lipase by limited proteolysis, *Prot. Engin.* **5:**105–111.

Alder, A., Lazarus, R. L., Dennis, M. S., and Wagner, G., 1991, Solution structure of kristin, a potent platelet aggregation inhibitor and GP IIb-IIa antagonist, *Science* **253:**445–448.

Ameis, P., Stahnke, G., Kobayashi, J., McLean, J., Lee, G., Buscher, M., and Schotz, M. C., 1990, Isolation and characterization of the human hepatic lipase gene, *J. Biol. Chem.* **265:**6552–6555.

Andersson, L., Carriere, F., Lowe, M. E., Nilsson, A., and Verger, R., 1996, Pancreatic lipase-related protein 2 but not classical pancreatic lipase hydrolyzes galactolipids, *Biochim. Biophys. Acta* **1302:**236–240.

Arvan, P., and Castle, J. D., 1987, Phasic release of newly synthesized secretory proteins in the unstimulated rat exocrine pancreas, *J. Cell Biol.* **104:**243–252.

Arvan, P., and Chang, A., 1987, Constitutive protein secretion from the exocrine pancreas of fetal rats, *J. Biol. Chem.* **262:**3886–3890.

Baskys, B., Klein, F., and Lever, W. F., 1963, Lipases of blood and tissue III. Purification and properties of pancreatic lipase, *Arch. Biochem. Biophys.* **102:**201–209.

Borgström, B., and Erlanson, C., 1971, Pancreatic juice co-lipase: Physiological importance, *Biochim. Biophys. Acta* **242:**509–513.

Borgström, B., and Erlanson-Albertsson, C., 1984, Pancreatic colipase, in: *Lipases* (B. Borgström and H. L. Brockman, eds.), Elsevier, Amsterdam, pp. 152–183.

Bourne, Y., Martinex, C., Kerfelec, B., Lombardo, D., Chapus, C., and Cambillau, C., 1994, Horse pancreatic lipase. The crystal structure refined at 2.3 A resolution, *J. Mol. Biol.* **238:**709–732.

Breg, J., Sarda, L., Cozone, P. J., Rugani, N., Boelens, R., and Kaptein, R., 1995, Solution structure of porcine pancreatic procolipase as determined from 1H homonuclear two-dimensional and three-dimensional NRM, *Eur. J. Biochem.* **227:**663–672.

Brockerhoff, H., 1970, Substrate specificity of pancreatic lipase. Influence of the structure of fatty acids on the reactivity of esters, *Biochim. Biophys. Acta* **212:**92–101.

Brockerhoff, H., and Jensen, R. G., 1974, Lipases, in: *Lipolytic Enzymes,* Academic Press, New York, pp. 25–100.

Brzozowski, A. M., Derewenda, U., Derewenda, Z. S., Dodson, G. G., Lawson, D. M., Turkenburg, J. P., Bjorkling, F., Huge-Jensen, B., Patkar, S. A., and Thim, L., 1991, A model for interfacial activation in lipases from the structure of a fungal lipase-inhibitor complex, *Nature* **251:**491–494.

Cai, S.-J., Wong, D. M., Chen, S.-H., and Chan, L., 1989, Structure of the human hepatic triglyceride lipase gene., *Biochemistry* **28:**8966–8971.

Carey, M. C., and Hernell, O., 1992, Digestion and absorption of fat, *Sem. Gastrointest. Dis.* **3:**189–208.

Carriere, F., Barrowman, J. A., Verger, R., and Laugier, R., 1993, Secretion and contribution to lipolysis of gastric and pancreatic lipases during a test meal in humans, *Gastroenterology* **105:**876–888.

Chaillan, C., Kerfelec, B., Foglizzo, E., and Chapus, C., 1992, Direct involvement of the C-terminal extremity of pancreatic lipase (403–499) in colipase binding, *Biochem. Biophys. Res. Comm.* **184:**206–211.

Chuat, J. C., Raisonnier, A., Etienne, J., and Galibert, F., 1992, The lipoprotein lipase-encoding human gene: Sequence from intron-6 to intron-9 and presence in intron-7 of a 40-million-year-old Alu sequence, *Gene* **110:**257–261.

Constantin, M. J., Pasero, L., and Desneulle, P., 1960, Quelques remarques complementaires sur L'hydrolyse des triglycerides par la lipase pancreatique, *Biochim. Biophys. Acta* **43:**103–109.

Cordle, R., and Lowe, M. E., 1998, The hydrophobic surface of colipase influences lipase activity at an oil–water interface., *J. Lipid Res.* **39:**1759–1767.

Cygler, M., Schrag, J. D., Sussman, J. L., Harel, M., Silman, I., Gentry, M. K., and Doctor, B. P., 1993, Relationship between sequence conservation and three-dimensional structure in a large family of esterases, lipases, and related proteins, *Prot. Sci.* **2:**366–382.

Davis, R. C., Diep, N., Hunziker, W., Klisak, I., Mohandas, T., Schotz, M. C., Sparkes, R. S., and Jusis, A. J., 1991, Assignment of human pancreatic lipase gene (PNLIP) to chromosome 10q24-q26, *Genomics.* **11:**1164–1166.

Deeb, S. S., and Peng, R., 1989, Structure of the human lipoprotein lipase gene, *Biochem.* **28:**4131–4135.

Derewenda, Z. S., 1995, A twist in the tale of lipolytic enzymes, *Struct. Biol.* **2:**347–349.

Derewenda, U., Swenson, L., Green, R., Wei, Y., Yamaguchi, S., Joerger, R., Haas, M. J., and Derewenda, Z. S., 1994, Current progress in crystallographic studies of new lipases from filamentous fungi, *Prot. Engin.* **7:**551–557.

Derewenda, U., Swenson, L., Wei, Y., Green, R., Kobos, P. M., Joergere, R., Haas, M. J., and Derewenda, Z. S., 1995, Conformational lability of lipases observed in the absence of an oil–water interface: Crystallographic studies of enzymes from the fungi Humocola lanuginosa and Rhizopus delemar, *J. Lipid Res.* **35:**524–534.

Desnuelle, P., 1972, The Lipases, in: *The Enzymes,* Volume 7, Academic Press, New York and London, pp. 575–603.

Egloff, M. P., Marguet, F., Buono, G., Verger, R., Cambillau, C., and van Tilbeurgh, H., 1995a, The 2.46 A resolution of the pancreatic lipase–colipase complex inhibited by a C11 alkyl phosphonate, *Biochem.* **24:**2751–2762.

Egloff, M. P., Sarda, L., Verger, R., Cambillau, C., and van Tilbeurgh, H., 1995, Crystallographic study of the structure of colipase and of the interaction with pancreatic lipase, *Prot. Sci.* **4:**44–57.

Erlanson-Albertsson, C., 1980, The importance of the tyrosine residues in pancreatic colipase for its activity., *FEBS Lett.* **117:**295–298.

Erlanson-Albertsson, C., 1992, Pancreatic colipase. Structural and physiological aspects, *Biochim. Biophys. Acta* **1125:**1–7.

Erlanson-Albertsson, C., 1994, Enterostatin—A peptide regulating fat intake, *Scand. J. Nutr.* **38:**11–14.

Erlanson-Albertsson, C., and Larsson, A., 1981, Importance of the N-terminal sequence in porcine pancreatic colipase, *Biochim. Biophys. Acta* **665:**250–255.

Figarella, C., De Caro, A., Leupoid, D., and Poley, J. R., 1980, Congenital pancreatic lipase deficiency, *Pediatrics* **96:**412–416.

Garton, G. A., Hilditch, T. P., and Meara, M. L., 1952, The composition of the depot fat of a pig fed on a diet rich in whale oil, *Biochem. J.* **50:**517–524.

Ghishan, F. K., Moran, J. R., Durie, P. R., and Greene, H. L., 1984, Isolated congenital lipase–colipase deficiency, *Gastroenterology* **86:**1580–1582.

Giller, T., Buchwald, P., Blum-Kaelin, D., and Hunziker, W., 1992, Two novel human pancreatic lipase related proteins, hPLRP1 and hPLRP2: Differences in colipase dependency and in lipase activity, *J. Biol. Chem.* **267:**16509–16516.

Grochulski, P., Li, Y., Schrag, J. D., Bouthillier, F., Smith, P., Harrison, D., Rubin, B., and Cygler, M., 1993, Insights into interfacial activation from an open structure of candida rugosa lipase, *J. Biol. Chem.* **268:**12843–12847.

Grochulski, P., Li, Y., Scharg, J. D., and Cygler, M., 1994, Two conformational states of candida rugosa lipase, *Prot. Sci.* **3:**82–91.

Grusby, M. J., Nabavi, N., Wong, H., Dick, R. F., Bluestone, J. A., Schotz, M. C., and Glimcher, L. H., 1990, Cloning of an Interleukin-4 inducible gene from cytotoxic T lymphocytes and its identification as a lipase, *Cell* **60:**451–459.

Hermoso, J., Pignol, D., Kerfelec, B., Crenon, I., Chapus, C., and Fontecilla-Camps, J. C., 1996, Lipase activation by nonionic detergents, *J. Biol. Chem.* **271:**18007–18016.

Hermoso, J., Pignol, D., Penel, S., Roth, M., Chapus, C., and Fontecilla-Camps, J. C., 1997, Neutron crystallographic evidence of lipase–colipase complex activation by a micelle, *The EMBO Journal,* 16:5531–5536.

Hide, W. A., Chan, L., and Li, W.-H., 1992, Structure and evolution of the lipase superfamily, *J. Lipid. Res.* **33:**167–178.

Hjorth, A., Carriere, F., Cudrey, C., Woldike, H., Boel, E., Lawson, D. M., Ferrato, F., Cambillau, C., Dodson, G. G., Thim, L., and Verger, R., 1993, A structural domain (the lid) found in pancreatic lipase is absent in the guinea pig (phospho)lipase, *Biochem.* **32:**4702–4707.

Hofmann, A. F., and Borgström, B., 1962, Physico-chemical state of lipids in intestinal contents during their digestion and absorption, *Am. J. Clin. Nutr.* **21**:43–50.

Hofmann, A. F., and Borgström, B., 1964, The intraluminal phase of fat digestion in man: The lipid content of the micellar and oil phases of intestinal content obtained during fat digestion and absorption, *J. Clin. Invest.* **12**:631–634.

Hwang, J., Tamm, L. K., Bohm, C., Ramalingam, T. S., Betzig, E., and Edidin, B., 1995, Nanoscale complexity of phsopholipid monolayers investigated by near-field scanning optical microscopy, *Science* **270**:610–614.

Janin, J., and chothia, C., 1990, The structure of protein–protein recognition sites., *J. Biol. Chem.* **265**:16027–16030.

Jennens, M. L., and Lowe, M. E., 1994, A surface loop covering the active site of human pancreatic lipase influences interfacial activation and lipid binding, *J. Biol. Chem.* **269**:25470–25474.

Jennens, M. L., and Lowe, M. E., 1995a, Rat GP-3 is a pancreatic lipase with kinetic properties that differ from colipase-dependent pancreatic lipase, *J. Lipid Res.* **36**:2374–2381.

Jennens, M. L., and Lowe, M. E., 1995b, The C-terminal domain of human pancreatic lipase is required for stability and maximal activity but not colipase reactivation, *J. Lipid Res.* **36**:1029–1036.

Kaplan, M. H., Boyer, S. N., and Grusby, M. J., 1996, Genomic organization of the murine CTL lipase gene, *Genomics* **35**:606–609.

Kerfelec, B., LaForge, K. S., Puigserver, A., and Scheele, G., 1986, Primary structures of canine pancreatic lipase and phospholipase A2 messenger RNAs, *Pancreas.* **1**:430–437.

Kirchgessner, T. G., Svenson, K. L., Lusis, A. J., and Schotz, M. C., 1987, The sequence of cDNA encoding lipoprotein lipase, *J. Biol. Chem.* **262**:8463–8466.

Kirchgessner, T. G., Chuat, J.-C., Heinzman, C., Etienne, J., Guilhot, S., Svenson, K., Ameis, D., Pilon, C., D'Auriol, L., Andalibi, A., Schotz, M. C., Galibert, F., and Lusis, A. J., 1989, Organization of the human lipoprotein lipase gene and evolution of the lipase gene family, *Nuc. Acid Res.* **86**:9647–9651.

Kullman, J., Gisi, C., and Lowe, M. E., 1996, Dexamethasone-regulated expression of pancreatic lipase and two related proteins in AR42J cells, *Am. J. Physiol.* **270**:G746–G751.

Larsson, A., and Erlanson-Albertsson, C., 1991, The effect of pancreatic procolipase and colipase on pancreatic lipase activation, *Biochim. Biophys. Acta* **1983**:283–288.

Li, L., and Lowe, M. E., 1993, Dexamethasone stimulation of human colipase gene expression, *Gastroenterol.* **104**:A317.

Lipp, M. M., Lee, K. Y. C., Zasadzinski, J. A., and Waring, A. J., 1996, Phase and morphology changes in lipid monolayers induced by SP-B protein and its amino-terminal peptide, *Science* **273**:1196–1199.

Lowe, M. E., 1992, The catalytic site residues and interfacial binding of human pancreatic lipase, *J. Biol. Chem.* **267**:17069–17073.

Lowe, M. E., 1996, Mutation of the catalytic site Asp177 to Glu177 in human pancreatic lipase produces an active lipase with increased sensitivity to proteases, *Biochim. Biophys. Acta* **1302**:177–183.

Lowe, M., 1997a, Colipase stabilizes the lid domain of pancreatic triglyceride lipase, *J. Biol. Chem.* **272**:9–12.

Lowe, M. E., 1997b, Molecular mechanisms of rat and human pancreatic triglyceride lipases, *J. Nutr.* **127**:539–557.

Lowe, M. E., 1997c, New pancreatic lipases: Gene expression, protein secretion, and the newborn, *Meth. Enzym.* **284**:285–297.

Lowe, M. E., Rosenblum, J. L., and Strauss, A. W., 1989, Cloning and characterization of human pancreatic lipase cDNA, *J. Biol. Chem.* **264**:20042–20048.

Lowe, M. E., Rosenblum, J. L., McEwen, P., and Strauss, A. W., 1990, Cloning and characterization of the human colipase cDNA, *Biochem.* **29**:823–828.

Lowe, M. E., Li, L., and Kullman, J., 1994, Dexamethasone regulates expression of the human colipase gene, *Pediatr. Res.* **35**:131A.

Lowe, M., Kaplan, M. H., Jackson-Grusby, L., D'Agostino, D., and Grusby, M., 1998, Decreased neonatal dietary fat absorption and T cell cytotoxicity in pancreatic lipase-related protein 2-deficient mice, *J. Biol. Chem.* **273**:31215–31221.

Mahe-Gouhier, N., and Leger, C. L., 1988, Immobilized colipase affinities for lipases B, A, C and their terminal peptide (336–449): The lipase recognition site lysine residues are located in the C-terminal region, *Biochim. Biophys. Acta* **962**:91–97.

Martinelle, M., Holmquist, M., and Hult, K., 1995, On the interfacial activation of Candida antarctica lipase A and B as compared with Humicola lanuginosa lipase, *Biochim. Biophys. Acta* **1258**:272–276.

McIntyre, J., Hundley, P., and Behnke, W., 1987, The role of aromatic side chain residues in micelle binding by pancreatic colipase., *Biochem. J.* **245**:821–829.

McIntyre, J., Schroeder, F., and Behnke, W., 1990a, Synthesis and characterization of the dansyltyrosine derivatives of porcine pancreatic colipase, *Biochem.* **29**:2092–2101.

McIntyre, J., Schroeder, F., and Behnke, W., 1990b, The interaction of bile salt micelles with the dansyltyrosine derivatives of porcine colipase, *Biophys. Chem.* **38**:143–154.

Mickel, F. S., Weidenbach, F., Swarovsky, B., LaForge, K., and Scheele, G. A., 1989, Structure of the canine pancreatic lipase gene, *J. Biol. Chem.* **264**:12895–12901.

Momsen, M. M., Dahim, M., and Brockman, H. L., 1997, Lateral packing of the pancreatic lipase cofactor, colipase, with phosphatidylcholine and substrates, *Biochem.* **36**:10073–10081.

Momsen, W. E., Momsen, M. M., and Brockman, H. L., 1995, Lipid structural reorganization induced by the pancreatic lipase cofactor, procolipase, *Biochem.* **34**:7271–7281.

Morgan, R. G. H., and Hoffman, N. F., 1971, The interaction of lipase, lipase cofactor and bile salts in triglyceride hydrolysis, *Biochim. Biophys. Acta* **248**:143–148.

Ollis, D. L., Cheah, E., Cygler, M., Dijkstra, B., Frolow, F., Franken, S. M., Harel, M., Remington, S. J., Silman, I., Schrag, J., Sussman, J. L., Verschueren, K. H. G., and Goldman, A., 1992, The α/β hydrolase fold, *Prot. Engin.* **5**:197–211.

Patton, J. S., Albertsson, P. A., Erlanson, C., and Borgström, B., 1978a, Binding of porcine pancreatic lipase and colipase in the absence of substrate studied by two-phase partition and affinity chromatography, *J. Biol. Chem.* **253**:4195–4202.

Patton, J. S., Donner, J., and Borgstrom, B., 1978b, Lipase–colipase interactions during gel filtration, *Biochim. Biophys. Acta* **529**:67–78.

Payne, R. M., Sims, H. F., Jennens, M. L., and Lowe, M. E., 1994, Rat pancreatic lipase and two related proteins: Enzymatic properties and mRNA expression during development, *Am. J. Physiol.* **266**:G914–G921.

Peters, G. H., Toxvaerd, S., Larsen, N. B., Bjornholm, T., Schaunburg, K., and Kjaer, K., 1995, Structure and dynamics of lipid monolayers: Implications for enzyme catalysed lipolysis, *Struct. Biol.* **2**:395–401.

Rinderknecht, H., 1993, Pancreatic secretory enzymes, in: *The Pancreas: Biology, Pathobiology, and Disease* (V. L. W. Go, E. P. Dimagno, J. D. Gardner, E. Lebenthal, H. A. Reber, and G. A. Scheele, eds.), Raven Press, Ltd., New York, pp. 219–251.

Rosenheim, O., 1910, On pancreatic lipase III. The separation of lipase from its co-enzyme, *J. Physiol.* **15**:14–16.

Roussel, A., deCaro, J., Bezzine, S., Gastinel, L., de Caro, A., Carriere, F., Leydier, S., Verger, R., and Cambillau, C., 1998a, Reactivation of the totally inactive pancreatic lipase RP1 by structure-predicted point mutations, *Proteins* **32**:523–531.

Roussel, A., Yang, Y., Ferrato, F., Verger, R., Cambillau, C., and Lowe, M. E., 1998b, Structure and activity of rat pancreatic lipase related protein 2, *J. Biol. Chem.* **273**:32121–32128.

Rugani, N., Carriere, F., Thim, L., Borgstrom, B., and Sarda, L., 1995, Lipid binding and activating properties of porcine pancreatic colipase split at the Ile19-Thr80 bond, *Biochim. Biophys. Acta* **1247**:185–194.

Sari, H., Granon, S., and Semeriva, M., 1978, Role of tyrosine residues in the binding of colipase to taurodeoxycholate micelles, *FEBS Lett.* **95**:229–234.

Saudek, V., Atkinson, A., and Pelton, J. T., 1991, Three-dimensional structure of echistatin, the smallest active RGD protein, *Biochem.* **30**:7369–7372.

Savary, P., 1971, The action of pure pig pancreatic lipase upon esters of long chain fatty acids and short chain primary alcohols., *Biochim. Biophys. Acta* **159**:206–303.

Sims, H. F., and Lowe, M. E., 1992, The human colipase gene: Isolation, chromosomal location, and tissue-specific expression, *Biochem.* **31**:7120–7125.

Sims, H. F., Jennens, M. L., and Lowe, M. E., 1993, The human pancreatic lipase-encoding gene: Structure and conservation of an Alu sequence in the lipase gene family, *Gene* **131**:281–285.

Sternby, B., and Borgström, B., 1979, Purification and characterization of human pancreatic colipase, *Biochim. Biophys. Acta* **572**:235–243.

Sternby, B., and Borgström, B., 1984, One-step purification of procolipase from human pancreatic juice by immobilized antibodies against human colipase, *Biochim. Biophys. Acta* **786**:109–112.

Sternby, B., and Erlanson-Albertsson, C., 1982, Measurement of the binding of human colipase to human lipase and lipase substrates, *Biochim. Biophys. Acta* **711**:193–195.

Sternby, B., Engstrom, A., Hellman, U., Vihert, A. M., Sternby, N. H., and Borgström, B., 1984, The primary sequence of human pancreatic colipase, *Biochim. Biophys. Acta* **784**:75–80.

Storlein, L. H., Kraegen, E. W., Chisholm, D. J., Ford, G. L., Bruce, D. G., and Pascoe, W. S., 1987, Fish oil prevents insulin resistance induced in high-fat feeding in rats, *Science* **237**:885–888.

Terada, T., and Nakanuma, Y., 1995, Expression of pancreatic enzymes (alpha-amylase, trypsinogen, and lipase) during human liver development and maturation, *Gastroenterology* **108**:1236–1245.

Terada, T., Kida, T., and Nakanuma, Y., 1993, Extrahepatic peribiliary glands express alpha-amylase isozymes, trypsin and pancratic lipase: An immunohistochemical analysis, *Hepatology* **18**:803–808.

Thirstrup, K., Verger, R., and Carriere, F., 1994, Evidence for a pancreatic lipase subfamily with new kinetic properties, *Biochem.* **33**:2748–2756.

van Tilbeurgh, H., Sarda, L., Verger, R., and Cambillau, C., 1992, Structure of the pancreatic lipase–procolipase complex, *Nature* **359**:159–162.

van Tilbeurgh, H., Egloff, M. P., Martinez, C., Rugani, N., Verger, R., and Cambillau, C., 1993, Interfacial activation of the lipase–procolipase complex by mixed micelles revealed by X ray crystallography, *Nature* **362**:814–820.

Verger, R., 1980, Enzyme kinetics of lipolysis, *Methods. Enzymol.* **64**:341–393.

Verger, R., 1984, Pancreatic lipase, in: *Lipases* (B. Borgström and H. L. Brockman, eds.), Elsevier, Amsterdam, pp. 84–150.

Verger, R., 1997, Interfacial activation of lipases facts and artifacts, *Trends in Biochemical Technology* **15**:32–38.

Von Zastrow, M., and Castle, J. D., 1987, Protein sorting among two distinct export pathways occurs from the content of maturing exocrine storage granules, *J. Cell Biol.* **105**:2675–2684.

Wagner, A. C. C., Wishart, M. J., Mulders, S. M., Blevins, P. M., Andrews, P. C., Lowe, A. W., and Williams, J. A., 1994, GP-3, a newly characterized glycoprotein on the inner surface of the zymogen granule membrane, undergoes regulated secretion, *J. Biol. Chem.* **269**:9099–9104.

Warden, C. H., Davis, R. D., Yoon, M.-Y., Hui, D. Y., Svenson, K., Xia, Y.-R., Diep, A., He, K.-Y., and Lusis, A. J., 1993, Chromosomal localization of lipolytic enzymes in the mouse: Pancreatic lipase, colipase, hormone-sensitive lipase, hepatic lipase, and carboxyl ester lipase, *J. Lipid. Res.* **34**:1451–1455.

Wicker-Planquart, C., and Puigserver, A., 1992, Primary structure of rat pancreatic lipase mRNA, *FEBS Lett.* **296**:61–66.

Wieloch, T., Borgstrom, B., Karl-Erik, F., and Forsen, S., 1979, High-resolution proton magnetic resonance study of porcine colipase and its interaction with taurodeoxycholate, *Biochem.* **18**:1622–1628.

Wieloch, T., Borgstrom, B., Pieroni, G., Pattus, F., and Verger, R., 1981, Porcine pancreatic procolipase and its trypsin-activated form, *FEBS Lett.* **128**:217–220.

Winkler, F., K., D'Arcy, A., and Hunziker, W., 1990, Structure of human pancreatic lipase, *Nature* **343**:771–774.

Wishart, M. J., Andrews, P. C., Nichols, R., Blevins, G. T., Logsdon, C. D., and Williams, J. A., 1993, Identification and cloning of GP-3 from rat pancreatic acinar zymogen granules as a glycosylated membrane-associated lipase, *J. Biol. Chem.* **268**:10303–10311.

Yang, L.-Y., Kuksis, A., and Myher, J. J., 1990, Lipolysis of menhaden oil triacylglycerols and the corresponding fatty acid alkyl esters by pancreatic lipase *in vitro:* A reexamination., *J. Lipid Res.* **31**:137–148.

CHAPTER 4

Pancreatic Lipase

Physiological Studies

HOWARD BROCKMAN

1. Introduction

In vertebrates there are at least three genetic families of lipases that regulate the distribution of ingested neutral ester lipids, that is, triacylglycerols and their degradation products, among tissues and among intracellular compartments. Each of the lipases involved in lipid homeostasis is adapted to function optimally in a specific environment. Even so, the major challenges faced by lipases are shared by all species. Most notably, the lipases of the families mentioned previously are separated from the bulk of their substrates by a monolayer or multilayer of amphipathic molecules. This occurs because the lipases are water-soluble proteins, whereas their triacylglycerol and diacylglycerol substrates are relatively apolar and form bulk oil phases in aqueous environments (Small, 1970). *In vivo,* amphipathic lipids and/or proteins are also present, depending on where the lipase functions. Thus, the lipase needs to remain soluble in water yet be able to attach itself to the appropriate substrate-containing lipid droplet, lipoprotein, or membrane. It must also avoid futile association with inert interfaces, that is, membranes that do not cover substrate-containing bulk phases, which would effectively inhibit the activity of the lipase by separating it from its intended substrate. The specific strategies used for adsorbing to the proper interface and gaining access to the substrate depend on the interfacial compositions of the substrate-containing and competing interfaces.

Perhaps some of the biggest challenges are faced by the lipases of the digestive tract. This is because they are confronted with a myriad of ingested substances in varying proportions. In addition, as the substances transit the digestive tract, their composition changes continuously as a consequence of the actions of the many other hydrolytic enzymes present (Guy and Figarella, 1981). In humans the initial stage of lipolysis is catalyzed by lingual/gastric lipase, which releases 10% of ingested triacylglycerol fatty acids (Carriere *et al.,* 1993). The pancreas secretes two lipases, colipase-dependent pancreatic triacylglycerol li-

---

HOWARD BROCKMAN • The Hormel Institute, University of Minnesota, Austin, Minnesota 55912.
Intestinal Lipid Metabolism, edited by Charles M. Mansbach II *et al.,* Kluwer Academic/Plenum Publishers, 2001.

pase (PTL) and pancreatic carboxylester lipase (Rudd and Brockman, 1984; Wang and Hartsuck, 1993; Cygler *et al.*, 1993; Hui, 1996; Chapus *et al.*, 1988; Lowe, 1994; Ransac *et al.*, 1996), that catalyze the hydrolysis of tri- and diacylglycerols in the intestine (Verger, 1984; Rudd and Brockman, 1984).

This chapter describes PTL from the standpoint of its role in intestinal lipid hydrolysis and how PTL is able to carry out that role. Of necessity, it must also include pancreatic colipase (Borgström and Erlanson-Albertsson, 1984) because understanding the interaction of this protein with lipids and with PTL is essential to understanding how PTL overcomes the challenges to lipolysis in the intestinal lumen. The goal of this chapter is to summarize our understanding of PTL function with emphasis on newer studies. Recently two volumes, 284 and 286, of *Methods in Enzymology* that review lipases from a more methodological perspective have appeared. These provide alternate and complementary sources for some of the subject matter covered in this chapter.

2. Role of PTL in Lipid Digestion

The importance of PTL to normal lipid digestion and absorption becomes apparent by the occurrence of fat malabsorption in its absence. This can occur because the enzyme activity is missing due to genetic defects (Figarella *et al.*, 1980; Ghishan *et al.*, 1984), is intentionally reduced by pharmacological intervention (Drent *et al.*, 1995), or is lowered by pancreatic insufficiency (DiMagno *et al.*, 1973). However, individuals with up to 90% deficiency in pancreatic enzyme output are sometimes able to maintain normal lipid digestion (DiMagno *et al.*, 1973). This may reflect the fact that in normal humans the amount of PTL secreted daily can digest 1000 times the typical 100 g of fat consumed per day (Patton, 1981). Alternately, compensatory mechanisms involving other enzymes may result in sufficient lipid hydrolysis (Verger *et al.*, 1996). Aging does not appear to compromise intestinal lipid hydrolysis and absorption (Holt and Balint, 1993), but neonates secrete only about 10% the adult level of PTL until 1 to 2 years of age (Lee and Lebenthal, 1993; Manson and Weaver, 1997).

Both PTL and carboxylester lipase are present in pancreatic juice. However, the ability of PTL to act on emulsified substrate and the preference of carboxylester lipase for micellar substrates suggests that they function sequentially in lipid digestion (Rudd and Brockman, 1984; Lindstrom *et al.*, 1988; Borgström, 1993). Thus, gastric lipase begins the hydrolysis of fat in the stomach. The lipids that enter the duodenum are emulsified into 1 to 50 μm diameter particles by mechanical action (Armand *et al.*, 1996). After mixing with bile and pancreatic enzymes and more colipase, lipolysis is catalyzed initially by PTL with the aid of colipase. As partial glycerides, fatty acids, and lysophospholipids accumulate in mixed bile salt micelles, hydrolysis is completed by carboxylester lipase.

3. Regulation of PTL and Colipase Levels

The synthesis of PTL and colipase are under both transcriptional and translational control by hormones. As for other pancreatic proteins, the levels of these hormones are regulated by dietary metabolites such that there is dietary adaptation, albeit asynchronous, of both PTL and colipase levels (Brannon, 1990; Hirschi *et al.*, 1991; Scheele, 1993). Recently, procolipase, colipase, and enterostatin, the pentapeptide produced when procolipase is proteolytically activated to colipase (Borgström *et al.*, 1979; Borgström and Erlanson-Albertsson, 1984), were all found in gastric juice (Sörhede *et al.*, 1996). This could reflect the fact that enterostatin is a satiety factor that reduces fat intake (Erlanson-Albertsson and York, 1997). This effect occurs both peripherally (Tian *et al.*, 1994) and centrally (Rössner *et al.*, 1995) on differing time scales. Alternately, the presence of colipase in gastric juice could aid gastric lipolysis (Bernbäck *et al.*, 1987) or prepare lipid emulsions for attack by pancreatic lipase in the duodenum, as is discussed in the following sections.

The secretion of PTL and other digestive enzymes is under hormonal, neural, and cephalic control (DiMagno and Layer, 1993). Enzyme secretion is low and periodic in the interdigestive period but increases markedly following ingestion of food. PTL then catalyzes the release of fatty acids that trigger the release of cholecystokinin, leading to further secretion of pancreatic enzymes and bile (Hildebrand *et al.*, 1998). In this acute phase, fat digestion is approximately 80% complete within the proximal duodenum (Holtmann *et al.*, 1997) and 90 to 95% complete within the proximal jejunum (Borgström *et al.*, 1957). At the end of this phase of lipid digestion, PTL activity has decreased by 70% as a consequence of proteolysis. Lipolytic activity during jejunoileal transit is decreased by the presence of bile acids and chymotrypsin, but this is ameliorated by an increase in dietary calories and carbohydrate (Holtmann *et al.*, 1997).

4. Regulation of PTL Adsorption to Interfaces

4.1. Adsorption to Substrate

To understand how PTL is able to attack emulsified substrates first requires knowledge of how this water-soluble enzyme effects the hydrolysis of water-insoluble lipids in the absence of either colipase or competing amphiphiles, such as detergents, proteins, and phospholipids. The context in which this must be approached is the general model for lipolysis that has evolved from the work of many investigators (Verger, 1980; Brockman, 1984), beginning with Pierre Desnuelle (Sémériva and Desnuelle, 1979). These studies have shown that the site of lipolysis is the lipid–water interface. The interface is best visualized as a separate phase only one lipid molecule thick that separates the bulk oil phase, containing the substrate, from the bulk aqueous phase, containing the enzyme (Brockman, 1984). For lipolysis to occur, both PTL and the substrate must reside in this quasi-two-dimensional phase. In emulsion systems, substrate partitioning from the bulk oil phase to the interfacial phase is regulated by the balance between the properties of the acyl chains and the polar head group. Triacylglycerols show very little tendency to occupy sites in interfaces because they are relatively apolar (Small *et al.*, 1983). Most studies of lipolysis have been done using systems in which substrate availability at the interface and the composition of the interface were unknown. However, with some model systems interfaces consisting solely of PTL substrate or products can be created (Brockman *et al.*, 1973; Labourdenne *et al.*, 1997; Brockman 1984; Verger and Pieroni, 1986; Pieroni *et al.*, 1990). These have allowed the individual steps of the catalytic process to be examined.

In a system utilizing monomolecular films of diacylglycerol, PTL partitioned to the interface with high affinity (Gargouri *et al.*, 1986b). As the packing density of the lipid molecules was increased to physiologically relevant surface pressures greater than 20 mN/m, PTL adsorption became independent of lipid packing density. This suggests that the protein does not significantly penetrate between the lipid molecules, but rather adsorbs to the interface, replacing water of hydration in the lipid head group region. Supporting this idea is the observation that almost a monolayer of PTL can adsorb to tightly packed fatty acid monolayers (Muderhwa and Brockman, 1992a). Finally, at a methylated glass surface, PTL adsorbs with a binding constant of 1.3×10^8 M (Brockman *et al.*, 1973). This shows that high affinity adsorption of PTL to interfaces does not require molecular recognition of the interface by the enzyme in the sense of classical enzyme–substrate binding. Instead, it appears to be a simple partitioning driven by the replacement of water of hydration of the lipid head groups by the enzyme.

The first three-dimensional structures of PTL without (Winkler *et al.*, 1990) and with (van Tilbeurgh *et al.*, 1992) procolipase showed that it consists of two domains connected by a flexible hinge. The N-terminal domain (residues 1–335) contains the active site and has an α/β-fold structure common to other hydrolases (Ollis *et al.*, 1992). The C-terminal domain (residues 336–449) is a β-barrel. As seen in the Figure 1A, there are no large hydrophobic faces exposed in the N-terminal domain on the face of the molecule above the active site. Thus, this structure does not reveal the reason for the affinity of PTL for hydrophobic surfaces. The answer to that question was provided by the structure of a complex of a human PTL-porcine procolipase in the presence of phosphatidylcholine, bile salts, and a nonionic detergent (van Tilbeurgh *et al.*, 1993a, 1993b) shown in Figure 1B. Relative to the "closed" conformation shown in Figure 1A, there is complex movement of residues 238–262 ("the lid"), 77–80 ("the β-5 loop") and 213–217 that both create an exposed hydrophobic surface near the active site and segregate it from charged residues. It is referred to as the "open" conformation. The exposed hydrophobic residues of the lid and β-5 loops (black) and the exposed charged residues (light gray) are shown as space filled. The filled residues are those that are exposed in either structure when viewed from above the active site, relative to the open conformation. Also filled (dark gray) in Figure 1B are the hydrophobic residues on the "fingers" of procolipase described in the following text. No charged residues are exposed on the hydrophobic surface of procolipase. An essentially identical structure was later obtained for porcine PTL-colipase in the presence of a nonionic detergent, indicating the lack of a specific role for the procolipase N-terminal pentapeptide, bile salts, or phospholipid in determining the structure (Hermoso *et al.*, 1996). Functional evidence for the role of the hydrophobic lid residues in driving the adsorption of PTL to substrates was obtained by mutation studies (Jennens and Lowe, 1994). Whereas wild-type enzyme partitioned almost exclusively to the substrate phase, mutants lacking all or part of the lid remained in the aqueous phase.

Teleologically, the sequestration of hydrophobic residues of the lid and β-5 loops in the closed conformation is probably necessary for effective function. In pancreatic juice and in the intestinal lumen, if the open conformation was assumed and the apolar surface was continuously exposed, PTL would likely dimerize or be more susceptible to proteolysis. However, if the closed conformation predominates in solution, what triggers it to open? It could be proximity of the enzyme to an interface, but no mechanism has been provided for how this could occur. A simpler explanation is that for PTL and other lipases, the open and closed conformations are in dynamic equilibrium (Rubin, 1994) with a significant energy

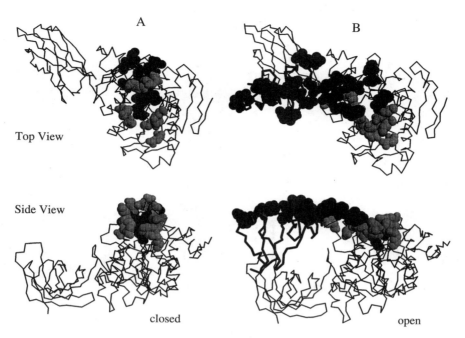

Figure 1. Structures of (A) human PTL in the closed conformation and (B) the human PTL–porcine procolipase complex in the open conformation. Shown space filled are selected (see text) hydrophobic (dark gray and black) and charged (light gray) residues. The backbone of procolipase is shown in bold and in each panel the N-terminal domain of PTL is on the right. Atomic coordinates for PTL in the closed PTL–procolipase complex were kindly provided by C. Cambillau and coordinates for the open form were from the Brookhaven protein database.

barrier between the two states. This is depicted in Figure 2A. The free energy required to hydrate hydrophobic residues suggests that in solution, the equilibrium will favor the closed conformation but that for a finite fraction of time, the open conformation is present. Collision of the enzyme with an appropriate interface while open will result in productive adsorption. Once PTL is adsorbed in the open conformation, the PTL–surface complex is stabilized by the energy required to hydrate both it and the apolar surface to which it is adsorbed. Consistent with this model (Figure 2A), the functional irreversibility of adsorption of PTL to apolar surfaces has been demonstrated by direct transfer experiments (Pieroni *et al.,* 1990).

4.2. Inhibition

The flatness of the surface containing the exposed hydrophobic residues of PTL in the open conformation implies that it must encounter a relatively large area of substrate-covered surface. On this basis it is easy to understand that the adsorption of PTL alone can be inhibited by surface obstacles, like proteins. As reviewed (Pieroni *et al.,* 1990), a number of common proteins, such as β-lactoglobulin and bovine serum albumin, inhibit the hy-

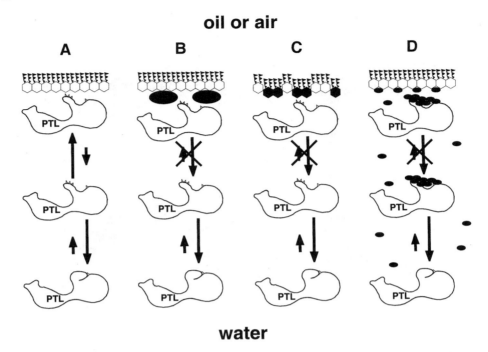

Figure 2. Adsorption of the open form of PTL to a substrate interface (A) and its inhibition by proteins (B) phosphatidylcholine (C) and bile salts (D).

drolysis of didecanoylglycerol monolayers. This inhibition occurs not from the effects of these proteins on interfacial tension, protein charge, or the masking of substrate, but from inhibition of PTL adsorption to the interface as illustrated in Figure 2B. A recently described protein isolated from wheat flour also inhibits PTL (Tani *et al.,* 1994). In contrast to the proteins described previously, this inhibitor binds directly to the enzyme. However, it is unknown if it is PTL adsorption or catalysis that is inhibited.

Phospholipids, such as diacyl phosphatidylcholine, are a major constituent of bile and cellular membranes and are present at about 1 mM in the intestinal lumen (Borgström, 1993). In a classic paper in which a phosphatidylcholine-stabilized triacylglycerol emulsion was used as substrate for PTL, Borgström (1980) showed that the hydrolysis of the triacylglycerol by PTL exhibited a long lag phase after which the reaction rate accelerated rapidly to a steady state. The reason for the lag was shown to be the inability of PTL to adsorb to the substrate-containing interface and it was later shown that adsorption of PTL was inversely correlated with the mole fraction of phosphatidylcholine in the interface (Pieroni and Verger, 1979). It was also demonstrated that adsorption of PTL to egg phosphatidylcholine monolayers ceased above a surface pressure of about 25 mN/m, in contrast to its adsorption to diacylglycerol monolayers (Pieroni *et al.,* 1990). These data suggest that the adsorption of PTL is blocked sterically by the phosphocholine head groups of the phospholipid, as depicted in Figure 2C. This again emphasizes the idea that PTL requires a relatively large area, clear of obstacles, to adsorb to substrate-containing interfaces.

Bile salts are a third component of the intestinal lumen that inhibit the adsorption of PTL to interfaces. They are present at a total concentration of about 15 mM, which is well above the critical micelle concentration for each of the various species (Hernell *et al.,* 1990). Thus, bile salts are present as mixed micelles with other lipids, as monomers in the aqueous phase in millimolar concentrations, and as adsorbed components of the emulsion interface. This will cause phospholipids, proteins, and other surface-active constituents to redistribute between micelles and emulsion particles (Nichols, 1986; Staggers *et al.,* 1990; Heuman *et al.,* 1996; de la Maza *et al.,* 1997). Taurodeoxycholate does not bind significantly to PTL until it is present in the range of its critical micelle concentration (Borgström and Donnér, 1976). At 0.8 mM and higher concentrations of taurodeoxycholate, mixed micelles form that contain 12 moles of taurodeoxycholate per mole of PTL. A more recent inhibition study showed that bile salt micelles were required to render the active site serine of PTL accessible to diethyl *p*-nitrophenyl phosphate (Moreau *et al.,* 1991). This suggests that interaction with bile salts stabilizes the lid in the open position as illustrated in Figure 2D. However, in the presence of bile salts PTL adsorption to apolar interfaces is blocked (Borgström, 1975; Momsen and Brockman, 1976). It is interesting that this effect is seen at 0.8 mM taurodeoxycholate, near the critical micelle concentration of bile salt, but not at 0.3 mM, consistent with mixed PTL–taurodeoxycholate micelles being the nonadsorbable species. In addition, at 0.3 mM, where the surface excess of taurocholate should be near maximal, lipase is not only adsorbed, but is stabilized against denaturation at the apolar interface (Momsen and Brockman, 1976). It has also been shown that the surface charge created by nonconjugated bile salts on the emulsion surface does not inhibit PTL (Wickham *et al.,* 1998) and that low concentrations of bile salts and of other detergents can activate PTL in emulsion assay systems (Gargouri *et al.,* 1983). These observations are consistent with the relatively low surface activity of bile salts (O'Connor *et al.,* 1983) and provide further evidence that it is mixed micelle formation with PTL lid residues, not obstruction of the substrate-containing interface by adsorbed bile salt, that is responsible for their inhibition of PTL adsorption.

4.3. Colipase

Considering that proteins, phospholipids, and bile salts are all present in high concentrations in the intestinal lumen, PTL should be unable to adsorb to substrate-containing interfaces, particularly before proteolysis is complete, phospholipolysis is complete, and bile salts are readsorbed. Being unable to adsorb, it should remain separated from its substrates and, therefore, be unable to function normally, if at all. Indeed, this is true, based on the steatorrhoea observed in patients with a congenital deficiency of the PTL cofactor, colipase (Hildebrand *et al.,* 1982). Colipase is a 10 kD protein, about one fifth the size of PTL. It was initially described as a heat-and-acid-stable factor required for PTL activity to be expressed in the presence of bile salts (Baskys *et al.,* 1963; Morgan *et al.,* 1969) and was later purified from porcine pancreas (Maylie *et al.,* 1971). Its role in enabling PTL function has been the subject of several reviews (Borgström and Erlanson-Albertsson, 1984; Chapus *et al.,* 1988; Erlanson-Albertsson, 1992; Lowe, 1997b). The picture that has emerged is of a protein with a higher affinity than PTL for interfaces in the presence of bile salts, phospholipids, and proteins. It also has an affinity for PTL, with which it forms a 1:1 complex (Figure 1B). Thus, it has become accepted that either procolipase or its trypsin-activated form, colipase, can serve as an anchor for PTL in interfaces to which it could otherwise not adsorb.

5. Regulation of Colipase Adsorption to Interfaces

How procolipase and colipase are each able to bind to interfaces to which PTL cannot has recently been reinvestigated using mixed monolayers of phosphatidylcholine and a lipolysis reactant (W. E. Momsen *et al.,* 1995; M. M. Momsen *et al.,* 1997). The reactant was either a substrate for PTL (e.g., diacylglycerol), or a product (e.g., fatty acid). In describing experiments in which similar results were obtained with either form of the cofactor, the term *(pro)colipase* is used. The results showed that (pro)colipase, in the absence of PTL, interacted preferentially with the reactants of lipolysis compared to the phosphatidylcholine. Adsorption affinity to the phospholipid alone was comparable to that measured earlier for adsorption of PTL to phosphatidylcholine (Pieroni *et al.,* 1990). Thus, (pro)colipase preferentially adsorbs to interfaces rich in lipolysis reactants with high affinity. Studies of PTL–(pro)colipase interactions under physiologically relevant conditions led Larsson and Erlanson-Albertsson (1991) to conclude that the trypsin-catalyzed conversion of procolipase to colipase in the intestinal lumen serves no immediate purpose in lipid digestion. Rather, it was proposed that proteolysis occurs to release the satiety factor, enterostatin. This hypothesis was tested by having procolipase and colipase compete for lipid interfaces (Schmit *et al.,* 1996). In all cases if any lipid, even phosphatidylcholine, was present in the interface, colipase quantitatively displaced procolipase from the interface. Thus, the activation of procolipase to colipase may regulate the affinity of the cofactor for surfaces *in vivo.*

The preference of (pro)colipase for interaction with lipolysis reactants implies that in the plane of the interface, lipolysis reactants will spend more time near (pro)colipase than does phosphatidylcholine. Stated differently, (pro)colipase will laterally concentrate lipolysis reactants in its vicinity and, thereby, will recruit substrate to PTL that is anchored to the interface via (pro)colipase. The extent of interaction between colipase and lipolysis reactants was titrated (M. M. Momsen *et al.,* 1997). This revealed that the maximal number of acyl chains that sense the influence of colipase is 20 to 25. Hence, the interaction between colipase and lipolysis reactants is not complex formation in the sense of specific binding at a defined site. Nor, because of the limited number of acyl chains involved, do the data suggest (pro)colipase-induced lateral phase separation of reactants and (pro)colipase. Rather, the results describe a two-dimensional adsorption process. Intriguing possibilities are that in the intestinal lumen the preferential interaction of colipase with reactant-containing interfaces helps to target it or the PTL–colipase complex to reactant-containing interfaces, prevents their nonproductive adsorption to substrate-deficient surfaces, such as cellular membranes, and regulates desorption of the PTL–colipase complex from substrate-depleted particles. The last possibility assumes that sufficient bile salt micelles are present to rapidly remove fatty acid and monoacylglycerol products from the particle undergoing lipolysis (Thomson and Dietschy, 1981).

Structural studies of PTL–(pro)colipase complexes (Egloff *et al.,* 1995a, 1995b; Ayvazian et al., 1996) and of colipase alone (Breg *et al.,* 1995) have identified the determinants of (pro)colipase interaction with lipids. (Pro)colipase is an amphipathic, wedge-shaped protein (Fig. 1B). The X ray structures for the complexes were determined in the presence of lipids and showed that one face of the molecule is comprised of four hydrophobic loops or "fingers," connected by β-turns and stabilized by five disulfide bridges among the different fingers. It is the hydrophobic residues on these fingers and three isoleucines near the N-terminus that presumably interact with the lipid–water interface. These are space filled (dark gray) in Figure 1B. The structures also showed no direct interactions between PTL and the colipase N-terminus. Thus, proteolytic cleavage of the pentapeptide, enterostatin, does not appear to play a direct role in PTL–(pro)colipase interactions. The structural studies did not address directly the higher affinity of colipase, relative to procolipase, for interfaces. However, the presence of charged residues in the N-terminal pentapeptide of procolipase is likely to interfere with interactions of the three adjacent isoleucines with lipids. These were not resolved in the X ray structures and are not shown in Fig. 1B.

Similar to PTL, colipase forms mixed micelles with bile salts, such as taurodeoxycholate (Borgström and Donnér, 1976; Charles *et al.*, 1980; McIntyre *et al.*, 1987) and the PTL–colipase complex binds as much taurocholate as do PTL and colipase separately (Borgström and Donnér, 1976). The X ray crystal structure of PTL–procolipase in the presence of bile salts shown in Figure 1B (van Tilbeurgh *et al.*, 1993b) does not reveal the location of the bile salt micelles. The structure of this open PTL–procolipase complex is essentially identical to that determined for PTL–colipase in the presence of a nonionic detergent (Hermoso *et al.*, 1996). For the latter structure it was possible to localize the detergent micelles with respect to the PTL–colipase complex (Hermoso *et al.*, 1997). Surprisingly, this showed that interaction between the complex and a detergent micelle does not involve the flat, hydrophobic face of the complex shown in Figure 1B.

6. The Functional PTL–Colipase–Lipid Complex

PTL–colipase binding studies in solution identified the C-terminal domain of PTL as the site of interaction with colipase (Bousset-Risso *et al.*, 1985; Abousalham *et al.*, 1992; Ayvazian *et al.*, 1996). This was confirmed by the first crystal structure of the closed PTL–procolipase complex (van Tilbeurgh *et al.*, 1992), which showed multiple interactions between procolipase and C-terminal residues of PTL. The ionic and hydrogen-bonding residues that interact with PTL are on the polar side of the molecule. This simple picture of the role of colipase and of PTL–procolipase interactions was challenged by the structure of the PTL–procolipase complex determined in the presence of phosphatidylcholine and bile salt (van Tilbeurgh *et al.*, 1993b). It showed PTL in the open conformation (Figure 1B) thought to represent the active form of the enzyme at interfaces. In this structure the orientation of the C-terminal domain of PTL was changed, but its contacts with procolipase remained. Three additional hydrogen bonds formed between procolipase and the lid domain of PTL. Thus, in the presence of lipids, the procolipase–PTL interactions not only strengthened the complex but appeared to stabilize the open conformation of PTL. The importance of colipase interactions with the N-terminal lid residues was reinforced by mutation studies of PTL (Jennens and Lowe, 1995). Single and multiple mutations of C-terminal residues involved in interactions with colipase failed to reduce the ability of the cofactor to reactivate PTL. Even deletion of the entire C-terminal domain resulted in an enzyme that had reduced activity but still required colipase for activity against tributyroylglycerol–bile salt emulsions. In contrast, mutation of Glu 15 of colipase, a residue involved in stabilizing the open conformation, reduced activity of the cofactor 175-fold (Lowe, 1997a).

Based on the biochemical, structural, and molecular biological studies of PTL–(pro)colipase interactions, it is clear that the proteins interact at multiple sites and that the formation of the binary complex coupled with the surface activity of (pro)colipase contributes to localizing PTL at substrate-containing interfaces. But do protein–protein interactions alone determine the stability of the complex? As discussed (Egloff *et al.*, 1995a), the strength of the PTL–(pro)colipase complex is relatively weak compared to other protein–protein interactions. Moreover, several biochemical studies have shown that lipids, especially free fatty acids, increase the stability of the complex and enhance PTL binding to and activity in phosphatidylcholine-rich interfaces (Wieloch *et al.*, 1982; Erlanson-Albertsson, 1983; Larsson and Erlanson-Albertsson, 1986). However, these studies did not indicate how the effect was manifested.

One possible consequence of the ability of colipase to laterally concentrate lipolysis reactants in its vicinity is facilitated adsorption of the hydrophobic lid residues of PTL to the interface in the presence of phosphatidylcholine. Evidence to support this hypothesis has been provided by looking at the role of fatty acid on the adsorption of PTL to phosphatidylcholine–colipase monolayers (Dahim *et al.*, 1998). It was found, surprisingly, that

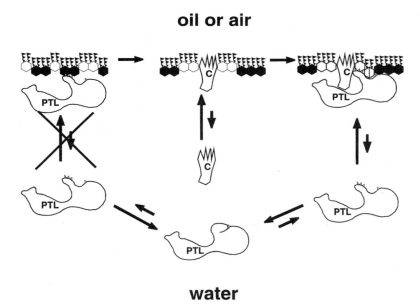

Figure 3. Facilitation of PTL adsorption to phosphatidylcholine-rich interfaces by colipase and lipolysis reactants. Phosphatidylcholine is shown filled and lipolysis reactant (e.g., fatty acid or diacylglycerol), is shown open. Left, inhibition of PTL adsorption; center, adsorption of colipase (C) and lateral rearrangement of lipid molecules; right, formation of the catalytically competent PTL–colipase–reactant complex. Adapted from Dahim *et al.*, 1998.

the presence of colipase in a phosphatidylcholine interface was insufficient to enable PTL adsorption. Nor could the preformed PTL–colipase complex adsorb to phosphatidylcholine alone. For any PTL adsorption to occur, the interface had to contain fatty acid. Moreover, the dependence of adsorption on increasing fatty acid mole fraction was exponential. Because substrates and fatty acids interact similarly with colipase in interfaces, these data suggest that substrates will also facilitate PTL adsorption to phospholipid-rich interfaces. These observations led to proposal of the model for the regulation of PTL adsorption shown in Fig. 3. It shows that in a phosphatidylcholine-rich interface it is the ability of colipase to create a fatty acid/substrate "nanodomain" around itself that kinetically facilitates formation of the functional ternary PTL lid–colipase–lipid complex in the interface. Although this model leaves many questions unanswered, it does explain qualitatively a body of data concerning phosphatidylcholine-induced lags in lipolysis. This autocatalytic model also underscores the importance of generating surface-active lipolysis products in the stomach for rapid and efficient intestinal lipolysis (Gargouri *et al.,* 1986a) and could explain why colipase is secreted into the stomach (Sörhede *et al.,* 1996).

7. Regulation of PTL Catalysis

7.1. Substrate Concentration and Kinetics

The preceding paragraphs described the regulation of the adsorption of PTL to interfaces both with and without colipase. The body of evidence on hand suggests that the transformation of PTL from the closed to the open conformation in either the presence or absence of colipase is a prerequisite for adsorption and is, in large part, responsible for the observed "interfacial activation" of lipases. This term describes the observation that many lipases, particularly PTL, exhibit very low activity on water-soluble, monomeric substrates compared to their activity at substrate–water interfaces. As described previously and recently discussed (Verger, 1997), the term "activation" is somewhat of a misnomer because the enzyme is really overcoming inhibition of its adsorption caused by surface obstacles and bile salts. On the other hand, during activation the enzyme comes to reside at an interface at which the effective concentration of substrate is, typically, orders of magnitude higher than can be attained for typical PTL substrates (Brockman *et al.,* 1973). For example, the effective three-dimensional substrate concentration for an insoluble triacylglycerol can be estimated at about 0.5M in a substrate-only interface and about 2mM in a phospholipid interface! This latter number is a minimum, because it accounts for neither the lateral substrate-concentrating effect of colipase nor the substrate-orientating effect of the interface. A number of kinetic models to describe lipolysis and derive essential activity and inhibition constants have been obtained, but their discussion is outside the purpose and beyond the scope of this review (Panaiotov *et al.,* 1997; Bernard *et al.,* 1996; Marangoni, 1994; Havsteen *et al.,* 1992; Jain and Berg, 1989; Ransac *et al.,* 1990). As emphasized previously, the essential feature that distinguishes these models from the classical Michaelis–Menten kinetic model is the notion of the nonspecific adsorption of the lipase from the bulk solution phase to the interface prior to its interaction with the substrate in the interfacial plane.

When PTL is adsorbed to a triacylglycerol interface it displays a high turnover rate of about 10^5/min. One reason for this is the high concentration of properly oriented substrate. Another is the weakness of the interaction between the adsorbed enzyme and the substrate in the interface, that is, the two-dimensional Michaelis constant. Practically, this means that the rate at which adsorbed enzyme turns substrate over is linearly dependent on the interfacial concentration of substrate (Momsen and Brockman, 1981). Because in a substrate-only interface the substrate is virtually neat, the interfacial Michaelis constant must be higher than the two-dimensional substrate concentration and, reasonably, the lifetime of the PTL–substrate complex will be short.

7.2. Structure and Catalysis

What is not revealed by kinetic studies is whether the opening of the PTL lid and adjacent loops has consequences for catalysis beyond simply getting the enzyme active site to a place where there is a high concentration of properly oriented substrate. However, structural data show that the transition from the closed to the open conformation that accompanies PTL adsorption has direct, mechanistic consequences. First and obvious from inspection of the three-dimensional structures of the enzyme, assumption of the open conformation by PTL allows substrate and solvent access to the buried residues of the active site, thereby enhancing catalysis. The active site triad of PTL consists of Ser 153, His 264, and Asp 177, based on both biochemical evidence (Verger, 1984) and the properties of PTL mutants (Lowe, 1992). Structurally, the catalytic triad of PTL resembles that found in trypsin and in other hydrolases (Winkler *et al.,* 1990), and the mechanism of hydrolysis is the same (Ransac *et al.,* 1996). The conformation of the catalytic triad in PTL is essentially the same in both the closed and opened forms. However, there is an important consequence of the opening of the lid (van Tilbeurgh *et al.,* 1993b; Derewenda and Sharp, 1993; Rubin, 1994). In the closed form of PTL there is no residue in the correct position to stabilize the oxyanion, whereas in the open form there is. Thus, the catalytic turnover of PTL should be enhanced in the open conformation relative to the closed from both by increased accessibility of the active site and by stabilization of the oxyanion intermediate.

The opening of the lid and other loops also regulates PTL specificity by creating the part of the site that interacts with the apolar parts of the substrate (Egloff *et al.,* 1995b). PTL hydrolyzes acyl moieties from a variety of natural and synthetic substrates, provided they are primary esters, such as those at the *sn*-1 and *sn*-3 positions of glycerol in natural substrates (Brockerhoff and Jensen, 1974). Among natural substrates, it shows some selectivity, about six fold, against esters of polyunsaturated fatty acids (Yang *et al.,* 1990; Chen *et al.,* 1994). The small extent of this discrimination, the excess of PTL in the intestine, and the presence of carboxylester lipase argue against this selectivity having any measurable effect on lipid adsorption. Lipid type specificity may also be influenced by residues in the movable loops (Withers-Martinez *et al.,* 1996; Carrière *et al.,* 1997a, 1997b), particularly with respect to phospholipase A_1 activity. Natural phospholipids normally do not serve as substrates, but acidic phospholipids with medium length chains can be attacked in the presence of calcium (Verger *et al.,* 1976). The relative lack of specificity of PTL extends to its stereospecificity (Ransac *et al.,* 1996). In contrast to other lipases, it shows little ability to distinguish between the *sn*-1 and *sn*3 positions of either 1,2-diacylglycerols or prochirial triacylglycerols. As discussed (Ransac *et al.,* 1996), however, some caution must be exercised in making absolute statements about lipase specificity because it is influenced by the substrate under study and its physical state in any particular interface.

7.3. Colipase

By enabling or assisting PTL adsorption to interfaces and by laterally concentrating substrates, colipase increases the efficiency of lipolysis. In conventional assay systems the specific activity of PTL measured against triacylglycerols alone is comparable to that measured for triacylglycerols in the presence of bile salts and colipase. The conventional wisdom has been that colipase plays no direct role in regulating the catalytic turnover of PTL. However, recent experiments with a novel oil drop tensiometer have indicated that the specific activity of PTL is enhanced 10- to 15-fold by colipase at a trioleoylglycerol–water interface (Labourdenne *et al.*, 1997). It is speculated that this effect may be due to the stabilization of the open lid by colipase or, possibly, to the lateral-substrate-concentrating effect of colipase described earlier. Another, more speculative possibility is that in addition to concentrating lipids, colipase stabilizes a particular surface lipid conformation that is a better substrate for PTL. Kinetic data for the hydrolysis of diacylglycerol monolayers by PTL were consistent with the substrate existing in two conformational states in the interface with areas of 7.5 nm^2 and 3.8 nm^2 (Momsen *et al.*, 1979). Only the more condensed state served as a substrate for the enzyme. More recently, a phase transition to a compact state of diacylglycerols with a molecular area of 3.8 nm^2 has been detected in molecular dynamics simulations of diacylglycerol monolayers (Peters *et al.*, 1995).

7.4. Inhibition by Proteins, Phosphatidylcholine, and Bile Salts

As described previously, proteins, phospholipids, and bile salts all impede the adsorption of PTL to interfaces. Once PTL is adsorbed, these substances can dilute the substrate in the interface, force it from the interface by altering interfacial tension, or interfere with substrate diffusion to the active site within the interfacial plane. With respect to the last point, if a significant portion of an interface is occupied by proteins, diffusion of lipids in the interface will be slowed and can become restricted in range (Saxton, 1994; Schram *et al.*, 1994; Muderhwa and Brockman, 1992a). In the absence of such large diffusional barriers, diffusion of substrate and phospholipid in the normally fluid interfaces encountered by PTL should be rapid. However, studies using mixed monomolecular films of a phosphatidylcholine and any of several substrates of either PTL (Muderhwa and Brockman, 1992a) or carboxylester lipase (Tsujita *et al.*, 1989; Muderhwa and Brockman, 1992b) showed that the access of adsorbed enzyme to substrate is severely restricted in the range of low substrate mole fractions. Moreover, for both enzymes there is an almost discontinuous increase in both substrate accessibility and apparent reaction mechanism when substrate mole fraction exceeds a critical value (Muderhwa and Brockman, 1992a, 1992b). From these data it was concluded that, at low mole fractions of substrate, there is a kinetic barrier in the accessibility of the enzymes to the substrate. Whether this is a barrier to lateral movement of substrate to the enzymes in the plane of the interface or to moving from the interface to the active site is unknown.

8. Perspective

Although the studies described previously provide a reasonably coherent picture of how PTL is able to function in the intestine, they are incomplete. For example, the site of intestinal lipolysis is not firmly established. It has traditionally been assumed to be the intestinal lumen, but some provocative data suggests an epithelial location (Bosner *et al.,* 1988, 1989). It is also not yet clear how colipase is able to overcome the inhibition of PTL adsorption by bile salts. Is it because the long hydrophobic fingers of colipase can intercalate with lipids but the relatively flat hydrophobic surface of PTL cannot, or do bile salt micelles bind away from the hydrophobic surface as recently proposed (Hermoso *et al.,* 1996)? Does colipase contribute to the specificity of lipolysis by preferentially concentrating substrates with saturated or monounsaturated acyl groups? Is it coincidence that the structure of the C-terminal domain of PTL resembles a fatty-acid-binding protein? In this regard it is of interest that the recently determined structure of mammalian 15-lipoxygenase contains a β-barrel domain similar to that of PTL (Gillmor *et al.,* 1997). Because both utilize or produce fatty acids, does this structure serve a common function? Answering these questions will provide new insights into how PTL and colipase overcome the challenges presented by the intestinal milieu.

References

Abousalham, A., Chaillan, C., Kerfelec, B., Foglizzo, E., and Chapus, C., 1992, Uncoupling of catalysis and colipase binding in pancreatic lipase by limited proteolysis, *Protein Eng.* **5:**105–111.

Armand, M., Borel, P., Pasquier, B., Dubois, C., Senft, M., Andre, M., Peyrot, J., Salducci, J., and Lairon, D., 1996, Physicochemical characteristics of emulsions during fat digestion in human stomach and duodenum, *Am. J. Physiol.* **34:**G:172–G 183.

Ayvazian, L., Crenon, I., Granon, S., Chapus, C., and Kerfelec, B., 1996, Recombinant C-terminal domain of pancreatic lipase retains full ability to bind colipase, *Protein Eng.* **9:**707–711.

Baskys, B., Klein, E., and Lever, W. F., 1963, Lipases of blood and tissues. III. Purification and properties of pancreatic lipase, *Arch. Biochem. Biophys.* **102:**201–209.

Bernard, C., Buc, J., and Piéroni, G., 1996, Lipolysis and heterogeneous catalysis. A new concept for expressing the substrate concentration, *Lipids* **31:**261–267.

Bernbäck, S., Hernell, O., and Bläckberg, L., 1987, Bovine pregastric lipase: A model for the human enzyme with respect to properties relevant to its site of action, *Biochim. Biophys. Acta* **922:**206–213.

Borgström, B., 1975, On the interactions between pancreatic lipase and colipase and the substrate, and the importance of bile salts, *J. Lipid Res.* **16:**411–417.

Borgström, B., 1980, Importance of phospholipids, pancreatic phospholipase A₂, and fatty acid for the digestion of dietary fat. *In vitro* experiments with the porcine enzymes, *Gastroenterology* **78:**954–962.

Borgström, B. 1993, Luminal digestion of fats, in: *The Pancreas: Biology, Pathobiology and Disease,* 2nd edition (V. L. W. Go, E. P. DiMagno, J. D. Gardner, E. Lebenthal, H. A. Reber, and G. A. Scheele, eds.), Raven Press, New York, pp. 475–488.

Borgström, B., and Donnér, J., 1976, On the binding of bile salt to pancreatic lipase, *Biochim. Biophys. Acta* **450:**352–357.

Borgström, B., and Erlanson-Albertsson, C., 1984, Pancreatic colipase, in: *Lipases* (B. Borgström and H. L. Brockman, eds.), Elsevier Science Publishers, Amsterdam, pp. 151– 183.

Borgström, B., Dahlquist, A., Lundh, G., and Sjovall, J., 1957, Studies of intestinal digestion and adsorption in the human, *J. Clin. Invest.* **36;** 1521–1536.

Borgström, B., Wieloch, T., and Erlanson-Albertsson, C., 1979, Evidence for a pancreatic procolipase and its activation by trypsin, *FEBS Lett.* **108:**407–410.

Bosner, M. S., Gulick, T., Riley, D. J. S., Spilburg, C. A., and Lange, L. G., III, 1988, Receptorlike function of heparin in the binding and uptake of neutral lipids, *Proc. Natl. Acad. Sci. U.S.A.* **85:**7438–7442.

Bosner, M. S., Gulick, T., Riley, D. J. S., Spilburg, C. A., and Lange, L. G., 1989, Heparin-modulated binding of pancreatic lipase and uptake of hydrolyzed triglycerides in the intestine, *J. Biol. Chem.* **264:**20261–20264.

Bousset-Risso, M., Bonicel, J., and Rovery, M., 1985, Limited proteolysis of porcine pancreatic lipase. Lability of the Phe 335–Ala 336 bond towards chymotrypsin, *FEBS Lett.* **182:**323–326.

Brannon, P. M., 1990, Adaptation of the exocrine pancreas to diet, *Annu. Rev. Nutr.* **10:**85–105.

Breg, J. N., Sarda, L., Cozzone, P. J., Rugani, N., Boelens, R., and Kaptein, R., 1995, Solution structure of porcine pancreatic procolipase as determined from ^{1}H homonuclear two-dimensional and three-dimensional NMR, *Eur. J. Biochem.* **227:**663–672.

Brockerhoff, H., and Jensen, R. G., 1974, Lipases, in; *Lipolytic Enzymes,* Academic Press, New York, pp. 25–175.

Brockman, H. L. 1984, General features of lipolysis: Reaction scheme, interfacial structure and experimental approaches, in: *Lipases* (B. Borgström and H. L. Brockman, eds.), Elsevier Science Publishers, Amsterdam, pp. 1–46.

Brockman, H. L., Law, J. H., and Kézdy, F. J., 1973, Catalysis by adsorbed enzymes. The hydrolysis of tripropionin by pancreatic lipase adsorbed to siliconized glass beads, *J. Biol. Chem.* **248:**4965–4970.

Carriere, F., Barrowman, J. A., Verger, R., and Laugier, R., 1993, Secretion and contribution to lipolysis of gastric and pancreatic lipases during a test meal in humans, *Gastroenterology* **105:**876–888.

Carrière, F., Bezzine, S., and Verger, R., 1997a, Molecular evolution of the pancreatic lipase and two related enzymes towards different substrate selectivities, *J. Mol. Catal. B: Enzym.* **3:**55–64.

Carrière, F., Thirstrup, K., Hjorth, S., Ferrato, F., Nielsen, P. F., Withers-Martinez, C., Cambillau, C., Boel, E., Thim, L., and Verger, R., 1997b, Pancreatic lipase structure–function relationships by domain exchange, *Biochem.* **36:**239–248.

Chapus, C., Rovery, M., Sarda, L., and Verger, R., 1988, Minireview on pancreatic lipase and colipase, *Biochimie* **70:**1223–1234.

Charles, M., Semeriva, M., and Chabre, M., 1980, Small-angle neutron scattering study of the association between porcine pancreatic colipase and taurodeoxycholate micelles, *J. Mol. Biol.* **139:**297–317.

Chen, Q., Bläckberg, L., Nilsson, Å., Sternby, B., and Hernell, O., 1994, Digestion of triacylglycerols containing long-chain polyenoic fatty acids in vitro by colipase-dependent pancreatic lipase and human milk bile salt-stimulated lipase, *Biochim. Biophys. Acta* **1210:**239–243.

Cygler, M., Schrag, J. D., Sussman, J. L., Harel, M., Silman, I., Gentry, M. K., and Doctor, B. P., 1993, Relationship between sequence conservation and three-dimensional structure in a large family of esterases, lipases, and related proteins, *Protein, Sci.* **2:**366–382.

Dahim, M., Momsen, W. E., and Brockman, H. L., 1998, How colipase–fatty acid interactions mediate adsorption of pancreatic lipase to interfaces. *Biochem.* **37:**8369–8377.

de la Maza, A., Manich, A. M., and Parra, J. L., 1997, Intermediate aggregates resulting in the interaction of bile salt with liposomes studied by transmission electron microscopy and light scattering techniques, *J. Microsc. (Oxford)* **186:**75–83.

Derewenda, Z. S., and Sharp, A. M., 1993, News from the interface: The molecular structures of triacylglyceride lipases, *Trends Biochem. Sci. (Pers. Ed.)* **18:**20–25.

DiMagno, E. P., and Layer, P., 1993, Human exocrine pancreatic enzyme secretion, in: *The Pancreas: Biology, Pathobiology and Disease,* 2nd edition (V. L. W. Go, E. P. DiMagno, J. D. Gardner, E. Lebenthal, H. A. Reber, and G. A. Scheele, eds.), Raven Press, New York, pp. 275–300.

DiMagno, E. P., Go, V. L. W., and Summerskill, W. H. J., 1973, Relations between pancreatic enzyme outputs and malabsorption in severe pancreatic insufficiency, *New Engl. J. Med.* **288:**813–815.

Drent, M. L., Larsson, I., William-Olsson, T., Quaade, F., Czubayko, F., von Bergmann, K., Strobel, W., Sjöström, I., and van der Veen, E. A., 1995, Orlistat (RO 18-0647), a lipase inhibitor, in the treatment of human obesity: A multiple dose study, *Int. J. Obes.* **19:**221–226.

Egloff, M.-P., Sarda, L., Verger, R., Cambillau, C., and van Tilbeurgh, H., 1995a, Crystallographic study of the structure of colipase and of the interaction with pancreatic lipase, *Protein Sci.* **4:**44–57.

Egloff, M.-P., Marguet, F., Buono, G., Verger, R., Cambillau, C., and van Tilbeurgh, H., 1995b, The 2.46 Å resolution structure of the pancreatic lipase–colipase complex inhibited by a C11 alkyl phosphonate, *Biochem.* **34:**2751–2762.

Erlanson-Albertsson, C., 1983, The interaction between pancreatic lipase and colipase: A protein–protein interaction regulated by a lipid, *FEBS Lett.* **162:**225–229.

Erlanson-Albertsson, C., 1992, Pancreatic colipase. Structural and physiological aspects, *Biochim. Biophys. Acta* **1125:**1–7.

Erlanson-Albertsson, C., and York, D., 1997, Enterostatin—A peptide regulating fat intake, *Obesity Res.* **5:**360–372.

Figarella, C., De Caro, A., Leupold, D., and Poley, j. R., 1980, Congenital pancreatic lipase deficiency, *J. Pediatr.* **96:**412–416.

Gargouri, Y., Julien, R., Bois, A. G., Verger, R., and Sarda, L., 1983, Studies on the detergent inhibition of pancreatic lipase activity, *J. Lipid Res.* **24:**1336–1342.

Gargouri, Y., Pieroni, G., Riviere, C., Lowe, P. A., Sauniere, J.-F., Sarda, L., and Verger, R., 1986a, Importance of human gastric lipase for intestinal lipolysis: An *in vitro* study, *Biochim. Biophys. Acta* **879**:419–423.

Gargouri, Y., Pieroni, G., Riviere, C., Sarda, L., and Verger, R., 1986b, Inhibition of lipases by proteins: A binding study using dicaprin monolayers, *Biochem.* **25**:1733–1738.

Ghishan, F. K., Moran, J. R., Durie, P. R., and Greene, H. L., 1984, Isolated congenital lipase–colipase deficiency, *Gastroenterology* **86**:1580–1582.

Gillmor, S. A., Villaseñor, A., Fletterick, R., Sigal, E., and Browner, M. F., 1997, The structure of mammalian 15-lipoxygenase reveals similarity to the lipases and the determinants of substrate specificity, *Nat. Struct. Biol.* **4**:1003–1009.

Guy, O., and Figarella, C., 1981, The proteins of human pancreatic external secretion, *Scand. J. Gastroenterol.* **16**:59–61.

Havsteen, B. H., Varón Castellanos, R., Molina, M., García Meseguer, M. J., Valero, E., and García-Moreno, M., 1992, Kinetic theory of the action of lipases, *J. Theor. Biol.* .**157**:523–533.

Hermoso, J., Pignol, D., Kerfelec, B., Crenon, I., Chapus, C., and Fontecilla-Camps, J. C., 1996, Lipase activation by nonionic detergents. The crystal structure of the porcine lipase–colipase–tetraethylene glycol monooctyl ether complex, *J. Biol. Chem.* **271**:18007–18016.

Hermoso, J., Pignol, D., Penel, S., Roth, M., Chapus, C., and Fontecilla-Camps, J. C., 1997, Neutron crystallographic evidence of lipase–colipase complex activation by a micelle, *EMBO J.* **16**:5531–5536.

Hernell, O., Staggers, J. E., and Carey, M. C., 1990, Physical–chemical behavior of dietary and biliary lipids during intestinal digestion and absorption. 2. Phase analysis and aggregation states of luminal lipids during duodenal fat digestion in healthy adult human beings, *Biochem.* **29**:2041–2056.

Heuman, D. M., Bajaj, R. S., and Lin, Q., 1996, Adsorption of mixtures of bile salt taurine conjugates to lecithin–cholesterol membranes: Implications for bile salt toxicity and cytoprotection, *J. Lipid Res.* **37**:562–573.

Hildebrand, H., Borgström, B., Békássy, A., Erlanson-Albertsson, C., and Helin, A., 1982, Isolated colipase deficiency in two brothers, *Gut* **23**:243–246.

Hildebrand, P., Petrig, C., Burckhardt, B., Ketterer, S., Lengsfeld, H., Fleury, A., Hadváry, P., and Beglinger, C., 1998, Hydrolysis of dietary fat by pancreatic lipase stimulates cholecystokinin release, *Gastroenterology* **114**:123–129.

Hirschi, K. K., Sabb, J. E., and Brannon, P. M., 1991, Effects of diet and ketones on rat pancreatic lipase in cultured acinar cells, *J. Nutr.* **121**:1129–1134.

Holt, P. R., and Balint, J. A., 1993, Effects of aging on intestinal lipid absorption, *Am. J. Physiol.* **264**:G1–G6.

Holtmann, G., Kelly, D. G., Sternby, B., and DiMagno, E. P., 1997, Survival of human pancreatic enzymes during small bowel transit. Effect of nutrients, bile acids, and enzymes, *Am. J. Physiol.* **36**:G553–G558.

Hui, D. Y., 1996, Molecular biology of enzymes involved with cholesterol ester hydrolysis in mammalian tissues, *Biochim. Biophys. Acta* **1303**:1–14.

Jain, M. K., and Berg, O. G., 1989, The kinetics of interfacial catalysis by phospholipase A2 and regulation of interfacial activation: Hopping versus scooting, *Biochim. Biophys. Acta* **1002**:127–156.

Jennens, M. L., and Lowe, M. E., 1994, A surface loop covering the active site of human pancreatic lipase influences interfacial activation and lipid binding, *J. Biol. Chem.* **41**:25470–25474.

Jennens, M. L., and Lowe, M. E., 1995, C-terminal domain of human pancreatic lipase is required for stability and maximal activity but not colipase reactivation, *J. Lipid Res.* **36**:1029–1036.

Labourdenne, S., Brass, O., Ivanova, M., Cagna, A., and Verger, R., 1997, Effects of colipase and bile salts on the catalytic activity of human pancreatic lipase. A study using the oil drop tensiometer, *Biochem.* **36**:3423–3429.

Larsson, A., and Erlanson-Albertsson, C., 1986, Effect of phosphatidylcholine and free fatty acids on the activity of pancreatic lipase–colipase, *Biochim. Biophys. Acta* **876**:543–550.

Larsson, A., and Erlanson-Albertsson, C., 1991, The effect of pancreatic procolipase and colipase on pancreatic lipase activation, *Biochim. Biophys. Acta* **1083**:283–288.

Lee, P. C., and Lebenthal, E., 1993, Prenatal and postnatal development of the human exocrine pancreas, in: *The Pancreas: Biology, Pathobiology and Disease*, 2nd edition (V. L. W. Go, E. P. DiMagno, J. D. Gardner, E. Lebenthal, H. A. Reber, and F. A. Scheele, Eds.), Raven Press, New York, pp. 57–73.

Lindstrom, M., Sternby, B., and Borgström, B., 1988, concerted action of human carboxyl ester lipase and pancreatic lipase during digestion *in vitro:*Importance of the physiochemical state of the substrate, *Biochim. Biophys. Acta* **959**:178–184.

Lowe, M. E., 1992, The catalytic site residues and interfacial binding of human pancreatic lipase, *J. Biol Chem.* **267**:17069–17073.

Lowe, M. E., 1994, Pancreatic triglyceride lipase and colipase: Insights into dietary fat digestion, *Gastroenterology* **107:**1524–1536.

Lowe, M. E., 1997a, Colipase stabilized the lid domain of pancreatic triglyceride lipase, *J. Biol. Chem.* **272:**9–12.

Lowe, M. E., 1997b, Structure and function of pancreatic lipase and colipase, *Annu. Rev. Nutr.* **17:**141–158.

Manson, W. G., and Weaver, L. T., 1997, Fat digestion in the neonate, *Arch. Dis. Child.* **76:**F 200–F 211.

Marangoni, A. G., 1994, Enzyme kinetics of lipolysis revisited: The role of lipase interfacial binding, *Biochem. Biophys. Res. Commun.* **200:**1321–1328.

Maylie, M. F., Charles, M., Gache, C., and Desnuelle, P., 1971, Isolation and partial identification of a pancreatic colipase, *Biochim. Biophys. Acta* **229:**286–289.

McIntyre, J. C., Hundley, P., and Behnke, W. D., 1987, The role of aromatic side chain residues in micelle binding by pancreatic colipase. Fluorescence studies of the porcine and equine proteins, *Biochem. J.* **245:**821–829.

Momsen, M. M., Dahim, M., and Brockman, H. L., 1997, Lateral packing of the pancreatic lipase cofactor, colipase, with phosphatidylcholine and substrates, *Biochem.* **36:**10073–10081.

Momsen, W. E., and Brockman, H. L., 1976, Effects of colipase and taurodeoxycholate on the catalytic and physical properties of pancreatic lipase B at an oil–water interface, *J. Biol. Chem.* **251:**378–383.

Momsen, W. E., and Brockman, H. L., 1981, The adsorption to and hydrolysis of 1,3-didecanoylglycerol monolayers by pancreatic lipase, *J. Biol. Chem.* **256:**6913–6916.

Momsen, W. E., Smaby, J. M., and Brockman, H. L., 1979, Interfacial structure and lipase action. Characterization of taurodeoxycholate–didecanoylglycerol monolayers by physical and kinetic methods, *J. Biol. Chem.* **254:**8855–8860.

Momsen, W. E., Momsen, M. M., and Brockman, H. L., 1995, Lipid structural reorganization induced by the pancreatic lipase cofactor, procolipase, *Biochem.* **34:**7271–7281.

Moreau, H., Moulin, A., Gargouri, Y., Noël, J.-P., and Verger, R., 1991, Inactivation of gastric and pancreatic lipases by diethyl *p*-nitrophenyl phosphate, *Biochem.* **30:**1037–1041.

Morgan, R. G. H., Barrowman, J., and Borgström, B., 1969, The effect of sodium taurodesoxycholate and pH on the gel filtration behaviour of rat pancreatic protein and lipases, *Biochim. Biophys. Acta* **175:**65–75.

Muderhwa, J. M., and Brockman, H. L., 1992a, Lateral lipid distribution is a major regulator of lipase activity. Implications for lipid-mediated signal transduction, *J. Biol. Chem.* **267:**24184–24192.

Muderhwa, J. M., and Brockman, H. L., 1992b, Regulation of fatty acid ^{18}O exchange catalyzed by pancreatic carboxylester lipase. 2. Effects of lateral lipid distribution in mixtures with phosphatidylcholine, *Biochem.* **31:**149–155.

Nichols, J. W., 1986, low concentrations of bile salts increase the rate of spontaneous phospholipid transfer between vesicles, *Biochem.* **25:**4596–4601.

O'Connor, C. J., Ch'ng, B. T., and Wallace, R. G., 1983, Studies in bile salt solutions. 1. Surface tension evidence for a stepwise aggregation model, *J. Colloid Interface Sci.* **95:**410–419.

Ollis, D. L., Cheah, E., Cygler, M., Dijkstra, B., Frolow, F., Franken, S. M., Harel, M., Remington, S. J., Silman, I., Schrag, J., Sussman, J. L., Verschueren, K. H. G., and Goldman, A., 1992, the α/β hydrolase fold, *Protein Eng.* **5:**197–211.

Panaiotov, I., Ivanova, M., and Verger, R., 1997, Interfacial and temporal organization of enzymatic lipolysis, *Curr. Opin. Colloid Interface Sci.* **2:**517–525.

Patton, J. S., 1981, Gastrointestinal lipid digestion, in: *Physiology of the Gastrointestinal Tract* (L. R. Johnson, ed.), Raven Press, New York, pp. 1123–1146.

Peters, G. H., Toxvaerd, S., Olsen, O. H., and Svendsen, A., 1995, Modeling of complex biological systems. 2. Effect of chain length on the phase transitions observed in diglyceride monolayers, Langmuir **11;** 4072–4081.

Pieroni, G., and Verger, R., 1979, Hydrolysis of mixed monomolecular films of triglyceride/lecithin by pancreatic lipase, *J. Biol. Chem.* **254:**10090–10094.

Pieroni, G., Gargouri, Y., Sarda, L., and Verger, R., 1990, Interactions of lipases with lipid monolayers. Facts and fictions, *Adv. Colloid Interface Sci.* **32:**341–378.

Ransac, S., Rivière, C., Soulié, J. M., Gancet, C., Verger, R., and de Haas, G. H., 1990, Competitive inhibition of lipolytic enzymes. I. A kinetic model applicable to water-insoluble competitive inhibitors, *Biochim. Biophys. Acta* **1043:**57–66.

Ransac, S., Carrière, F., Rogalska, E., Verger, R., Marguet, F., Buono, G., Melo, E. P., Cabral, J. M. S., Egloff, M.-P., van Tilbeurgh, H., and Cambillau, C., 1996, The kinetics, specificities and structural features of lipases *NATO ASI Ser., Ser. H:***96:**265–303.

Rössner, H., Barkeling, B., Erlanson-Albertsson, C., Larsson, P., and Wählin-Boll, E., 1995, Intravenous enterostatin does not affect single meal food intake in man, *Appetite* **24**:37–42.

Rubin, B., 1994, Grease pit chemistry exposed, *Nat. Struct. Biol.* **1**:568–572.

Rudd, E. A., and Brockman, H. L., 1984, Pancreatic carboxyl ester lipase (cholesterol esterase), in: *Lipases,* (B. Borgström and H. L. Brockman, eds.), Elsevier Science Publishers, Amsterdam, pp. 185–204.

Saxton, M. J., 1994, Anomalous diffusion due to obstacles: A Monte Carlo study, *Biophys. J.* **66**:394–401.

Scheele, G. A., 1993, Regulation of pancreatic gene expression in response to hormones and nutritional substrates, in: *The Pancreas: Biology, Pathobiology and Disease,* 2nd edition (V. L. W. Go, E. P. DiMagno, J. D. Gardner, E. Lebenthal, H. A. Reber, and G. A. Scheele, eds.), Raven Press, New York, pp. 103–120.

Schmit, G. D., Momsen, M. M., Owen, W. G., Naylor, S., Tomlinson, A., Wu, G., Stark, R. E., and Brockman, H. L., 1996, The affinities of procolipase and colipase for interfaces are regulated by lipids, *Biophys. J.* **71**:3421–3429.

Schram, V., Tocanne, J.-F., and Lopez, A., 1994, Influence of obstacles on lipid lateral diffusion: Computer simulation of FRAP experiments and application to proteoliposomes and biomembranes, *Eur. Biophys. J.* **23**:337–348.

Sémériva, M., and Desnuelle, P., 1979, Pancreatic lipase and colipase. An example of heterogeneous biocatalysis, *Adv. Enzymol. Rel. Areas Mol. Biol.* **48**:319–371.

Small, D. M., 1970, Surface and bulk interactions of lipids and water with a classification of biologically active lipids based on these interactions, *Fed. Proc.* **29**:1320–1326.

Small, D. M., Miller, K., Cistola, D., Ginsburg, G., Parks, J., Atkinson, D., and Hamilton, J. A., 1983, Physicochemical studies on the position of molecules in emulsions and membranes, *Falk Symp.* **33**; 25–30.

Sörhede, M., Mulder, H., Mei, J., Sundler, F., and Erlanson-Albertsson, C., 1996, Procolipase is produced in the rat stomach—A novel source of enterostatin, *Biochim. Biophys. Acta* **1301**:207–212.

Staggers, J. E., Hernell, O., Stafford, R. J., and Carey, M. C., 1990, Physical–chemical behavior of dietary and biliary lipids during intestinal digestion and adsorption. 1. Phase behavior and aggregation states of model lipid systems patterned after aqueous duodenal contents of healthy adult human beings, *Biochem.* **29**:2028–2040.

Tani, H., Ohishi, H., and Watanabe, K., 1994, Purification and characterization of proteinous inhibitor of lipase from heat flour, *J. Agric. Food Chem.* **42**:2382–2385.

Thomson, A.B.R., and Dietschy, J. M., 1981, Intestinal lipid absorption:Major extracellular and intracellular events, in: *Physiology of the Gastrointestinal Tract* (L. R. Johnson, ed.), Raven Press, New York, pp. 1147–1220.

Tian, Q., Nagase, H., York, D. A., and Bray, G. A., 1994, Vagal–central nervous system interactions modulate the feeding response to peripheral enterostatin, *Obesity Res.* **2**:527–534.

Tsujita, T., Muderhwa, J. M., and Brockman, H. L., 1989, Lipid–lipid interactions as regulators of carboxylester lipase activity, *J. Biol. Chem.* **264**:8612–8618.

van Tilbeurgh, H., Sarda, L., Verger, R., and Cambillau, C., 1992, Structure of the pancreatic lipase–colipase complex, *Nature* **359**:159–162.

van Tilbeurgh, H., Gargouri, Y., Dezan, C., Egloff, M.-P., Nésa, M. P., Ruganie, N., Sarda, L., Verger, R., and Cambillau, C., 1993a, Crystallization of pancreatic procolipase and of its complex with pancreatic lipase, *J. Mol. Biol.* **229**:552–554.

van Tilbeurgh, H., Egloff, M.-P., Martinez, C., Rugani, N., Verger, R., and Cambillau, C., 1993b, Interfacial activation of the lipase–procolipase complex by mixed micelles revealed by X ray crystallography, *Nature* **362**:814–820.

Verger, R., 1980, Enzyme kinetics of lipolysis, *Methods Enzymol.* **64**:340–392.

Verger, R. 1984, Pancreatic lipase, in: *Lipases* (B. Borgström and H. L. Brockman, eds.), Elsevier Science Publisher, Amsterdam, pp. 83–150.

Verger, R., 1997, Interfacial activation of lipases: Facts and artifacts, *Trends in Biotechnol.* **15**:32–38.

Verger, R., and Pieroni, G., 1986, Monomolecular layers: A bio-topology in the past, present and future, in: *Lipids and Membranes: Past, Present and Future* (J. A. F. op den Kamp, B. Roelofsen, and K. W. A. Wirtz, eds.), Elsevier Science Publishers B.V., Amsterdam, pp. 153–170.

Verger, R., Rietsch, J., van Dam-Mieras, M. C. E., and de Haas, G. H., 1976, Comparative studies of lipase and phospholipase A_2 acting on substrate monolayers, *J. Biol. Chem.* **251**:3128–3133.

Verger, R., Aoubala, M., Carrière, F., Ransac, S., Dupuis, L., De Caro, J., Ferrato, F., Douchet, I., Laugier, R., and de Caro, A., 1996, Regulation of lumen fat digestion: Enzymic aspects, *Proc. Nutr. Soc* **55**:5–18:7.

Wang, C.-S., and Hartsuck, J. A., 1993, Bile salt-activated lipase. A multiple function lipolytic enzyme, *Biochim. Biophys. Acta* **1166:**1–19.

Wickham, M., Garrood, M., Leney, J., Wilson, P. D. G., and Fillery-Travis, A., 1998, Modification of a phospholipid stabilized emulsion interface by bile salt: Effect on pancreatic lipase activity, *J. Lipid Res.* **39:**623–632.

Wieloch, T., Borgström, B., Pieroni, G., Pattus, F., and Verger, R., 1982, Product activation of pancreatic lipase. Lipolytic enzymes as probes for lipid/water interfaces, *J. Biol. Chem.* **257:**11523–11528.

Winkler, F. K., D'Arcy, A., and Hunziker, W., 1990, Structure of human pancreatic lipase, *Nature* **343:**771–774.

Withers-Martinez, C., Carrière, F., Verger, R., Bourgeois, D., and Cambillau, C., 1996, A pancreatic lipase with a phospholipase A1 activity: Crystal structure of a chimeric pancreatic lipase-related protein 2 from guinea pig, *Structure* **4:**1363–1374.

Yang, L.-Y., Kuksis, A., and Myher, J. J., 1990, Lipolysis of menhaden oil triacylglycerols and the corresponding fatty acid alkyl esters by pancreatic lipase *in vitro:* A reexamination, *J. Lipid Res.* **31:**137–147.

CHAPTER 5

Biology, Pathology, and Interfacial Enzymology of Pancreatic Phospholipase A_2

REYNOLD HOMAN and MAHENDRA KUMAR JAIN

1. Introduction

The phospholipase A_2 (PLA_2) secreted by the exocrine pancreas is one of the earliest known members of a large class of enzymes that catalyze hydrolysis of the ester-bond linking fatty acids to the sn-2 position of glycerophospholipids (Fig. 1). The phospholipase A_2 enzyme class has been intensely scrutinized, in large part because the sn-2-specific deacylation of glycerophospholipids is a critical regulatory step in a variety of metabolic pathways, ranging from digestion and defense mechanisms to eicosanoid synthesis and signal transduction.

Pancreatic phospholipase A_2 ($pPLA_2$) has long been the archetype for elucidating PLA_2 structure and mechanism. The ground work for the biochemical characterization of $pPLA_2$ was laid largely through efforts of de Haas and coworkers (Verheij *et al.*, 1981). Features of $pPLA_2$ that have proven favorable for such studies include stability, ease of isolation from a relatively accessible and abundant source, and, more significantly, the fact that the active site architecture and 14 kDa size are common to a wide range of PLA_2 family members produced in venoms or secreted by inflammatory cells (Scott and Sigler, 1994). Pancreatic PLA_2 has also served as a paradigm for characterizing the unique features of enzyme catalysis at the lipid–water interface (Jain and Berg, 1989; Verger and De Haas, 1976), which has resulted in the development of a comprehensive model of interfacial kinetics shown in Fig. 4 (Berg *et al.*, 1991, 1997; Jain *et al.*, 1995). The kinetic and mechanistic results are complemented by an abundance of detailed structural data (Dijkstra *et al.*, 1983; Scott and Sigler, 1994; Sekar *et al.*, 1997a; van den Berg *et al.*, 1995; Sekar and Sundaralingam, 1999), an example of which is shown in Figure 2. These data are further augmented by site-directed mutagenesis studies that have helped resolve the contributions of

REYNOLD HOMAN • Cardiovascular Therapeutics, Parke-Davis Pharmaceutical Research Division, Ann Arbor, Michigan 48105.   MAHENDRA KUMAR JAIN • Department of Chemistry and Biochemistry, University of Delaware, Newark, Delaware 19716.
Intestinal Lipid Metabolism, edited by Charles M. Mansbach II *et al.,* Kluwer Academic/Plenum Publishers, 2001.

individual amino acid residues to $pPLA_2$ structure and function (Dua *et al.*, 1995; Dupureur *et al.*, 1992a, 1992b; Huang *et al.*, 1996; Li *et al.*, 1994; Liu *et al.*, 1995; Sekar *et al.*, 1997b; Verheij, 1995). Such extensive characterization of $pPLA_2$ has produced a detailed framework for understanding PLA_2 catalytic mechanisms (Figure 3) and for developing PLA_2 inhibitors (Gelb *et al.*, 1994, 1997) of the type shown in Fig. 5.

In this chapter we review the biology of $pPLA_2$, noting the biological milieu in which this enzyme operates, and highlighting some novel facets of its function including regulation of intestinal cholesterol absorption and cell activation—possibly through a specific receptor. This expanding body of evidence has broadened the physiological significance of $pPLA_2$ beyond simple hydrolysis of phosphatidylcholine in the digestive tract. In the later part of the chapter we summarize key interfacial enzymological findings relating to $pPLA_2$ mechanism and function.

2. Distribution of Pancreatic Phospholipase A_2

2.1. Pancreas

Pancreatic PLA_2 is one of the many hydrolases produced by the acinar cells of the exocrine pancreas for release into the duodenum via pancreatic juice. The acinar cells of the human pancreas are prolific in hydrolase manufacture, producing 6 to 20 g of protein each day consisting primarily of proteases (Lowe, 1994). The pancreatic lipases, which include $pPLA_2$, triglyceride lipase, carboxyl ester hydrolase, and colipase, are a relatively minor component of the acinar cell output comprising less than 5% of the total proteins secreted (Scheele *et al.*, 1981). Pancreatic PLA_2 makes up an even smaller fraction because the bulk of the lipase mass consists of carboxyl ester hydrolase and triglyceride lipase (Sternby *et al.*, 1991). The relative abundance of hydrolases in pancreatic juice, however, does not necessarily correspond to the relative importance or content of the respective substrates in the diet because the lipases are highly efficient in converting the complex dietary lipids to more water soluble and absorbable hydrolytic products (Arnesjö *et al.*, 1969; B. Borgström *et al.*, 1957).

Many of the digestive hydrolases, including $pPLA_2$, are produced and released by the acinar cells as inactive zymogens containing peptide extensions at the amino-termini. The zymogens are subsequently activated in the duodenum by proteolytic cleavage of the peptide extensions. Activation involves a series of proteolytic events initiated by the hydrolysis of trypsinogen to trypsin by enterokinase (enteropeptidase), a product of the intestinal epithelial cells. The remaining zymogens, including prophospholipase A_2 (pro-$pPLA_2$), are subsequently activated by trypsin (Lowe, 1994). Pro-$pPLA_2$, which is one of the more rapidly activated zymogens *in vitro* (A. Borgström *et al.*, 1993), is activated by tryptic cleavage of a heptapeptide from the amino-terminus. Removal of the heptapeptide results in exposure of an interface recognition site that is necessary for $pPLA_2$ adsorption to the lipid–water interface of aggregated lipids such as micelles and bilayers (Verheij *et al.*, 1981).

The initial proteolytic event converting zymogens to active hydrolases is counterbalanced by subsequent inactivation due to additional proteolysis of the hydrolases during transit through the small intestine (Layer *et al.*, 1986). The lipases, particularly $pPLA_2$ (A. Borgström *et al.*, 1993) and triglyceride lipase (Layer *et al.*, 1986), appear to be more susceptible to proteolytic degradation than the proteases. In the case of triglyceride lipase, only 1% of the lipase activity initially detected in the proximal duodenum remains in human intestinal contents retrieved from the ileum.

2.2. Non-Pancreatic Tissues

The significance of the pancreas as the chief source of pPLA$_2$ has long been manifest in the abundance of PLA$_2$ activity found almost exclusively in pancreatic tissues and juices from a wide variety of mammalian species. More recent evidence, however, indicates that pPLA$_2$ is not solely produced in the pancreas. Messenger RNA encoding pPLA$_2$ has been detected in mouse lung, liver and spleen (Valentin *et al.*, 1999), in the gastric mucosa of guinea pigs (Ying *et al.*, 1993), in mouse keratinocytes (Li-Stiles *et al.*, 1998), and in rat (Sakata *et al.*, 1989), guinea pig (Ying *et al.*, 1993), and human (Johnson *et al.*, 1990) lung tissues as well as in human liver, ovary, prostrate gland, and spleen (Cupillard *et al.*, 1997). Further substantiation of pPLA$_2$ gene expression in lung tissue is provided by a report of pPLA$_2$ detection in human lung tissue by immunostaining (Matsuda *et al.*, 1987). Immunodetection and protein purification studies have also yielded evidence of pPLA$_2$ in rat spleen and gastric mucosa (Tojo *et al.*, 1988b; Kortesuo *et al.*, 1993). With the exception of gastric mucosa, the levels of pPLA$_2$ expression in nonpancreatic tissues are quite small, compared to the pPLA$_2$ content in the pancreas.

The discovery of pPLA$_2$ production in the guinea pig gastric mucosa solved a long-standing enigma that contradicted the general observation that the pancreas was the principal source of pPLA$_2$. The enigma was created by the discovery that the guinea pig pancreas is essentially devoid of pPLA$_2$ synthesis and secretion (Fauvel *et al.*, 1981). Despite this apparent anomaly, phospholipid hydrolysis and absorption in the guinea pig gut appeared to be normal (Diagne *et al.*, 1987). The isolation of a novel phospholipase from microvillous membranes of guinea pig intestinal epithelium that exhibited both PLA$_2$ and lyosphospholipase activity suggested that an alternate pathway for phospholipid hydrolysis existed in the guinea pig gut (Gassama-Diagne *et al.*, 1989; Pind and Kuksis, 1991). A more complete resolution of this unusual situation was achieved with the discovery that pPLA$_2$ was indeed supplied to the digestive tract of the guinea pig, not by the pancreas, but by the gastric mucosa (Tojo *et al.*, 1993). A cDNA cloned from the gastric mucosa demonstrated that the mucosal PLA$_2$ exhibited all the primary sequence characteristics of pPLA$_2$ and was 71% homologous to the rat pPLA$_2$ (Ying *et al.*, 1993). Although the cDNA indicated that a proenzyme was expressed in the mucosa, no proenzyme could be isolated, suggesting the mucosa secreted the active enzyme. The pPLA$_2$ activity in guinea pig gastric juice was not evident until the pH was raised to values above 7, which is consistent with the established pH dependence of pPLA$_2$ enzymes (Tojo *et al.*, 1993). Furthermore, immunohistochemical staining indicated that pPLA$_2$ was located in the chief cells, which are the exocrine cells of the gastric mucosa that also produce the digestive hydrolases pepsin (Plebani, 1993) and gastric lipase (Hamosh, 1990).

The search for pPLA$_2$ in guinea pig gastric mucosa was spurred by the prior discovery of pPLA$_2$ production in rat gastric mucosa. Although the rat pancreas produces significant quantities of pPLA$_2$, immunostaining of stomach tissue with antibodies created against rat pPLA$_2$ isolated from the pancreas revealed that the rat gastric mucosa contained considerable amounts of pPLA$_2$ quite comparable to the pancreatic pPLA$_2$ content (Kortesuo *et al.*, 1993; Tojo *et al.*, 1988a). Furthermore, unlike the guinea pig, the rat gastric mucosa appeared to contain significant amounts of both pro-$pPLA_2$ zymogen and active $pPLA_2$ (Tojo *et al.*, 1988a). Immunohistochemical staining confirmed that the $pPLA_2$ was located in both the pancreatic acinar cells and the gastric chief cells, particularly in the zymogen granules of both cell types (Kortesuo *et al.*, 1993; Tasumi *et al.*, 1990).

The observation of $pPLA_2$ production in rat and guinea pig gastric mucosa suggests that gastric $pPLA_2$ production may also occur in other species. Preliminary evidence of gastric $pPLA_2$ production in other species may come, as it did in the case of rat and guinea pig, from the detection $pPLA_2$-like activity in gastric contents that is calcium dependent and exhibits a pH optimum at 7 or higher. Keeping in mind that duodenal reflux could also account for such findings, it is intriguing to note a report of the detection of PLA_2 activity in stomach tissue homogenates from several species that yielded an activity ranking of rat > mouse >> rabbit > dog (Hirohara *et al.*, 1988). Pancreatic PLA_2-like activity has also been reported to occur in human gastric mucosa biopsies (Grataroli *et al.*, 1987). This result is complemented by the detection of immunoreactive $pPLA_2$ activity in human stomach tissue (Matsuda *et al.*, 1987) and in human gastric juice (Huhtinen *et al.*, 1999). We have also observed very significant $pPLA_2$-like activity in hamster gastric contents neutralized to pH 7. Whether any of these various observations resulted from gastric $pPLA_2$ production remains to be proved more definitively.

3. Pancreatic Phospholipase A_2 Function

3.1. Nondigestive Functions

The detection of $pPLA_2$ in nondigestive tissues implies that metabolic functions exist for $pPLA_2$ outside the digestive tract but the identities of those functions remain to be discovered. The recent elucidation of a specific, high-affinity membrane receptor for $pPLA_2$ may provide some clues (for recent reviews see Ohara *et al.*, 1995; Lambeau and Lazdunski, 1999). The receptor, which is structurally related to the macrophage mannose receptor, has been located in a variety of cell types and tissues from several species. The human receptor, however, exhibits significantly lower affinity for $pPLA_2$ than the analogous receptor in other species (Cupillard *et al.*, 1999). The $pPLA_2$ receptor in rat tissues exhibits significant specificity for $pPLA_2$ because pro-$pPLA_2$ and other PLA_2 types do not effectively compete with $pPLA_2$ for binding. In several cases it has been shown that $pPLA_2$ need not be catalytically active for binding to take place. A number of cellular responses including proliferation, chemokinesis, smooth muscle cell contraction and prostanoid synthesis are associated with $pPLA_2$ addition to cells expressing $pPLA_2$ receptors. The cellular responses exhibit the same selectivity for $pPLA_2$ over other PLA_2 types as observed in the binding results.

The evidence of cell activation via a $pPLA_2$ receptor and of $pPLA_2$ expression in nondigestive tissues suggests that $pPLA_2$ may function in a hormonelike manner. Although the importance of $pPLA_2$ catalytic activity in cell activation remains to be more clearly resolved, the significance of receptor-dependent activation is enhanced by a recent report showing that $pPLA_2$ receptor-induced proliferation of human pancreatic cancer cells involves stimulation of the mitogen-activated protein kinase cascade, independent of $pPLA_2$

catalytic activity (Kinoshita *et al.,* 1997). Similar resolution of receptor binding and catalytic activity was observed for pPLA$_2$-induced cellular invasion of extracellular matrices (Kundu and Mukherjee, 1997) and with a catalytically inactive pPLA$_2$ mutant (G30S) that bound rat mesangial cells and activated prostaglandin E$_2$ production in the same manner as normal pPLA$_2$ (Kishino *et al.,* 1995). More dramatic evidence of a physiological function for the pPLA$_2$ receptor was recently obtained with pPLA$_2$-receptor-deficient mice that were shown to be resistant to endotoxic shock (Hanasaki *et al.,* 1997). Receptor-deficient mice that were sensitized with sublethal doses of bacterial lipopolysaccharide were resistant to the normally lethal effect of subsequent pPLA$_2$ injection.

3.2. Digestive Functions

3.2.1. Pancreatic Phospholipase A$_2$ Substrates In Vivo

Pancreatic PLA$_2$ is one of several lipases released at various points along the alimentary tract to facilitate lipid absorption by hydrolysis of the complex acylated lipids that enter the tract from the diet and the bile. The triglycerides, which constitute the major dietary lipid component and are consumed by adult humans on a Western diet at an average rate of 150 g per day (Carey *et al.,* 1983) are hydrolyzed by lingual, gastric, and pancreatic lipases (Hamosh, 1990). The glycerophospholipids on which pPLA$_2$ acts are the second most abundant dietary lipid class and enter the digestive tract from both the diet and the bile. The daily ingestion of glycerophospholipids via the Western diet is 2 to 7 g, of which 70% is in the form of phosphatidylcholine and the bulk of the remainder is in the form of phosphatidylethanolamine (Åkesson, 1982). An additional 7 to 22 g of almost exclusively phosphatidylcholine is added to the digestive tract contents from the bile (Carey *et al.,* 1983). A third source of luminal glycerophospholipids arises from desquamation of intestinal epithelium. Determinations of the amounts of glycerophospholipids introduced by this process have not been reported, but, based on estimates of cholesterol amounts released by desquamation (Homan and Krause, 1997) and the typical cholesterol to glycerophospholipid ratio in intestinal epithelium (Simons and van Meer, 1988), it can be estimated that 0.6 to 1.5 g of glycerophospholipid enter the intestinal lumen through desquamation.

The distribution of glycerophospholipid substrate in the intestinal contents is more complex than the relatively homogenous and uniformly dispersed substrate preparations required for obtaining interpretable kinetics *in vitro* (see section 6). Pancreatic PLA$_2$ substrate in the intestinal contents is distributed among multiple lipid phases of varying organization and composition. For example, an examination of intestinal contents from human subjects fed a fatty meal identified several distinct lipid phases (Hernell *et al.,* 1990). The predominant phase was an oil–water emulsion rich in triglyceride with lesser amounts of diglyceride and cholesteryl ester that together formed the emulsion core. The core lipids were stabilized by a surface layer of more polar lipids including monoglycerides, fatty acids, phosphatidylcholine, and bile acids (Fig. 1). Coexisting with the emulsion phase were micellar, vesicular, and crystalline lipid phases composed of the same lipids but those phases were greatly enriched in the polar lipid constituents compared to the emulsion phase.

Lipid phase heterogeneity may have significant effects on pPLA$_2$ activity *in vivo* resulting in variable glycerophospholipid hydrolysis rates according to the nature of the lipid phase. The significance of phase heterogeneity is indicated by an analysis of pPLA$_2$ ad-

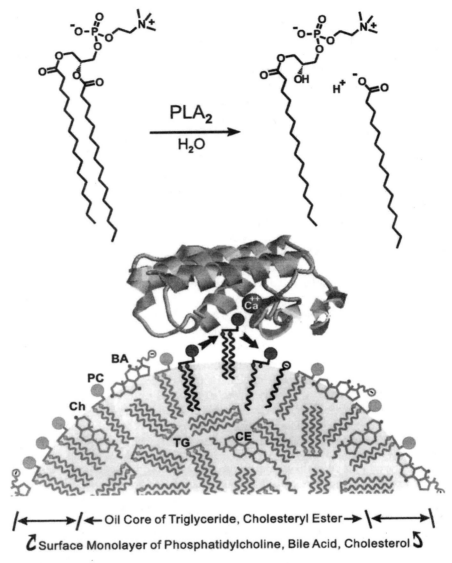

|←——→|←— Oil Core of Triglyceride, Cholesteryl Ester —→|←——→|

ᘓ Surface Monolayer of Phosphatidylcholine, Bile Acid, Cholesterol ᘔ

Figure 1. Hydrolysis of phosphatidylcholine catalyzed by PLA_2 at the lipid–water interface of an oil emulsion particle as might be found in the intestinal contents.

sorption to model lipid dispersions, which showed that $pPLA_2$ adsorbed preferentially to mixed-lipid micelles in the presence of competing triglyceride emulsions (Nalbone *et al.,* 1983). From the kinetic standpoint and as developed in more detail in section 6, three fundamental effects of substrate heterogenity may apply to PLA_2 kinetics in the intestinal contents. First, the phase organization does not change the interfacial kinetic constants K_M^* or k^*_{cat}, and, therefore, the substrate specificity. Moreover, $pPLA_2$ does not show any substrate specificity for choline as the head group versus other head group types, and, the fatty acyl chain preference is modest at best. Second, the anionic charge at the interface, introduced by bile salts and the free fatty acids released by hydrolysis, enhances the binding of the enzyme to the interface and also increases k^*_{cat}. Third, the interfacial substrate mole fraction, that is, the substrate concentration that the bound enzyme "sees" for the formation of the interfacial Michaelis complex, changes with incorporation of other additives in the interface.

3.2.2. Interdependencies of Phosphatidylcholine Hydrolysis and Lipid Absorption

Considering the abundance of glycerophospholipids entering the digestive tract, it might be concluded that pPLA$_2$ simply functions to convert a very significant nutrient pool of glycerophospholipids to absorbable products. However, evidence that several stages in the lipid absorption process are directly coupled to the hydrolysis of phosphatidylcholine, which is the predominant glycerophospholipid in the lumenal contents, suggests the absorptions of other lipid classes are dependent on pPLA$_2$ activity as well (Homan and Hamelehle, 1998; Tso and Scobey, 1986).

Emulsification of water-insoluble lipids is one stage of lipid digestion for which phosphatidylcholine is essential and thus potentially significantly impacted by pPLA$_2$ activity. As an amphiphile, phosphatidylcholine acts to stabilize microdroplets of triglyceride, cholesterol, and other nonpolar dietary lipids that are otherwise insoluble in the aqueous environment of the intestinal contents and the bile (Carey *et al.*, 1983). As lipid digestion progresses, the surfactant pool is augmented by the fatty acid, monoacylglycerol, and lysophospholipid products of lipid hydrolysis, but initially, phosphatidylcholine is the only surfactant available for dietary lipid dispersion in tandem with the bile acids.

Stabilization of neutral lipid microdroplets by emulsification enhances lipid absorption, in part, by increasing the surface area over which water-insoluble lipids like triglycerides can approach the lipid–water interface where the neutral lipid lipases are located. *In vitro* studies, however, indicate that the activity of neutral lipid lipases at the interface is restricted unless the phosphatidylcholine monolayer on the droplet surface is partially hydrolyzed, suggesting that neutral lipid absorption may be facilitated by pPLA$_2$ activity. Such dependence of neutral lipid hydrolysis on pPLA$_2$ activity has been demonstrated for lipolysis of emulsified triglyceride by pancreatic lipase–colipase complex (B. Borgström, 1980; Young and Hui, 1999) and lipolysis of cholesteryl ester by carboxyl ester lipase *in vitro* (Lindström *et al.*, 1991). It remains to be demonstrated that such concerted action between pPLA$_2$ and neutral lipid lipases exists *in vivo*.

3.2.2a. Cholesterol Absorption.

A growing body of evidence indicates that cholesterol absorption is coupled to phosphatidylcholine turnover in the intestinal contents and, consequently, to phospholipase activity. Like phosphatidylcholine, the majority of cholesterol in the intestinal contents originates from the bile. Cholesterol solubility in the bile is absolutely dependent on emulsification by biliary phosphatidylcholine because the aqueous solubility of cholesterol is extremely low and the bile acids alone are poor solubilizers of cholesterol (Carey *et al.*, 1983). Generally, emulsification and dispersion of lipids into smaller particles in the intestinal contents leads to enhanced absorption due to increased interfacial area between the aqueous and lipid phases to which lipases can adsorb and from which lipolytic products can desorb. However, as recent experiments indicate, cholesterol solubilization by phosphatidylcholine does not necessarily lead to enhanced cholesterol absorption.

Evidence of restricted cholesterol absorption in the presence of phosphatidylcholine has been observed both *in vitro* and *in vivo*. For example, Caco-2 cell monolayers, which are a cultured cell model of intestinal epithelium, absorbed significantly less cholesterol from bile acid micelles containing phosphatidylcholine than from phospholipid-free micelles or from micelles in which 50% or more of the phosphatidylcholine was replaced by lysophosphatidylcholine (Homan and Hamelehle, 1998). The reduction in cholesterol absorption by phosphatidylcholine was reversed by $pPLA_2$ addition and the $pPLA_2$-dependent increase in cholesterol absorption could be blocked with the competitive PLA_2 inhibitor FPL 67047XX (Beaton *et al.*, 1994). The same inhibitor effect has been observed *in vivo* in rats treated with FPL 67047XX (Homan and Krause, 1997). The absorption inhibition *in vivo* was specific for cholesterol because FPL 67047XX had no effect on triglyceride absorption. Further *in vivo* evidence of the significance of $pPLA_2$ in cholesterol absorption has been obtained in cholesterol-fed hamsters treated with the potent $pPLA_2$ inhibitor MJ99 (Jain *et al.*, 1991a). We found that oral coadministration of a cholesterol-rich coconut oil emulsion and 50 mg/kg MJ99 resulted in a 52% reduction in the elevated plasma cholesterol levels that otherwise occur when hamsters are given emulsion alone. The significance of $pPLA_2$ in cholesterol absorption is further supported by the observation that $pPLA_2$ is the component of porcine pancreas extract responsible for activating Caco-2 cell absorption of cholesterol from phosphatidylcholine-containing micelles (Mackay *et al.*, 1997). These studies strongly implicate the importance of $pPLA_2$ activity for cholesterol absorption but they do not exclude the possible involvement of other phospholipases in the digestive tract.

3.2.2b. Enterocyte Lipid Secretion Another mode in which lipid absorption may be linked to $pPLA_2$ activity derives from the fact that the flux of absorbed lipid through the enterocytes of the intestinal epithelium depends directly on cellular phosphatidylcholine synthesis, which is limited by the supply of synthetic precursors available from hydrolysis of phosphatidylcholine in the lumenal contents (Mansbach, 1977). It is well established that transport of absorbed lipids through enterocytes requires reacylation of the absorbed lipids followed by assembly of reacylated lipids into chylomicron particles composed of a triglyceride-rich core surfaced by apolipoproteins and a phospholipid monolayer consisting primarily of phosphatidylcholine (Tso, 1994). A direct dependence of intestinal chylomicron production on phosphatidylcholine synthesis from lumenally derived synthetic precursors is indicated by several *in vivo* studies showing that induction of intestinal triglyceride output in chylomicrons by intestinal lipid infusion is blunted if the supply of phosphatidylcholine in the intestinal contents is limited (for a review see Tso and Scobey, 1986). The results suggest that the capacity of enterocytes to transport absorbed lipids into the circulation via lipoproteins depends on $pPLA_2$ hydrolysis of phosphatidylcholine in the lumenal contents.

3.3. Autodigestion

The apical membranes of the intestinal epithelial cells represent another substrate pool that is potentially accessible to $pPLA_2$. Yet, no evidence of apical membrane degradation by $pPLA_2$ under normal circumstances has been reported. Although there is potential for autodigestion of the gastrointestinal epithelium by any of the variety of digestive hydrolases released into the digestive tract, several factors may serve to create barriers to such hydrolase activity. The most significant barrier may be the continuous layer of protective mucus that covers the entire gastrointestinal epithelium. The mucus is a 5% hydrated gel of the glycoprotein mucin that forms a permeability barrier to high-molecular-weight molecules. It is estimated that a 10 to 20 kDa protein would require several hours to freely diffuse through a 180 μm layer of mucus gel as is typically found adhering to the human intestinal epithelium (Allen *et al.,* 1993). Furthermore, based on the fact that the oligosaccharide sidechains of mucin are highly anionic due to an abundance of sulfate and sialic acid residues and evidence that pPLA$_2$ binds to heparin, a polyanionic, sulfated glycosaminoglycan (Diccianni *et al.,* 1991; Yu *et al.,* 1997a), it is also possible that mucin may further inactivate and restrict pPLA$_2$ migration to the epithelial surface by direct binding interactions.

Several additional factors may act to further attenuate the hydrolytic activity of any pPLA$_2$ that does reach the apical membrane surface. One level of protection may result from the fact that the apical membranes of intestinal epithelial cells are greatly enriched in glycosphingolipids at the expense of phosphatidylcholine and other glycerophospholipids (Simons and van Meer, 1988), thereby reducing the interfacial substrate concentration. Further reduction in pPLA$_2$ activity may result from the unique structure of the interface recognition domain in pPLA$_2$ that contains specific cationic lysyl residues that greatly enhance pPLA$_2$ adsorption to anionic surfaces and diminish pPLA$_2$ affinity for the type of neutral lipid interface that is characteristic of the apical membrane surface (see section 6.5.; Dua *et al.,* 1995; Rogers *et al.,* 1998; Snitko *et al.,* 1999). Finally, the pH at the apical membrane surface of the intestinal epithelium is in the range of 5 to 6, which is at least a full unit lower than the pH in the lumenal contents (Fawcus *et al.,* 1997) and, more important, significantly below the pH optimum of pPLA$_2$ (Tojo *et al.,* 1993). The increased acidity would significantly attenuate pPLA$_2$ activity at the membrane surface. Collectively, these multiple factors can act to effectively neutralize epithelial membrane hydrolysis by pPLA$_2$.

4. Pathology

Pathological conditions arise when pPLA$_2$ and the accompanying pancreatic hydrolases migrate beyond the protective barriers of the digestive tract and gain access to cells in surrounding tissues. Such can occur in acute pancreatitis when the tissue integrity of the exocrine pancreas is disrupted by infection, inflammation, alcoholism, ductal obstruction, or cancer (Agarwal and Pitchumoni, 1993; Banks, 1998). In these situations, the pancreatic hydrolases escape the normal path of pancreatic juice flow and infiltrate the surrounding tissue as well as enter the systemic circulation. Partial activation of zymogens occurs during the tissue infiltration, presumably by trypsin that is cleaved from trypsinogen by a protease other than enterokinase. The ensuing autodigestion greatly exacerbates the tissue disruption and necrosis associated with pancreatitis.

A direct correlation has been observed between episodes of acute pancreatitis and elevated serum levels of both pro-pPLA$_2$ and active pPLA$_2$ in humans (Funakoshi *et al.,* 1993). The consequences of active pPLA$_2$ in the circulation are unknown but, in addition to the potential deleterious effects of phospholipid hydrolysis, further toxic responses may be elicited by activation of cells expressing pPLA$_2$ receptors. The elevated serum levels of

$pPLA_2$ are also linked to increased levels of other nonpancreatic phospholipase A_2s, which may further exacerbate the inflammatory effects associated with pancreatitis (Nevalainen *et al.,* 1999).

Pancreatic insufficiency caused by cystic fibrosis, chronic pancreatitis, or cancer can result in the loss of digestive enzyme output. This is accompanied by lipid malabsorption as manifest by steatorrhea (Banks, 1998). The significance of $pPLA_2$ in this situation is not clear because output of all the pancreatic hydrolases is diminished or absent. One study of phospholipid absorption in cystic fibrosis patients, where $pPLA_2$ was found to be totally absent, detected significant phospholipid hydrolysis, suggesting that alternate sources of phospholipase activity may exist in the digestive tract (Roy *et al.,* 1988).

The contributions of $pPLA_2$ hydrolytic activity to tissue disruption potentially involve degradation of cell membrane phospholipids with release of fatty acids and lysophosphatidylcholine, some of which may be bioactive. Arachidonic acid is one such bioactive fatty acid that leads to the production of proinflammatory eicosanoids (for a recent review see Tischfield 1997). Specific bioactive properties have also been ascribed to lysophosphatidylcholine (for a recent review see Sakai *et al.* 1998). Only limited amounts of phospholipid hydrolysis need occur for changes in cell activation by such pathways to become significant. Thus, even though $pPLA_2$ has a lower affinity for the neutral phospholipid interfaces characteristic of the external side of cell membranes, reduced $pPLA_2$ activity may still have a considerable impact on cell function.

It is unlikely that phospholipid lipolysis leads to significant cytotoxicity as a result of membrane structure disruption and increased permeability. It has been shown for both phospholipid vesicles (Wilschut *et al.,* 1979) and cells (Wilbers *et al.,* 1979) that more than half of the phospholipids in the outer membrane leaflet can be hydrolyzed by externally applied PLA_2 without significant changes in membrane structure or permeability to water or to low-molecular-weight solutes. Membrane integrity is largely maintained because the hydrolysis products do not diffuse away from the membrane, even though they are sufficiently water soluble to exchange through the aqueous phase. Lipolytic product retention results from the lipophilicity and, consequently, high membrane partitioning of the hydrolysis products and the energetically unfavorable changes in interfacial tension that would occur with the loss of the lipids from one leaflet of the membrane bilayer. The thermodynamic stability of phospholipid bilayer membranes and a lack of lipid exchange after treatment with PLA_2 has been demonstrated (Jain *et al.,* 1995; Yu *et al.,* 1997b). The notion that the fatty acids and lysophosphatidylcholine disrupt the bilayer is based on the observation that externally added products force reorganization of the two halves of the bilayer. In contrast, when lipolysis products are formed *in situ,* they replace the phospholipid in the same bilayer, in which case the PLA_2-treated membranes remain intact and nonleaky with no noticeable transbilayer movement of phospholipids.

5. Biochemistry of Pancreatic Phospholipase A_2

5.1. Classification and Structure

Since the first observations of PLA_2 activity in pancreatic juice in the late 20th century (Wittcoff, 1951), the number of unique phospholipase A_2 enzymes identified has grown into a list of several dozen that has been organized into 10 distinct types (Dennis, 1997; Tischfield, 1997; Valentin *et al.,* 1999). Pancreatic PLA$_2$s are classified as type Ib enzymes, which are evolutionarily related to other type I as well as type II, III, V, and X PLA$_2$s. Features common to this collection of PLA$_2$s include the 13 to 15 kDa size, heat stability, 6 to 8 disulfide bridges, a remarkably conserved active site architecture, and a catalytic mechanism that is based on a His–Asp diad with a role for calcium in the chemical step (Yu *et al.,* 1998).

An extra peptide segment made up of residues 62 to 66 and the presence of certain lysyl residues are the only significant features that distinguish the Type Ib pPLA$_2$ from the type Ia PLA$_2$s found in *Elapidae* (cobra, krait) and *Hydrophiiae* (sea snake) venoms (Davidson and Dennis, 1990). Due to the absence of these lysyl residues, the type Ia enzymes are distinguished from most other PLA$_2$ types, including pPLA$_2$ by the lack of increased affinity for anionic interfaces (see the following paragraphs).

Although the sequence homology between types I, II and III PLA$_2$ is only 20% and the gross architectures differ significantly, the architecture for the catalytic residue and for the calcium and substrate binding sites is reasonably well conserved (Scott and Sigler, 1994; Sekar *et al.,* 1997a). The active site consists of heptacoordinated calcium with a neighboring catalytic histidine. X-ray analysis indicates that calcium is also coordinated to the *sn-2* carbonyl of substrate mimics cocrystallized with pPLA$_2$ (Fig. 2). Substrate access to the active site is through a slot, the wall of which is lined with hydrophobic residues for the bind-

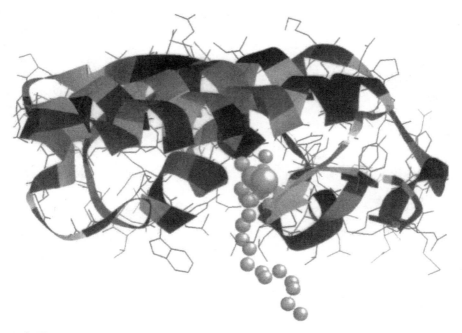

Figure 2. The backbone–ribbon rendition of the crystal structure of bovine pPLA$_2$ with MJ33 in the active site (Sekar *et al.,* 1997a). The methyl end of the inhibitor is protruded down, and it is likely that the plane perpendicular to the chain is the interfacial binding face (i-face) of the enzyme.

ing of the acyl chains. The collar of this slot is also lined with hydrophobic residues and is further surrounded by several cationic residues that are somewhat farther out from the entry point of the slot. The hydrophobic collar and the cationic ring presumably form the i-face of the protein that comes in contact with the interface in such a way that the polar charged region is within 0.5 nm of the interface while the glycerol backbone and the acyl chain regions of the interfacial lipids form closer contacts (Ramirez and Jain, 1991). The i-face has apparently evolved to make a watertight seal with the interface without a significant penetration into the acyl chain region of the interface. Such a contact must permit access of the substrate and inhibitor from the interface to the active site without exposing the chains to the aqueous environment.

Components of the i-face that are unique to the pancreatic enzyme include the extra surface loop at residues 62 to 66 and cationic residues at positions 53, 56, and 120. The function of the loop is not yet evident but the cationic residues control the anionic charge preference and are thus responsible for the k^*_{cat} activation (Rogers *et al.*, 1998; Berg *et al.*, 1997). The i-face also involves the N-terminus and the C-terminus regions. Thus the extra seven residues at the N-terminus of pro-pPLA$_2$ zymogen, which is catalytically inert at the interface, presumably block close contact of the i-face with the lipid surface.

5.2. Catalytic Mechanism

Site-directed mutagenesis studies have provided insights into the catalytic mechanism and interfacial recognition of pPLA$_2$ (Yu *et al.*, 1999a). The role of calcium and the His-48/Asp-99 couple in the chemical step of esterolysis by PLA$_2$ is now well established (Sekar *et al.*, 1997b). Recent studies suggest that the nucleophilic water may also be coordinated, along with the carbonyl oxygen of the substrate, to calcium in a near-attack conformation in an expanded octacoordinated shell (Figure 3). Although a concerted mechanism for attack by water and cleavage of the ester could account for the PLA$_2$ catalyzed esterolysis, formation of a tetrahedral intermediate is invoked in analogy with serine pro-

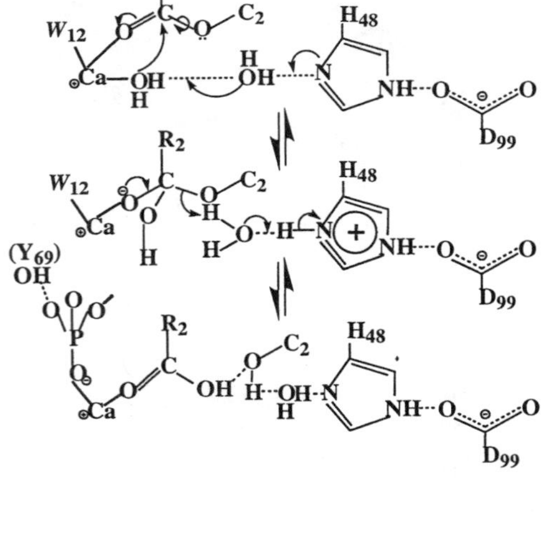

Figure 3. Proposed sequence of events during the chemical step of the catalytic cycle of secreted 14 kDa PLA$_2$. Initial binding of the substrate carbonyl occurs with an expansion of the calcium coordination shell (first step), where W_5 remains coordinate and acts as a nucleophile. The rate-limiting conversion of the tetrahedral intermediate is initiated as the oxygen of *sn*-3-phosphate displaces W_{12} (Sekar *et al.*, 1997b).

teases, in which case formation of the intermediate is unlikely to be rate limiting. In either case, the energetically demanding transition state probably lies in the step where the *sn-3*-phosphate shifts in the coordination shell of calcium (Rogers *et al.*, 1996; Yu *et al.*, 1998)

6. Interfacial Enzymology

Kinetics of interfacial catalysis can be, and should be, interpreted and evaluated in the broader context of the extended Michaelis paradigm as adopted in Fig. 4 for the interfacial kinetic turnover. The turnover in the aqueous phase is not relevant for the hydrolysis of natural phospholipid because the monomer substrate concentration is exceedingly low. Moreover, k_{cat} for pPLA2 through the ES path in the aqueous phase is also negligible in comparison to the interfacial turnover through the interfacial K_M^* and k^*_{cat} (Yu *et al.*, 1999b). Therefore, as outlined in this section, the main task of interfacial kinetic analysis (Berg *et al.*, 1991, 1997; Jain *et al.*, 1995) is to eliminate, or to quantitatively evaluate (which is often not possible), the contribution of the various exchange reactions that ultimately determine the concentration of the substrate as seen by the enzyme at the interface for the processive turnover.

6.1. The Steady-State Condition

A fundamental assumption for kinetic interpretation of ensemble behavior is that all enzyme molecules at a given time point must be in the identical environment, and that, except for the change in the concentration of the substrate and products, the overall environment must remain unchanged during the initial steady state where the steps of the catalytic turnover cycle determine the rate. This condition is readily satisfied in solution enzymolo-

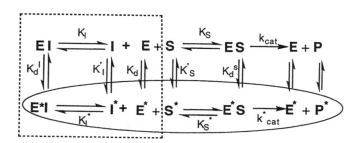

Figure 4. General kinetic model for interfacial catalysis. The species and the constants have their standard enzymological significance, and those marked with an asterisk indicate species at the interface. The scheme can be analytically resolved under three steady-state conditions where the chemical step remains rate limiting: the scooting mode of infinite processivity where any of the species do not exchange (Jain *et al.*, 1986), rapid exchange of the substrate and products (Berg *et al.*, 1997), and on an interface for a rapidly exchangeable substrate and product which does not form the interface (Berg *et al.*, 1998). Also the thermodynamic box in the square has been analyzed in detail to establish the conditions for allosteric K_S^*-activation (Jain *et al.*, 1993a). The anionic charge at the interface also induces allosteric k^*_{cat}-activation (Berg *et al.*, 1997).

gy because all the components are monodisperse and in rapid thermal motion. Imagine the complications that are created if one of the components is in some sort of monomer–aggregate equilibrium that is shifted with the reaction progress. Such results will be virtually uninterpretable in terms of the interfacial Michaelis rate and equilibrium constants because the microscopic steady-state condition for the turnover is not the same as the bulk averaged condition through the measured reaction progress.

It can hardly be overemphasized that for obtaining interpretable kinetic results with interfacial enzymes, one must pay special attention to what an interfacial enzyme "sees" at the microscopic level in the assay mixture, and it must be assured that all enzyme molecules see identical environments at any given time point during the reaction progress. The microscopic environment of an interfacial enzyme may differ significantly from the bulk average environment derived on the basis of the concentrations of the added components. It is also necessary to pay special attention to the structures and properties of the phases that are formed in aqueous dispersions of the substrate, products, and other lipidic additives. This brings into focus the variables controlled by partition coefficient, ideal mixing, and Poisson distribution of the components in the mixture. Such factors are of special kinetic significance under the conditions where the heterogeneity of the interface must be taken into account to establish the steps of the turnover cycle and therefore the mechanism. Space limitation does not permit detailed discussion of such problems (Jain and Berg, 1989) and their resolution (Berg *et al.*, 1997, 1998; Jain *et al.*, 1995), however, major pitfalls are identified in the discussion that follows. With such an appreciation of the conditions for the processivity and of the microscopic steady state, it is possible to obtain functional and mechanistic information about $pPLA_2$ in terms of the primary rate and equilibrium parameters.

6.2. Key Considerations for the Assay of PLA_2 Activity

PLA_2 enzymes have evolved to carry out hydrolysis of phospholipid at organized interfaces. Natural phospholipids are found exclusively in organized interfaces, such as micelles and bilayers. The monomer phospholipid concentration is insignificant because the critical micelle concentration is in the subnanomolar range. Release of fatty acid by PLA_2 has been monitored by several techniques. The real-time, continuous measurements by pH-stat titrations are very reliable for the determination of absolute activities as well as for the full kinetic characterization of the reaction progress in terms of well-defined primary rate, equilibrium parameters, and definitive mechanistic significance (Berg *et al.*, 1991, 1997, 1998). A more sensitive assay for detection of residual activity during protein modification is based on the use of pyrene-labeled phospholipid (Bayburt *et al.*, 1995). Virtually all other methods, such as those employing radiolabeled substrate, rely on indirect analysis of the assay volume. In addition, protocols to eliminate extraneous surface effects that often plague assays have been developed (Yu *et al.*, 1999b; Cajal *et al.*, 2000).

Interpretation of the observed rates is not possible without the knowledge of the microscopic environment of the enzyme. For example, use of the monodisperse premicellar substrates for kinetic purposes is misleading because in such cases the site of reaction is the surfaces of the container and air bubbles (Berg *et al.*, 1998). In fact, the estimated "monomer rate" for the action of pig $pPLA_2$ is less than 0.1 sec^{-1}, compared to the apparent rate of 10 sec^{-1} obtained when contributions from such extraneous surfaces is not eliminated (Yu et al., 1999b). Similar complications from the exposed hydrophobic surface of the trough and the unstirred layer make the air/water-monolayer-based assays quantitatively unreliable (Jain and Berg, 1989; Cajal *et al.,* 2000). As discussed next, the substrate replenishment, rather than the chemical step, becomes rate limiting in assays with mixed micelles (Jain *et al.,* 1993b).

6.3. Quality of the Substrate Interface Influences Enzyme Binding

An enigma of interfacial enzymology is that the observed rate changes with the way substrate is presented to the enzyme. According to the scheme in Fig. 4, it is clear that such "quality of interface effects" are due to the rate-limiting exchange rates that determine the residence time for the enzyme at the interface and the availability of the substrate for the steady-state turnover cycles. A lack of appreciation for the fact that these parallel processes introduce artifacts has fueled much controversy during the last three decades. Under most experimental conditions, the binding and desorption of the enzyme from the interface is a relatively slow process. For the maximal expression of the interfacial turnover rate, not only must enzyme exchange between interfaces be absent, but the substrate replenishment and the product release should be nonexistent, as in the case of scooting kinetics (see the following text). A pseudo-scooting mode condition for the microscopic steady state is also established with a rapid substrate replenishment of short-chain phosphatidylcholine substrates on the submillisecond time scale, which permits not only determination of rates in the range of 5000 sec^{-1} but also analysis of the apparent rate parameters (Berg *et al.,* 1997, 1998). On the other hand, the microscopic steady-state condition can not be identified during the hydrolysis of monolayers (Cajal *et al.,* 2000) or mixed-micelles (Jain *et al.,* 1993b). For example, with a few dozen molecules of long-chain phospholipid per micelle, which will be hydrolyzed in less than 0.1 sec by an enzyme with the intrinsic catalytic turnover of 300 sec^{-1}. Under these conditions the observed kinetics for the long-chain natural phospholipid substrates become limited by the replenishment rate because the intrinsic phospholipid monomer exchange rate between vesicles is less than 1 per hour even after a substantial portion of the phospholipid is hydrolyzed. Because this is a thermodynamic limit for the monomer exchange, it should also hold for mixed micelles. The reports of fast exchange rates result from the exchange of the detergent molecules. Even the half-times for the exchange through the fusion–fission of mixed micelles are in seconds. Thus, hydrolysis by an enzyme on a mixed micelle containing a few dozen substrate molecules becomes limited by the exchange rate, which depends on the bulk concentration of the micelles. In short, exchange-limited kinetics in mixed micelles make them unsuitable for kinetic and mechanistic analysis.

Kinetics in the scooting mode takes advantage of the virtually infinite processivity of tightly bound pPLA$_2$. The tight binding eliminates enzyme hopping between vesicles and, thus, permits monitoring of hydrolysis only in vesicles to which enzyme is initially bound (Berg *et al.,* 1991; Jain *et al.,* 1986). As described in the following section, this permits unequivocal determination of the interfacial kinetic parameters k*$_{cat}$ for the turnover in the interface in units of per second, and K$_M$* in units of mole fraction of the substrate that the enzyme sees at the interface. Under these conditions the phospholipid exchange can be accelerated by protein or peptide-mediated contacts between vesicles (Cajal *et al.,* 1996). Coupled with the fact that the intrinsic catalytic rate of pPLA$_2$ and lipase is typically in the range of 500 to 5000 sec^{-1}, the exchange constraints for the reactants and products provide insights into the *in vivo* kinetics.

Based on the model for an emulsion particle (Fig. 1), it is estimated that *in vivo* the enzyme operates below the K_M^*, that is at less than 50% k^*_{cat}. If enough enzyme is secreted to distribute over each particle, or if particles fuse and exchange their contents in the presence of bile salts and products of hydrolysis, virtually all the substrate for $pPLA_2$ and lipases can be readily acted on during transit through the digestive tract.

6.4. Intrinsic Equilibrium and Kinetic Parameters

Figure 4 provides a minimal kinetic scheme for the interfacial catalytic turnover and the exchange processes associated with it. For the turnover in the scooting mode with virtually infinite processivity, following the initial pre-steady-state binding of the enzyme to the interface (E to E*), the steady-state reaction progress occurs on the interface through the steps shown in the oval. Due to higher affinity of $pPLA_2$ for anionic interfaces, the residence time of the enzyme at the interface can be up to hours. The turnover by E* obeys the Michaelis–Menten paradigm adopted for the interface, as characterized by interfacial K_M^* and k^*_{cat}.

The substrate concentration relates to what the bound enzyme sees during the interfacial turnover, rather than the bulk substrate concentration. Only the fraction of the total enzyme at the interface is related to the bulk substrate concentration through K_d and K_d^s. A disregard for this basic physical fact has created considerable confusion in the literature. The two-dimensional substrate concentration e.g., (moles per cm^2), is best approximated as the mole fraction, and thus the maximum attainable substrate mole fraction, X_S, is 1. Obviously, X_S remains constant with the bulk substrate concentration. However, X_s, and therefore the observed rate, changes during the course of hydrolysis due to substrate depletion and product accumulation or by the introduction of other surface additives. The presence of other surface components (cholesterol, glycerides) will lower X_s for $pPLA_2$, and thus will lower the rate even though these additives may not bind to the active site. Similarly, X_I for a competitive inhibitor is related to its partitioning equilibrium dissociation constant K_I', which will correspondingly lower the substrate mole fraction to $1-X_I$ in a two-component system. If the inhibitor is to compete with the substrate, it is the mole fraction of the inhibitor with respect to the substrate that ultimately determines the degree of inhibition. Thus, even a "submicromolar inhibitor" can be a meaningless artifact if the bulk substrate concentration is also in this range, that is, both X_I and $1-X_I$ are nearly equal.

Relationships between substrate structure and catalytic parameters are well established. Interfacial $K_M^* = (k_{-1} + k^*_{cat})/k_1$ for bovine and pig $pPLA_2$ are in the 0.2 to 0.8 mole fraction range for most phospholipids regardless of headgroup type, and the effect of the chain length on k^*_{cat} is only modest. The catalytic turnover rate at $X_S = 1$ mole fraction gives $k^*_{cat} = 400$ sec^{-1} for DMPM. Because k^*_{cat} for the *sn*-2-thio analogue of the substrate is about 10% of the rate with the oxy-analogue, the chemical step, rather than a physical step preceding the chemical step, is rate limiting in the turnover cycle, that is, $k_2 = k^*_{cat}$ (Jain *et al.*, 1992a). Independent measurements show that $K_S^* (= k_{-1}/k_1)$ for the long-chain substrates are near 0.03 mole fraction, that is, the substrate once bound to the active site is committed to catalysis (i.e., $k^*_{cat} > k_{-1}$). For the short-chain substrates K_M^* and K_S^* are comparable (i.e., $k^*_{cat} < k_{-1}$), that is, the Michaelis complex reverts readily. It appears that the origin of the chain-length dependence may be in Leu-2, because the L2W mutant shows a large discrimination between the short- and the long-chain substrates (Liu *et al.*, 1995).

The cofactor role for calcium has been resolved. Calcium is not required for the binding of the enzyme to the interface; however, it is required for the substrate or mimic (competitive inhibitor or product) binding to the active site and also for the chemical step (Yu *et al.*, 1993). Several divalent and trivalent cations replace calcium for the substrate binding, but fewer replace calcium in the chemical step (Yu *et al.*, 1998).

6.5. Interfacial Activation in the Substrate Binding and the Chemical Step

The interface has a significant effect on the interfacial kinetic parameters of pPLA$_2$. Enhanced affinity of the enzyme at the interface for the substrate (mimics), K_S^* activation, is clearly demonstrated in terms of the thermodynamic box (square box in Fig. 4) by the fact that the affinity of E for the interface is about 40 to 100 times lower than the affinity of ES (Jain *et al.*, 1993a). This effect does not seem to depend on the micellar or bilayer organization or on the presence of the charge at the interface. On the other hand, the interfacial anionic charge has a dramatic effect on k^*_{cat}, that is, the rate of hydrolysis of zwitterionic substrates at anionic interface is about 50 times higher than at the zwitterionic interface (Berg *et al.*, 1997). It appears that pPLA$_2$ is designed for the anion-induced k^*_{cat}-activation, because the K53M and K56M mutants show a significant increase in hydrolysis rate at the zwitterionic interface, and the anion-induced increase is less than fivefold for either of these mutants (Rogers *et al.*, 1998). Thus, pPLA$_2$ does not readily hydrolyze phosphatidylcholine-rich host membranes, whereas the venom enzymes with hydrophobic residues at positions 53 and 56 are designed to cause maximum damage at the zwitterionic interface.

6.6. Kinetic Interpretation of the Role of Bile Salts in PLA$_2$ Catalyzed Hydrolysis

Pancreatic PLA$_2$ is designed to work on phospholipids in the presence of bile salts. Phospholipid mixed micelles with 0.3 to 0.5 mole fraction bile salt show maximum rates of hydrolysis (Jain *et al.*, 1993b). These particles probably exist as large disks. Besides solubilization of the substrate bilayers, two kinetic effects of bile salts are noteworthy: the anionic interface promotes the binding of the enzyme, and only the enzyme bound to the anionic interface is catalytically functional due to the k^*_{cat} activation by the anionic interface.

7. Inhibitors

The substrate and calcium-binding domains offer sites for two types of competitive inhibitors. Several classes of substrate mimics are now available as competitive inhibitors (Fig. 5). The more efficient ones include primary acylamides (I; Jain *et al.*, 1992a), *sn*-2-acylamido-phospholipid analogues (II; de Hass *et al.*, 1990), and *sn*-2-phospho- or phosphono-derivatives, such as MG14 (Jain *et al.*, 1989). Data from such inhibitors coupled with crystallographic and molecular simulations suggest considerable flexibility in the active site, which has provided further insights for inhibitor design. The MJ series of inhibitors, which include MJ33 and MJ99, were designed to coordinate with calcium in the

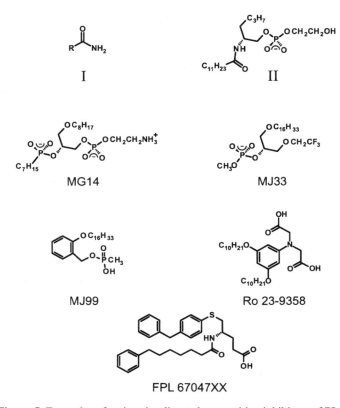

Figure 5. Examples of active site-directed competitive inhibitors of PLA_2.

active site without the need for the *sn*-3 phosphate (Jain *et al.*, 1991a). The structures of these inhibitors are less similar to the glycerophospholipid substrate. Ro 23–9358 (LeMaheiu *et al.*, 1993) and FPL 67047XX (Beaton *et al.*, 1994) are examples of potent inhibitors derived by molecular modeling with structures that also depart considerably from that of the substrate

X ray crystallograpic data for FPL 67047XX (Cha *et al.*, 1996) and an *sn*-2-acylamido-phospholipid analogue (II; Thunnissen *et al.*, 1990; Tomoo *et al.*, 1997) indicate the acylamide group occupies the same position as the *sn*-2-acyl chain with the carbonyl coordinated to calcium and the $-NH_2$ group H-bonded to catalytic His-48. Acylamides (I) are probably bound to $pPLA_2$ in the same position.

Covalent inhibitors of $pPLA_2$ include phenacyl bromides, which alkylate the catalytic histidine (Verheij *et al.*, 1981), manoalogue, which modifies lysine residues (Ghomashchi *et al.*, 1991), and gossypol, which appears to modify the N-terminus residue of the interfacial recognition region (Yu *et al.*, 1997a).

The calcium site is inhibited by micromolar concentrations of Zn^{2+} and Cu^{2+}. The inhibitory concentration depends on the substrate because such cations support substrate binding. The calcium requirement for substrate binding to the active site is supported by crystal structures (Fig. 2).

Examples of nonspecific inhibitors include anesthetics, solvents, and cationic anti-malarials that lower pPLA$_2$ activity by altering the charge or organization of the substrate interface and thus lower the binding of the enzyme (Jain *et al.*, 1991b). Similarly, heparin and other sulfated glycoconjugates trap pPLA$_2$ away from the substrate interface (Yu *et al.*, 1997a).

8. Conclusions

Many scientific disciplines have converged on pPLA$_2$ to produce one of the most comprehensive biochemical characterizations available for an enzyme. The scientific impetuses for these efforts have been multifarious, ranging from the interest in pPLA$_2$ as a surrogate for the PLA$_2$s involved in second-messenger and eicosanoid metabolism pathways to understanding and developing a kinetic formalism for interfacial catalysis. In contrast to the diverse interests in the biochemical attributes of pPLA$_2$, perspectives on the physiological significance of pPLA$_2$ have, until recently, remained focused on the function of pPLA$_2$ as a member of a group of lipases required for hydrolysis of the respective dietary lipid substrates to more absorbable products. It is now becoming evident, however, that pPLA$_2$ performs a broader physiological function than simply catalyzing the recovery of glycerophospholipids from the digestive tract. More recent evidence suggests pPLA$_2$ is the rate-limiting factor for the absorption of a range of lipid classes, in addition to glycerophospholipids. These observations provide a basis for the therapeutic use of pPLA$_2$ inhibitors in the treatment of hyperlipidemias (Homan and Krause, 1997). Even more intriguing are the indications that pPLA$_2$ has functions outside the digestive tract involving cells expressing a specific pPLA$_2$ receptor. Further study is needed to elucidate the physiological significance of this receptor.

Much effort has been applied to resolving the actions of pPLA$_2$ *in vitro,* but further work is needed to determine how the *in vitro* results apply to pPLA$_2$ in the milieu of the intestinal contents. For example, it would be beneficial to determine whether or not there is an interdependence in the action of the various hydrolases whose substrates reside in emulsified lipid aggregates. Meager knowledge of the substrate and phase preference for these interfacial enzymes does not permit speculation, however it is quite likely that once an emulsion droplet is attacked by pPLA$_2$, the products of its hydrolysis may promote the binding of lipases. A consequence of a long interfacial residence time of the lipases adsorbed to the droplet surface and a high processivity may be the near completion of substrate hydrolysis before the lipases move on to the next droplet. Also, as the polar amphiphilic products are formed, larger droplets may break into smaller ones to accomodate the polar amphiphiles at the interface and thus increase the surface area/volume ratio. The phase properties of particles thus formed are difficult to predict. Further work is needed in the characterization of such multicomponent phases.

References

Agarwal, N., and Pitchumoni, C. S., 1993, Acute pancreatitis: A multisystem disease, *Gastroenterologist.* **1**:113–128.

Åkesson, B., 1982, Content of phospholipids in human diets studied by the duplicate-portion technique, *Br. J. Nutr.* **47**:223–229.

Allen, A., Flemström, G., Garner, A., and Kivilaakso, E., 1993, Gastroduodenal mucosal protection, *Physiol. Rev.* **73:**823–857.

Arnesjö, B., Nilsson, Å., Barrowman, J., and Borgström, B., 1969, Intestinal digestion and absorption of cholesterol and lecithin in the human, *Scand. J. Gastroenterol.* **4:**653–665.

Banks, P. A., 1998, Acute and chronic pancreatitis, in: *Gastrointestinal and Liver Disease,* Volume 1 (M. Feldman, B. F. Scharschmidt, and M. H. Sleisenger, eds.), W. B. Saunders Company, Philadelphia, pp. 809–862.

Bayburt, T., Yu, B. Z., Street, I., Ghomashchi, F., Laliberte, F., Perrier, H., Wang, Z., Homan, R., Jain, M. K., and Gelb, M. H., 1995, Continuous, vesicle-based fluorimetric assays of 14- and 85-kDA phospholipase A_2, *Anal. Biochem.* **232:**7–23.

Beaton, H. G., Bennion, C., Connolly, S., Cook, A. R., Gensmantel, N. P., Hallam, C., Hardy, K., Hitchin, B., Jackson, C. G., and Robinson, D. H., 1994, Discovery of new non-phospholipid inhibitors of the secretory phospholipases A_2, *J. Med. Chem.* **37:**557–559.

Berg, O. G., Yu, B. Z., Rogers, J., and Jain, M. K., 1991, Interfacial catalysis by phospholipase A_2: Determination of the interfacial kinetic rate constants, *Biochem.* **30:**7283–7297.

Berg, O. G., Rogers, J., Yu, B. Z., Yao, J., Romsted, L. S., and Jain, M. K., 1997, Thermodynamic and kinetic basis of interfacial activation: Resolution of binding and allosteric effects on pancreatic phospholipase A_2 at zwitterionic interfaces, *Biochem.* **36:**14512–14530.

Berg, O. G., Cajal, Y., Butterfoss, G. L., Grey, R. L., Álsina, M. A., Yu, B. Z., and Jain, M. K., 1998, Interfacial activation of triglyceride lipase from Thermomyces (Humicola) lanuginosa: Kinetic parameters and a basis for control of the lid, *Biochem.* **37:**6615–6627.

Borgström, A., Erlanson-Albertsson, C., and Borgström, B., 1993, Human pancreatic proenzymes are activated at different rates *in vitro, Scand. J. Gastroenterol.* **28:**455–459.

Borgström, B., 1980, Importance of phospholipids, pancreatic phospholipase A_2, and fatty acid for the digestion of dietary fat, *Gastroenterol.* **78:**954–962.

Borgström, B., Dahlqvist, A., Lundh, G., and Sjövall, J., 1957, Studies of intestinal digestion and absorption in the human, *J. Clin. Invest.* **36:**1521–1536.

Cajal, Y., Rogers, J., Berg, O. G., and Jain, M. K., 1996, Intermembrane molecular contacts by polymyxin B mediate exchange of phospholipids, *Biochem.* **35:**299–308.

Cajal, Y., Berg, O. G., and Jain, M. K., 2000, Product accumulation during the lag phase as the basis for the activation of phospholipase A_2 on monolayers, *Langmuir* **16:**252–257.

Carey, M. C., Small, D. M., and Bliss, C. M., 1983, Lipid digestion and absorption, *Ann. Rev. Physiol.* **45:**651–677.

Cha, S. S., Lee, D., Adams, J., Kurdyla, J. T., Jones, C. S., Marshall, L. A., Bolognese, B., Abdel-Meguid, S. S., and Oh, B. H., 1996, High-resolution X-ray crystallography reveals precise binding interactions between human nonpancreatic secreted phospholipase A_2 and a highly potent inhibitor (FPL67047XX), *J. Med. Chem.* **39:**3878–3881.

Cupillard, L., Koumanov, K., Lazdunski, M., and Lambeau, G., 1997, Cloning, chromosomal mapping, and expression of a novel human secretory phospholipase A_2, *J. Biol. Chem.* **272:**15745–15752.

Cupillard, L., Mulherkar, R., Gomez, N., Kadam, S., Valentin, E., Lazdunski, M., and Lambeau, G., 1999, Both Group IB and Group IIA secreted phospholipases A_2 are natural ligands of the mouse 180-kDA M-type receptor, *J. Biol. Chem.* **274:**7043–7051.

Davidson, F. F., and Dennis, E. A., 1990, Evolutionary relationships and implications for the regulation of phospholipase A_2 from snake venom to human secreted forms, *J. Mol. Evol.* **31:**228–238.

de Hass, G. H., Dijkman, R., Ransac, S., and Verger, R., 1990, Competitive inhibition of lipolytic enzymes. IV. Structural details of acylamino phospholipid analogues important for the potent inhibitory effects on pancreatic phospholipase A_2, *Biochim. Biophys. Acta* **1046:**249–257.

Dennis, E. A., 1997, The growing phospholipase A_2 superfamily of signal transduction enzymes, *Trends in Biochem. Sci.* **22:**1–2.

Diagne, A., Mitjavila, S., Fauvel, J., Chap, H., and Douste-Blazy, L., 1987, Intestinal absorption of ester and ether glycerophospholipids in guinea pig. Role of phospholipase A_2 from brush border membrane, *Lipids* **22:**33–40.

Diccianni, M. B., Lilly-Stauderman, M., McLean, L. R., Balasubramaniam, A., and Harmony, J. A. K., 1991, Heparin prevents the binding of phospholipase A_2 to phospholipid micelles: Importance of the amino-terminus, *Biochem.* **30:**9090–9097.

Dijkstra, B. W., Renetseder, R., Kalk, K. H., Hol, W. G. J., and Drenth, J., 1983, Structure of porcine pancreatic phospholipase A$_2$ at 2.6 Å resolution and comparison with bovine phospholipase A$_2$, *J. Mol. Biol.* **168:**163–179.

Dua, R., Wu, S. K., and Cho, W., 1995, A structure–function study of bovine pancreatic phospholipase A$_2$ using polymerized mixed liposomes, *J. Biol. Chem.* **270:**263–268.

Dupureur, C. M., Yu, B. Z., Jain, M. K., Noel, J. P., Deng, T., Li, Y., Byeon, I. J. L., and Tsai, M. D., 1992a, Phospholipase A$_2$ engineering. Structural and functional roles of highly conserved active site residues tyrosine-52 and tyrosine-73, *Biochem.* **31:**6402–6413.

Dupureur, C. M., Yu, B. Z., Mamone, J. A., Jain, M. K., and Tsai, M. D., 1992b, Phospholipase A$_2$ engineering. The structural and functional roles of aromaticity and hydrophobicity in the conserved phenylalanine-22 and phenylalanine-106 aromatic sandwich, *Biochem.* **31:**10576–10583.

Fauvel, J., Bonnefis, M.-J., Chap, H., Thouvenot, J.-P., and Douste-Blazy, L., 1981, Evidence for the lack of classical secretory phospholipase A$_2$ in guinea pig pancreas, *Biochim. Biophys. Acta* **666:**72–79.

Fawcus, K., Gorton, V. J., Lucas, M. L., and McEwan, G. T. A., 1997, Stimulation of three distinct guanylate cyclases induces mucosal surface alkalinisation in rat small intestine *in vitro, Comp. Biochem. Physiol.* **118A:**291–295.

Funakoshi, A., Yamada, Y., Migita, Y., and Wakasugi, H., 1993, Simultaneous determinations of pancreatic phospholipase A$_2$ and prophospholipase A$_2$ in various pancreatic diseases, *Dig. Dis. Sci.* **38:**502–506.

Gassama-Diagne, A., Fauvel, J., and Chap, H., 1989, Purification of a new, calcium-independent, high molecular weight phospholipase A$_2$/lysophospholipase (phospholipase B) from guinea pig intestinal brush-border membrane, *J. Biol. Chem.* **264:**9470–9475.

Gelb, M. H., Jain, M. K., and Berg, O. G., 1994, Inhibition of phospholipase A$_2$, *FASEB J.* **8:**916–924.

Gelb, M. H., Jain, M. K., and Berg, O. G., 1997, Principles of inhibition of phospholipase A$_2$ and other interfacial enzymes, in: *Phospholipase A$_2$. Basic and Clinical Aspects in Inflammatory Diseases,* (W. Uhl, T. J. Nevalainen, and M. W. Büchler, eds.), Kargel, Basel, pp. 123–129.

Ghomashchi, F., Yu, B. Z., Mihelich, E. D., Jain, M. K., and Gelb, M. H., 1991, Kinetic characterization of phospholipase A$_2$ modified by manoalogue, *Biochem.* **30:**9559–9569.

Grataroli, R., Charbonnier, M., Léonardi, J., Grimaud, J.-C., Lafont, H., and Nalbone, G., 1987, Phospholipase A$_2$ activity in rat stomach, *Arch. Biochem. Biophys.* **258:**77–84.

Hamosh, M., 1990, Lingual and gastric lipases, *Nutrition* **6:**421–428.

Hanasaki, K., Yokota, Y., Ishizaki, J., Itoh, T., and Arita, H., 1997, Resistance to endotoxic shock in phospholipase A$_2$ receptor-deficient mice, *J. Biol. Chem.* **272:**32792–32797.

Hernell, O., Staggers, J. E., and Carey, M. C., 1990, Physical–chemical behavior of dietary and biliary lipids during intestinal digestion and absorption. 2. Phase analysis and aggregation states of luminal lipids during duodenal fat digestion in healthy adult human beings, *Biochem.* **29:**2041–2056.

Hirohara, J., Sugatani, J., Okumura, T., Sameshima, Y., and Saito, K., 1988, Properties and localization of phospholipase A$_2$ activity in rat stomach, *Biochim. Biophys. Acta* **919:**231–238.

Homan, R., and Hamelehle, K. L., 1998, Phospholipase A$_2$ relieves phosphatidylcholine inhibition of micellar cholesterol absorption and transport by human intestinal cell line Caco-2, *J. Lipid Res.* **39:**1197–1209.

Homan, R., and Krause, B. R., 1997, Established and emerging strategies for inhibition of cholesterol absorption, *Curr. Pharm. Design* **3:**29–44.

Huang, B., Yu, B. Z., Rogers, J., Byeon, I. J. L., Sekar, K., Chen, X., Sundaralingam, M., Tsai, M. D., and Jain, M. K., 1996, Phospholipase A$_2$ engineering. Deletion of the C-terminus segment changes substrate specificity and uncouples calcium and substrate binding at the zwitterionic interface, *Biochem.* **35:**12164–12174.

Huhtinen, H. T., Gronroos, J. M., Haapamaki, M. M., and Nevalainen, T. J., 1999, Phospholipases A$_2$ in gastric juice of Helicobacter pylori-positive and negative individuals, *Clin. Chem. Lab. Med.* **37:**61–64.

Jain, M. K., and Berg, O. G., 1989, The kinetics of interfacial catalysis by phospholipase A$_2$ and regulation of interfacial activation: hopping versus scooting, *Biochim. Biophys. Acta* **1002:**127–156.

Jain, M. K., Rogers, J., Jahagirdar, D. V., Marecek, J. F., and Ramirez, F., 1986, Kinetics of interfacial catalysis by phospholipase A$_2$ in intravesicular scooting mode, and heterodiffusion of anionic zwitterionic vesicles, *Biochim. Biophys. Acta* **860:**435–447.

Jain, M. K., Yuan, W., and Gelb, M. H., 1989, Competitive inhibition of phospholipase A$_2$ in vesicles, *Biochem.* **28:**4135–4139.

Jain, M. K., Tao, W., Rogers, J., Arenson, C., Eibl, H., and Yu, B. Z., 1991a, Active-site-directed specific competitive inhibitors of phospholipase A$_2$: Novel transition-state analogues, *Biochem.* **30:**10256–10268.

Jain, M. K., Yu, B. Z., Rogers, J., Ranadive, G. N., and Berg, O. G., 1991b, Interfacial catalysis by phospholipase

A_2: Dissociation constants for calcium, substrate, products, and competitive inhibitors, *Biochem.* **30**:7306–7317.

Jain, M. K., Ghomashchi, G., Yu, B. Z., Bayburt, T., Murphy, D., Houck, D., Brownell, J., Reid, J. C., Solowiej, J. E., Jarrell, R., Sasser, M., and Gelb, M. H., 1992a, Fatty acid amides: Scooting mode-based discovery of tight-binding competitive inhibitors of secreted phospholipases A_2, *J. Med. Chem.* **35**:3584–3586.

Jain, M. K., Yu, B. Z., Rogers, J., Gelb, M. H., Tsai, M. D., Hendrickson, E. K., and Hendrickson, H. S., 1992b, Interfacial catalysis by phospholipase A_2: the rate-limiting step for enzymatic turnover, *Biochem.* **31**:7841–7847.

Jain, M. K., Yu, B. Z., and Berg, O. G., 1993a, Relationship of interfacial equilibria to interfacial activation of phospholipase A_2, *Biochem.* **32**:11319–11329.

Jain, M. K., Rogers, J., Hendrickson, H. S., and Berg, O. G., 1993b, The chemical step is not rate-limiting during the hydrolysis by phospholipase A_2 of mixed micelles of phospholipid and detergent, *Biochem.* **32**:8360–8367.

Jain, M. K., Gelb, M. H., Rogers, J., and Berg, O. G., 1995, Kinetic basis for interfacial catalysis by phospholipase A_2, *Methods Enzymol.* **249**:567–614.

Johnson, L. J., Frank, S., Vades, P., Pruzanski, W., Lusis, A. J., and Seilhamer, J. J., 1990, Localization and evolution of two human phospholipase A_2 genes and two related genetic elements, *Adv. Exp. Med. Biol.* **275**:17–34.

Kinoshita, E., Handa, K., Kajiyama, G., and Sugiyama, M., 1997, Activation of MAP kinase cascade induced by human pancreatic phospholipase A_2 in a human pancreatic cancer cell line, *FEBS Lett.* **407**:343–346.

Kishino, J., Kawamoto, K., Ishizaki, J., Verheij, H. M., Ohara, O., and Arita, H., 1995, Pancreatic-type phospholipase A_2 activates prostaglandin E_2 production in rat mesangial cells by receptor binding reaction, *J. Biochem.* **117**:420–424.

Kortesuo, P. T., Hietaranta, A. J., Jämiä, M., Hirsimäki, P., and Nevalainen, T. J., 1993, Rat pancreatic phospholipase A_2, *Int. J. Pancreatology* **13**:111–118.

Kundu, G. C., and Mukherjee, A. B., 1997, Evidence that porcine pancreatic phospholipase A_2 via its high affinity receptor stimulates extracellular matrix invasion by normal and cancer cells, *J. Biol. Chem.* **272**:2346–2353.

Lambeau, G., and Lazdunski, M., 1999, Receptors for a growing family of secreted phospholipases A_2, *Trends Pharmacol. Sci.* **20**:162–170.

Layer, P., Go, V. L. W., and DiMagno, E. P., 1986, Fate of pancreatic enzymes during small intestinal aboral transit in humans, *Am. J. Physiol.* **251**:G475-G480

LeMaheiu, R. A., Carson, M., Han, R. J., Madison, V. S., Hope, W. C., Chen, T., Morgan, D. W., and Hendrickson, H. S., 1993, N-(Carboxymethyl-N-[3,5-bis(decyloxy)phenyl]glycine (Ro 23–9358): A potent inhibitor of secretory phospholipases A_2 with antiinflammatory activity, *J. Med. Chem.* **36**:3029–3031.

Li, Y., Yu, B. Z., Zhu, H., Jain, M. K., and Tsai, M. D., 1994, Phospholipase A_2 engineering. Structural and functional roles of the highly conserved active site residue aspartate-49, *Biochem.* **33**:14714–14722.

Lindström, M. B., Persson, J., Thurn, L., and Borgström, B., 1991, Effect of pancreatic phospholipase A_2 and gastric lipase on the action of pancreatic carboxyl ester lipase against lipid substrates *in vitro, Biochim. Biophys. Acta* **1084**:194–197.

Li-Stiles, B., Lo, H., and Fischer, S. M., 1998, Identification and characterization of several forms of phospholipase A_2 in mouse epidermal keratinocytes, *J. Lipid Res.* **39**:569–582.

Liu, X., Zhu, H., Huang, B., Rogers, J., Yu, B. Z., Kumar, A., Jain, M. K., Sundaralingam, M., and Tsai, M. D., 1995, Phospholipase A_2 engineering. Probing the structural and functional roles of N-terminal residues with site-directed mutagenesis, X-ray, and NMR, *Biochem.* **34**:7322–7334.

Lowe, M. E., 1994, The structure and function of pancreatic enzymes, in: *Physiology of the Gastrointestinal Tract,* (L. R. Johnson, ed.), Raven Press, New York, pp. 1531–1542.

Mackay, K., Starr, J. R., Lawn, R. M., and Ellsworth, J. L., 1997, Phosphatidylcholine hydrolysis is required for pancreatic cholesterol esterase- and phospholipase A_2-facilitated cholesterol uptake into intestinal Caco-2 cells, *J. Biol. Chem.* **272**:13380–13389.

Mansbach, C. M., 1977, The origin of chylomicron phosphatidylcholine in the rat, *J. Clin. Invest.* **60**:411–420.

Matsuda, Y., Ogawa, M., Shibata, T., Nakaguchi, K., Nishijima, J., Wakasugi, C., and Mori, T., 1987, Distribution of immunoreactive pancreatic phospholipase A_2 (IPPL-2) in various human tissues, *Res. Commun. Chem. Pathol. Pharmacol.* **58**:281–284.

Nalbone, G., Charbonnier-Augeire, M., Lafont, H., Grataroli, R., Vigne, J. L., Lairon, D., Chabert, C., Leonardi,

J., Hauton, J. C., and Verger, R., 1983, Adsorption of pancreatic (pro)phospholipase A$_2$ to various physiological substrates, *J. Lipid Res.* **24:**1441–1450.

Nevalainen, T. J., Hietaranta, A. J., and Gronroos, J. M., 1999, Phospholipase A$_2$ in acute pancreatitis: New biochemical and pathological aspects, *Hepatogastroenterology* **46:**2731–2735.

Ohara, O., Ishizaki, J., and Arita, H., 1995, Structure and function of phospholipase A$_2$ receptor, *Prog. Lipid Res.* **34:**117–138.

Pind, S., and Kuksis, A., 1991, Further characterization of a novel phospholipase B (phospholipase A$_2$–lysophospholipase) from intestinal brush-border membranes, *Biochem. Cell Biol.* **69:**346–357.

Plebani, M., 1993, Pepsinogens in health and disease, *Crit. Rev. Clin. Lab. Sci.* **30:**273–328.

Ramirez, F., and Jain, M. K., 1991, Phospholipase A$_2$ at the bilayer interface, *Proteins* **9:**229–239.

Rogers, J., Yu, B. Z., Serves, S. V., Tsivgoulis, G. M., Sotiropoulos, D. N., Ioannou, P. V., and Jain, M. K., 1996, Kinetic basis for the substrate specificity during hydrolysis of phospholipids by secreted phospholipase A$_2$, *Biochem.* **35:**9375–9384.

Rogers, J., Yu, B. Z., Tsai, M. D., Berg, O. G., and Jain, M. K., 1998, Cationic residues 53 and 56 control the anion-induced interfacial k*$_{cat}$-activation of pancreatic phospholipase A$_2$, *Biochem.* **37:**9549–9556.

Roy, C. C., Weber, A. M., Lepage, G., Smith, L., and Levy, E., 1988, Digestive and absorptive phase anomalies associated with the exocrine pancreatic insufficiency of cystic fibrosis, *J. Pediatr. Gastroenterol. Nutr.* **7:**S1–S7

Sakai, M., Shichiri, M., Hakamata, H., and Horiuchi, S., 1998, Endocytosed lysophosphatidylcholine, through the scavenger receptor, plays an essential role in oxidized low-density lipoprotein-induced macrophage proliferation, *Trends Cardiovasc. Med.* **8:**119–124.

Sakata, T., Nakamura, E., Tsuruta, Y., Tamaki, M., Teraoka, H., Tojo, H., Ono, T., and Okamoto, M., 1989, Presence of pancreatic type phospholipase A$_2$ mRNA in rat gastric mucosa and lung, *Biochim. Biophys. Acta* **1007:**124–126.

Scheele, G., Bartelt, D., and Bieger, W., 1981, Characterization of human exocrine pancreatic proteins by two-dimensional isoelectric focusing/sodium dodecyl sulfate gel electrophoresis, *Gastroenterol.* **80:**461–473.

Scott, D. L., and Sigler, P. B., 1994, Structure and catalytic mechanism of secretory phospholipases A$_2$, *Adv. Protein Chem.* **45:**53–88.

Sekar, K., Eswaramoorthy, S., Jain, M. K., and Sundaralingam, M., 1997a, Crystal structure of the complex of bovine pancreatic phospholipase A$_2$ with the inhibitor 1-hexadecyl-3-(trifluoroethyl)-*sn*-glycero-2-phosphomethanol, *Biochem.* **36:**14186–14191.

Sekar, K., Yu, B. Z., Rogers, J., Lutton, J., Liu, X., Chen, X., Tsai, M. D., Jain, M. K., and Sundaralingam, M., 1997b, Phospholipase A$_2$ engineering. Structural and functional roles of the highly conserved active site residue aspartate-99, *Biochem.* **36:**3104–3114.

Sekar, K., and Sundaralingam, M., 1999, High-resolution refinement of orthorhombic bovine pancreatic phospholipase A$_2$, *Acta Crystallogr. D Biol. Crystoallogr.* **55:**46–50.

Simons, K., and van Meer, G., 1988, Lipid sorting in epithelial cells, *Biochem.* **27:**6197–6202.

Snitko, Y., Han, S. K., Lee, B. I., and Cho, W., 1999, Differential interfacial and substrate binding modes of mammalian pancreatic phospholipase A$_2$: A comparison among human, bovine and porcine enzymes, *Biochem.* **38:**7803–7810.

Sternby, B., Nilsson, Å., Melin, T., and Borgström, B., 1991, Pancreatic lipolytic enzymes in human duodenal contents, *Scand. J. Gastroenterol.* **26:**859–866.

Tasumi, H., Tojo, H., Senda, T., Ono, T., Fujita, H., and Okamoto, M., 1990, Immunocytochemical studies on the localization of pancreatic-type phospholipase A$_2$ in rat stomach and pancreas, with special reference to the stomach cells, *Histochemistry* **94:**135–140.

Thunnissen, M. M. G. M., Eiso, A. B., Kalk, K. H., Drenth, J., Dijkstra, B. W., Kuipers, O. P., Dijkman, R., de Hass, G. H., and Verheij, H. M., 1990, X-ray structure of phospholipase A$_2$ complexed with a substrate-derived inhibitor, *Nature* **347:**689–691.

Tischfield, J. A., 1997, A reassessment of the low molecular weight phospholipase A$_2$ gene family in mammals, *J. Biol. Chem.* **272:**17247–17250.

Tojo, H., Ono, T., and Okamoto, M., 1988a, A pancreatic-type phospholipase A$_2$ in rat gastric mucosa, *Biochem. Biophys. Res. Comm.* **151:**1188–1193.

Tojo, H., Ono, T., Kuramitsu, S., Kagamiyama, H., and Okamoto, M., 1988b, A phospholipase A$_2$ in the supernatant fraction of rat spleen, *J. Biol. Chem.* **263:**5724–5731.

Tojo, H., Ying, Z., and Okamoto, M., 1993, Purification and characterization of guinea pig gastric phospholipase A_2 of the pancreatic type, *Eur. J. Biochem.* **215:**81–90.

Tomoo, K., Yamane, Y., Ishida, T., Fujii, S., Ikeda, K., Iwama, S., Katsumura, S., Sumiya, S., Miyagawa, H., and Kitamura, K., 1997, X-ray crystal structure determination and molecular dynamics simulation of prophospholipase A_2 inhibited by amide-type substrate analogues, *Biochim. Biophys. Acta* **1340:**178–186.

Tso, P., 1994, Intestinal lipid absorption, in: *Physiology of the Gastrointestinal Tract,* (L. R. Johnson, ed.), Raven Press, New York, pp. 1867–1908.

Tso, P., and Scobey, M., 1986, The role of phosphatidylcholine in the absorption and transport of dietary fat, in: *Fat Absorption,* (A. Kuksis, ed.), CRC Press, Boca Raton, pp. 177–195.

Valentin, E., Ghomashchi, F., Gelb, M. H., Lazdunski, M., and Lambeau, G., 1999, On the diversity of secreted phospholipases A_2, *J. Biol. Chem.* **274:**31195–31202.

van den Berg, B., Tessari, M., De Haas, G. H., Verheij, H. M., Boelens, R., and Kaptein, R., 1995, Solution structure of porcine pancreatic phospholipase A_2, *EMBO J.* **14:**4123–4131.

Verger, R., and De Haas, G. H., 1976, Interfacial enzyme kinetics of lipolysis, *Annu. Rev. Biophys. Bioeng.* **5:**77–117.

Verheij, H. M., 1995, Structure and mechanism of pancreatic phospholipase A_2—A molecular biology approach, in: *Phospholipase A_2 in Clinical Inflammation, Molecular Approaches to Pathophysiology,* (K. B. Glaser and P. Vadas, eds.), CRC Press, Boca Raton, pp. 3–24.

Verheij, H. M., Slotboom, A. J., and De Haas, G. H., 1981, Structure and function of phospholipase A_2, *Rev. Physiol. Biochem. Pharmacol.* **91:**91–203.

Wilbers, K. H., Haest, C. W., von Bentheim, M., and Deuticke, B., 1979, Influence of enzymatic phospholipid cleavage on the permeability of the erythrocyte membrane: I. Transport of non-electrolytes via the lipid domain, *Biochim. Biophys. Acta* **554:**388–399.

Wilschut, J. C., Regts, J., and Scherphof, G., 1979, Action of phospholipase A_2 on phospholipid vesicles, *FEBS Lett.* **98:**181–186.

Wittcoff, H., 1951, *The lysophosphatides and lecithinases,* Reinhold Publishing Corp., New York.

Ying, Z., Tojo, H., Nonaka, Y., and Okamoto, M., 1993, Cloning and expression of phospholipase A_2 from guinea pig gastric mucosa, its induction by carbachol and secretion *in vitro, Eur. J. Biochem.* **215:**91–97.

Young, S. C., and Hui, D. Y., 1999, Pancreatic lipase/colipase-mediated triacylglycerol hydrolysis is required for cholesterol transport from lipid emulsions to intestinal cells, *Biochem. J.* **339:**615–620.

Yu, B. Z., Berg, O. G., and Jain, M. K., 1993, The divalent cation is obligatory for the binding of ligands to the catalytic site of secreted phospholipase A_2, *Biochem.* **32:**6485–6492.

Yu, B. Z., Rogers, J., Ranadive, G. N., Baker, S., Wilton, D. C., Apitz-Castro, R., and Jain, M. K., 1997a, Gossypol modification of Ala-1 of secreted phospholipase A_2: A probe for the kinetic effects of sulfate glycoconjugates, *Biochem.* **36:**12400–12411.

Yu, B. Z., Ghomashchi, F., Cajal, Y., Annand, R. R., Berg, O. G., Gelb, M. H., and Jain, M. K., 1997b, Use of an imperfect neutral diluent and outer vesicle layer scooting mode hydrolysis to analyze the interfacial kinetics, inhibition, and substrate preferences of bee venom phospholipase A_2, *Biochem.* **36:**3870–3881.

Yu, B. Z., Rogers, J., Nicol, G. R., Theopold, K. H., Seshadri, K., Vishweshwara, S., and Jain, M. K., 1998, Catalytic significance of the specificity of divalent cations as K_s^* and k^*_{cat} cofactor for secreted phospholipase A_2, *Biochem.* **37:**12576–12587.

Yu, B. Z., Rogers, J., Tsai, M. D., Pidgeon, C., and Jain, M. K., 1999a, Contributions of residues of pancreatic phospholipase A_2 to interfacial binding, catalysis, and activation, *Biochem.* **38:**4875–4884.

Yu, B. Z., Berg, O. G., and Jain, M. K., 1999b, Hydrolysis of monodisperse substrate by phospholipase A_2 occurs at vessel walls and air bubbles, *Biochem.* **38:**10449–10456.

CHAPTER 6

Enterostatin/Procolipase

A Peptide System Regulating Fat Intake

CHARLOTTE ERLANSON-ALBERTSSON

1. Introduction

Man is constructed to eat regular meals. Moreover, man is constructed to prefer food that is energy-rich and tasty. Dietary lipids are both energy-rich and tasty, with several spices being lipid soluble. In the early history of man dietary lipids were scarce but gradually increased with the development of an agricultural society. Since the 1950s dietary fat intake in the Western world has dramatically increased from around 30 energy percent to 40 energy percent (Dreon *et al.*, 1988). This has as a consequence an increased frequency of obesity and insulin resistance, eventually leading to Type 2 diabetes at the moment of failure of the islets of Langerhans (Bray *et al.*, 1990; Steffens *et al.*, 1991; Shafrir and Gutman, 1993). The reason that a high fat intake has serious metabolic implications is that fat taken in is not automatically oxidized in proportion to its consumption as is dietary carbohydrate and protein (Thomas *et al.*, 1992). Instead an increased fat intake leads to the accumulation of fat in skeletal muscle and adipose tissue. The insulin resistance following high-fat feeding is due to an impairment of the insulin receptor signaling events downstream in the target cell, the insulin receptor and its insulin receptor substrates, for instance Insulin receptor substrate 1 (IRS-1) and Phosphatidylinositol-3 kinase (PI3-kinase), being phosphorylated at several serine/threonine residues instead of the normal tyrosine phosphorylation (Paz *et al.*, 1996). In the muscle that is the first tissue to become insulin resistant following high-fat feeding, long-chain acyl-CoA has been shown to raise the levels of protein kinase C, which in turn activates serine/threonine residues (Schmitz Peiffer *et al.*, 1997). In light of the multiple metabolic disturbances of a high dietary fat intake, great interest has been taken in the regulation of appetite, especially with the interest of a specific macronutrient appetite regulation. Such studies were prompted by the finding that overfeeding with one type of macronutrient lead to a reduced intake of that particular macronutrient, whereas the

CHARLOTTE ERLANSON-ALBERTSSON • Department of Cell and Molecular Biology, Section for Molecular Signalling, University of Lund, S-221 00 Lund, Sweden.
Intestinal Lipid Metabolism, edited by Charles M. Mansbach II *et al.,* Kluwer Academic/Plenum Publishers, 2001.

intake of other nutrients was uneffected (van Amelsvoort *et al.,* 1988; Leibowitz, 1994; Shor-Posner *et al.,* 1994). A number of peptides and neurotransmitters have been shown to selectively regulate the intake of a specific macronutrient (Bray, 1992). Hence, carbohydrate intake has been found to be preferentially stimulated by Neuropeptide Y (NPY) (Stanley *et al.,* 1985) and by the α_2 effects of noradrenaline, whereas *kappa* opioids have been found to selectively stimulate fat intake (Barton *et al.,* 1995). Certain peptides or neurotransmittors are partially specific in that they are influenced by the background preferences of the animal studied. Thus galanin has been found to stimulate fat intake in certain animals (Tempel *et al.,* 1988; Lin *et al.,* 1993a) and carbohydrate in others (Smith *et al.,* 1997). The same is true for serotonin, which has been found to inhibit carbohydrate intake (J. Wurtman *et al.,* 1993) or fat intake (Blundell and Clawton, 1995) depending on the situation. Inhibition of protein intake has been claimed to occur through the peptide glucagon (Vanderweele and Macrum, 1986). A peptide that has been found to selectively inhibit fat intake is enterostatin (Okada *et al.,* 1991). Enterostatin is formed through the processing of procolipase, a protein necessary for intestinal fat digestion (Erlanson-Albertsson, 1992). This chapter provides an overview of the biological actions of enterostatin.

2. Biological Activity of Enterostatin

2.1. Feeding Response

The first experiments involving enterostatin demonstrated its ability to decrease food intake in the rat when given either peripherally or centrally (Erlanson-Albertsson and Larsson, 1988; Shargill *et al.,* 1991). In the following experiments rats were given a three-choice macronutrient diet of fat, carbohydrate, and protein and with this regimen enterostatin was found to reduce selectively the intake of the fat macronutrient (Okada *et al.,* 1991, Mizuma *et al.,* 1994). On a two-choice diet of high fat and low fat, enterostatin was subsequently found to reduce the intake of the high-fat diet only (Erlanson-Albertsson *et al.,* 1991a). It was also found that enterostatin inhibited high-fat food intake in rats stimulated to eat high-fat food with galanin, whereas with a low-fat diet stimulated with NPY there was no effect (Lin *et al.,* 1993a). The introduction of satiety by enterostatin occurred through an early onset of satiety, with a reduced time spent feeding and a prolonged time spent resting (Lin *et al.,* 1993b). The behavior of the animals following injection of enterostatin was thus similar to natural satiety, suggesting that enterostatin was not mediating its effects through nausea. This was supported by the finding that enterostatin did not initiate any conditioned aversion (Mei and Erlanson-Albertsson, 1992). Further experiments demonstrated that enterostatin was effective in reducing high-fat feeding, also when given intraduodenally (Mei and Erlanson-Albertsson, 1996b), orally (York, personal communication) and intravenously (Mei and Erlanson-Albertsson, 1992), in addition to the previously reported intraperitoneal and intracerebroventricular routes. By these routes enterostatin was shown to have a rapid effect (< 30 min) on food intake after administration of peptide with the exception of the intravenous administration, where the response time was found to be 60 to 120 min.

The enterostatin-induced satiety had a long duration of action, lasting up to 6 hours after a single injection in rats adapted to a 6-hour feeding schedule (Mei and Erlanson-Albertsson, 1992) or lasting up to 24 hours after a single injection in *ad-libitum* fed rats (Okada *et al.*, 1992). During chronic intracerebroventricular (ICV) administration of enterostatin for 11 days in Osborne-Mendel, for rats fed a high-fat diet, there was a decrease in daily food intake, fat deposition, and body weight gain (Okada *et al.*, 1993b). In a similar way Sprague–Dawley rats chronically treated with enterostatin ICV during 9 days on a two-choice high-fat and low-fat regime reduced the intake of the high-fat diet with the maximum suppression at Day 4 (Lin *et al.*, 1997). There was no compensatory overeating of the low-fat food, which hence resulted in a diminshed body weight. The reduction in body weight at the end of the experiment was however too large to be explained only through a reduction in food intake, suggesting that enterostatin might have effects on energy expenditure as well, increasing thermogenesis (Lin *et al.*, 1997). Similar results were found in Sprague–Dawley rats treated with enterostatin through intraperitoneal pumps and fed a high-fat diet for 7 days, losing body weight compared to control high-fat fed animals in excess of what could be expected from the reduction in food intake (Sörhede-Winzell, unpublished observations). In low-fat fed Sprague-Dawley rats treated with enterostatin intraperitoneally there was, however, no significant change in body weight gain (Mei and Erlanson-Albertsson, 1996a), emphasizing the importance of a lipid factor in order for enterostatin to be effective.

2.2. Metabolic Effects

In addition to appetite-regulating properties enterostatin has been found to have significant metabolic effects by modulating the secretion of two important energy-regulating hormones—insulin and the glucocorticoid hormone. Enterostatin was hence found to reduce insulin secretion (Mei *et al.*, 1993a; Okada *et al.*, 1993b; Erlanson-Albertsson *et al.*, 1994; Silvestre *et al.*, 1996; Lin *et al.*, 1997) and to increase the glucocorticoid secretion (Okada *et al.*, 1993b; Mei and Erlanson-Albertsson, 1996a; Lin *et al.*, 1997). The inhibition of enterostatin on insulin secretion has been demonstrated on isolated rat islets from Sprague–Dawley rats (Mei *et al.*, 1993a; Erlanson-Albertsson *et al.*, 1994) as well as in a perfused pancreas (Silvestre *et al.*, 1996). This inhibiton by enterostatin was observed after insulin secretion had been stimulated with either glucose (Mei *et al.*, 1993a; Erlanson-Albertsson *et al.*, 1994) or tolbutamide and arginine (Silvestre *et al.*, 1996). There was no inhibition of insulin secretion by enterostatin in mouse islets isolated from the *ob/ob* mouse (Mei, 1998, personal communication), suggesting the involvement of leptin in the reponse of enterostatin. The inhibition of insulin secretion by enterostatin in Sprague–Dawley rat islets was not related to any *beta* cell damage; furthermore, the inhibition was found to be specific to insulin secretion with no influence on glucagon or somatostatin secretion (Silvestre *et al.*, 1996).

Inhibition of insulin release by enterostatin was also observed *in vivo* following either acute or chronic administration of enterostatin (Okada *et al.*, 1993b; Mei and Erlanson-Albertsson, 1996a; Lin *et al.*, 1997). In these experiments there was only a slight or no associated change in blood glucose, suggesting that insulin sensitivity was improved either directly or indirectly by enterostatin (Mei and Erlanson-Albertsson, 1996a; Lin *et al.*, 1997). The insulin-lowering effect of enterostatin was also observed after intraintestinal administration of the peptide (Mei *et al.*, 1997), which might have significant physiological importance, as the intestinal lumen is the natural site of production of enterostatin. The mechanism through which enterostatin inhibits insulin secretion is unclear, the intestinal absorption of enterostatin being low due to proteolytic processing of the enterostatin molecule in the intestine (Huneau *et al.,* 1994; Bouras *et al.,* 1995) to suggest a direct effect of enterostatin on the islets. Rather a modulation of the enteric or autonomic nervous system output to the islets of Langerhans may be a mechanism of action for enterostatin. The importance of the lowering of insulin secretion by enterostatin may be to prevent hypoglycemia during fat feeding. In addition, lowering of insulin levels leads to an increased activation of the adipose tissue hormone-sensitive lipase (Belfrage, 1985), through which fatty acids are released from the adipose tissue ready for β-oxidation. In this way enterostatin might increase the utilization of lipids and decrease adipose tissue mass as has also been observed after chronic administration of enterostatin (Okada *et al.,* 1993b; Lin *et al.,* 1997).

The other hormone affected by enterostatin is corticosterone. The enhanced release of corticosterone by enterostatin occurred both after central and peripheral administration (Okada *et al.,* 1993b; Mei and Erlanson-Albertsson, 1996a; Lin *et al.,* 1997). The observation that serum corticosterone levels were greatly elevated in rats chronically infused with enterostatin suggested that enterostatin might stimulate corticotropin-releasing hormone (CRH) secretion. By using a CRH antagonist it was, however, demonstrated that enterostatin did not mediate its anorectic effect through CRH (Okada *et al.,* 1993b). An increased CRH release during enterostatin treatment might however prevent the rats from overeating the low-fat food as described in the two-choice feeding situation of low-fat and high-fat food (Lin *et al.,* 1997).

The physiological significance of the activation of the hypothalamic–pituitary–adrenal axis by enterostatin is not known. It is, however, well known that high-fat feeding activates the hypothalamic–pituitary–adrenal axis (York, 1992). Because the production of enterostatin increases following high-fat feeding (Mei *et al.,* 1993b), it could be argued that this activation is mediated through enterostatin. The release of corticosterone by enterostatin during high-fat feeding could enhance the sensitivity of hormone-sensitive lipase to catecholamines and so stimulate adipose tissue lipolysis (Briendley, 1995). Thus enterostatin, by both lowering insulin levels and by raising corticosterone levels, may act to increase the utilization of fat as energy substrate.

In addition to its endocrine effects, enterostatin has been found to influence energy metabolism through its activation of the sympathetic drive to brown adipose tissue, hence increasing thermogenesis (Nagase *et al.,* 1997). This stimulation was observed only in rats that had been adapted to a high-fat diet, in a similar way as the appetite-suppressing effect of enterostatin. The increase in firing rate of sympathetic nerves was observed 5 minutes after the injection of enterostatin ICV in anesthetized rats, the effect being independent of a feeding situation (Nagase *et al.,* 1997). The effects of an increased thermogenesis evoked by enterostatin was also suggested in Sprague–Dawley rats during 9 days of high-fat feeding compared to control high-fat fed animals (Lin *et al.,* 1997).

2.3. Gastrointestinal Effects

Hunger and satiety are strongly correlated with changes in gastrointestinal motility. Whereas hunger is associated with contractions and hyperactivity of the intestine, satiety is associated with the opposite state, a state of low activity. Thus, inhibition of gastric emptying has been found to be an essential target mechanism for several gastrointestinal satiety peptides, like cholecystokinin and amylin. In rat it was found that enterostatin reduced gastric emptying of a methylcellulose test meal after intracerebroventricular administration (Lin and York, 1997). However, this response was not related to the inhibitory effect of enterostatin on the consumption of a high-fat diet in comparisons made across three strains of rat, hence distinct from the appetite-regulating effect of enterostatin.

In pigs, intraintestinal administration of enterostatin was found to significantly inhibit pancreatic protein secretion after stimulation with cholecystokinin (Erlanson-Albertsson *et al.*, 1991b). This inhibition of pancreatic secretion was further shown to be linked to an inhibition of the motility of the intestine, with enterostatin acting to prolong the non-spike activity phase, a phase with no electrical activity recorded from the intestine (Pierzynowsky *et al.*, 1994). This effect was observed only after intraintestinal administration of enterostatin and was not observed after intravenous infusion of the peptide or in isolated intestinal segments (Kiela *et al.*, 1994), suggesting the enterostatin response to be dependent on intact nervous or blood circulatory transmission.

3. Mechanism of Action of Enterostatin

3.1. Central Mechanisms

The mechanism of action of enterostatin has suggested a central as well as a peripheral mode of action. Using crude brain membranes, a specific binding of enterostatin was demonstrated with a high-affinity binding, kd = 0.5 nM, and a low-affinity binding, kd = 30 nM (Sörhede *et al.*, 1993). In further studies a human neuroblastoma cell line, the SK-N-MC-cell line, was identified, which specifically bound enterostatin, also with a two-site binding of 1nM and 30 nM, respectively (Berger *et al.*, 1998). The intracellular events following this binding are not known. The neurochemical pathways affected by enterostatin after the peptide has been centrally administered have been identified as being κ-opioidergic (Barton *et al.*, 1995; Ookuma *et al.*, 1997). These studies showed that a κ-opioid agonist (U50488) when administered to rats stimulated fat intake, whereas with enterostatin there was an inhibition of fat intake (Ookuma *et al.*, 1997). In combination the κ-agonist U50488 prevented the enterostatin inhibition of high-fat food intake, suggesting the enterostatin to act through a κ-opioid receptor-mediated pathway (Ookuma *et al.*, 1997). It was also shown that the effect of enterostatin to inhibit fat intake was additive to a κ-antagonist, norbinaltorphimine, demonstrating the interaction of enterostatin with an opioid pathway during regulation of fat intake (Ookuma *et al.*, 1997). Numerous studies have demonstrated the role of opioids in macronutrient intake (Levine *et al.*, 1985). It is thus clear that the stimulation of fat intake occurs through an opioidergic pathway, and that enterostatin interferes in an inhibitory way with this pathway.

3.2. Brain–Gut Interactions

Because enterostatin is released peripherally either from the pancreas or the gastrointestinal tract, a relevant question is how the response of peripheral enterostatin is transmitted to the brain. Tian *et al.* (1994) found that the vagal–central nervous system was important for the feeding response to peripheral enterostatin. Transection of the hepatic vagus completely blocked the normal response to intraperitoneal enterostatin of inhibiting high-fat diet consumption in rats after overnight starvation. The importance of neuronal transmission for the feeding response of enterostatin was also suggested by the attenuation of the feeding response to intraintestinal enterostatin after tetracain administration to block peripheral nerve endings (Mei and Erlanson-Albertsson, 1996b). These studies thus suggest a neuronal transmission of enterostatin response from the intestine to the brain. Further support for a neuronal pathway was provided by the demonstration of a specific activation of certain hypothalamic nuclei including the nucleus tractus solitarius, parabrachial, paraventricular, and supraoptic nuclei following peripheral administration of enterostatin and that this effect was absent in rats with selective hepatic vagotomy (Tian *et al.,* 1994).

4. Enterostatin/Procolipase Production

4.1. Site of Enterostatin/Procolipase Synthesis

Procolipase, the parent molecule of enterostatin, is present in the exocrine pancreatic cells, where it is secreted into pancreatic juice (Erlanson-Albertsson, 1992). Following tryptic activation, procolipase is split into colipase and its activation peptide, named enterostatin (Borgström *et al.,* 1979; Fig. 1). The role of colipase is to activate pancreatic lipase, to which it binds forming a 1:1 molar complex (Hermoso *et al.,* 1997). The ratio between lipase and colipase is variable. In man, the ratio is 1:1; in pigs there is an abundance of colipase relative to lipase, whereas in rats and mice there is a deficiency of colipase relative to lipase (Rippe and Erlanson-Albertsson, 1998).

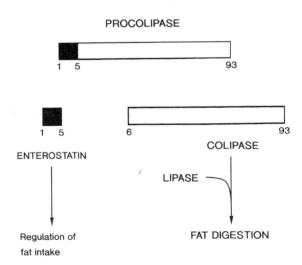

Figure 1. Colipase is a protein cofactor for pancreatic lipase to which it binds in a 1:1 molar complex that is subsequently bound to the lipid substrate to hydrolyze the dietary triglycerides in the presence of bile salt micelles. All colipase molecules known are produced in the procolipase form. The sequence of the enterostatin is APGPR in man, rabbit, chicken, and rat, and VPDPR in dog, cat, horse, pig, and cow (Erlanson-Albertsson, 1992; Rippe and Erlanson-Albertsson, 1998).

The relative deficiency of colipase relative to lipase lead to investigations of other possible localizations of colipase in the gastrointestinal tract. Such studies demonstrated the presence of procolipase mRNA in the stomach of rats (Okada *et al.*, 1993c), the procolipase being localized to the chief cells in the fundus region visulized both through *in situ* hybridization and immune reactivity (Sörhede *et al.*, 1996b). Because colipase is an acid-stable molecule containing five disulfide bridges the colipase molecule might well tolerate the acidic conditions of the stomach for further passage into the duodenum. This leads to the hypothesis of procolipase/colipase being anchored to the lipid substrate, waiting for the appearance of pancreatic lipase in the duodenum. At this time, it is not clear if procolipase is cleaved to yield colipase and enterostatin already in the gastric juice. At acidic conditions there is, however, no difference between procolipase and colipase in the interaction of the lipid substrate, making the prebinding of colipase to its substrate a possible scenario. In neither the acinar cells of the pancreas nor the chief cells of the stomach was there any free enterostatin peptide present visualized through immune reactivity (Sörhede *et al.*, 1996b).

The existence of free enterostatin without colipase or procolipase was however demonstrated in enterochromaffin cells of the gastrointestinal tract, these cells being most abundant in the antral part of the stomach, with fewer in the duodenum and jejunum, and only a few cells in the ileum (Sörhede *et al.*, 1996a). In some of these enterochromaffin cells there was, moreover, a co-localization of enterostatin and serotonin (Sörhede *et al.*, 1996a). The significance of this observation is not known. It may be that enterostatin and serotonin are both released during fat feeding, with both factors having a role to selectively inhibit fat intake in a situation of macronutrient choice (Mei and Erlanson-Albertsson, 1996b; Smith *et al.*, 1997).

4.2. Relationship between Colipase and Fat Intake

Various rat strains have been shown to have a natural preference for either dietary fat or dietary carbohydrate. The Osborne–Mendel rat has a strong dietary preference for fat, whereas the S5B/PL rat prefers dietary carbohydrate and is highly restrictive in dietary fat intake (Bray *et al.*, 1990; Fisler and Bray, 1995). Because dietary fat intake could be modulated with exogenous enterostatin (Okada *et al.*, 1991), it was of interest to study the endogenous levels of enterostatin or its precursor molecule, procolipase. Such investigations revealed a two- to threefold higher level of procolipase in the S5B/Pl rats compared to the Osborne–Mendel rats (Okada *et al.*, 1992). Indeed, voluntary fat intake was reciprocally related to pancreatic procolipase levels, both within and across rat species when rats were allowed to choose their dietary macronutrients (Okada *et al.*, 1992). Higher endogenous pancreatic procolipase activities were associated with lower intakes of dietary fat, and vice versa.

Additionally, in Sprague–Dawley rats an inverse relationship was established between dietary fat intake and pancreatic colipase activities (Sörhede-Winzell, unpublished results), again suggesting that endogenous production of enterostatin may be important in regulating fat intake. In a similar way the obese fa/fa rat was found to have a low colipase activity (Erlanson-Albertsson and Larsson, 1988) and a low expression of pancreatic procolipase mRNA (Okada *et al.*, 1993a) compared to its lean fa/? counterpart. In addition adrenalectomy of the obese fa/fa rat increased the levels of procolipase mRNA concomitant with an inhibition of further obesity (Okada *et al.*, 1993a). These correlations suggest, but do not prove, an endogenous role of enterostatin/procolipase in the regulation of dietary fat intake.

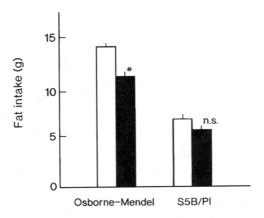

Figure 2. Enterostatin reduces fat intake in the high-fat-consuming Osborne–Mendel rat, whereas enterostatin has no effect in the fat-restrictive S5B/Pl rat (Okada *et al.,* 1992). The nonresponsiveness of the S5B/PL rat may either be due to a high endogenous production of procolipase/enterostatin in this animal or the absence of an opioidergic pathway to stimulate fat intake.

The correlational studies also demonstrated that the effect of exogenous enterostatin was dependent on the presence of endogenous procolipase—the animals with high levels of procolipase showed no response to enterostatin. Thus, the S5B/Pl rats with a high pancreatic procolipase activity did not respond to exogenously administered enterostatin, in contrast to the Osborne–Mendel rats, which readily decreased dietary fat intake on exogenous administration of enterostatin (Okada *et al.,* 1992; Fig. 2). The obese fa/fa rat with low endogenous pancreatic procolipase production responded to enterostatin by decreasing food intake but was unresponsive after the procolipase levels were incresased after adrenalectomy (Okada *et al.,* 1993a). Cook *et al.* (1994) found, in a group of Sprague–Dawley rats, that only fat-preferring animals responded to enterostatin, whereas the carbohydrate-preferring animals were nonresponding. In these experiments it should be of clear interest to assay the pancreatic procolipase amounts.

4.3. Regulation of Enterostatin/Procolipase Synthesis

In understanding the role of enterostatin in mediating a feeding pattern, the expression of enterostatin and/or its precursor protein, procolipase, is of physiological interest. The most significant finding is the increased production of enterostatin/procolipase during fat feeding (Wicker and Puigserver, 1987; Mei *et al.,* 1993b). This increased procolipase synthesis occurred in proportion to the amount of fat ingested 24 hours after the onset of a high-fat diet. An increased enterostatin production might hypothetically form a long-term feedback signal regulating the levels of fat ingested. Gabert *et al.* (1996) found that polyunsaturated fat was more effective in stimulating pancreatic procolipase production and release than saturated fat. Such a stimulation of polyunsaturated fat suggests that the quality of fat is important for the regulation and activity of enterostatin, as polyunsaturated fat is known to improve insulin resistance during high-fat feeding (Pan *et al.,* 1994).

The mechanism through which enterostatin/procolipase production is increased by high-fat diets is not known. One peptide of significance could be gastric inhibitory polypeptide, which is released in the gastrointestinal tract during fat feeding and which has been shown to stimulate procolipase synthesis (Duan and Erlanson-Albertsson, 1992). Opposed to this stimulation of procolipase synthesis two other key hormones in energy metabolism, insulin and corticosterone, inhibit procolipase synthesis (Duan and Erlanson-Albertsson, 1990; Duan *et al.*, 1991; Okada *et al.*, 1993a). Both hormones inhibit procolipase mRNA production, procolipase mRNA translation proceeding at a lower rate than for pancreatic lipase (Duan *et al.*, 1991). The significance of the reciprocal regulation of procolipase synthesis and corticosterone was demonstrated by adrenalectomy, which was shown to increase procolipase expression (Okada *et al.*, 1993a) as well as procolipase synthesis (Duan and Erlanson-Albertsson, 1990) concomitant with reduced intake of high-fat diet and body weight. Likewise the reciprocal regulation of procolipase and insulin was demonstrated after experimentally induced diabetes, which increased procolipase synthesis (Duan and Erlanson-Albertsson, 1989).

Enterostatin has been found in the intestinal content of humans after a test meal in a 1:1 molar relationship with pancreatic colipase, suggesting a complete cleavage of procolipase to colipase and enterostatin (Erlanson-Albertsson, 1994). The presence of immunoreactive enterostatin (APGPR) has also been demonstrated in the intestinal content of the rat, being increased following cholecystokinin (CCK) stimulation as well after one day of high-fat feeding (Mei *et al.*, 1993b). The rise in intestinal enterostatin levels following high-fat feeding occurred in parallel with an increase in pancreatic colipase synthesis (Mei *et al.*, 1993b), suggesting a close correlation between intestinal enterostatin and procolipase originating from the pancreas. Enterostatin has also been identified in lymph, where it was increased fourfold following ingestion of a cream meal (Townsley *et al.*, 1996), suggesting that enterostatin may associate with chylomicrons for delivery to the lymph rather than direct diffusion into the blood. Such a route of absorption would suggest that enterostatin appearing in the circulation after a meal does not act as a satiety signal to reduce intake of that immediate meal.

In humans enterostatin has also been identified in the circulating blood, where it appeared in response to a satiating meal (Boywer *et al.*, 1993). Two peaks of immunoreactive enterostatin were identified; one peak being maximal after 40 min, and a second 80 min after the onset of eating. The reason for the biphasic response is not known, but it may correspond to an early, vagally stimulated peak of pancreatic secretion and a second, late intestinally CCK-mediated peak of pancreatic secretion (Boywer *et al.*, 1993).

Immunoreactive enterostatin (APGPR) was also identified in urine, suggesting that enterostatin indeed survived intact in the circulation for a sufficient time to allow systemic dispersal. A second conclusion of this investigation was that the excretion of enterostatin by the kidney was rapid (Boywer *et al.*, 1993). The possibility that a postprandial rise in enterostatin is an important satiety signal needs further investigations. Bowyer *et al.* (1993) investigated the urine of three morbidly obese and normal-weighted individuals after a satiating meal. The authors found enterostatin to be secreted in the urine in two patients at a significantly lower level than the normal-weight subjects. Enterostatin immunoreactivity could not be detected in serum of the obese individuals but was detectable in the normal-weighted individuals, however, with a large standard deviation. The reason for a large variation may be that the method used was not sensitive enough to detect serum enterostatin, the detection limit being in the nanomolar range (Boywer *et al.*, 1991). Measurement of procolipase activity in intestinal content following stimulation with CCK in normal-weight and in obese patients demonstrated a three- to fourfold reduction in the production of pancreatic procolipase in the obese subjects compared to the normal-weight patients (Erlanson-Albertsson and York, 1997). The reduction was not so low as to impair fat digestion and fat absorption. One explanation of the reduced pancreatic secretion in obese patients is a vagal dysfunction, which will lead to decreased secretion of both gastric and pancreatic juice. It is, however, not known if this is a consequence or a cause of the obesity. These preliminary observations prompt the need for further studies to establish a possible relationship of enterostatin secretion to feeding behavior in normal and in obese subjects.

5. Conclusions

High levels of dietary fat are known to convey health risks. The ability to regulate fat intake is important for the regulation of body weight and energy balance, because fat oxi-

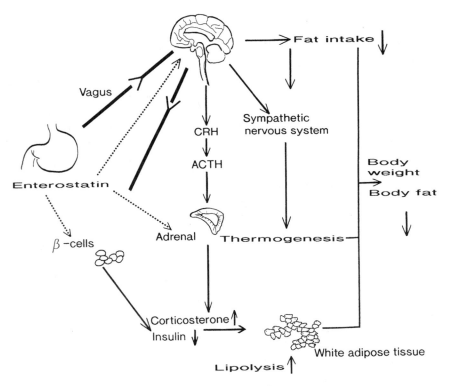

Figure 3. Hypothetical scheme for the mode of action of enterostatin in the regulation of fat intake and body weight. Enterostatin is produced in the gastrointestinal tract. From here enterostatin may interact with vagal afferent nerves to signal satiety in the hypothalamus of the brain through the blocking of opioidergic pathways. From the brain various efferent pathways may be affected, leading to stimulation of sympathetic activity and thermogenesis, inhibition of insulin release, and stimulation of glucocorticoid hormone. It is not known if these signals are also transmitted through the circulating blood (modified after Erlanson-Albertsson and York, 1997).

dation is not closely coupled to fat intake. Understanding the mechanisms that determine fat preference and fat intake may thus provide new approaches toward reducing the consumption of dietary fat. Experimental work with animal models has shown that enterostatin will selectively inhibit fat intake. Both peripheral (gastrointestinal) and central sites of action have been identified as summarized in Fig. 3. Future work will show if the interesting biology of enterostatin is a reflection of a physiological role of this peptide to modulate fat intake and body composition.

Acknowledgments

In parts this work was made possible through grants from the Swedish Medical Research Council, K98-03X-07904-12A.

References

Barton, C., Lin L., York D. A., and Bray G. A., 1995, Differential effects of enterostatin, galanin and opioids on high-fat diet consumption, *Brain Res.* **702**:55–60.

Belfrage, P., 1985, Hormonal control of lipid degradation, in: *New Perspectives in Adipose Tissue* (A. Cryer, and R. L. R. Van, eds), Butterworths, London, pp. 121–144.

Berger, K., Sörhede-Winzell, M., and Erlanson-Albertsson, C., 1998, Binding of enterostatin to the human neuroepithelioma cell line SK-N-MC, *Peptides* **19**:1525–1531.

Blundell, J. E., and Clawton, C. L., 1995, Serotonin and dietary fat intake: Effects of dexfenfluramine, *Metabolism* **44** (Suppl. 2):33–37.

Borgström, B., Wieloch, T., and Erlanson-Albertsson, C., 1979, Evidence for a pancreatic procolipase and its activation by trypsin, *FEBS Lett.* **108**:407–410.

Bouras, M., Huneau, J. F., Luengo, C., Erlanson-Albertsson, C., and Tomé, D., 1995, Metabolism of enterostatin in rat intestine, brain membranes and serum: Differential involvement of proline-specific peptidases, *Peptides* **16**:399–405.

Bowyer, R. C., Jehanli, A. M. T., Patel, G., and Hermon-Taylor, J., 1991, Development of enzyme-linked immunosorbent assay for free human procolipase activation peptide (APGPR), *Clin. Chim. Acta* **200**:137–152.

Bowyer, R. C., Rowston, W. M., Jehanli, A. M. T., Lacey, J. H., and Hermon-Taylor, J., 1993, The effect of a satiating meal on the concentrations of procolipase activation peptide in the serum and urine of normal and morbidly obese individuals, *Gut* **34**:1520–1525.

Bray, G. A., 1992, Peptides affect the intake of specific nutrients and the sympathetic nervous system, *Am. J. Clin. Nutr.* **55**:2655–271S.

Bray, G. A., Fisler, J. S., and York, D. A.,1990, Neuroendocrine control of the development of obesity: Understanding gained from studies of experimental models of obesity, *Prog. Neuroendocrinol.* **4**:128–181.

Briendley, D. N., 1995, Role of glucocorticoids and fatty acids in the impairement of lipid metabolism observed in the metabolic syndrome, *Int. J. Obes.* **19** (Suppl. 1): S69–S75.

Cook, C., Gatchair-Rose, A., Herminghuysen, D, Nair, R., Prasad, A., Mizuma, H., and Prasad, C., 1994, Individual differences in the macronutrient profile of outbred rats: implications for nutritional, metabolic and pharmacological studies, *Life Sci.* **55**:1463–1470.

Dreon, D. M., Frey-Hewitt, B., Ellsworth, N., Williams, P. T., Terry, R. B., and Wood, P. D., 1988, Dietary fat: Carbohydrate ratio and obesity in middle aged men, *Am. J. Clin. Nutr.* **47**:995–1000.

Duan, R., and Erlanson-Albertsson, C., 1989, Pancreatic lipase and colipase activity increase in pancreatic acinar tissue of diabetic rats, *Pancreas* **4**:329–334.

Duan, R., and Erlanson-Albertsson, C., 1990, The anticoordinate changes of pancreatic lipase and colipase activity to amylase activity by adrenalectomy in normal and diabetic rats, *Int. J. Pancreatol.* **6**:271–279.

Duan, R., and Erlanson-Albertsson, C., 1992, Gastric inhibitory polypeptide stimulates pancreatic lipase and colipase synthesis in rats, *Amer. J. Physiol.* **262** :G779-G784.

Duan, R., Wicker, C., and Erlanson-Albertsson, C., 1991, Effect of insulin administration on contents, secretion and synthesis of pancreatic lipase and colipase in rats, *Pancreas* **6**:595–602.

Erlanson-Albertsson, C., 1992, Pancreatic colipase. Structural and physiological aspects. *Biochim. Biophys. Acta* **1125**:1–7.

Erlanson-Albertsson, C., 1994, Pancreatic lipase, colipase and enterostatin—A lipolytic triad, in: *Esterases, Lipases and Phospholipases: From Structure to Clinical Significance* (M. I. Mackness and M. Clerc, eds.), Plenum Press, New York, pp. 159–168.

Erlanson-Albertsson, C., and Larsson, A., 1988, A possible physiological function of procolipase activation peptide in appetite regulation, *Biochimie* **70**:1245–1250.

Erlanson-Albertsson, C., and York, D. A., 1997, Enterostatin—A peptide regulating fat intake, *Obes. Res.* **5**:360–372.

Erlanson-Albertsson, C., Mei, J., Okada, S., York, D., and Bray, G. A., 1991a, Pancreatic procolipase propeptide, enterostatin, specifically inhibits fat intake, *Physiol. Behav.* **49**:1191–1194.

Erlanson-Albertsson, C., Weström, B., Pierzynowski, S., Karlsson, S., and Ahrén B., 1991b, Pancreatic procolipase activation peptide—enterostatin—inhibits pancreatic enzyme secretion in the pig, *Pancreas* **6**:619–624.

Erlanson-Albertsson, C., Hering, B. J., Bretzler, B. G., and Federlin, K., 1994, Enterostatin inhibits insulin secretion from isolated perifused rat islets, *Acta Diabetol.* **31**:160–164.

Fisler, J. S., and Bray, G. A., 1995, Dietary obesity: Effects of drugs on food intake in S5B/PL and Osborne–Mendel rats, *Physiol. Behav.* **34**:225–231.

Gabert, V. M., Jensen, M. S., Jörgensen, H:, Engberg, R. M., and Jensen, S. K., 1996, Exocrine pancreatic secretion in growing pigs fed diets containing fish oil, rapeseed oil or coconut oil, *J. Nutr.* **126**:2076–2082.

Hermoso, J., Pignol, D., Penel, S., Roth, M., Chapus, C., and Fontecilla-Camps, J. C., 1997, Neutron crystallographic evidence of lipase–colipase complex activation by a micelle, *EMBO J.* **16**:5531–5536.

Huneau, J. F., Erlanson-Albertsson, C., Beauvallet, C., and Tomé, D., 1994, The *in vitro* absorption of enterostatin is limited by brush border membrane peptidases, *Regul. Pept.* **54**:495–503.

Kiela, P., Pierzynowski, S. G., Oldak, T., Ceregrzyn, M., Erlanson-Albertsson, C., Wiechetek, M., and Garwacki, S., 1994, The effect of enterostatin on pig duodenal, jejunal and ileal motility *in vitro, Biomed. Res.* **15** (Suppl. 2):303–307.

Leibowitz, S. F., 1994, Specificity of hypothalamic peptides in the control of behavioural and physiological processes, *Ann. N. Y Acad. Sci.* **739**:12–35.

Levine, A. S., Morley, J. E., Gosnell, B. A., Billington, C. J., and Bartness, T. J., 1985, Opioids and consummatory behaviour, *Brain Res. Bull.* **14**:663–672.

Lin, L., and York, D. A., 1997, Comparisons of the effects of enterostatin on food intake and gastric emptying in rats, *Brain Res.* **745**:205–209.

Lin, L., Gehlert, D. R., York, D. A., and Bray, G. A., 1993a, Effect of enterostatin on the feeding responses to galanin and NPY, *Obes. Res.* **1**:186–192.

Lin, L., McClanahan, S., York, D. A., and Bray, G. A., 1993b, The peptide enterostatin may produce early satiety, *Physiol. Behav.* **53**:789–794.

Lin, L., Chen, J., and York, D. A., 1997, Chronic ICV enterostatin preferentially reduced fat intake and lowered body weight, *Peptides* **18**:651–655.

Mei, J., and Erlanson-Albertsson, C., 1992, Effect of enterostatin given intravenously and intracerebroventricularly on high-fat feeding in rats, *Regul. Pept.* **41**:209–218.

Mei, J., and Erlanson-Albertsson, C., 1996a, Plasma insulin response to enterostatin and effect of adrenalectomy in rat, *Obes. Res.* **4**:513–519.

Mei, J., and Erlanson-Albertsson, C., 1996b, Role of intraduodenally administered enterostatin in rat: Inhibition of food intake, *Obes. Res.* **4**:161–165.

Mei, J., Cheng, Y., and Erlanson-Albertsson, C., 1993a, Enterostatin—Its ability to inhibit insulin secretion and to decrease high-fat food intake, *Int. J. Obes.* **17**:701–704.

Mei, J., Bowyer, R. C., Jehanli, A.M.T., Patel, G., and Erlanson-Albertsson, C., 1993b, Identification of enterostatin, the pancreatic procolipase activation peptide, in the intestine of rat: Effect of CCK-8 and high-fat feeding, *Pancreas* **8**:488–493.

Mei, J., Bouras, M., and Erlanson-Albertsson, C., 1997, Inhibition of insulin release by intraduodenally infused enterostatin-VPDPR in rats, *Peptides* **18**:651–655.

Mizuma, H., Abadie, J., and Prasad, C., 1994, Corticosterone facilitation of inhibition of fat intake by enterostatin (Val-Pro-Asp-Pro-Arg), *Peptides* **15**:447–452.

Nagase, H., Bray, G. A., and York, D. A., 1997, Effect of galanin and enterostatin on sympathetic nerve activity to intrascapular brown adipose tissue, *Brain Res.* **709**:44–50.

Okada, S., York, D. A., Bray, G. A., and Erlanson-Albertsson, C., 1991, Enterostatin (Val-Pro-Asp-Pro-Arg) the activation peptide of procolipase selectively reduces fat intake, *Physiol. Behav.* **49**:1185–1189.

Okada, S., York, D. A., Bray, G. A., Mei, J., and Erlanson-Albertsson, C., 1992, Differential inhibition of fat intake in two strains of rat by the peptide enterostatin, *Amer. J. Physiol.* **262**:R1111–R1116.

Okada, S., Onai, T., Kilroy, G., York, D. A., and Bray, G. A., 1993a, Adrenalectomy of the obese Zucker rat: Effects on the feeding response to enterostatin and specific mRNA levels, *Amer. J. Physiol.* **265**:R21–R27.

Okada, S., Lin, L., York, D. A., and Bray, G. A., 1993b, Chronic effects of intracerebral ventricular enterostatin in Osborne–Mendel rats fed a high-fat diet, *Physiol. Behav.* **54**:325–330.

Okada, S., York, D. A., and Bray, G. A., 1993c, Procolipase mRNA: Tissue localization and effects of diet and adrenalectomy, *Biochem. J.* **292**:787–789.

Ookuma, K., Barton, C., York, D. A., and Bray, G. A., 1997, Effect of enterostatin and kappa-opioids on macronutrient selection and consumption, *Peptides* **18**:785–791.

Pan, D. H., Hulbert, A. J., and Storlien, L. H., 1994, Dietary fats, membrane phospholipids and obesity, *J. Nutr.* **124**:1555–1565.

Paz, K., Voliovitch, H., Hadari, Y. R., Roberts, C. T., LeRoith, D., and Zick, Y., 1996, Interaction beween the insulin receptor and its downstream effectors. Use of individually expressed receptor domains for structure/function analysis, *J. Biol. Chem.* **271**:6998–7003.

Pierzynowski, S. G., Erlanson-Albertsson, C., Podgurniak, P., Kiela, P., and Weström B., 1994, Possible integration of the electrical activity of the duodenum and pancreas secretion through enterostatin, *Biomed. Res.* **15** (Suppl. 2):257–260.

Rippe, C., and Erlanson-Albertsson, C., 1998, Identification of enterostatin and the relationship between lipase and colipase in various species, *Nutritional Neuroscience* **1**:111–117.

Schmitz Peiffer, C., Browne, C. L., Oakes, N. D., Watkinson, A., Chisholm, D. J., Kraegen, E. W., and Biden, T. J., 1997, Alterations in the expression and cellular localization of protein kinase C isozymes ε and ϕ are assocoated with insulin resistance in skeletal muscle of the high-fat fed rat, *Diabetes* **46**:169–178.

Shafrir, E., and Gutman, A., 1993, *Psammomys obesus* of the Jerusalem colony: A model for nutritionally induced, non-insulin dependent diabetes, *J. Basic Clin. Physiol. Pharmacol.* **4**:83–99.

Shargill, N. S., Tsujii, S., Bray, G. A., and Erlanson-Albertsson, C., 1991, Enterostatin suppresses food intake following injection into the third ventricle of rats, *Brain Res.* **544**:137–140.

Shor-Posner, G., Brennan, G., Ian, C., Jasaitis, R., Madhu, K., and Leibowitz, S. F., 1994, Meal patterns of macronutrient intake in rats with particular dietary preferences, *Am. J. Physiol.* **266**:R1395–R1402.

Silvestre, R. A., Rodrigeux-Gallardo, J., and Marco, J., 1996, Effect of enterostatin on insulin, glucagon and somatostatin secretion in the perfused rat pancreas, *Diabetes* **45**:1157–1160.

Smith, B. K., York, D. A., and Bray, G. A., 1997, Chronic d-fenfluramine treatment reduces fat intake and increases carbohydrate intake in rats, *Pharmacol. Biochem. Behav.* **18**:207–211.

Sörhede, M., Mei, J., and Erlanson-Albertsson, C., 1993, Enterostatin—a gut–brain peptide—regulating fat intake in rat, *J. Physiol. (Paris)* **87**:273–275.

Sörhede, M., Erlanson-Albertsson, C., Mei, J., Nevalainen, T., Aho, A., and Sundler, F., 1996a, Enterostatin in gut endocrine cells—Immunocytochemical evidence, *Peptides* **17**:609–614.

Sörhede, M., Mulder, H., Mei, J., Sundler, F., and Erlanson-Albertsson, C., 1996b, Procolipase is produced in the rat stomach—a novel source of enterostatin. *Biochim Biophys Acta.* **1301**:207–212.

Sörhede-Winzell, M., 1998, Enterostatin in the gastrointestinal tract: Production and possible mechanism of action. Ph.D. diss., University of Lund.

Stanley, B. G., Daniel, D. R., Chin, A. S., and Leibowitz, S. F., 1985, Paraventricular nucleus injection of peptide YY and neuropeptide Y preferentially enhance carbohydrate ingestion, *Peptides* **6**:1205–1211.

Steffens, A. B., Strubbe, J. H., Balkan, B., and Scheurink, A. J., 1991, Neuroendocrine factors regulating blood glucose, plasma FFA and insulin in the development of obesity, *Brain Res. Bull.* **27**:505–510.

Tempel, D. L., Leibowitz, K. J., and Leibowitz, S. F., 1988, Effects of PVN galanin on macronutrient selection, *Peptides* **9**:309–314.

Thomas, C. D., Peters, J. C., Reed, G. W., Abumrad, N. N., Sun, M., and Hill, J. O., 1992, Nutrient balance and energy expernditure during *ad libitum* feeding of high-fat and high-carbohydrate diets in humans, *Am. J. Clin. Nutr.* **55**:934–942.

Tian, Q., Nagase, H., York, D. A., and Bray, G. A., 1994, Vagal–central nervous system interactions modulate the feeding response to peripheral enterostatin, *Obes. Res.* **2**:527–534.

Townsley, M. I., Erlanson-Albertsson, C., Ohlsson, A., Rippe, C., and Reed, R. K., 1996, Enterostatin efflux in cat intestinal lymph: Relation to lymph flow, hyaluronan and fat absorption, *Am. J. Physiol.* **271**:G714–G721.

Van Amelsvoort, J. M., Van Stratum, P., Kraal, J. H., Lussenberg, R. N., and Houtsmuller, V. M. T., 1988, Effects of varying the carbohydrate: fat ratio in a hot lunch on postprandial variables in male volunteers, *Br. J Nutr.* **61**:267–283.

Vanderweele, D. A., and Macrum, B. L., 1986, Glucagon, satiety from feeding and liver pancreatic interactions, *Brain Res. Bull.* **17**:539–543.

Wicker, C., and Puigserver, A., 1987, Effects of inverse changes in dietary lipid and carbohydrate on the synthesis of some pancreatic secretory proteins, *Eur. J. Biochem.* **162**:25–30.

Wurtman, J., Wurtman, R., Berry, E., Gleason, R., Goldberg, H., McDermott, J., Kahne, M., and Tsay, R., 1993, Dexfenfluramine, fluoxetine, and weight loss among female carbohydrate cravers, *Neuropsychopharmacol.* **9**:201–210.

York, D. A., 1992, Central regulation of appetite and autonomic activity by CRH, glucocorticoids and stress, *Prog. Neuroendoc. Immunol.* **5**:153–165.

CHAPTER 7

Cholesterol Esterase

PHILIP N. HOWLES and DAVID Y. HUI

1. Introduction

The name pancreatic cholesterol esterase is ascribed to the only enzyme in the pancreas that hydrolyzes cholesterol esters to unesterified cholesterol and free fatty acids. However, extensive investigations over a period of more than 30 years revealed that a protein with similar properties can also be purified from homogenates of several other tissues and body fluids and that the enzyme is a nonspecific lipase capable of hydrolyzing cholesteryl esters, vitamin esters, triacylglycerol, phospholipids, and lysophospholipids. At the onset of these investigations, it was not clear whether these various enzyme activities were properties of the same protein. Thus, this enzyme was also named nonspecific lipase, phospholipase A_1, lysophospholipase, bile-salt-stimulated lipase, bile-salt-dependent lipase, carboxyl ester lipase, and carboxyl ester hydrolase. (Please see Rudd and Brockman, 1984; Wang and Hartsuck, 1993, for a historic perspective of this protein.) It required the cloning of cDNA based on different enzyme activities from different tissues and different species to conclusively demonstrate that the same gene product has several distinct but related enzymatic activities (Han *et al.*, 1987; Kissel *et al.*, 1989; Kyger *et al.*, 1989; Hui and Kissel, 1990; Reue *et al.*, 1991; Colwell *et al.*, 1993; Mackay and Lawn, 1995). Sequence comparison with other proteins also revealed that this enzyme is responsible for the lipoamidase activity in milk (Hui *et al.*, 1993), which may account for its ability to hydrolyze the physiological lipoamide substrate ceramide (Nyberg *et al.*, 1998). Due to the various substrates for this enzyme, the best nomenclature for its name has been widely debated over the years. The more commonly used names include carboxyl ester lipase (CEL), which is based on its general reactivity with lipids containing carboxyl ester bonds; cholesterol esterase or cholesterol ester lipase, which is based on its documented physiological function as a cholesteryl ester hydrolase; and bile-salt-stimulated lipase (BSSL) or bile salt-dependent lipase (BSDL), which is based on the unique bile-salt-dependency of this enzyme. Because CEL can be used to refer to the enzymatic properties and physiological function of this protein, we have adopted this terminology throughout this chapter.

PHILIP N. HOWLES and DAVID Y. HUI • Department of Pathology and Laboratory Medicine, University of Cincinnati College of Medicine, Cincinnati, Ohio 45267–0529.

Intestinal Lipid Metabolism, edited by Charles M. Mansbach II *et al.,* Kluwer Academic/Plenum Publishers, 2001.

The major tissues for CEL protein biosynthesis are the acinar cells of exocrine pancreas and lactating mammary glands. The enzyme synthesized by the pancreas is stored in zymogen granules and is secreted with pancreatic juice in a process stimulated by gastric hormones such as cholecystokinin, secretin, and bombesin (Huang and Hui, 1991, 1994). In the lumen of the digestive tract, CEL is mixed with bile salt, becoming an active enzyme where it may catalyze nutrient digestion and absorption through the gastrointestinal tract (Rudd and Brockman, 1984). The CEL protein synthesized by lactating mammary glands is secreted as a major constituent of milk proteins (Hernell and Olivecrona, 1974; Wang and Johnson, 1983), thereby reaching the digestive tract of infants where it is also postulated to play a role in nutrient digestion and absorption (Hernell and Olivecrona, 1974). Low but significant levels of CEL are also synthesized in other tissues, such as the liver (Zolfaghari *et al.,* 1989; Camulli *et al.,* 1989), eosinophils (Holtsberg *et al.,* 1995), macrophages (Li and Hui, 1997), and endothelial cells (Li and Hui, 1998). The physiological function of the CEL synthesized outside the digestive tract is unknown. Preliminary data have suggested its possible involvement in remodeling of plasma lipoproteins (Brodt-Eppley *et al.,* 1995; Shamir *et al.,* 1996) and modulation of atherosclerosis in the vessel wall (Shamir *et al.,* 1996; Li and Hui, 1997, 1998). Because the focus of this book is on intestinal lipid metabolism, this chapter focuses on the structure–function relationship of CEL and its physiological role in lipid metabolism and transport in the digestive tract. The potential role of CEL in the circulation and in the vessel wall is discussed briefly where appropriate.

2. Protein Structure–Function Relationship

The primary structure of CEL, as deduced from nucleotide sequencing of its cDNA from various species, indicates that this enzyme is highly conserved and is a member of the α/β-hydrolase family of enzymes. As is the case with other serine esterases in this family, CEL utilizes a catalytic triad of Ser-His-Asp/Glu to form the charge relay network required for substrate hydrolysis. Site-specific mutagenesis experiments documented that Ser^{194} is the key residue in CEL responsible for nucleophilic attack on the substrate carboxyl ester bond (DiPersio *et al.,* 1990). The reaction is assisted by His^{435} (DiPersio *et al.,* 1991) through a general acid–base catalysis reaction on the substrate carbonyl, eventually leading to an acylenzyme intermediate (Stout *et al.,*1985; Feaster *et al.,* 1997). Mutagenesis studies also revealed the participation of Asp^{320} in the catalytic triad (DiPersio and Hui, 1993), which probably serves by providing a better acid–base reaction through modulation of the pK_a (Feaster *et al.,* 1997).

The CEL enzyme is active alone in hydrolyzing carboxyl esters containing short-chain fatty acids, but it requires bile salt activation for the hydrolysis of carboxyl esters with long-chain fatty acyl groups (Rudd and Brockman, 1984; Wang and Hartsuck, 1993; Gjellesvik, 1991). The mechanism for bile-salt activation of CEL has been the subject of numerous studies. The consensus finding is that bile salt interacts with two sites on the protein, each producing a different effect (Blackberg and Hernell, 1993). One site is termed the nonspecific site, which is capable of binding both di- and trihydroxylated bile salts (Blackberg and Hernell, 1993). At this site, the negatively charged side chain of the bile salt interacts with one or more arginine residues in the enzyme. This interaction protects CEL from proteolysis and also promotes the binding of CEL to the surface of lipid emulsions prior to its hydrolysis of emulsified substrates (Blackberg and Hernell, 1993). Bile-salt binding to this nonspecific site has no effect on CEL hydrolysis of water-soluble substrates. In contrast, the second bile salt binding site on CEL is specific for trihydroxylated bile salts. The binding of bile salts such as cholate and taurocholate to this site induces a conformational change of the enzyme (Jacobson *et al.*, 1990) and increases hydrolytic activity against both water-soluble and lipid-emulsified substrates (Blackberg and Hernell, 1993; Jacobson *et al.*, 1990).

The size of the CEL protein from various species differs primarily due to the number of proline-rich repeating sequences near the carboxyl terminus of the protein. The largest CEL protein is the human enzyme at 100 kDa, which contains 16 repeating units with the consensus sequence of PVPPTDDSQ. The rat, mouse, bovine, and rabbit enzymes are smaller proteins at approximately 74 kDa and contain four, three, and two such repeating units, respectively. These proline-rich repeating units have been shown to be the site of *O*-glycosylation, which is apparently important for maintaining the stability of the protein (Bruneau *et al.*, 1997). In addition to differences in the number of proline-rich repeating units, the carboxyl terminus of CEL also contains a domain that is required for normal intracellular processing and secretion of the protein. Recent data from Lombardo's laboratory suggested that this domain is responsible for CEL interaction with a membrane-folding complex containing a 94 kDa protein that is similar to the glucose-regulated protein 94 (Grp94 Bruneau and Lombardo, 1995). This interaction is apparently important for proper sorting of CEL from the endoplasmic reticulum to the *trans*-Golgi prior to its secretion from cells (Bruneau *et al.*, 1997). The interaction between CEL and the Grp94-containing membrane-folding complex does not require the presence of the proline-rich repeats, because deletion of all the repeating units resulted in a stable protein that could be secreted from transfected cells (DiPersio *et al.*, 1994). The truncated CEL without any proline-rich repeating units or the C-terminal domain also retains hydrolytic activities against both water-soluble substrates and lipid substrates, suggesting that the repeating units do not directly participate in the catalytic activity of the protein (DiPersio *et al.*, 1994). Interestingly, mutagenized enzymes without the C-terminal domain, including the deletion of all proline-rich repeating units, were more active than the native enzyme in substrate hydrolysis at low bile salt concentrations (DiPersio *et al.*, 1994). These data suggest that the C-terminal domain and the proline-rich repeating units are important in regulating substrate accessibility to the active site of the protein (DiPersio *et al.*, 1994).

The recent reports from two different laboratories of the X ray crystal structure of bovine CEL provided additional support for the importance of the C-terminal domain in regulating substrate delivery to the active site domain of CEL (X. Wang *et al.*, 1997; Chen *et al.*, 1998). Wang and colleagues crystallized a truncated version of bovine CEL without the C-terminal repeats and showed a structure with 13 β-strands and 14 α-helices at .28 nm resolution (X. Wang *et al.*, 1997). In contrast, Chen *et al.* (1998) reported a structure with 11 β-strands and 15 α-helices at .16 nm resolution using a mutagenized full length protein without *N*-linked glycosylation. Both structures showed the central location of the active site triad similar to that observed in other lipases and esterases. Based on homology with acetylcholinesterase, Ala^{195}, Gly^{107}, and Ala^{108} are predicted to be involved with coordinating the oxyanion intermediate in the reaction (Chen *et al.*, 1998). This prediction is consistent with predictions made from molecular modeling of the human and rat CEL (Feaster *et al.*, 1997). Both crystal structures as well as predictions made from molecular modeling of the protein indicate that CEL lacks the amphipathic helical lid domain of other lipases that is important for interfacial activation. In contrast, CEL contains a truncated lid with a pair of antiparallel β-strands, overlapping the N-terminal disulfide loop at residues 64–80 (Chen *et al.*, 1998). Interestingly, crystal structure of the full-length CEL showed that the carboxyl terminus of the protein forces this oxyanion hole away from the active site domain (Chen *et al.*, 1998). Thus, movement of the C-terminal peptide domain away from the hydrophobic pocket is necessary in order for the substrate to access the active site domain of the protein. This can be achieved by bile-salt binding to the proximal site near the catalytic triad of the protein, which leads to increased ordering of residues 116–125 (Wang *et al.*, 1997; Chen *et al.*, 1998) and swinging the binding loop away from the active site to allow substrate access (Chen *et al.*, 1998). Bile-salt binding to the distal site was also shown to be involved with opening the substrate-binding pocket to accommodate the bulky cholesteryl ester moiety (Wang *et al.*, 1997; Chen *et al.*, 1998).

3. Cholesterol Absorption

The physiological role of CEL, especially in adult pancreatic secretion, has been debated since the 1950s. This debate has centered around the role of this enzyme in the absorption and esterification of dietary cholesterol. Hydrolysis of cholesteryl esters by this enzyme has never been questioned but, because cholesteryl esters represent only about 10% of dietary cholesterol, this function is not likely to play a major role in overall cholesterol absorption. However, the ability of CEL to re-esterify cholesterol *in vitro* (Gallo, 1981), and its localization within intestinal epithelial cells in immunohistochemistry studies (Gallo *et al.*, 1980; Lechene de la Porte *et al.*, 1987), led to the hypothesis that CEL may serve as either a docking or a carrier protein for free cholesterol uptake by enterocytes. It has been postulated that CEL mediates the re-esterification of cholesterol after it traverses through the membrane bilayer and this step is required prior to cholesterol secretion into the lymph as part of chylomicrons and very low–density lipoproteins (VLDL) (Bhat and Brockman, 1982).

In early studies, such as that described by Borja *et al.* (1964), cholesterol absorption and esterification were measured in rats with a lymph fistula and diverted pancreatic and/ or biliary secretions. Free cholesterol administered as oleic acid/bile salt/cholesterol micelles was absorbed only 20% to 50% as efficiently in the absence of pancreatic secretions as compared to controls. This reduced absorption correlated roughly with reduced levels of CEL activity in homogenates of mucosal scrapings. The absorbed cholesterol was esterified with equal efficiency in control and experimental animals. These results suggested that CEL plays a role in the absorption but not the esterification of dietary cholesterol. Using a similar animal model, Watt and Simmonds (1981) generated data that led them to quite different conclusions. They found similar rates and extents of secretion of total cholesterol and cholesteryl ester into mesenteric lymph with or without pancreatic diversion, arguing against any role for CEL in these processes. In a later study, however, Gallo *et al.* (1984) refined the pancreatic diversion model by using a CEL-specific antibody to deplete the enzyme from donor pancreatic juice and then measuring cholesterol absorption and esterification in pancreatobiliary-diverted lymph fistula rats supplemented with pancreatic juice with or without CEL. They found an 80% reduction in cholesterol absorption in the absence of CEL, and thus argued that this enzyme is critical for cholesterol absorption.

Various inhibitors that are specific for either CEL or acyl-CoA:cholesterol acyltransferase (ACAT) have also been used in animal model systems to determine the relative role of these two enzymes in cholesterol absorption. Using the Sandoz 58–035 ACAT inhibitor, Bennett Clark and Tercyak (1984) reported a 50% to 60% reduction in lymphatic output of an infused dose of cholesterol. Furthermore, the reduction was entirely constituted by decreased cholesteryl ester output. In these experiments, there was no diversion of pancreatic or biliary secretions. The investigators concluded that ACAT and not CEL was the rate-limiting enzyme for cholesterol absorption and esterification. A similar result was reported for rabbits fed a related compound, but only when they were fed a high-cholesterol diet (Heider *et al.,* 1983). This latter study suggested that ACAT may be significant only when the cholesterol load is high. Other studies utilizing the same inhibitors, however, reported opposite results and showed that even when ACAT was inhibited up to 90%, there was no difference in cholesterol absorption in control versus treated animals (Gallo *et al.,* 1987).

Cholesterol esterase inhibitors have also yielded conflicting results. Tetrahydrolipstatin (THL) is a lipase inhibitor derived from a natural product of *Streptomyces toxytricini,* which is a potent inhibitor of CEL, pancreatic lipase, and gastric lipase. Thus, when administered to rats as part of an emulsion comprised of triglyceride, phospholipid, cholesterol, and bile acids, both triglyceride and cholesterol absorption were significantly reduced (Fernandez and Borgström, 1989). Under these conditions, the cholesterol remained associated with the triglyceride of the undigested emulsion particles. However, when THL was administered as part of a mixed micelle solution of oleic acid, lysophospholipid, monoolein, cholesterol, and bile salts, cholesterol absorption was unchanged by the presence of the inhibitor (Fernandez and Borgström, 1989). Inhibition of the CEL was verified by the finding that cholesteryl oleate absorption was essentially abolished under these conditions. Other investigators have used phenoxyphenyl carbamate inhibitors of CEL (WAY 121,751 and 121,898) and reported reduced cholesterol absorption while retaining full ACAT activity in rats and in dogs (McKean *et al.,* 1992). In a large study that compared the efficacy of a CEL inhibitor (WAY 121,898) and an ACAT inhibitor (CI-976) for reducing serum cholesterol in rats, rabbits, and hamsters, Krause *et al.* (1998) found that, under dietary conditions of increased cholesterol load, both these inhibitors could have an effect and that when administered together, the two inhibitors were synergistic. The specificity of these inhibitors was questioned and the possibility of species-specific effects was also suggested by the data.

In vitro experimental systems have also been used to assess the role of CEL in free cholesterol absorption. The earliest studies utilized isolated intestinal epithelial cells and reported modest stimulation of cholesterol uptake but significant increase in esterification in the presence of exogenous CEL (Gallo *et al.,* 1977). In a different study, cholesterol absorption and esterification was followed in everted intestinal sacs prepared from rats (Bhat and Brockman, 1982). In this study, cholesterol uptake was increased 40% to 50% in the presence of CEL and esterification was increased threefold. Later experiments were performed with the colon carcinoma cell line Caco-2, which can be induced to differentiate into a monolayer resembling the absorptive epithelium of the small intestine in morphology and biochemistry. In an initial study with this system, Huang and Hui (1990) demonstrated that CEL had no effect on free cholesterol uptake by these cells, whereas cholesteryl ester hydrolysis and absorption was stimulated approximately eightfold. The concentration of bile salt used in these experiments was low, however, and when similar studies were performed by another group, CEL was found to stimulate the uptake of free cholesterol in a manner that was dependent on both bile-salt concentration and CEL concentration (Lopez-Candales *et al.,* 1993). The latter study was criticized for having a low concentration of cholesterol in the vesicular substrate. However, a fairly definitive set of experiments was later reported by Shamir *et al.* (1995) in which cholesterol concentration was varied 500-fold while the bile salt concentration was kept constant at 6 mM, near its critical micellar concentration. In addition, the cholesterol was presented as micelles containing either phospholipid or monoolein along with the bile salt and cholesterol. In these experiments, the Caco-2 cells were transfected with a CEL gene and secreted the protein into the medium at high concentrations. Addition of exogenous enzyme was also tested. Under none of the several conditions tested were these investigators able to demonstrate any role for CEL in the uptake or the esterification of free cholesterol in the medium. These studies supported the concept that CEL was not involved with free cholesterol uptake by intestinal cells. However, a shortcoming of these studies with Caco-2 cells is that none of the experiments were performed with cells grown on semi-permeable membrane supports. Under such conditions, the cells not only differentiate but also secrete lipoproteins into the underlying medium. Whether CEL plays a role in cholesterol flux through the enterocytes remains unclear.

It is difficult to directly compare the results reported in these various *in vitro* and *in vivo* experimental systems. Differences in the substrate preparation—emulsions versus micelles, composition of the micelle—cholesterol load, method of assay, and so on may have substantial effects on the outcome. It seemed plausible that under certain conditions, CEL may in fact play a role in the absorption of dietary cholesterol. The conflicting results and experimental inconsistencies in these previously described reports led us to take a different approach for determining the physiological role of CEL. We used gene targeting in embryonic stem cells to generate lines of mice in which the CEL gene was no longer functional (Howles *et al.,*1996b). Pancreatic extracts from these knockout mice were assayed by both Western blot and enzymatic assays to demonstrate that no CEL or CEL-derived peptide fragments were being synthesized. We then measured cholesterol absorption in homozygous knockout animals as compared to wild-type and heterozygous littermates. Absorption was measured by gavage feeding a test meal consisting of an emulsion of triglyceride, phospholipid, and free cholesterol. Also included were radiolabeled [^{3}H]-cholesterol and [^{14}C]-sitosterol. Sitosterol is a plant sterol that is minimally absorbed through the gastrointestinal tract (Borgström, 1968) and thus served as a marker for nonabsorbable sterol in this study. Fecal material was collected for 24 hr, homogenized, and the sterols were extracted with chloroform and methanol (2:1). The fraction of the cholesterol that was absorbed was calculated by comparing the ratio of the radiolabels in the starting emulsion to their ratio in the fecal extract using the formula

$$\% \text{ absorbed} = 100 \times [1 - (^3H_{fecal}/^{14}C_{fecal})/(^3H_{initial}/^{14}C_{initial})]$$

The results of this study demonstrated that free cholesterol was absorbed equally well in the presence or absence of CEL (Howles *et al.,* 1996b). Different emulsion and phospholipid vesicle formulations were tested but in no case was a significant difference in absorption observed between control and CEL knockout mice. We also tested whether or not a high cholesterol load would reveal an effect of the absence of CEL. Animals were fed a diet high in fat and cholesterol (1.25%) for 6 weeks and then dosed with the test meal. Although the high-cholesterol load decreased the percent of cholesterol absorbed from the test meal, there was no difference in cholesterol absorption between control and CEL knockout mice. Because some of the earlier studies had suggested that CEL affected the rate of cholesterol absorption, the appearance of the radiolabeled cholesterol in the serum of the gastrically infused animals was measured. In this assay also, no difference was observed between cholesterol and CEL knockout mice. Absorption of esterified cholesterol, on the other hand, was reduced approximately 60% in the absence of CEL (Howles *et al.,* 1996b). Because the majority of dietary cholesterol and all of the biliary cholesterol exist in the unesterified form, this finding does not point to a major physiological role for CEL in mediating cholesterol absorption through the gastrointestinal tract.

Although the data generated with the CEL knockout mice are quite compelling, there are several points worth noting before dismissing the observations of several investigators since 1970. First, the studies done with the CEL knockout mice used quite different assays than those performed with the lymph fistula rats. We defined cholesterol absorption as that portion of a gastric bolus dose that was not excreted within 24 hr whereas the studies with lymph fistula rats defined absorption as the cholesterol that was transported from the lumen to the mesenteric lymph as chylomicrons and VLDL particles. Additionally, the lymph studies varied in the time of collection, and cholesterol was administered either as a bolus dose or as a continuous infusion. The mass of cholesterol delivered also varied greatly from study to study. Our studies with the CEL knockout mice used about one tenth of the dose (in mg/kg body weight) than previous studies. It is possible that there are subtle difference between normal and CEL knockout mice that would be revealed only by lymph fistula techniques. For example, a potential difference is that the lymph fistula technique measures the rate of cholesterol absorption whereas the gastric infusion technique measures the total amount of cholesterol absorbed from a single dose over a 24 hr period. Whether CEL plays a role in determining the rate of cholesterol absorption has not been resolved. The use of the lymph fistula technique to study cholesterol absorption in the mouse is technically difficult. Recently, Tso and his colleagues have reported the successful use of lymph fistulae to study cholesterol absorption in mice (Zheng *et al.,* 1998). The procedure can now be employed with the CEL knockout mice to assess the potential role of this enzyme in mediating the rate of cholesterol transport through the intestine.

A second potential explanation for the different observations is that mice and rats are fundamentally different in how they absorb cholesterol. It is possible that CEL does play a role in this process in rats and in other species but not in mice. Precedence for differences between rats and mice already exists. An intriguing difference with respect to the processes of digestion and absorption is the fact that rats lack a gallbladder and thus have continuous bile flow whereas mice have a gallbladder that stores the liver secretions and releases them into the lumen only when stimulated by secretagogues such as cholecystokinin. Species differences for the expression of the CEL gene has been reported (Zolfaghari *et al.,* 1992). In rabbits, for example, CEL is synthesized by the small intestine as well as the pancreas, but not by the liver. In rats, the gene is never expressed in the intestine but is expressed in the liver. Though the physiological significance of these differences has not yet been explored, it has been speculated that lack of circulating CEL in the rabbit may play a role in the greater tendency of rabbits to accumulate cholesteryl esters as compared to the rat (Zolfaghari *et al.,* 1992) Thus, CEL expression differences in the digestive tract may also suggest potential species differences in the role of this protein in cholesterol absorption.

4. Vitamin Absorption

A natural CEL substrate present in the diet is vitamin A, which is present in the form of long-chain fatty acyl esters of retinol. Absorption of this natural form of vitamin A requires hydrolysis of the ester in much the same way as cholesteryl ester absorption. Because CEL is able to recognize retinyl esters as a substrate, it has been proposed that this enzyme may be important for absorption of dietary vitamin A. Some *in vivo* evidence in support of this hypothesis was first suggested in a report that showed reduced retinyl palmitate absorption when rats were fed tetrahydrolipstatin (Fernandez and Borgström, 1990). When retinyl palmitate was delivered in a micellar form with oleic acid, monoolein, and bile salts, the inhibitor-treated animals transferred 70% less retinyl palmitate to the lymph than the untreated control rats. Delivery of the substrate in micellar form precluded the possibility that the effect was due to poor triglyceride hydrolysis because of pancreatic lipase inhibition. Thus, these data suggested a role for CEL in retinyl palmitate hydrolysis and in vitamin A absorption.

Although pancreatic CEL is clearly capable of mediating vitamin ester hydrolysis in the digestive tract, both rats and humans have another retinyl ester hydrolase (REH) that is synthesized by the intestine and located on the brush border membrane (Rigtrup and Ong, 1992; Rigtrup *et al.,* 1994b). In an elegant set of experiments by Rigtrup and Ong (1992), two retinyl ester hydrolytic activities could be distinguished in rat small intestine. The two activities are distinct in bile acid requirement, in acyl chain preference, and in location along the length of the intestine (Rigtrup and Ong,1992; Rigtrup *et al.,* 1994a). The intrinsic REH, although activated by taurocholate, was six times more active with deoxycholate. On the other hand, as noted earlier, CEL is activated only by trihydroxylated bile salts such as taurocholate. The intrinsic REH is found exclusively in the distal half of the small intestine and displays a substrate preference for long-chain acyl esters of retinol, especially lauryl, myristoyl, and palmitoyl esters. In contrast, CEL is found primarily in the proximal half of the small intestine. It prefers short-chain acyl esters in the absence of bile salt but has no chain length preference in the presence of trihydroxylated bile salts. The relative contribution of CEL and REH toward vitamin ester absorption through the digestive tract remains unclear. However, Herr *et al.* (1993) showed that mucosal expression of CRBP-II and enzymes required for retinol re-esterification, such as lecithin:retinol acyltransferase and retinal reductase, are highest in the proximal small intestine. In addition, Takase *et al.* (1993) showed that serum retinol decreased by approximately 20% after a partial jejunum bypass and suggested that the distal small intestine was unable to adequately deliver retinyl esters from the diet to the lymph. Taken together, these results along with observations showing the highest level of CEL in proximal portions of the intestine suggest that CEL is important for retinyl ester hydrolysis and vitamin absorption under physiological conditions.

Direct evidence to support the hypothesis that CEL is involved with vitamin A absorption has yet to be reported. Although the studies using THL remain suggestive, it is possible that the intrinsic REH is also inhibited by this compound because CEL and REH are functionally related proteins. We have initiated studies with the CEL knockout mice, using the retinyl palmitate load test (Aalto-Setala *et al.*, 1994), to address this issue. Mice were administered with 1.7 mg (3000 IU) of retinyl palmitate in 100 &1 of triolein by stomach gavage. The mice were sacrificed after 2 hr, blood was collected, and the serum levels of retinyl ester were determined by high-performance liquid chromatography (HPLC). Results showed a 30% to 40% lower level of serum retinyl palmitate in the CEL knockout mice as compared to control mice (Howles *et al.*, 1999). High variability was observed in these experiments and the resulting differences did not reach statistical significance in every experiment. This variability may result from problems inherent in the assay procedure or may result from the heterogenous genetic background of these animals (129/sv x Black Swiss). Repeating the experiments with animals of a pure genetic background will allow us to determine if the variation is due to genetic heterogeneity of the animals.

The modest differences in retinyl ester absorption observed between wild-type and $CEL(-/-)$ mice may also be attributed to the presence of the additional REH in the small intestine. It is possible that the intrinsic REH located in the distal intestine provides sufficient hydrolytic activity to compensate for the absence of retinyl ester hydrolysis in the proximal intestine of CEL knockout mice. Hence, hydrolysis and absorption of retinyl esters in $CEL(-/-)$ mice may be delayed, resulting in minor differences in total absorption over a 24 hr period. This study is currently being repeated using the mouse lymph fistula assay to determine if CEL gene deletion alters the rate of vitamin ester transport from intestinal lumen to the lymph.

5. Triglyceride Absorption

In the adult, triglyceride hydrolysis is adequately accomplished by the high concentration of pancreatic lipase. Monoacylglycerol and free fatty acids, the end products of this digestive process, are both efficiently absorbed by the small intestine. Although CEL may enhance the digestion and absorption by hydrolyzing the monoacylglycerol, the physiological implication of this is not clear. In preliminary experiments, we used the dual isotope, fecal extraction assay method described earlier and detected no difference between control and CEL knockout mice in the absorption of triolein. Thus, the overall contribution of CEL to triglyceride hydrolysis and absorption may be minimal. However, this initial study does not rule out its involvement in the absorption of certain long-chain fatty acids. *In vitro* studies by Hernell, Bläckburg, and Nilsson demonstrated that CEL significantly enhanced both the rate and the extent of pancreatic lipase-catalyzed hydrolysis of triglycerides containing long-chain polyunsaturated fatty acids, such as arachidonate, eicosapentaenoate, and docosahexaenoate (Chen *et al.*, 1989, 1990; Hernell *et al.*, 1993; Chen *et al.*, 1994). In the absence of CEL, pancreatic lipase was unable to digest 1,2-diacylglycerol containing these fatty acyl esters. Thus, CEL may play a role in polyunsaturated fatty acid absorption. This hypothesis needs to be tested in *in vivo* experiments with the CEL knockout mice.

6. Phospholipid Digestion

The physiological importance of the phospholipase A_1 and lysophospholipase activities of CEL have not been addressed in the literature. The digestion and absorption of phospholipids is poorly understood. It is likely that this process is mediated by the type 1 phospholipase A_2 secreted from the pancreas and/or the phospholipase B synthesized by the small intestine (Rigtrup *et al.,* 1994b; Gassama-Diagne *et al.,* 1989; Pind and Kuksis, 1991). The products liberated from these enzyme activities, lysophospholipids, and free fatty acids, are efficiently taken up by enterocytes. Thus, it is unlikely that CEL plays any significant role in phospholipid absorption. One exception to this hypothesis is that the cod lacks both pancreatic lipase and the type 1 phospholipase A_2 (Gjellesvik *et al.,* 1992). Thus, CEL is the only lipolytic enzyme in the intestinal lumen of the cod and may function in lipid absorption in this species. Hence, CEL in other species may also serve a backup function to other lipolytic enzymes for transport and uptake of these vital nutrients. The availability of CEL knockout mice along with type 1 phospholipase A_2 knockout mice currently being produced in our laboratory will be useful to test this latter hypothesis.

The phospholipase and lysophospholipase activities of CEL are probably more important for lipid metabolism and cell regulation in the vasculature. Studies from Fisher's laboratory (Shamir *et al.,* 1996) as well as from ours (Li and Hui, 1998) have demonstrated the presence of CEL in the vessel wall, especially in areas with pronounced atherosclerotic lesions. Fisher's group documented the ability of CEL to hydrolyze lysophosphatidylcholine produced as a consequence of low-density lipoprotein oxidation (Shamir *et al.,* 1996). Because many of the atherogenic properties of oxidized LDL can be attributed to the lysophosphatidylcholine moiety in the modified lipoprotein (Kohno *et al.,* 1998), these investigators postulated that CEL in the vasculature is protective against the deleterious effects of oxidized LDL. Our data showing that CEL biosynthesis by macrophages and by endothelial cells can be increased by oxidized LDL also suggested that CEL activity in the vessel wall may be a cellular response to challenge by atherogenic agents such as oxidized LDL (Li and Hui, 1997, 1998). The observation that CEL protects endothelial cells against the cytotoxic effects of lysophosphatidylcholine provided additional support for this hypothesis. However, all these observations were made from *in vitro* experiments. Additional *in vivo* experiments need to be performed to determine the physiological significance of CEL in the vessel wall and the significance of the lysophospholipase activity of this protein.

7. Neonatal Nutrition

A considerable amount of direct and indirect evidence has accumulated to support the hypothesis that milk-derived CEL plays an important role in neonatal nutrition, especially in the digestion and absorption of milk fat (as an energy source) and fat-soluble vitamins. In an early study, Williamson *et al.* (1978) compared the fat absorption and excretion in very low birthweight infants fed either raw, pasteurized, or boiled human milk. Fecal fat output doubled when the infants were fed heat-treated milk and their fat absorption rates were decreased by 30% to 40%. It was also found that the amount of free fatty acids in the fecal fat was significantly greater after feeding the infants with raw milk, indicating that the malabsorption was a reflection of reduced digestion of the milk fat. Furthermore, weight gain was greater during the period the infants were being fed the raw milk. These findings suggested that a heat-labile factor in milk was important for proper digestion and absorption of fat by preterm infants. However, an alternate explanation for the data is that heat treatment may alter the structure of the milk globule so that it is less digestible by either the gastric or the pancreatic lipase. In a later study with human subjects, Alemi *et al.* (1981) studied the effects of supplementing low-birthweight infant formula or a mixture of 40% human milk plus 60% formula. Fecal fat excretion was followed for 3 days. Infants receiving the milk/formula mixture excreted 50% less fat than those receiving formula alone. These results suggest that human milk contains factor(s) important for fat digestion and absorption in infants. One of these factors may be CEL. However, several other factors in human milk may also account for the positive effect on the growth and health of these preterm infants.

The first animal model used to directly test the importance of milk CEL in neonatal fat absorption was reported by Wang *et al.* (1989). These investigators refined the supplementation experiments described previously by adding purified human CEL to standard kitten formula. They measured the differences in weight gain of pair-fed kittens receiving either formula alone or formula supplemented with CEL. Kittens receiving the supplemented formula were observed to gain weight at almost twice the rate as those fed the formula alone. The CEL-supplemented formula supported growth at essentially the same rate as those nursed by their mothers. Because CEL was the only factor added to the supplemented formula, these results provided strong evidence that the milk CEL plays an important role in neonatal fat absorption.

Our experiments with CEL knockout mice have confirmed the role of CEL in triglyceride absorption in neonates. First, our data confirmed an earlier report that noted the presence of CEL in mouse milk (Stromqvist *et al.*, 1996). Our data showed that CEL constitutes 1% to 6% of the whey protein in mouse milk. There is also evidence for variability in enzyme level based either on mouse strain or on the time of lactation (Howles *et al.*, 1996a; Howles *et al.*, 1999). Pups nursed by CEL($-/-$) females, and thus receiving no CEL in milk, have serum triglyceride levels that are at least 30% lower than pups nursed by CEL($+/-$) females, which contain CEL in their milk. Furthermore, this effect is not dependent on the genotype of the pups, indicating that endogenous synthesis of the enzyme by the pancreas of the pups is not sufficient to compensate for the lack of CEL in milk. Interestingly, this effect was observed only in pups up to 4 or 5 days of age, suggesting that maturation of pancreatic function was incomplete until this time (Howles *et al.*, 1998). Concomitant with the reduced serum triglycerides, there was a dramatic increase in fecal lipids. In a parallel study, the intestinal epithelial cells of pups without CEL were shown to accumulate large neutral lipid droplets, especially in the distal third of the small intestine (Howles *et al.*, 1996a; Howles *et al.*, 1999). This study suggested that fat digestion and absorption may be delayed in the absence of CEL. Because the distal intestine has been shown to be less efficient at chylomicron synthesis and secretion (Bennett Clark *et al.*, 1973; Wu *et al.*, 1975), these data provided additional support to suggest a critical role of CEL in determining the rate and extent of milk fat digestion and absorption in neonates. In view of the documented contribution of lumenal lipids to intestinal diseases such as necrotizing enterocolitis and ischemia/reperfusion gut injury (Crissinger *et al.*, 1994), these results also suggest a role for CEL in preventing small bowel diseases in premature and low birthweight infants.

Finally, as discussed earlier in the chapter, pancreatic and preduodenal lipases are inefficient at hydrolyzing fatty acyl esters of arachidonate, docosahexaenoate, and docosapentaenoate, and CEL greatly enhances the release of these polyunsaturated fatty acids from triglycerides. These fatty acids are essential components of cell membranes of the nervous system, and their absorption from the neonatal diet may be necessary for optimal neural development (Carlson *et al.,* 1996; Jorgensen *et al.,* 1996). There is evidence that these fatty acids are more readily absorbed from human milk containing CEL than from formula (Boehm *et al.,* 1997; Morgan *et al.,* 1998). Supplementation of formula with fish oils can result in sufficient absorption of these fatty acids (Foreman-van Drongelen *et al.,* 1996; Carlson *et al.,* 1996; Boehm *et al.,* 1996). However, their distribution between lymph triglycerides and phospholipids may be different than in human milk (Koletzko *et al.,* 1995). Thus, the combined effect of sufficient levels of these fatty acids in breast milk and the ability of milk CEL to digest them suggests the importance of CEL in this aspect of neonatal nutrition.

8. Prospectus

It is clear from this chapter that recent advances in the field have contributed significant information regarding the structure and function relationship of CEL. The availability of inhibitors for the enzyme and CEL gene knockout mice provided additional information regarding the physiological function of this protein. It is clear from recent studies that CEL does not play a significant role in determining the amount of fat and cholesterol absorption from a single meal in adults, as postulated some time ago. However, CEL may be important for the absorption of lipid-soluble vitamins, such as vitamins A and E, and possibly for the absorption of essential fatty acids. Alternately, CEL may serve only as an auxiliary protein to back-up the functions of other digestive enzymes such as pancreatic lipase and REH in facilitating nutrient transport and absorption in the digestive tract. Therefore, the significance of the high level of CEL found in the adult digestive tract remains unclear. It is possible that CEL may alter the rate of absorption of lipid nutrients through the gastrointestinal tract. The application of the lymph fistula model to study the rate and extent of absorption of specific lipids and lipid-soluble nutrients may shed new light on these questions.

In contrast to the adults, increasing information is becoming available to support the hypothesis that CEL plays an important role in nutrient absorption during the neonatal period. Further studies to define the requirement of CEL in neonatal nutrient digestion and prevention of potentially life-threatening diseases in the gut are warranted. These studies will be helpful in designing infant formula supplementation for preterm infants, especially in cases where breast feeding is not an option. The discovery of CEL in tissues outside the digestive tract is also an exciting finding with potential clinical significance. Understanding its role in lipoprotein remodeling and atherogenesis will be the next challenge.

Acknowledgments

Research in authors' laboratories are supported by NIH grants HD33926 (to PNH) and DK40917 (to DYH).

References

Aalto-Setala, K., Bisgaier, C. L., Ho, A., Kieft, K. A., Traber, M. G., Kayden, H. J., Ramakrishnan, R., Walsh, A., Essenburg, A. D., and Breslow, J. L., 1994, Intestinal expression of human apolipoprotein A-IV in transgenic mice fails to influence dietary lipid absorption or feeding behavior, *J. Clin. Invest.* **93**:1776–1786.

Alemi, B., Hamosh, M., Scanlon, J. W., Salzman-Mann, C., and Hamosh, P., 1981, Fat digestion in very low birthweight infants: Effect of addition of human milk to low birthweight formula, *Pediatrics* **68**:484–489.

Bennett Clark, S., and Tercyak, A. M., 1984, Reduced cholesterol transmucosal transport in rats with inhibited mucosal acyl CoA:cholesterol acyltransferase and normal pancreatic function, *J. Lipid Res.* **25**:148–159.

Bennett Clark, S., Lawergren, B., and Martin, J. V.,1973, Regional intestinal absorptive capacities for triolein: An alternative to markers, *Am. J. Physiol.* **225**:574–585.

Bhat, S. G., and Brockman, H. L., 1982, The role of cholesteryl ester hydrolysis and synthesis in cholesterol transport across rat intestinal mucosal membrane. A new concept, *Biochem. Biophys. Res. Commun.* **109**:486–492.

Blackberg, L., and Hernell, O., 1993, Bile salt-stimulated lipase in human milk. Evidence that bile salt induces lipid binding and activation via binding to different sites, *FEBS Letters* **323**:207–210.

Boehm, G., Borte, M., Bohles, H. J., Muller, H., Kohn, G., and Moro, G., 1996, Docosahexaenoic and arachidonic acid content of serum and red blood cell membrane phospholipids of preterm infants fed breast milk, standard formula or formula supplemented with *n*-3 and *n*-6 PUFA, *Eur. J. Pediatrics* **155**:410–416.

Boehm, G., Muller, H., Kohn, G., Moro, G., Minoli, I., and Bohles, H. J., 1997, Docosahexaenoic and arachidonic acid absorption in preterm infants fed LCP-free or LCP-suplemented formula in comparison to infants fed fortified breast milk, *Ann. Nutrit. Met.* **41**:235–241.

Borgström, B., 1968, Quantitative aspects of the intestinal absorption and metabolism of cholesterol and B-sitosterol in the rat, *J. Lipid Res.* **9**:473–481.

Borja, C. R., Vahouny, G. V., and Treadway, C. R., 1964, Role of bile and pancreatic juice in cholesterol absorption and esterification, *Am. J. Physiol.* **206**:223–228.

Brodt-Eppley, J., White, P., Jenkins, S., and Hui, D. Y., 1995, Plasma cholesterol esterase level is a determinant for an atherogenic lipoprotein profile in normolipidemic human subjects, *Biochim. Biophys. Acta* **1272**:69–72.

Bruneau, N., and Lombardo, D., 1995, Chaperone function of a Grp94-related protein for folding and transport of the pancreatic bile salt-dependent lipase, *J. Biol. Chem.* **270**:13524–13533.

Bruneau, N., Nganga, A., Fisher, E. A., and Lombardo, D., 1997, *O*-glycosylation of C-terminal tandem repeated sequences regulates the secretion of rat pancreatic bile salt dependent lipase, *J. Biol. Chem.* **272**:27353–27361.

Camulli, E. A., Linke, M. J., Brockman, H. L., and Hui, D. Y., 1989, Identity of a cytosolic neutral cholesterol esterase in rat liver with the pancreatic bile salt stimulated cholesterol esterase, *Biochim. Biophys. Acta* **1005**:177–182.

Carlson, S. E., Ford, A. J., Werkman, S. H., Peeples, J. M., and Koo, W. W., 1996, Visual acuity and fatty acid status of term infants fed human milk and formulas with and without docosahexaenoate and arachidonate from egg yolk lecithin, *Ped. Res.* **39**:882–888.

Chen, Q., Sternby, B., and Nilsson, A., 1989, Hydrolysis of triacylglycerol arachidonic and linoleic acid ester bonds by human pancreatic lipase and carboxyl ester lipase, *Biochim. Biophys. Acta* **1004**:372–385.

Chen, Q., Sternby, B., Akesson, B., and Nilsson, A., 1990, Effects of human pancreatic lipase-colipase and carboxyl ester lipase on eicosapentaenoic and arachidonic acid ester bonds of triacylglycerols rich in fish oil fatty acids. *Biochim. Biophys. Acta.* **1044**:111–117.

Chen, Q., Blackberg, L., Nilsson, A., Sternby, B., and Hernell, O., 1994, Digestion of triacylglycerols containing long-chain polyenoic fatty acids *in vitro* by colipase-dependent pancreatic lipase and human milk bile salt-stimulated lipase, *Biochim. Biophys. Acta* **1210**:239–243.

Chen, J .C. H., Miercke, L. J. W., Krucinski, J., Starr, J. R., Saenz, G., Wang, X., Spilburg, C. A., Lange, L. G., Ellsworth, J. L., and Stroud, R. M., 1998, Structure of bovine pancreatic cholesterol esterase at 1.6 angstrom: Novel structural features involved in lipase activation, *Biochem.* **37**:5107–5117.

Colwell, N. S., Aleman-Gomez, J. A., and Kumar, B. V., 1993, Molecular cloning and expression of rabbit pancreatic cholesterol esterase, *Biochim. Biophys. Acta* **1172**:175–180.

Crissinger, K. D., Burney, D. L., Velasquez, O. R., and Gonzalez, E., 1994, An animal model of necrotizing enterocolitis induced by infant formula and ischemia in developing piglets, *Gastroenterol.* **106**:1215–1222.

DiPersio, L. P., and Hui, D. Y., 1993, Aspartic acid 320 is required for optimal activity of rat pancreatic cholesterol esterase, *J. Biol. Chem.* **268**:300–304.

DiPersio, L. P., Fontaine, R. N., and Hui, D. Y., 1990, Identification of the active site serine in pancreatic cholesterol esterase by chemical modification and site-specific mutagenesis, *J. Biol. Chem.* **265**:16801–16806.

DiPersio, L. P., Fontaine, R. N., and Hui, D. Y., 1991, Site-specific mutagenesis of an essential histidine residue in pancreatic cholesterol esterase, *J. Biol. Chem.* **266**:4033–4036.

DiPersio, L. P., Carter, C. P., and Hui, D. Y., 1994, Exon 11 of the rat cholesterol esterase gene encodes domains important for intracellular processing and bile salt-modulated activity of the protein, *Biochem.* **33**:3442–3448.

Feaster, S. R., Quinn, D. M., and Barnett, B. L., 1997, Molecular modeling of the structures of human and rat pancreatic cholesterol esterases, *Protein Science* **6**:73–79.

Fernandez, E., and Borgström, B., 1989, Effects of tetrahydrolipstatin, a lipase inhibitor, on absorption of fat from the intestine of the rat, *Biochim. Biophys. Acta* **1001**:249–255.

Fernandez, E., and Borgström, B., 1990, Intestinal absorption of tetinol and retinyl palmitate in the rat, *Lipids* **25**:549–552.

Foreman-van Drongelen, M. M., van Houwelingen, A. C., Kester, A. D., Blanco, C. E., Hasaart, T. H., and Hornstra, G., 1996, Influence of feeding artificial formula milks containing docosahexaenoic and arachidonic acids on the postnatal long-chain polyunsaturated fatty acid status of healthy preterm infants, *Br. J. Nutrit.* **76**:649–667.

Gallo, L. L., 1981, Sterol ester hydrolase from rat pancreas, *Methods in Enzymol.* **71**:664–674.

Gallo, L. L., Newbill, T., Hyun, J., Vahouny, G. V., and Treadwell, C. R., 1977, Role of pancreatic cholesterol esterase in the uptake and esterification of cholesterol by isolated intestinal cells, *Proc. Soc. Expt. Biol. Med.* **156**:277–281.

Gallo, L. L., Chiang, Y., Vahouny, G. V., and Treadwell, C. R., 1980, Localization and origin of rat intestinal cholesterol esterase determined by immunocytochemistry, *J. Lipid Res.* **21**:537–545.

Gallo, L. L., Clark, S. B., Myers, S., and Vahouny, G. V., 1984, Cholesterol absorption in rat intestine: Role of cholesterol esterase and acyl coenzyme A:cholesterol acyltransferase, *J. Lipid Res.* **25**:604–612.

Gallo, L. L., Wadsworth, J. A., and Vahouny, G.V., 1987, Normal cholesterol absorption in rats deficient in intestinal acyl coenzyme A:cholesterol acyltransferase activity, *J. Lipid Res.* **28**:381–387.

Gassama-Diagne, A., Fauvel, J., and Chap, H., 1989, Purification of a new, calcium-independent, high molecular weight phospholipase A-2/lysophospholipase (Phospholipase B) from guinea pig intestinal brush-border membrane, *J. Biol. Chem.* **264**:9470–9475.

Gjellesvik, D. R., 1991, Fatty acid specificity of bile salt-dependent lipase: Enzyme recognition and super-substrate effects, *Biochim. Biophys. Acta.* **1086**:167–172.

Gjellesvik, D. R., Lombardo, D., and Walther, B. T., 1992, Pancreatic bile salt dependent lipase from cod (Gadus morhua): Purification and properties, *Biochim. Biophys. Acta* **1124**:123–134.

Han, J. H., Stratowa, C., and Rutter, W. J., 1987, Isolation of full-length putative rat lysophospholipase cDNA using improved methods for mRNA isolation and cDNA cloning, *Biochem.* **26**:1617–1625.

Heider, J. G., Pickens, C. E., and Kelly, L. A., 1983, Role of acyl CoA:cholesterol acyltransferase in cholesterol absorption and its inhibition by 57–118 in the rabbit, *J. Lipid Res.* **24**:1127–1134.

Hernell, O., and Olivecrona, T., 1974, Human milk lipases. II. Bile salt stimulated lipase, *Biochim. Biophys. Acta* **369**:234–244.

Hernell, O., Blackberg, L., Chen, Q., Sternby, B., and Nilsson, A., 1993, Does the bile salt-stimulated lipase of human milk have a role in the use of the milk long-chain polyunsaturated fatty acids, *J. Ped. Gastroenterol. Nutrit.* **16**:426–431.

Herr, F. M., Wardlaw, S. A., Kakkad, B., Albrecht, A., Quick, T. C., and Ong, D. E., 1993, Intestinal vitamin A metabolism: coordinate distribution of enzymes and CRBP(II), *J. Lipid Res.* **34**:1545–1554.

Holtsberg, F. W., Ozgur, L. E., Garsetti, D. E., Myers, J., Egan, R. W., and Clark, M. A., 1995, Presence in human eosinophils of a lysophospholipase similar to that found in the pancreas, *Biochem. J.* **309**:141–144.

Howles, P., Stemmerman, G., Fenoglio-Preiser, C., and Hui, D., 1996a, Bile salt stimulated lipase activity prevents intestinal damage in mice, *FASEB J.* **10**:A190.

Howles, P. N., Carter, C. P., and Hui, D. Y., 1996b, Dietary free and esterified cholesterol absorption in cholesterol esterase (bile salt-stimulated lipase) gene-targeted mice, *J. Biol. Chem.* **271**:7196–7202.

Howles, P., Wagner, R., and Davis, L., 1998, Bile salt stimulated lipase is required for proper digestion and absorption of milk triglycerides in neonatal mice, *FASEB J.* **12**:A851.

Howles, P. N., Stemmerman, G. N., Fenoglio-Preiser, C. M., and Hui, D. Y., 1999, Cholesterol ester lipase activity in milk prevents fat-derived intestinal injury in neonatal mice. *Am. J. Physiol.* **277**:G653–G661.

Huang, Y., and Hui, D. Y., 1990, Metabolic fate of pancreas-derived cholesterol esterase in intestine. An *in vitro* study using Caco-2 cells, *J. Lipid Res.* **31**:2029–2037.

Huang, Y., and Hui, D. Y., 1991, Cholesterol esterase biosynthesis in rat pancreatic AR42J cells. Post-transcriptional activation by gastric hormones, *J. Biol. Chem.* **266**:6720–6725.

Huang, Y., and Hui, D. Y., 1994, Synergistic effects of bombesin and cholecystokinin on cholesterol esterase biosynthesis and secretion by AR42J cells, *Arch. Biochem. Biophys.* **310**:54–59.

Hui, D. Y., and Kissel, J .A., 1990, Sequence identity between human pancreatic cholesterol esterase and bile salt-stimulated milk lipase, *FEBS Letters.* **276**:131–134.

Hui, D. Y., Hayakawa, K., and Oizumi, J., 1993, Lipoamidase activity in normal and mutagenized pancreatic cholesterol esterase (bile salt-stimulated lipase), *Biochem. J.* **291**:65–69.

Jacobson, P. W., Wiesenfeld, P. W., Gallo, L. L., Tate, R. L., and Osborne, J. C., 1990, Sodium cholate-induced changes in the conformation and activity of rat pancreatic cholesterol esterase. *J. Biol. Chem.* **265**:515–521.

Jorgensen, M. H., Hernell, O., Lund, P., Holmer, G., and Michaelsen, K. F., 1996, Visual acuity and erythrocyte docosahexaenoic acid status in breast fed and formula fed term infants during the first four months of life, *Lipids* **31**:99–105.

Kissel, J. A., Fontaine, R. N., Turck, C., Brockman, H. L., and Hui, D.Y., 1989, Molecular cloning and expression of cDNA for rat pancreatic cholesterol esterase, *Biochim. Biophys. Acta* **1006**:227–236.

Kohno, M., Yokokawa, K., Yasunari, K., Minami, M., Kano, H., Hanehira, T., and Yoshikawa, J., 1998, Induction by lysophosphatidylcholine, a major phospholipid component of atherogenic lipoproteins, of human coronary artery smooth muscle cell migration, *Circulation* **98**:353–359.

Koletzko, B., Edenhofer, S., Lipowsky, G., and Reinhardt, D., 1995, Effects of a low birthweight infant formula containing human milk levels of docosahexaenoic and arachidonic acids, *J. Ped. Gastroenterol. Nutrit.* **21**:200–208.

Krause, B. R., Sliskovic, D. R., Anderson, M., and Homan, R., 1998, Lipid-lowering effects of WAY-121,898, an inhibitor of pancreatic cholesteryl ester hydrolase, *Lipids* **33**:489–498.

Kyger, E. M., Wiegand, R. C., and Lange, L. G., 1989, Cloning of the bovine pancreatic cholesterol esterase/lysophospholipase, *Biochem. Biophys. Res. Commun.* **164**:1302–1309.

Lechene de la Porte, P., Abouakil, N., Lafont, H., and Lombardo, D., 1987, Subcellular localization of cholesterol ester hydrolase in the human intestine, *Biochim. Biophys. Acta* **920**:237–246.

Li, F., and Hui, D. Y., 1997, Modified low density lipoprotein enhances the secretion of bile salt-stimulated cholesterol esterase by human monocyte-macrophages. Species-specific difference in macrophage cholesteryl ester hydrolase, *J. Biol. Chem.* **272**:28666–28671.

Li, F., and Hui, D. Y., 1998, Synthesis and secretion of the pancreatic-type carboxyl ester lipase by human endothelial cells, *Biochem. J.* **329**:675–679.

Lopez-Candales, A., Bosner, M. S., Spilburg, C. A., and Lange, L. G., 1993, Cholesterol transport function of pancreatic cholesterol esterase: Directed sterol uptake and esterification in enterocytes, *Biochem.* **32**:12085–12089.

Mackay, K., and Lawn, R. M., 1995, Characterization of the mouse pancreatic/mammary gland cholesterol esterase-encoding cDNA and gene, *Gene* **165**:255–259.

McKean, M. L., Commons, T. J., Berens, M. S., Hsu, P. L., Ackerman, D. M., Steiner, K. E., and Adelman, S. J., 1992, Effects of inhibitors of pancreatic cholesterol ester hydrolase (PCEH) on ^{14}C-cholesterol absorption in animal models, *FASEB J.* **6**:A1388.

Morgan, C., Stammers, J., Colley, J., Spencer, S. A., and Hull, D., 1998, Fatty acid balance studies in preterm infants fed formula milk containing long-chain polyunsaturated fatty acids (LCP) II, *Acta Paediatrica* **87**:318–324.

Nyberg, L., Farooqi, A., Blackberg, L., Duan, R. D., Nilsson, A., and Hernell, O., 1998, Digestion of ceramide by human milk bile salt stimulated lipase, *J. Ped. Gastroenterol. Nutrit.* **27**:560–567.

Pind, S., and Kuksis, A., 1991, Further characterization of a novel phospholipase B (Phospholipase A2-lysophospholipase) from intestinal brush border membranes, *Biochem. Cell Biol.* **69**:346–357.

Reue, K., Zambaux, J., Wong, H., Lee, G., Leete, T. H., Ronk, M., Shively, J. E., Sternby, B., Borgstrom, B., Ameis, D., and Schotz, M. G., 1991, cDNA cloning of carboxyl ester lipase from human pancreas reveals a unique proline-rich repeat unit, *J. Lipid Res.* **32**:267–276.

Rigtrup, K. M., and Ong, D. E., 1992, A retinyl ester hydrolase activity intrinsic to the brush border membrane of rat small intestine, *Biochem.* **31**:2920–2926.

Rigtrup, K. M., Kakkad, B., and Ong, D. E., 1994a, Purification and partial characterization of a retinyl ester hydrolase from the brush border of rat small intestine mucosa: Probable identity with brush border phospholipase B, *Biochem.* **33**:2661–2666.

Rigtrup, K. M., McEwen, L. R., Said, H. M., and Ong, D. E., 1994b, Retinyl ester hydrolytic activity associated with human intestinal brush border membranes, *Am. J. Clin. Nutri.* **60**:111–116.

Rudd, E. A., and Brockman, H. L., 1984,   Pancreatic carboxyl ester lipase (cholesterol esterase), in: *Lipases* (B. Borgstrom and H. L. Brockman, eds.), Elsevier Science Publishers, Amsterdam, pp. 185–204.

Shamir, R., Johnson, W. J., Zolfaghari, R., Lee, H. S., and Fisher, E. A., 1995, Role of bile salt-dependent cholesteryl ester hydrolase in the uptake of micellar cholesterol by intestinal cells, *Biochem.* **34:**6351–6358.

Shamir, R., Johnson, E. J., Morlock-Fitzpatrick, K., Zolfaghari, R., Li, L., Mas, E., Lambardo, D., Morel, D. W., and Fisher, E. A., 1996, Pancreatic carboxyl ester lipase: A circulating enzyme that modifies normal and oxidized lipoproteins *in vitro*, *J. Clin. Invest.* **97:**1696–1704.

Stout, J. S., Sutton, L. D., and Quinn, D. M., 1985, Acylenzyme mechanism and solvent isotope effects for cholesterol esterase-catalyzed hydrolysis of p-nitrophenyl butyrate, *Biochim. Biophys. Acta* **837:**6–12.

Stromqvist, M., Tornell, J., Edlund, M., Edlund, A., Johansson, T., Lindgren, K., Lundberg, L., and Hansson, L., 1996, Recombinant human bile salt-stimulated lipase: An example of defective O-glycosylation of a protein produced in milk of transgenic mice, *Transgenic Res.* **5:**475–485.

Takase, S., Goda, T., and Shinohara, H., 1993, Adaptive changes of intestinal cellular retinol-binding protein, type II following jejunum-bypass operation in the rat, *Biochim. Biophys. Acta* **1156:**223–231.

Wang, C. S., and Hartsuck, J. A., 1993, Bile salt-activated lipase: a multiple function lipolytic enzyme, *Biochim. Biophys. Acta* **1166,**1–19.

Wang, C. S., and Johnson, K., 1983, Purification of human milk bile salt-activated lipase, *Anal. Biochem.* **133:**457–461.

Wang, C. S., Martindale, M. E., King, M. M., and Tang, J., 1989, Bile-salt-activated lipase: Effect on kitten growth rate, *Am. J. Clin. Nutr.* **49:**457–463.

Wang, X., Wang, C. S., Tang, J., Dyda, F., and Zhang, X. C., 1997, The crystal structure of bovine bile salt activated lipase: Insights into the bile salt activation mechanism, *Structure* **5:**1209–1218.

Watt, S. M., and Simmonds, W. J., 1981, The effect of pancreatic diversion on lymphatic absorption and esterification of cholesterol in the rat, *J. Lipid Res.* **22:**157–165.

Williamson, S., Finucane, E., Ellis, H., and Gamsu, H. R., 1978, Effect of heat treatment of human milk on absorption of nitrogen, fat, sodium, and phosphorus by preterm infants, *Arch. Dis. Child.* **53:**555–563.

Wu, A. L., Bennett Clark, S., and Holt, P. R., 1975 Transmucosal triglyceride transport rates in proximal and distal rat intestine *in vivo*, *J. lipid Res.* **16:**251–257.

Zheng, S., Ee, L., Yao, L., Hui, D. Y., and Tso, P., 1998, A study of the uptake and lymphatic transport of lipid in the mouse, *Gastroenterology* **114:**A916.

Zolfaghari, R., Harrison, E. H., Ross, A. C., and Fisher, E. A., 1989, Expression in Xenopus oocytes of rat liver mRNA coding for a bile salt-dependent cholesteryl ester hydrolase, *Proc. Natl. Acad. Sci. USA* **86:**6913–6916.

Zolfaghari, R., Harrison, E. H., Han, J. H., Rutter, W. J., and Fisher, E. A., 1992, Tissue and species differences in bile salt-dependent neutral cholesteryl ester hydrolase activity and gene expression, *Arteriosclerosis and Thrombosis* **12:**295–301.

CHAPTER 8

The Influence of the Intestinal Unstirred Water Layers on the Understanding of the Mechanisms of Lipid Absorption

ALAN B. R. THOMSON and GARY WILD

1. Introduction

The accurate measurement of the rates of uptake from the intestinal lumen into the enterocyte is central to the elucidation of the cellular processes associated with intestinal absorption of lipids. From the knowledge of the correct rates of uptake, it is possible to determine the permeability properties of the intestinal brush border membrane (BBM), and therefore its lipophilic properties. The permeability of the BBM is subject to adaptation in health and disease (Thomson and Wild, 1997a, 1997b). For example, the uptake of lipids increases with the feeding of a saturated as compared with a polyunsaturated diet, with aging, and with diabetes mellitus. For the treatment of obesity or hyperlipidemia, it may be desirable to develop therapeutic strategies that will reduce the absorption of lipids. Therefore, it becomes necessary to ensure that the experimentally determined rates of lipid uptake are correct, and that the mechanisms of lipid uptake are correctly understood.

If we assume that most lipid uptake is by a process of passive diffusion, then the measured rates of uptake used to calculate the BBM permeability coefficients (P) may be in error because of two pre-epithelial barriers, the intestinal unstirred water layer (UWL) and the acid microclimate. In this chapter, the nature of these two nonhomogeneous compartments is described, together with a consideration of the potential controversies and the relative impact of these two compartments on nutritional versus physiological aspects of lipid uptake. It should be stressed that a component of lipid uptake may also be carrier mediated. Failure to correct for the effective resistance of the UWL may again lead to errors in the assessment of the kinetic constants of this component of lipid uptake, and thus may intro-

ALAN B. R. THOMSON • Cell and Molecular Biology Collaborative Network in Gastrointestinal Physiology, Nutrition and Metabolism Research Group, Division of Gastroenterology, Department of Medicine, University of Alberta, Edmonton, Alberta T6G 2C2 Canada.   GARY WILD • Department of Medicine, Division of Gastroenterology, Department of Anatomy and Cell Biology, McGill University, Montreal H3A 2T5 Canada.
Intestinal Lipid Metabolism, edited by Charles M. Mansbach II *et al.,* Kluwer Academic/Plenum Publishers, 2001.

duce major errors in our appreciation of the true contribution of passive versus mediated uptake of lipids.

2. What Is the Unstirred Water Layer?

In all biological membranes there is a layer of relatively unstirred water through which solute must move by diffusion (Dainty, 1963). In the intestine, this UWL consists of a series of water lamellae that extends out from the BBM, each layer of which is progressively more stirred until they blend in with the bulk water phase in the intestinal contents (Wilson and Dietschy, 1974). The UWL is an operational term, because the boundary between the bulk water phase and the UWL is not well defined. Therefore, the dimensions of UWL thickness and surface area give rise to a value of "effective resistance."

It is unclear whether the UWL is the physical structure of water, mucus, glycocalyx, villi, and microvilli (Desimione, 1983; Smithson et al., 1981; Westergaard et al., 1986); laminar flow of the luminal perfusate (M. D. Levitt et al., 1987); or the intervillous space (IVS; Westergaard and Dietschy, 1974; Harris and Kennedy, 1988; Harris et al., 1988). Like all membranes in contact with an aqueous phase, the intestinal mucosa is covered with an unstirred layer of water molecules. Its thickness has been estimated to be approximately 30 to 1000 μm, depending on the methods used to make the measurement, the use of in vitro or in vivo tissue preparations to obtain varying degrees of stirring or mixing of the luminal fluid, the shape of the villi, and the extent of villous motility observed in the different species studied (Debnam and Levin, 1975; Read et al., 1976; Westergaard and Dietschy, 1974; Winne, 1978a, 1978b, 1978c; Winne and Markgraf, 1979).

The effective resistance of the UWL is determined not just by its thickness, but also by its surface area and by the diffusion coefficient of the probe under consideration. The diffusion coefficient may vary with the viscosity of the bulk phase or of the mucus in the UWL. Earlier physiological studies have shown that the uptake of most nutrients occurs from the upper portion of the villus (King et al., 1981; Kinter and Wilson, 1965; Ryu and Grim, 1982; Westergaard and Dietschy, 1974; Lee, 1963; Padykula, 1962; Westergaard et al., 1986; Levitt, 1985). The "recruitment" of additional transporters or surface area available for diffusion-based uptake will depend on the rate of uptake of solute from the enterocytes at the top of the villus, their access to the intervillus space (IVS), and the movement of the villi. Thus, the surface area of the BBM used for nutrient uptake is much less than the total membrane surface area.

3. What Are the Dimensions of the Unstirred Water Layer?

A variety of approaches have been developed to estimate UWL resistance. These include techniques that are based on the following:

1. the morphometric analysis of small bowel dimensions (Winne, 1978a, 1978b, 1978c);
2. rates of carbon monoxide diffusion out of the intestinal lumen (Levitt et al., 1984;
3. effects of changes in the viscosity of the perfusate on solute absorption (Andreoli and Troutman, 1971);

4. the time required for the development of diffusion potentials (Diamond, 1966);
5. the kinetics of entry of macromolecules into the intervillus spaces (Ryu and Grim, 1982);
6. the comparison of the K_m values for solute transport or peptide digestion with the values observed in the intact intestine (Smithson *et al.*, 1981).

Measurements of UWL resistance *in vitro* have varied from 35 to 1000 μm, depending on the method used to make these measurements. The osmotic transient technique was first used to measure the UWL thickness in the gallbladder (Diamond, 1966), in which the bulk lumenal fluid could be switched rapidly from a fluid of one osmolality to a second of different osmolality, leaving a homogeneous boundary layer of initial fluid in the gallbladder. This technique was then applied to the small intestine (Harris *et al.*, 1986; Westergaard *et al.*, 1986) and man (Flourie *et al.*, 1984; Frase *et al.*, 1985; Read *et al.*, 1977; Sparso *et al.*, 1984). In this setting there would be no boundary layer, and the entire contents of the intestinal lumen must be drained rapidly by the replacing solution in order to establish a new steady-state potential difference (PD). Technical considerations limit the usefulness of this approach.

The osmotic transient technique has been used *in vitro*, and it assumes the presence of a planar surface (Westergaard and Dietschy, 1974). The osmotic transient technique has also been used in intestinal perfusion studies in animals (Gerencser *et al.*, 1984) and in humans (Flourie *et al.*, 1984; Frase *et al.*, 1985; Read *et al.*, 1974, 1977; Sparso *et al.*, 1984). UWL thickness estimates of 300 to 800 μm have been obtained from estimates of the half-time of changes in the potential difference in a perfused gut segment when one solution is rapidly substituted for another with a different osmolarity. This method probably does not yield accurate estimates of UWL thickness, because the osmolarity of the luminal bulk solution is achieved only slowly. Furthermore, fluid perfused through the rat intestine moves by laminar flow, rather than segregating into a UWL and well-mixed bulk luminal contents (Amidon *et al.*,. 1980; Elliot *et al.*, 1980; Kneip *et al.*, 1987; M. D. Levitt *et al.*, 1987; Miyamoto *et al.*, 1982; Winne and Markgraf, 1979; Yuasa *et al.*, 1984, 1986).

The rate of movement of a solute across the UWL is represented by the following equation (Thomson and Dietschy, 1984, p. 165):

$$Jd = (C_1 - C_2)(D/R)$$

Where Jd is the rate of uptake, C_1 is the concentration of the solute in the bulk phase, C_2 is the concentration of the BBM, D is the aqueous diffusion coefficient of the probe, and R is the effective resistance of the UWL composed of the thickness of the UWL (d) divided by the functional surface area (Sw; i.e., $R = d/Sw$).

The rate of absorption of a solute by passive mechanisms is given by the following equation (Barry and Diamond 1984; Dietschy, 1978; Thomson and Dietschy, 1984):

$$Jd = (PC_1)/(1 + PR/D)$$

The mathematical representation of the rate of uptake of a solute by carrier-mediated mechanisms is given by Thomson and Dietschy, 1977, 1984: 70

$$Jd = 0.5 \, D/R \, [C_1 + K_m + RV_{max}/D - \sqrt{(C_1 + K_m + RV_{max}/D)^2 - 4C_1 \, RV_{max}/D}]$$

Thus, the higher the value of R, the lower the rate of transport. The greater the value of the resistance term, the greater the error in estimating the K_m value for the transport process (Thomson and Dietschy, 1977).

Even *in vivo,* the UWL thickness varies depending on the method used to access this dimension: in rat jejunum the UWL is 700 to 800 μm thick in a 30 cm segment perfused in the conventional fashion on the abdominal cavity, falling to 200 to 400 μm when the segment is placed in the abdominal wall, and falling to 32 to 68 μm with shaking of the intact rat (M. D. Levitt *et al.,* 1992a). The values of the effective thickness of the UWL in rat perfusion experiments have been obtained by various methods: 410 to 430 μm by measuring the development of osmotically induced potential difference (Debnam and Levin, 1975), 460 to 486 μm by the segmented flow technique (Winne, 1978; Winne *et al.,* 1979a, 1979b, 1979c), and 212 to 708 μm by changing the flow rate (Komiya *et al.,* 1980).

Another method used to assess the effective resistance of the UWL is to determine the rate of intestinal uptake of a homologous series of probes, such as saturated fatty acids (Westergaard and Dietschy, 1974). The values of the apparent passive permeability coefficients (P^*) increase with chain length of the fatty acids, until the value of P^* becomes proportional to their free diffusion coefficients. This point indicates that uptake was limited by the rate of diffusion across the UWL up to the BBM. From the rates of uptake of such diffusion-limited probes, the UWL resistance in rat jejunum *in vivo* may be calculated, and it closely approximates the value of the UWL resistance calculated from the half-time of change in the diffusion potential (Westergaard *et al.,* 1986). Over a range of perfusion rates seen *in vivo* (1.5–15 ml/min), the diffusion barrier resistance falls by about 45%, and the value of the apparent Michaelis constant for *d*-glucose transport falls by about the same percentage.

A method used recently to obtain measures of UWL thickness *in vivo* involves the measurement of the rate of hydrolysis of probes *in vitro* and *in vivo;* from the difference in these rates, the UWL effect is calculated (M. D. Levitt *et al.,* 1992a; Strocchi and Levitt, 1991). Such estimates have yielded UWL thickness values of 48 μm in the human jejunum. This method assumes that the K_m of the disaccharides, measured *in vitro,* accurately reflects the K_m of these enzymes in the intact intestine. This is a reasonable assumption, given that the kinetics of these enzymes are not altered by dissociation from the BBM (Chang, 1977). This method also assumes that all disaccharide hydrolysis occurs in the BBM rather than in the lumenal contents, again a reasonable assumption (Dahlqvist and Borgström, 1961).

4. The Role of the Intervillus Space

Increasing the rate of stirring of the bulk phase increases the *in vitro* uptake of nutrients (Andreoli and Troutman, 1971; Barry and Diamond, 1984; Bindslev and Wright, 1976; Diamond, 1966; Diamond and Wright, 1969a, 1969b). Faster rates of perfusion through the intestine enhance the *in vivo* uptake of nutrients by a process that involves a reduction in the thickness of the UWL (Wilson and Dietschy, 1974) and an enhancement of the mucosal surface available for uptake (Lewis and Fordtran, 1975; Winne, 1978a, 1978b, 1978c; Eslenhans *et al.,* 1984; Read, 1984). Increasing the rate of perfusion through the rat ileum *in vivo* decreases the thickness of the UWL and augments the neutral NaCl absorption (Harris *et al.,* 1986). Distention of the rat ileum leads to a reduction in villous height and a marked increase in the width of the intervillous space (IVS) in both the transverse and longitudinal dimensions (Harris *et al.,* 1988). The net effect of this distention, however, is that there is no absolute change in the total mucosal surface area. However, there may have been a greater functional surface area along the villus occurring with distension, thereby exposing transport sites that were not exposed previously. The rapid absorption of many nutrients from the upper portion of the villus would reduce the availability of solute for uptake into the IVS and would result in a longitudinal gradient of decreasing concentration from villus tip to villus base.

Distention might also increase uptake as the result of changes in intestinal permeability (i.e., pore size; Swabb *et al.*, 1982; Hakim and Lifson, 1969), blood flow (Ohman, 1984), villous motility (Womack *et al.*, 1987), hydrostatic pressure (Hakim and Lifson, 1969), or the release of a variety of neurohumoral factors that might affect absorption (Caren *et al.*, 1974; Bulbring and Crema, 1959; Beubler and Juan, 1978). Distention enhances the absorption of passively absorbed probes (such as water and urea), perhaps by an augmentation of functional absorptive surface area, but has no effect on the absorption of solutes transported by carrier mechanisms (such as *d*-glucose and *l*-alanine; Harris and Kennedy, 1988). The uptake of some nutrients only at the upper portion of the villus might be due to the localization of carriers only at this site, or alternately may be due to the inaccessibility of solute to the IVS, limiting the participation of these enterocytes in the absorptive processes (Ryu and Grim, 1982; Westergaard and Dietschy, 1974).

When uptake occurs largely from cells along the upper portion of the villus, the thickness of the UWL can be approximated by simple laminar flow analysis (Yuasa *et al.*, 1984; M. D. Levitt *et al.*, 1987), and for modeling purposes the intestine can then be assumed to be a smooth cylinder. The IVS may become an important site of uptake when flow rate is high or when there is intestinal distention. However, the increase in the absorption rate after distention is smaller than the enlargement of the inner cylindrical surface area (Holzeimer and Winne, 1989), presumably because of a change in the supravillous diffusion resistance.

In canine jejunum *in vivo*, the substances absorbed into the villus tips must penetrate an unstirred layer of 500 to 1000 μm, and for those substances absorbed into the lateral surfaces of the villi, an additional barrier of as much as 800 μm exists (Ryu and Grim, 1982). The fluid in the IVS is poorly stirred, and presumably movement of the villi has little impact on mixing of the IVS fluid. Westergaard and Dietschy (1974) and Winne (1978a, 1978b, 1978c) also concluded that villus movement produced little stirring of the UWL. Accordingly, when water absorption is minimal, solutes must diffuse through this relatively thick UWL. However, when water absorption is occurring, then penetration of the UWL by the process of solvent drag may enhance absorption into the tip region of the villus. However, because water absorption occurs largely in the uppermost portion of the villi, diffusion is likely the only process for permeation across the BBM of enterocytes further down the villus.

How do the changes in lumenal flow rate produced in animals relate to the situation in normal humans? The flow rate of fluid through the proximal small intestine varies widely from an average of 2.5 mL/min in fasting subjects (Soergel, 1969) to as high as 20 ml/min after meals (Fordtran and Locklear, 1966; Fordtran and Ingelfinger, 1967). Increasing the flow rate increases the absorption of tritiated water and *d*-glucose (Lewis and Fordtran, 1975) and opens the IVS more widely (Harris *et al.*, 1988). Increasing jejunal flow rate from 5 to 20 ml/min decreases the permeability ratio of *l*-xylose/urea by approximately 30% and decreases the average calculated pore radius of the diffusion pathway from 1.3 nm to .8 nm. Presumably there is increased exposure of the lumenal fluid to the less permeable cells in the IVS (Fine *et al.,* 1995). The change in the ratio of the uptake of xylose and urea could not be explained just by an alteration in UWL resistance or by a change in the laminar flow properties of the perfused fluid.

According to the principles of laminar flow, the flow rate of fluid is most rapid and most perfectly stirred in the center of the intestinal lumen, with progressively slower flow rates and poorer stirring in fluid near the BBM (Amidon *et al.,* 1980; Levitt *et al.,* 1987, 1988). Laminar flow occurs in the gut (Levitt *et al.,* 1987). The two-dimensional laminar flow model is valid for determining the kinetic parameters of carrier-mediated transport *in situ* and for predicting the absorption rate *in situ* from the uptake rate *in vitro* (Yuasa *et al.,* 1986). Thus, in the constantly perfused intestine, fluid moves with laminar flow, which results in a lumen that is totally unstirred in the radial direction. This is different from the conventional unstirred layer model, which proposes a thickness of totally unstirred water adjacent to the lumen with well-mixed contents in the center of the lumen.

Most studies of perfused bowel *in vivo* have been performed in anesthetized laparotomized rats, where there is little motility and near-perfect laminar flow (Amidon *et al.,* 1980; Elliot *et al.,.* 1980; Kneip *et al.,* 1987; Levitt *et al.,* 1987; Miyamoto *et al.,* 1982; Winne and Markgraf, 1979; Yuasa *et al.,* 1984, 1986), just as if the perfused fluid mass were moving through a pipe. In conscious rats the maximal pre-epithelial resistance is equivalent to an UWL thickness of only about 100 μm, with anesthesia doubling this resistance, and anesthesia and laparotomy increasing resistance to about 600 μm (Anderson *et al.,* 1988).

5. What Is the Importance of Intestinal Unstirred Water Layers?

The intestine is subject to adaptation (Thomson and Wild, 1997a, 1997b), and the passive and carrier-mediated transport processes undergoing this adaptation are described kinetically on the basis of the passive permeability coefficient (P), the incremental Gibbs free-energy values $(\delta\Delta F_{w \to 1})$ associated with the addition of specific substituent groups to a probe molecule, the Michaelis affinity constant (K_m), and the maximal transport rate (V_{max}). These measurements provide critical information on such important characteristics of the BBM as its relative hydrophobicity and permeability, and on the number and characteristics of specific transporters in the BBM. Although it has been assumed that uptake is determined by the properties of the BBM, it is critical to correct uptake for potentially serious qualitative and quantitative errors arising from the failure to account for the effect of the UWL resistance. Failure to correct for the effect of the UWL may lead to underestimation of passive permeability coefficients (Barry and Diamond, 1984; Thomson and Dietschy, 1984, p. 165), incremental free-energy values (Diamond *et al.,* 1969a, 1969b), and reflection coefficients for carrier-mediated processes (Barry and Diamond, 1984); overestimation of Michaelis constants for carrier-mediated processes (Thomson and Dietschy, 1980; Wilson and Dietschy, 1974; Winne, 1977a); or erroneous identification of "transition" temperatures in the membrane Bindslev and Wright, 1976; Dietschy, 1978). Failure to correct for the UWL may lead to overestimation of the value of V_{max} if the contribution of concurrent passive uptake is not taken into account or if recruitment of carrier along the intervillus space is not taken into consideration. These potential errors may invalidate the use of Michealis–Menten kinetics for analysis of carrier-mediated transport processes (Dietschy, 1978; Thomson and Dietschy, 1977; Wilson and Dietschy, 1974; Winne, 1973).

There is now evidence to suggest that the dimensions of the UWL may be specific for individual experimental conditions and cannot be predicted (Thomson *et al.,* 1993). Thus, in order to correct for the effective resistance of the UWL and thereby to obtain valid estimates of the kinetic properties of the tissue in question, this resistance factor must be measured. Only in this way will it be possible to establish whether an adaptation in transport that occurs as a result of a dietary, pharmacological, or other experimental manipulation designed to mimic a disease state influences nutrient uptake as a result of alterations in the value of P, K_m, or V_{max}. Once the kinetic mechanism of altered transport is known, then the mechanisms responsible for this adaptation can be determined (Thomson and Wild, 1997a, 1997b). Thus, the value of the UWL resistance has to be measured for each change in experimental design, unless the appropriate equations for the laminar flow model are used. In this instance, the resistance of the lumenal contents moving with laminar flow is theoretically predictable from knowledge of the perfusion rate, gut length, and the aqueous diffusion coefficient.

The previous discussion pertains to the importance of the UWL from the physiological perspective—to establish valid estimates of transport constants, and thereby to understand the possible mechanisms responsible for adaptation. On the other hand, the recent evidence that suggests that the effective resistance of the UWL *in vivo* is much less than *in vitro* raises the possibility that the UWL may have less importance than previously considered from the perspective of retarding nutrient absorption to a degree that is nutritionally important. Measurements of UWL thickness made in the jejunum of conscious dogs by assessing the absorption rate of two rapidly absorbed probes, glucose and ^{14}C-warfarin, gave values of only approximately 35 and 50 μm for perfusion rates of 26 and 5 ml/min, respectively (Levitt *et al.,* 1990). Measurements of the maximal UWL thickness for the human jejunum calculated from previous studies of glucose absorption yielded a mean value of only 40 μm. These measurements are less than one tenth of previously reported values obtained using the osmotic transient technique.

If the thickness of the UWL *in vivo* is low, then what will be its impact on nutrient uptake? Levitt and colleagues (1990) have estimated that for a rapidly absorbed nutrient such as glucose, if the ratio of the cross-sectional surface area of the IVS to the luminal mucosa surface is 1:3, then 50% of the glucose will be absorbed within 9 μm of the villous tips, and greater than 90% within 40 μm of the tips. With an UWL of 40 μm over the villous tips, this preepithelial diffusion barrier may remain the rate-limiting step for rapidly permeating probes such as glucose. A UWL of 35 μm would still produce about 75% of the total resistance to glucose absorption, because the total resistance (UWL plus epithelial cell) to transport of low concentrations of glucose is equivalent to a UWL of 48 μm (Levitt *et al.,* 1990). As the infusate concentration increases, the carriers at the villous tip become saturated and the probe must diffuse down the IVS.

Because the proportion of the IVS that is involved in solute uptake will depend on the extent to which the cells at the villous tip exceed their capacity to transport the solute, the V_{max} or permeability coefficient may vary depending on the extent of the IVS used for uptake. With increasing "recruitment" of carriers along the IVS, or recruitment of increasing amounts of cell membrane for passive uptake, the estimated value of these kinetic constants may vary. The value of the V_{max} predicted from Michaelis–Menten kinetics would underestimate the true V_{max} (Levitt *et al.*, 1996).

Because the IVS is only about 50 μm wide in dogs (Levitt *et al.*, 1990) and 15 μm in rats (Levitt *et al.*, 1996; Harris *et al.*, 1988), 7 to 25 μm is the maximum UWL that could separate solute from the absorptive epithelium in this space. The remaining UWL must be confined to the villous tips. Using measurements of maltose hydrolysis in the rat jejunum, hydrolysis was accurately predicted by a model in which the unstirred fluid, extended from 20 μm over the villous tips throughout the IVS (Levitt *et al.*, 1996). In this model, the depth of diffusion into the IVS is inversely proportional to the efficiency of epithelial handling of the solute. As a result, both the aqueous barrier and the functional surface area are variables rather than constants.

6. Villous Motility

The functional surface area of the intestine may be controlled by regulation of solute access to the IVS. Widening of the IVS could occur with distention of the intestine, narrowing could occur with villous swelling, and access to the IVS could be limited by secretion of fluid from the crypts and convection of fluid up the IVS. Access to the IVS could also be influenced by villous movement. The villi of rats are nonmotile, whereas the villi of dogs, cats, and humans undergo movement. This movement varies in response to certain lumenal solutes or drugs (Womack *et al.*, 1987). This motility could thereby influence access to solute.

Intestinal villi exhibit spontaneous movement in the living animal. The villous movements are thought to be due to contraction of smooth muscle fibers from the muscularis mucosae that run longitudinally within the villous core. There is a pistonlike retraction and extension of the villi, pendular side-to-side movement, and tonic contraction of several or all villi (Elliot *et al.*, 1980; Frase *et al.*, 1985; Gerencser *et al.*, 1984; Levitt *et al.*, 1987; Miyamoto *et al.*, 1982). Local, neural, and hormonal factors modulate villous motility (Elliot *et al.*, 1980; Flourie *et al.*, 1984; Frase *et al.*, 1985; Kneip *et al.*, 1987; Levitt *et al.*, 1987).

A videomicroscopic method was used to analyze quantitatively villous motility in the dog intestine (Womack *et al.*, 1987): The villous retractions were most frequent and of longest duration in the duodenum, followed by the jejunum and the ileum. It was predicted that the villi are in the retracted state for approximately 30%, 13%, and 6% of the time in these three sites, respectively. The frequency of the pendular movements was greatest in the jejunum, followed by the ileum and the duodenum. The lumenal pH or the presence of glucose had no effect on villous motility, whereas amino acid and free fatty acids increased villous contraction frequency by 30% to 50% and 90%, respectively. It is unclear whether these changes in villous contraction achieved by feeding alone, or by specific nutrients, actually enhance nutrient absorption. Increasing villous motility by fluid expansion in the intact canine intestine actually decreases the absorption of water and lauric acid (Mailman *et al.*, 1990). Thus, is it unclear whether the villous motility is sufficient to perturb the UWL sufficiently to actually modify nutrient absorption.

The viscosity of fluid influences diffusion resistance by altering the diffusion velocity (diffusion coefficient). Dietary fiber or its components such as guar gum may be used to induce an increase in the viscosity of the lumenal contents (Cerda *et al.*, 1987). Such changes in viscosity may reduce absorption by their effect on UWL resistance (Johnson and Lee, 1981; Elsenhans *et al.*, 1980; Blackburn and Johnson, 1983). Fiber may also increase the pressure in the intestinal lumen, distend the bowel, and thereby increase the uptake of some substances such as antipyrine (a weak base, almost completely undissociated; Holzheimer and Winne, 1986). Therefore, there is a balance that can be achieved by increasing the pre-epithelial diffusion resistance and by increasing the mucosal surface area. The effect of dietary fiber depends on the specific experimental conditions, including the nature of the solute under study. For highly permeable nutrients, when UWL resistance may have a major effect on uptake, a polymer that increases lumenal viscosity would be expected to reduce absorption. For nutrients with low permeation through the BBM, or when the polymer distends the bowel and increases the available surface area uptake may actually be increased. *In vivo*, of course, other aspects of fiber may a play a role on nutrient absorption such as a slowing effect on gastric emptying or a prolonged mouth-to-cecum transit time (Blackburn *et al.*, 1984; Daumerie and Henquin, 1982; Holt *et al.*, 1979; Jenkins *et al.*, 1978; Lembecke *et al.*, 1984).

7. Acid Microclimate and the Unstirred Water Layer

A layer juxtaposed to the mucosal surface where the proton concentration is higher than in the bulk phase of the intestinal lumen is designated as the "acid microclimate." This is contained in the same region as the UWL, and presumably a portion of the UWL is acidic and is therefore a part of the acid microclimate. The existence of the acid microclimate was postulated on the basis of differences in the steady-state distribution of weak acids and bases from the values predicted by the pH-partition hypothesis (Hogben *et al.*, 1959). In a theoretical study, Winne (1977a, 1977b) concluded that the absorption of weak electrolytes could be modified by the existence of unstirred layers. Although the presence of the acid microclimate was initially contested (Jackson *et al.*, 1974, 1975, 1978), these apparent discrepancies were felt to be due to technical differences. The presence of the acid microclimate was demonstrated indirectly on the basis of acidification of the incubation media by isolated segments of rat jejunum (Blair *et al.*, 1975; Lucas, 1976), as well as by direct measurements using surface microelectrodes or tip microelectrodes on tissue *in vitro* or *in vivo* (Lucas *et al.*, 1975; Daniel *et al.*, 1985; Iwatsubo *et al.*, 1986; Shimada, 1987; Shimada and Hoshi, 1988). Point-by-point determinations using 50 μm tip diameter antimony microelectrodes show that in rat jejunum *in vitro*, the highest proton concentration (214–224 nmol/l = pH 6.67–6.65) was found 10 to 100 μm below the tip of the villus (Daniel *et al.*, 1985). No gradient is seen along the villi in the ileum (Daniel *et al.*, 1989). Toward the intestinal crypt, there is a steep decrease in proton concentration, and in the bulk phase there is a decrease in proton concentration in the UWL. The thickness of the acid microclimate in rat jejunum *in vitro*, 700 μm, is similar to the thickness of the UWL (Shiau *et al.*, 1985).

Metabolizable sugars increase the proton generation by the epithelial cells (Blair *et al.*, 1975; Shiau *et al.*, 1985; Shimada, 1987), and thereby at least in part contribute to the maintenance of the acid microclimate by way of H^+ exchanged for Na^+ across the BBM (Said *et al.*, 1986). Sodium in the incubation medium is important to produce the acid microclimate. Dipeptides (Shimada, 1987) and epidermal growth factor (Iwatsubo *et al.,* 1989) reduce the microclimate pH, whereas ouabain and amiloride increase the microclimate pH (Iwatsubo *et al.,* 1986), presumably due to their effect on the Na^+/H^+ exchanges in the BBM.

Mucus glycoproteins may contribute to this pH gradient by inhibiting the diffusion of protons into the bulk phase (Pfeifer, 1981; Shiau *et al.,* 1985). The mucus may act as an ampholyte, restricting H^+ movement in its matrix (Shiau *et al.,* 1985).

In many of the *in vitro* experiments, the pH microclimate is slightly more acidic than *in vivo* (Rechkemmer *et al.,* 1986). The pH in this microclimate remains relatively constant despite wide variations in the pH in the intestinal lumen (Shiau *et al.,* 1985; Rechkemmer *et al.,* 1986). Such an acid microclimate has been described in human intestinal biopsy material (Kitis *et al.,* 1982) and may change in disease states such as celiac disease (Lucas *et al.,* 1978). According to the pH-partition theory (Jacobs 1940), a dissociable substance permeates through a cell membrane in the unionized form (nonionic diffusion), and the permeation rate will depend on the degree of ionization of the probe, which is determined by the pH of the medium and the pK_a of the probe. Thus, the acid microclimate may be important for the absorption of weak acids such as folic acid and fatty acids (Lucas *et al.,* 1978).

8. Overview of Lipid Absorption

The topic of lipid digestion and absorption has been previously discussed (Bergholz *et al.,* 1991; Potter *et al.,* 1989; Tso, 1985, 1994; Levy, 1992; Thomson *et al.,* 1989, 1993; Ponich *et al.,* 1990; M. T. Clandinin *et al.,* 1991; Carey *et al.,* 1983; Bisgaier and Glickman, 1983). Dietary lipids include triglycerides, phospholipids, and cholesterol. These are integral components of cell membranes and are important for prostaglandin and leukotriene synthesis and for cellular metabolic processes (Thomson and Dietschy, 1981; Thomson *et al.,* 1989, 1993). Lipids are also involved in brain development, inflammatory processes, atherosclerosis, carcinogenesis, aging, and cell renewal (Shiau, 1986).

The average Western adult consumes about 100 g of triacylglycerol (TG) and 4 to 8 g of phospholipid [mostly phosphatidyl choline (PC), also know as lecithin], of which about two thirds are of animal origin (M. T. Clandinin *et al.,* 1991). In water, TG forms crude unstable emulsion droplets, and PC disperses to form relatively stable concentric lamellar structures, called liposomes. When these molecules of TG and PC interact physically in their dietary proportions of about 30:1, they form stable emulsions. Endogenous phospholipid of hepatic origin is also secreted into the intestinal lumen via the bile in quantities of approximately 7 to 22 g/day of lecithin (Northfield and Hofmann, 1975).

Lingual and gastric lipases begin the digestive process (Hamosh, 1986; Moreau *et al.,* 1988), with duodenal hydrolysis carried out as a result of pancreatic lipase, colipase, phospholipase A_2, and cholesterol esterase (Borgström *et al.,* 1957). The lipids are solubilized in bile acids, mixed micelles, and liposomes in the gel and liquid crystal phases (Patton and Carey, 1979). Bile salt micelles are not absorbed as intact structures, so that the lipolytic products solubilized in the micelle dissociate either into an aqueous phase and then across the BBM (the "disassociation–aqueous" model), or directly from the micelle into the BBM (the "collision" model; Westergaard and Dietschy, 1974, 1976; Wilson and Dietschy, 1972;

Shiau and Levine, 1980; Shiau, 1981, 1990; Shiau *et al.*, 1985). The lipids diffuse across the intestinal UWL, become protonated in the acid microclimate, and then transverse the BBM either by passive diffusion or by protein-mediated mechanisms (Dietschy *et al.*, 1971).

Protein-mediated movement of lipid across the BBM has been postulated to include the existence of a "flippase" (Herrmann *et al.*, 1990). Diacylglycerols may be catalysts of the transmembrane movement of lipids (Zachowski and Devaux, 1990). Plasma membrane lipid-binding proteins have been described for long and medium-chain fatty acids and cholesterol (Schoeller *et al.*, 1995b; Stremmel *et al.*, 1988), for cholesterol (Thurnhofer and Hauser, 1990a, 1990b), and for phospholipids (Thurnhofer *et al.*, 1991). A BBM fatty-acid-binding protein in the membrane may facilitate transport of fatty acids, cholesterol, and phospholipids (Clark and Armstrong, 1989). In the cytosol, the long-chain fatty acids (LCFA) are bound to an intestinal-type and to a liver-type fatty acid binding protein (*i*- and *l*-FABP), and are then transported to the endoplasmic reticulum (ER; Ockner and Manning, 1974). A variety of cytosolic binding proteins have also been described (Veerkamp *et al.*, 1991; Thomson *et al.*, 1993). A polyclonal antibody to $FABP_m$ inhibits intestinal oleate uptake in a dose-dependent, noncompetitive fashion, with a reduction in the uptake of fatty acid, cholesterol, monoglyceride, and lysophosphatidyl choline. (Stremmel *et al.*, 1988). The $FABP_m$ acts as a LCFA receptor and possibly also as a translocase (Nunn *et al.*, 1986). Cholesterol has a sterol carrier protein (Scallen *et al.*, 1985a, 1985b).

Re-esterification of monoglyceride and fatty acid takes place by the triglyceride synthase complex and the glycerol phosphate pathway, with complex interactions (Brindley, 1974). A portion of the LCFA may pass directly into the portal circulation (McDonald *et al.*, 1980), with the remainder being formed into lipoproteins from the apolipoproteins synthesized by the enterocyte (Apo A-1, Apo A-IV and Apo B), leading the enterocyte across the basolateral membrane.

A novel model has been proposed to explain the uptake of LCFAs in the intestine (Schoeller *et al.*, 1995b): the sodium–hydrogen exchangers (NHE) in the intestinal BBM are responsible for acidifying the UWL adjacent to the BBM. This facilitates the partitioning of fatty acids out of the bile salt micelles, their protonation, and hence their greater permeation across the BBM. In addition, some fatty acid uptake is facilitated by the fatty-acid-binding protein in the BBM. NHE appears to play a more important role when there is a H^+/Na^+ gradient across the BBM, whereas $FABP_m$ is important when there is less of such a gradient. Inhibition of NHE or $FABP_m$ results in an approximately 30% to 40% decline in the uptake of fatty acids (Schoeller *et al.*, 1995a, 1995b).

The kinetic constants for passive and active transport in the intestine vary as a function of the site of the intestine, the species and age of the animal, the presence of disease such as diabetes mellitus, bowel resection or radiation treatment, or in response to dietary changes (Thomson, 1984). In these various situations in which intestinal absorption varies in health and in disease, there is no relationship between changes in the effective resistance of the UWL and the incremental change in free energy associated ($\delta\Delta F_{w \to 1}$) with the uptake of a homologous series of fatty acids. Thus, the effective resistance of the UWL changes in health and disease, but the basis for this alteration is unclear and is not related to the permeability of the BBM.

In their extensive review of adaptive regulation of sugar and amino acid transport by vertebrate intestine, Karasov and Diamond (1983) remarked how the major kinetic change in carrier-mediated transport is in the value of the maximal transport (V_{max}) rate rather than in the Michaelis affinity constant (K_m). Alterations in nutrient transport as a result of changes in the effective resistance of the UWL would be more likely to occur at lower concentrations of nutrients absorbed by a carrier-mediated transport, because it is at this range that the value of the Michaelis constant (K_m) is most influenced by UWL resistance (Thomson and Dietschy, 1977). It is likely, therefore, that the alterations in the value of the V_{max} described in various animal models of intestinal adaptation (Thompson and Wild, 1997a, 1997b) are likely due to alterations in the total quantity of carrier protein, either per enterocyte or per villus.

The presence of the BBM lipid transport protein raises the suggestion that once lipids have partitioned out of the bile acid micelle, their uptake may occur by this carrier-mediated transport or by passive diffusion. This raises the possibility therefore that the alterations in lipid uptake observed in health and in disease may be a function of the basis of changes in the UWL resistance, the permeability properties of the membrane, or the contribution of the membrane lipid-binding proteins. As new therapeutic agents are developed to achieve a reduction in lipid uptake for the purpose of treating obesity or hyperlipidemia, it is clear that the definition of the mechanism of action of these agents on passive or mediated lipid uptake will require appropriate corrections to be made for the effective resistance of the intestinal UWL.

Acknowledgments

The authors wish to express their appreciation for the excellent word processing skills of Ms. Rachel S. Jacobs.

References

Amidon, G. L., Kou, J., Elliot, R. L., and Lightfoot, E. N., 1980, analysis of models for determining intestinal wall permeabilities, *J. Pharm. Sci.* **69**:1369–1373.

Anderson, B. W., Levine, A. S., Levitt, D. G., Kneip, J. M., and Levitt, M. D., 1988, Physiological measurement of luminal stirring in perfused rat jejunum, *Am. J. Physiol.* **254**:G843–G848.

Andreoli, T. E., and Troutman, S. L., 1971, An analysis of unstirred layers in series with "tight" and "porous" lipid bilayer membranes, *J. Gen. Physiol.* **57**:464–478.

Barry, P. H., and Diamond, J. M., 1984, Effects of unstirred layers on membrane phenomena. *Physiol. Rev.* **64**:763–872.

Bergholz, C. M., Jandacek, R. J., and Thomson, A. B. R., 1991, Review of laboratory and clinical studies of olestra, a nonabsorbable lipid, *Can. J. Gastroenterol.* **5**:137.

Beubler, E., and Juan, H., 1978. PGE-release, blood flow, and transmucosal water movement after mechanical stimulation of the rat jejunal mucosa, *Naunyn Schmiedebergs Arch. Pharmacol.* **305**:91–95.

Bindslev, N., and Wright, E. M., 1976, Effect of temperature on nonelectrolyte permeation across the toad urinary bladder, *J. Membr. Biol.* **29**:265–288.

Bisgaier, C., and Glickman, R. M., 1983, Intestinal synthesis, secretion, and transport of lipoproteins, *Annu. Rev. Physiol.* **45**:625.

Blackburn, N. A., and Johnson, I. T., 1983, The influence of guar gum on the movements of insulin, glucose and fluid in the rat intestine during perfusion *in vivo*, *Pflugers Arch.* **397**:144–148.

Blackburn, N. A., Redfern, J. S., Jarjis, H., Holgate, A. M., Hanning, I., Scarpello, J. H. B., Johnson, I. T., and Read, N. W., 1984, The mechanism of action of guar gum in improving glucose tolerance in man, *Clin. Sci.* **66**:329–336.

Blair, J. A., and Matty, A. J., 1974, Acid microclimate in intestinal absorption, *Clin. Gastroenterol.* **3**:183–197.

Blair, J. A., Lucas, M. L., and Matty, A. J., 1975, Acidification in the rat proximal jejunum, *J. Physiol. London* **245**:333–350.

Borgström, B., Dahlqvist, A., Lund, G., and Sjovall, J., 1957, Studies of intestinal digestion and absorption in the human, *J. Clin. Invest.* **36:**1521–1536.

Brindley, D. N., 1974, The intracellular phase of fat absorption, *Biomembranes* **4B(0):**621–671.

Bulbring, E., and Crema, A., 1959, The release of S-hydroxytryptamine in relation to pressure exerted on the intestinal mucosa, *J. Physiol. (London)* **146:**18–28.

Caren, J. F., Meyer, J. H., and Grossman, M. I., 1974, Canine intestinal secretion during and after rapid distention of the small bowel, *Am. J. Physiol.* **227:**183–188.

Carey, M. C., Small, D. M., and Bliss, C. M., 1983, Lipid digestion and absorption, *Annu. Rev. Physiol.* **45:**651–677.

Cerda, J. J., Robbins, F. L., Burgin, C. W., and Gerenscer, G. A., 1987, Unstirred water layers in rabbit intestine: Effects of guar gum, *J. Parenteral. and Enteral. Nut.* **11(1):**63–66.

Chang, T. M. S., 1977, *Biomedical Applications of Immobilized Enzymes and Proteins,* Volumes 1 and 2, Plenum Press, New York.

Clandinin, M. T., Cheema, S., Field, C. J., Garg, M. L., Venkatraman, J., and Clandinin, T. R., 1991, Dietary fat exogenous determination of membrane structure and cell function. *FASEB J.* **5:**2761.

Clark, S. D., and Armstrong, M. K., 1989, Cellular lipid binding proteins: Expression, function and nutritional regulation, *FASEB J.* **3(13):**2480–2487.

Dahlqvist, A., and Borgström, B., 1961, Digestion and absorption of disaccharides in man, *Biochem. J.* **81:**411–418.

Dainty, J., 1963, Water relations of plant cells, *Adv. Botan. Res.* **1:**279–326.

Daniel, H., Neugebauer, B., Kratz, A., and Rehner, G., 1985, Localization of acid microclimate along intestinal villi of rat jejunum, *Am. J. Physiol.* **248:**G293–G298.

Daniel, H., Fett, C., and Kratz, A., 1989, Demonstration and modification of intervillous pH profiles in rat small intestine *in vitro, Am. J. Physiol.* **257:**G489–G495.

Daumerie, C., and Henquin, J. C., 1982, Acute effects of guar gum on glucose tolerance and intestinal absorption of nutrients in rats, *Diabete et Metabolisme* **8:**1–5.

Debnam, E. S., and Levin, R. J., 1975, Effects of fasting and semistarvation on the kinetics of active and passive sugar absorption across the small intestine *in vivo, J. Physiol. (London)* **252:**681–700.

Desimione, J. A., 1983, Diffusion barrier in the small intestine, *Science* **220:**221–222.

Diamond, J. M., 1966, A rapid method for determining voltage-concentration relations across membranes, *J. Physiol. (London)* **183:**83–100.

Diamond, J. M., and Wright, E. M., 1969a, Biological membranes: The physical basis of iron and nonelectrolyte selectivity, *Annu. Rev. Physiol.* **31:**581–646.

Diamond, J. M., and Wright, E. M., 1969b, Molecular forces governing nonelectrolyte permeation through cell membranes, *Proc. R. Soc. Lond. B. Biol. Sci.* **172:**273–316.

Dietschy, J. M., 1978, Effect of diffusion barriers on solute uptake into biological systems, in: *Microenvironments and Metabolic Compartmentation.* Academic Press, New York, pp. 401–418.

Dietschy, J. M., Sallee, V. L., and Wilson, F. A., 1971, Unstirred water layers and absorption across the intestinal mucosa, *Gastroenterol.* **61:**932–934.

Elliot, R. L., Amidon, G. L., and Lightfoot, E. N., 1980, A convective mass transfer model for determining intestinal wall permeabilities: Laminar flow in a circular tube, *J. Theor. Biol.* **87:**757–771.

Elsenhans, B., Sufke, U., Blume, R., and Caspary, W. F., 1980, The influence of carbohydrate gelling agents on rat intestinal transport of monosaccharides and neutral amino acids *in vitro, Clin. Sci.* **59:**373–380.

Elsenhans, B., Zenker, D., and Caspary, W., 1984, Guaran effect on rat intestinal absorption. A perfusion study, *Gastroenterol.* **86:**645–653.

Fine, K. D., Santa Ana, C. A., Porter, J. L., and Fordtran, J. S., 1995, Effect of changing intestinal flow rate on a measurement of intestinal permeability, *Gastroenterol.* **108:**983–989.

Flourie, B., Vidon, N., Florent, C. H., and Bernier, J. J., 1984, Effect of pectin on jejunal glucose absorption and unstirred layer thickness in normal man, *Gut* **25:**936–941.

Fordtran, J. S., and Ingelfinger, F. J., 1967, Absorption of water, electrolytes, and sugars from the human gut, in: *Handbook of Physiology—Alimentary canal.* American Physiological Society, Washington, DC, p. 1457.

Fordtran, J. S., and Locklear, T. W., 1966, Ionic constituents and osmolality of gastric and small-intestinal fluids after eating, *Am. J. Dig. Dis.* **11:**503–521.

Frase, L., Strickland, A. D., Kachel, G. W., and Krejs, G. H., 1985, Enhanced glucose absorption in the jejunum of patients with cystic fibrosis, *Gastroenterol.* **88:**478–484.

Gerencser, G. A., Cerda, J., Burgin, C., Baig, M. M., and Guild, R., 1984, Unstirred water layers in rabbit intestine: Effects of pectin, *Proc. Soc. Exp. Biol. Med.* **176**:183–186.

Hakim, A. A., and Lifson, N., 1969, Effects of pressure on water and solute transport by dog intestinal mucosa *in vivo, Am. J. Physiol.* **216**:276–284.

Hamosh, M., 1986, Lingual lipase, *Gastroenterol.* **90**:1290–1292.

Harris, M. S., and Kennedy, J. G., 1988, Relationship between distention and absorption in rat intestine. II. Effects of volume and flow rate on transport, *Gastroenterol.* **94**:1172–1179.

Harris, M. S., Dobbins, J. W., and Binder, H. J., 1986, Augmentation of neutral sodium chloride absorption by increased flow rate in rat ileum *in vivo, J. Clin. Invest.* **78**:431–438.

Harris, M. S., Kennedy, J. G., Siegesmund, K. A., and Yorde, D. E., 1988, Relationship between distention and absorption in rat intestine. I. Effect of luminal volume on the morphology of the absorbing surface, *Gastroenterol.* **94**:1164–1171.

Herrmann, A., Zachowski, A., and Devaux, P. E., 1990, The protein mediated phospholipid translocation of the endoplasmic reticulum has a low lipid specificity, *Biochem.* **29**:2023–2027.

Hogben, C. A. M., Tocco, D. T., Brodie, B. B., and Schanker, L. A., 1959, On the mechanism of intestinal absorption of drugs, *J. Pharmacol. Exp. Ther.* **125**:275–282.

Holt, S., Heading, R. C., Carter, D. C., Prescott, L. F., and Tothill, P., 1979, Effect of gel fiber on gastric emptying and absorption of glucose and paracetamol. *Lancet* **1**:636–639.

Holzheimer, G., and Winne, D., 1986, Influence of dietary fiber and intraluminal pressure on absorption and preepithelial diffusion resistance (unstirred layer) in rat jejunum *in situ, Naunyn Schmiedebergs Arch. Pharmacol.* **334**:514–524.

Holzheimer, G., and Winne, D., 1989, Influence of distention on absorption and villous structure in rat jejunum, *Am. J. Physiol.* **256**:G188–G197.

Iwatsubo, T., Miyamoto, Y., Sugiyama, Y., Yuasa, H., Iga, T., and Hanano, M. 1986, Effects of potential damaging agents on the microclimate-pH in the rat jejunum, *J. Pharm. Sci.* **75(12)**: 1162–1165.

Iwatsubo, T., Yamazaki, M., Sugiyama, Y., Suzuki, H., Yanai, S., Kim, D. C., Satoh, H., Miyamoto, Y., Iga, T., and Hanano, M., 1989, Epidermal growth factor as a regulatory hormone maintaining a low pH microclimate in the rat small intestine, *J. Pharm. Sci.* **78(6)**: 457–459.

Jackson, M. J., and Morgan, B. N., 1975, Relations of weak electrolyte transport and acid–base metabolism in rat small intestine *in vitro, Am. J. Physiol.* **228**:482–487.

Jackson, M. J., Shiau, Y.-F., Bane, S., and Fox, M., 1974, Intestinal transport of weak electrolytes: Evidence in favor of a three compartment system, *J. Gen. Physiol.* **63**:187–213.

Jackson, M. J., Williamson, A. M., Dombrowski, W. A., and Garner, D. E., 1978, Intestinal transport of weak electrolytes, determinants of influx at the luminal surface, *J. Gen. Physiol.* **71**:301–327.

Jacobs, H. M., 1940, Some aspects of cell permeability to weak electrolytes, *Cold Spring Harbor Symp. Quant. Biol.* **8**:30–39.

Jenkins, D. J. A., Wolever, T. M. S., Leeds, A. R., Gassull, M. A., Haisman, P., Dilawari, J., Goff, D. V., Metz, G. L., and Alberti, K. G. M. M., 1978, Dietary fibres, fibre analogues, and glucose tolerance: Importance of viscosity, *Brit. Med. J.* **1**:1392–1394.

Johnson, I. T., and Lee, J. M., 1981, Effect of gel-forming gums on the intestinal unstirred layer and sugar transport *in vitro, Gut* **22**:398–403.

Karasov, W. H., and Diamond, J. M., 1983, Adaptive regulation of sugar and amino acid transport by vertebrate intestine, *Am. J. Physiol.* **8**:G442–G462.

King, I. S., Sepulveda, F. V., and Smith, M. W., 1981, Cellular distribution of neutral and basic amino acid transport systems in rabbit ileal mucosa, *J. Physiol. (London)* **319**:355–368.

Kinter, W. B., and Wilson, T. H., 1965, Autoradiographic study of sugar and amino acid absorption by everted sacs of hamster intestine, *J. Cell. Biol.* **25**:19–39.

Kitis, G., Lucas, M. L., Bishop, H., Sargent, A., Schneider, R. E., Blair, J. A., and Allan, R. N., 1982, Altered jejunal surface pH in coeliac disease: its effect on propanolol and folic acid absorption, *Clin. Sci.* **63**:373–380.

Kneip, J. M., Wickman, B. E., and Levitt, M. D., 1987, Predicting the influence of gut distension on absorption, *Gastroenterol.* **92**:1472.

Komiya, I., Park, J. Y., Kamani, A., Ho, N. F. H., and Higuchi, W. I., 1980, Quantitative mechanistic studies in simultaneous fluid flow and intestinal absorption using steroids as model solutes, *Int. J. Pharm.* **4**:249–262.

Lee, J. S., 1963, Role of mesenteric lymphatic system in water absorption from rat intestine *in vivo, Am. J. Physiol.* **204**:92–96.

Lembcke, B., Ebert, R., Ptok, M., Caspary, W. F., Creutzfeldt, W., Schicha, H., and Emrich, D., 1984, Role of gastrointestinal transit in the delay of absorption by viscous fibre (guar), *Hepatogastroenterol.* **31:**183–186.

Levitt, M. D., 1985, Use of carbon monoxide to measure luminal stirring in the rat gut, *J. Clin. Invest.* **74:**2056–2064.

Levitt, M. D., Aufderheide, T., Fetzer, C. A., Bond, J. H., and Levitt, D. G., 1984, Use of carbon monoxide to measure luminal stirring in the rat gut, *J. Clin. Invest.* **74:**2056–2064.

Levitt, M. D., Fetzer, C. A., Kneip, J. M., Bond, J. H., and Levitt, D. G., 1987, Quantitative assessment of luminal stirring in the perfused small intestine of the rat, *Am. J. Physiol.* **252:**G325–G332.

Levitt, M. D., Kneip, J. M., and Levitt, D. G., 1988, Use of laminar flow and unstirred layer models to predict intestinal absorption in rat, *J. Clin. Invest.* **81:**1365–1369.

Levitt, M. D., Furne, J. K., Strocchi, A., Anderson, B. W., and Levitt, D. G., 1990, Physiological measurements of luminal stirring in the dog and human small bowel, *J. Clin. Invest.* **86:**1540–1547.

Levitt, M. D., Strocchi, A., and Levitt, D. G., 1992a, Human jejunal unstirred layer: Evidence for extremely efficient luminal stirring, *Am. J. Physiol.* **262:**G593–G596.

Levitt, M. D., Fine, C., Furne, J. K., and Levitt, D. G., 1996, use of maltose hydrolysis measurements to characterize the interaction between the aqueous diffusion barrier and the epithelium in the rat jejunum, *J. Clin. Invest.* **97(10):**2308–2315.

Levy, E., 1992, The 1991 Borden Award lecture. Selected aspects of intraluminal and intracellular phases of intestinal fat absorption, *Can. J. Physiol. Pharmacol.* **70(4):**413–419.

Lewis, L. D., and Fordtran, J. S., 1975, Effect of perfusion rate on absorption, surface area, unstirred water layer thickness, permeability, and intraluminal pressure in the rat ileum *in vivo, Gastroenterol.* **68:**1509–1516.

Lucas, M. L., 1976, The association between acidification and electrogenic events in the rat jejunum, *J. Physiol. (London)* **275:**645–662.

Lucas, M. L., Schneider, W., Haberich, F. J., and Blair, J. A., 1975, Direct measurement by pH-microelectrode of the Ph microclimate in rat proximal jejunum, *Proc. R. Soc. London Ser. B.* **192:**39–48.

Lucas, M. L., Cooper, M. T., Lei, F. H., Johnson, I. T., Holmes, G. K. T., Blair, J. A., and Cooke, W. T., 1978, Acid microclimate in coeliac and Crohn's disease: A model for folate malabsorption, *Gut* **19:**735–742.

Mailman, D., Womack, W. A., Kvietys, P. R., and Granger, D. N., 1990, Villous motility and unstirred water layers in canine intestine, *Am. J. Physiol.* **258:**G238–G246.

McDonald, G. B., Saunders, D. R., Weidman, M., and Fisher, L., 1980, Portal venous transport of long-chain fatty acids absorbed from rat intestine, *Am. J. Physiol.* **239:**717–723.

Miyamoto, Y., Hanano, M., Iga, T., and Ishikawa, M., 1982, A drug absorption model of the intestinal tract based on the two-dimensional laminar flow in a circular tube, *J. Pharmacobio-Dyn.* **5:**445–447.

Moreau, H., Laugier, R., Gargouri, Y., Ferrato, F., and Verger, R., 1988, Human preduodenal lipase is entirely of gastric fundic origin, *Gastroenterol.* **95:**1221–1226.

Northfield, T. C., and Hofmann, A. F., 1975, Biliary lipid output during three meals and an overnight fast. I. Relationship to bile acid pool size and cholesterol saturation of bile in gallstone and control subjects, *Gut* **16(1):**12–17.

Nunn, W. D., Colburn, R. W., and Black, P. N., 1986, Transport of long-chain fatty acids in *Escherichia coli, J. Biol. Chem.* **261:**167–171.

Ockner, R. K., and Manning, J. 1974, Fatty acid-binding protein in small intestine. Identification, isolation and evidence for its role in intracellular fatty acid transport, *J. Clin. Invest.* **54:**326–338.

Ohman, U., 1984, The effect of luminal distention and obstruction on the intestinal circulation, in: *Physiology of the Intestinal Circulation* (A. P. Shepard and D. N. Granger, eds.), Raven Press, New York, pp. 321–334.

Padykula, H. A., 1962, Recent functional interpretations of intestinal morphology, *Fed. Proc.* **21:**873–879.

Patton, J. S., and Carey, M. C., 1979, Watching fat digestion, *Science* **204:**145–148.

Pfeifer, C. J., 1981, Experimental analysis of hydrogen ion diffusion in gastrointestinal mucus glycoprotein, *Am. J. Physiol.* **240:**G176–G183.

Ponich, T., Fedorak, R. N., and Thomson, A. B. R., 1990, The small intestine, in: *Current Gastroenterology,* Houghton Mifflin, Boston, pp. 69–106.

Potter, B. J., Sorrentino, D., and Berk, P. D., 1989, Mechanisms of cellular uptake of free fatty acids, *Annu. Rev. Nutr.* **9:**253–270.

Read, N. W., 1984, The relationship between colonic motility and transport, *Scand. J. Gastroenterol.* **19(Suppl 93).**35–42.

Read, N. W., Holdsworth, C. D., and Levin, R. J., 1974, Electrical measurement of intestinal absorption of glucose in man, *Lancet* 2:624–627.

Read, N. W., Levin, R. J., and Holdsworth, C. D., 1976, Measurement of the functional unstirred layer thickness in the human jejunum *in vivo, Gut* 17:387.

Read, N. W., Barber, D. C., Levin, R. J., and Holdsworth, C. D., 1977, Unstirred layer and kinetics of electrogenic glucose absorption on the human jejunum *in situ, Gut* 18:865–876.

Rechkemmer, G., Wahl, M., Kuschinsky, W., and von Engelhardt, W., 1986, pH-Microclimate at the luminal surface of the intestinal mucosa of guinea pig and rat, *Pflugers. Arch.* 407:33–40.

Ryu, K. H., and Grim, E., 1982, Unstirred water layer in canine jejunum, *Am. J. Physiol.* 242:G364–G369.

Said, H. M., Blair, J. A., Lucas, M. L., and Hilburn, M. E., 1986, Intestinal surface acid microclimate *in vitro* and *in vivo* in the rat, *J. Lab. Clin. Med.* 107:420–426.

Scallen, T. J., Pastuszyk, A., Noland, B. J., Chanderbhan, R., Kharoubi, A., and Vahouny, G. V., 1985a, Sterol carrier and lipid transfer proteins, *Chem. Phys. Lipids* 38:239–261.

Scallen, T. J., Noland, B. J., Gavey, K., Bass, N. M., Ockner, R. K., Chanderbhan, R., and Vahouny, G. V., 1985b, Sterol carrier protein 2 and fatty acid-binding protein. Separate and distinct physiological functions, *J. Biol. Chem.* 260:4733–4739.

Schoeller, C., Keelan, M., Mulvey, G., Stremmel, W., and Thomson, A. B. R., 1995a, Oleic acid uptake into rat and rabbit jejunal brush border membrane, *Biochimica. Biophys. Acta* 1236(1):51–64.

Schoeller, C., Keelan, M., Mulvey, G., Stremmel, W., and Thomson, A. B. R., 1995b, Role of a brush border membrane fatty acid binding protein in oleic acid uptake into rat and rabbit jejunal brush border membrane, *Clin. Invest. Med.* 18(5):380–388.

Shiau, Y.-F., and Levine, G. M., 1980, PH dependence of micellar diffusion and dissociation, *Am. J. Physiol.* 239:G177–G182.

Shiau, Y.-F., 1981, Mechanism of fat absorption, *Am. J. Physiol.* 240:G1–G9.

Shiau, Y.-F., 1986, *Physiology of the Gastrointestinal Tract, Volume II. Lipid Digestion and Absorption.* Raven Press, New York, pp. 1527–1556.

Shiau, Y.-F., 1990, Mechanism of intestinal fatty acid uptake in the rat: The role of an acidic microclimate, *J. Physiol. (London)* 421:463–474.

Shiau, Y.-F. Fernandez, P., Jackson, M. J., and McMonagle, S., 1985, Mechanisms maintaining a low-pH microclimate in the intestine, *Am. J. Physiol.* 248:G608–G617.

Shimada, T., 1987, Factors affecting the microclimate pH in rat jejunum, *J. Physiol.* 392:113–127.

Shimada, T., and Hoshi, T., 1988, Na^+-dependent elevation of the acidic surface pH (microclimate pH) of rat jejunum villus cells induced by cyclic nucleotides and phorbol ester: Possible mediators of the regulation of the Na^+/H^+ antiporter, *Biochim. Biophys. Acta* 937:328–334.

Smithson, K. W., Milar, D. B., Jacobs, L. R., and Gray, G. M., 1981, Intestinal diffusion barrier: Unstirred water layer or membrane surface mucous coat, *Science* 214:1241–1244.

Soergel, K. H., 1969, Flow measurements of test meals and fasting contents in the human small intestine. Gastrointestinal motility. *International Symposium on Motility of the GI Tract.* Erlangen, West Germany, July 15 and 16, Academic Press, New York.

Sparso, B. H., Luke, M., and Wium, E., 1984, Electrogenic transport of glucose in the normal upper duodenum. II. Unstirred water layer and estimation of real transport constant, *Scand. J. Gastroenterol.* 19:568–574.

Stremmel, W., Lotz, G., Strohmeyer, G., and Berk, P. D., 1988, Identification, isolation and partial characterization of a fatty acid binding protein from rat jejunal microvillus membranes, *J. Clin. Invest.* 75:1068–1076.

Strocchi, A., and Levitt, M. D., 1991, A reappraisal of the magnitude and implications of the intestinal unstirred layer, *Gastroenterol.* 101:843–847.

Swabb, E. A., Hynes, R. A., and Donowitz, M., 1982, Elevated intraluminal pressure alters rabbit small intestinal transport *in vivo, Am. J. Physiol.* 242:G58–G64.

Thomson, A. B. R., and Dietschy, J. M., 1977, Derivation of the equations that describe the effects of unstirred water layers on the kinetic parameters of active transport processes in the intestine. *J. Theor. Biol.* 4:277–294.

Thomson, A. B. R., 1984, Mechanisms of intestinal adaptation: Unstirred layer resistance and membrane transport, *Can. J. Physiol. Pharmacol.* 62:678–682.

Thomson, A. B. R., and Dietschy, J. M., 1980, Intestinal kinetic parameters: Effects of unstirred layers and transport preparation, *Am. J. Physiol.* 239:G372–G377.

Thomson, A. B. R., and Dietschy, J. M., 1981, Intestinal lipid absorption: Major extracellular and intracellular events, in: *Physiology of the Gastrointestinal Tract* (L. R. Johnson, ed.), Raven Press, New York: pp. 1147–1220.

Thomson, A. B. R., and Dietschy, J. M., 1984, The role of the unstirred water layer in intestinal permeation, in: *Handbook of Experimental Pharmacology,* Volume 70, Part II (T. Z. Csaky, ed.), Springer-Verlag, Berlin, pp. 165–269.

Thomson, A. B. R., and Wild, G., 1997a, Adaptation of intestinal nutrient transport in health and disease—Part I, *Dig. Dis. Sci.* **42(3)**:453–469.

Thomson, A. B. R., and Wild, G., 1997b, Adaptation of intestinal nutrient transport in health and disease—Part II, *Dig. Dis. Sci.* **42(3)**:470–488.

Thomson, A. B. R., Keelan, M., Garg, M. L., and Clandinin, M. T., 1989, Intestinal aspects of lipid absorption: In review. *Can. J. Physiol. Pharmacol.* **67**:179–191.

Thomson, A. B. R., Schoeller, C., Keelan, M., Smith, L., and Clandinin, M. T., 1993, Lipid absorption: Passing through the unstirred layers, brush-border membrane, and beyond, *Can. J. Physiol. Pharmacol.* **71**:531–555.

Thurnhofer, H., and Hauser, H., 1990a, The uptake of phosphatidycholine by small intestinal brush border membrane is protein-mediated, *Biochim. Biophys. Acta* **1024**:249–262.

Thurnhofer, H., and Hauser, H., 1990b, Uptake of cholesterol by small intestinal brush border membrane is protein-mediated, *Biochem.* **29**:2142–2148.

Thurnhofer, H., Schnabel, J., Betz, M., Lipka, G., Pidgeon, C., and Hauser, H., 1991, Cholesterol-transfer protein located in the intestinal brush-border membrane. Partial purification and characterization, *Biochim. Biophys. Acta* **1064**:275–286.

Tso, P., 1985, Gastrointestinal digestion and absorption of lipid, *Adv. Lipid Res.* **21**:143–186.

Tso, P., 1994, Intestinal lipid absorption, in: *Physiology of the Gastrointestinal Tract* (L. R. Johnson, D. H. Alpers, J. Christensen, E. D. Jacobson, and J. H. Walsh, eds.), Raven Press, New York, pp. 1867–1907.

Veerkamp, J. H., Peeters, R. A., and Maatman, R. G., 1991, Structural and functional features of different types of cytoplasmic fatty acid binding proteins, *Biochim. Biophys. Acta* **1081**:1–24.

Westergaard, H., and Dietschy, J. M., 1974, Delineation of the dimensions and permeability characteristics of the two major diffusion barriers to passive mucosal uptake in the rabbit intestine, *J. Clin. Invest.* **54**:718–732.

Westergaard, H., and Dietschy, J. M., 1976, The mechanism whereby bile acid micelles increase the rate of fatty acid and cholesterol uptake into the intestinal mucosal cell, *J. Clin. Invest.* **58**:97–108.

Westergaard, H., Holtermuller, K. H., and Dietschy, J. M., 1986, Measurement of resistance of barriers to solute transport *in vivo* in rat jejunum, *Am. J. Physiol.* **250**:G727–G735.

Wilson, F. A., and Dietschy, J. M., 1972, Characterization of bile acid absorption across the unstirred water layer and brush border of the rat jejunum, *J. Clin. Invest.* **51**:3015–3025.

Wilson, F. A., and Dietschy, J. M., 1974, The intestinal unstirred layer: Its surface area and effect on active transport kinetics, *Biochim. Biophys. Acta* **363**:112–126.

Winne, D., 1973, Unstirred layer, source of biased Michaelis constant in membrane transport, *Biochim. Biophys. Acta* **298**:27–31.

Winne, D., 1977a, Shift of pH-absorption curves, *J. Pharmacokinet. Biopharm.* **5**:53–94.

Winne, D., 1977b, The influence of unstirred layers on intestinal absorption, in: *Intestinal Permeation* (M. Kramer and F. Lauterbach, eds.) Excerpta Med, Amsterdam, pp. 58–64.

Winne, D., 1978a, Dependence of intestinal absorption *in vivo* on the unstirred layer, *Naunyn-Schmiedebergs Arch. Pharmacol.* **304**:175–181.

Winne, D., 1978b, Rat jejunum perfused *in situ:* Effect of perfusion rate and intraluminal radius on absorption rate and effective unstirred layer thickness, *Naunyn Schiedebergs Arch. Pharmacol.* **307**:265–274.

Winne, D., 1978c, The permeability coefficient of the wall of a villous membrane, *J. Math. Biol.* **6**:95–108.

Winne, D., and Markgraf, I., 1979, The longitudinal intraluminal concentration gradient in the perfused rat jejunum and the appropriate mean concentration for calculation of the absorption rate, *Naunyn-Schmiedeberg's Arch. Pharmacol.* **309**:271–279.

Winne, D., Kopf, S., and Ulmer, M.-L., 1979, Role of unstirred layer in intestinal absorption of phenylalanine *in vivo, Biochim. Biophys. Acta* **550**:120–130.

Womack, W. A., Barrowman J. A., Graham, W. H., Benoit, J. N., Kvietys, P. R., and Granger, D. N., 1987, Quantitative assessment of villous motility, *Am. J. Physiol.* **252**:G250–256.

Yuasa, H., Miyamoto, Y., Iga, T., and Hanano, M., 1984, A laminar flow absorption model for a carrier-mediated transport in the intestinal tract, *J. Pharm. Dyn.* **7:**604–606.

Yuasa, H., Miyamoto, Y., Iga, T., and Hanano, M., 1986, Determination of kinetic parameters of a carrier-mediated transport in the perfused intestine by two-dimensional laminar flow model: Effects of the unstirred water layer, *Biochim. Biophys. Acta* **856:**219–230.

Zachowski, A., and Devaux, P. F. Transmembrane movements of lipids, *Experientia* **46:**644–656.

CHAPTER 9

The Role of Fatty Acid Binding Proteins in Enterocyte Fatty Acid Transport

JUDITH STORCH

1. Introduction

The gastrointestinal tract is enormously efficient at digesting and absorbing the large quantities of lipid with which it is regularly presented. Dietary triacylglycerol (TG) intake on a Western-style diet can average 90 to 100 gm/day, with additional lipid input to the intestinal lumen coming from biliary lipids as well as from cellular and bacterial membrane debris. Despite the large substrate load, more than 95% of these lipids are effectively absorbed (Carey and Hernell, 1992). For TG, it is clear that digestive lipase redundancy accounts, in part, for this efficiency, with at least three distinct gastrointestinal lipases participating in the lumenal hydrolysis of TG to its major products, unesterified fatty acids and sn-2 monoacylglycerol. In addition to this robust digestive capacity, the intestine must also possess large absorptive and intracellular processing capacities in order to effectively take up and utilize the lipase products. It is generally thought that fatty acid binding proteins (FABPs) are key cellular proteins involved in the intracellular transport of fatty acids, and, in some cases, other hydrophobic ligands. It is of interest that the absorptive enterocyte abundantly expresses two types of FABPs, whereas most other cell types, even those that also process large quantities of fatty acids such as the adipocyte or the skeletal muscle myocyte, express only a single predominant FABP form.

The two intracellular FABPs that are present in the proximal intestinal enterocyte are the 15.1 kDa intestinal FABP (IFABP; Ockner and Manning, 1974) and the 14.1 kDa liver-type FABP (LFABP; Mishkin et al., 1972; Ockner et al., 1972). Other FABP family members are also expressed in the gastrointestinal tract; for example the heart-type FABP is found in the stomach (Kanda et al., 1989), and the distal small intestine contains the so-called ileal lipid-binding protein (originally termed gastrotropin), which has been reported to bind bile acids as well as fatty acids (Sacchettini et al., 1990; Lucke et al., 1996). Fur-

JUDITH STORCH • Department of Nutritional Sciences, Rutgers University, New Brunswick, New Jersey 08901–8525.
Intestinal Lipid Metabolism, edited by Charles M. Mansbach II et al., Kluwer Academic/Plenum Publishers, 2001.

thermore, cellular retinol binding protein II, also a member of the FABP gene superfamily, is highly expressed in the proximal enterocyte (Herr *et al.,* 1993). It has been estimated that the IFABPs and the LFABPs together represent approximately 5% of messenger RNA in the absorptive cell (Alpers *et al.,* 1984; Gordon *et al.,* 1982). Such abundant expression leads to approximately equivalent cellular concentrations of 0.1 to 0.3 mM (Bass, 1985; Alpers *et al.,* 1984). It has long been hypothesized that these two proteins participate in the intracellular fatty acid transport process, and recent efforts have focused on defining the functions and mechanisms of actions by which this may occur.

This chapter compares and contrasts the biochemical and structural characteristics of the two FABPs and then reviews the major factors that regulate their expression. Each of these topics affords important albeit indirect information regarding the fatty acid transport functions of these FABPs and the likely possibility that each of the proteins plays a different role in the cellular lipid transport process. The experimental evidence that more directly addresses the roles of the intestinal FABPs in fatty acid transport is then discussed, followed by a consideration of the potential structural mechanisms by which IFABP and LFABP may transport fatty acids within the enterocyte.

2. Comparative Properties of LFABP and IFABP: Functional Implications

2.1. Identification and Distribution

A fatty-acid binding activity was initially identified in intestinal mucosa by Ockner and co-workers (Ockner *et al.,* 1972), who proposed that it participates in the intracellular transport of fatty acid. They later demonstrated that inhibition of fatty acid binding to FABP resulted in the inhibition of fatty acid incorporation into triacylglycerol (Ockner and Manning, 1976), further supporting a role in the transport and metabolic trafficking of fatty acids. Similar molecular weight FABPs had also been identified in liver and in other tissues (Ockner *et al.,* 1972; Mishkin *et al.,* 1972), and it became clear that there existed a family of 14 to 15 kDa intracellular proteins that bound fatty acids and other hydrophobic ligands in a wide variety of mammalian cell types (Bass, 1985). The small intestine was found to contain high levels of both the intestinal FABP and the liver FABP, the latter thus termed because of its initial site of identification (Table 1). Indeed, the levels of expression of LFABP in the hepatocyte and the enterocyte are similar (Bass *et al.,* 1985; Gordon *et al.,* 1985). In the mouse, it was shown that mRNA levels for LFABP and IFABP are highest in the jejunum, with the peak of LFABP expression in the proximal jejunum and maximal IFABP mRNA levels in the more distal jejunum (Sweetser *et al.,* 1988a, 1988b). Further, the extent of IFABP expression along the duodenal to ileal axis was shown to be broader than that of LFABP, with high expression of the former extending from the duodenum through the ileum, in contrast to the somewhat narrower distribution of LFABP in the duodenum and jejunum (Sacchettini *et al.,* 1990). As tissue levels of FABP protein are almost uniformly reported to be regulated at the transcriptional level, it is likely that IFABP and LFABP levels in the intestinal tract closely reflect these mRNA levels. Expression of LFABP begins at the crypt–villus junction and is highest in villus cells, with a small decline as cells reach the villus tip (Iseki and Kondo, 1990). Indeed it has been shown that

Table 1. Comparison of Liver and Intestinal Fatty Acid Binding Proteins

PROPERTY	LFABP	IFABP	References
Molecular weight	14.1 kDa	15.1 kDa	Mishkin *et al.,* 1972; Ockner *et al.,* 1972; Ockner and Manning, 1974
Endogenous ligand(s)	Long chain FA, lyso-PL, MG, specific eicosanoids, bile salts, fatty acyl CoA's, (cholesterol?)	Long chain FA, (fatty acyl CoA's?)	Bass, 1988; Storch, 1993; Thumser and Wilton, 1995; Burrier and Brecher, 1986; Takikawa and Kaplowitz, 1986; Dutta-Roy *et al.,* 1987; Wilkinson and Wilton, 1987; Burrier *et al.,* 1987; Hubbell *et al.,* 1994; Thumser and Wilton, 1996; Rolf *et al.,* 1995; Bass, 1985; Nemecz *et al.,* 1991a
FA:FABP stoichiometry	2:1	1:1	Thompson *et al.,* 1997; Sacchettini *et al.,* 1989 Eads *et al.,* 1993
K_D (nM) of FABP for:			Richieri *et al.,* 1994
Palmitate	23 ± 2	30 ± 2	
Oleate	9 ± 2	39 ± 9	
Sites of high expression	Liver, SI	SI	Bass *et al.,* 1985; Gordon et *et al.,* 1985; Bass, 1985; Alpers *et al.,* 1984
Intestinal distribution	Duodenum and jejunum	Duodenum, jejunum, and ileum	Sacchettini *et al.,* 1990
Site of maximal intestinal expression	Proximal jejunum	Distal jejunum	Sweetser *et al.,* 1988a; Sweetser *et al.,* 1988b
Mechanism of transfer of fluorescent FA from FABP to PL membranes	Diffusion through aqueous phase	Direct interaction with membranes	Kim and Storch, 1992; Hsu and Storch, 1996

Abbreviations: SI, small intestine; FA, fatty acid; PL, phospholipid; MG, monoacylglycerol

LFABP is not expressed in villus tip cells at all, in contrast to IFABP which is (Hallden and Aponte, 1997). At the subcellular level, IFABP and LFABP have been found to be have similar cytoplasmic distributions (Shields *et al.,* 1986). Other immunohistochemical studies suggested that LFABP levels are higher in basolateral relative to apical cytosol (Iseki *et al.,* 1989). It has often been pointed out that the expression of LFABP and IFABP in the intestine closely tracks the distribution of lipid uptake and intracellular packaging of intestinal lipoproteins, as both these processes are maximal in the villus cells of the proximal jejunum (Kaikaus *et al.,* 1990; Sweetser *et al.,* 1987; Glatz and van der Vusse, 1996). The slightly shifted profiles of IFABP and LFABP expression, on the other hand, suggest that the relationship is not precisely quantitative, implying that each protein may have somewhat different functional properties. Nevertheless, such coordinate expression on a qualitative basis provides indirect support for FABPs as mediators of intestinal lipid assimilation.

2.2. Structure and Ligand Binding

IFABP was among the first of the FABP family of proteins whose tertiary structure was obtained at high resolution (Sacchettini *et al.,* 1988). It is now known that the structure of IFABP is in fact fully representative of the entire FABP gene superfamily, including the cellular retinoid-binding proteins (Sacchettini *et al.,* 1993). The proteins are composed of 10 antiparallel β-strands that form a barrel-like structure containing the ligand binding cavity, with the barrel capped by two short α-helical segments (Banaszak *et al.,* 1994). The two helices and the closely appositioned β-turns are generally referred to as the portal region of the FABP, where it is hypothesized that ligands may enter and exit the binding cavity. The structural and dynamic properties of IFABP and, more recently, LFABP, are being explored in detail using X ray crystallography and NMR spectroscopy (Thompson *et al.,* 1997; Hodsdon *et al.,* 1996; Sacchettini and Gordon, 1993; Wang *et al.,* 1998). Although the overall fold of the two proteins is quite similar, inspection of their holostructures demonstrates unique fatty acid binding characteristics. Previous biochemical and spectroscopic studies had suggested differential ligand binding stoichiometries for IFABP and LFABP (Haunerland *et al.,* 1984; Cistola *et al.,* 1989; Schulenberg-Schell *et al.,* 1988; Keuper *et al.,* 1985), as well as the potential for different fatty acid orientations within the respective binding pockets (Cistola *et al.,* 1989; Kim and Storch, 1992). Thus, cocrystallization of IFABP with various fatty acids resulted in a 1:1 protein:ligand complex (Sacchettini *et al.,* 1989; Eads *et al.,* 1993), whereas the LFABP crystal structure revealed a 2:1 ratio of oleic acid:LFABP (Thompson *et al.,* 1997) (Figure 1). For IFABP, the carboxylate group of the fatty acid ligand forms part of a five-membered hydrogen bonding network that includes the interior Arg^{106} residue. For LFABP, the carboxylate of one of its bound oleates is also part of an extensive hydrogen bonding network, including the interior Arg^{122}, with the acyl chain largely shielded from the exterior of the protein. In contrast, the carboxylate of the other LFABP-bound oleate has a more solvent-exposed position, interacting with several residues near the portal region of the protein (Thompson *et al.,* 1997). Recent NMR studies indicate that the two fatty acid binding sites of LFABP become occupied in a sequential fashion (He *et al.,* 1999).

The two enterocyte FABPs are distinguished by their ligand specificities as well as by their fatty acid binding capacities. Both LFABP and IFABP bind longer chain length fatty acids only, and do not interact with medium- or short-chain-length fatty acids ($\leq C_{12}$). LFABP binds a number of other endogenous hydrophobic ligands such as lysophospholipids, monoacylglycerol, certain eicosanoids, and bile salts (Bass, 1988; Storch, 1993; Thumser and Wilton, 1995; Burrier and Brecher, 1986; Takikawa and Kaplowitz, 1986; Dutta-Roy *et al.,* 1987; Riehl *et al.,* 1990), whereas IFABP appears to bind exclusively long-chain fatty acids (Lowe *et al.,* 1987). The fact that LFABP can bind both the fatty acid and MG products of lumenal TG hydrolysis is noteworthy, providing indirect evidence for a role in dietary lipid assimilation. It has also been shown that LFABP binds fatty acyl CoAs (Wilkinson and Wilton, 1987; Burrier *et al.,* 1987; Hubbell *et al.,* 1994), and IFABP may also bind fatty acyl CoAs (Hubbell *et al.,* 1994). Although it is generally found that LFABP does not exhibit high-affinity binding of cholesterol (Thumser and Wilton, 1996; Rolf *et al.,* 1995; Bass, 1985), this remains unresolved (Nemecz and Schroeder, 1991).

Several comparative studies have not revealed consistent or substantial differences in fatty acid equilibrium-binding affinities between LFABP and IFABP (Nemecz *et al.,* 1991a,

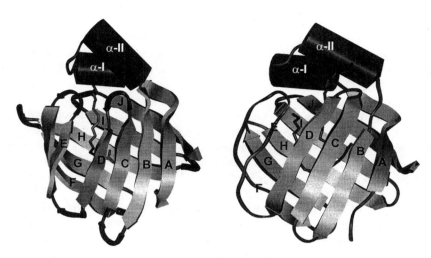

Figure 1. Tertiary structures of IFABP (left) and LFABP (right) complexed with one mol. of palmitic acid and two mol. oleic acid, respectively (Sacchettini *et al.*, 1989; Thompson *et al.*, 1997). α-helices I and II and β-strands A through J are indicated.

1991b; Lowe *et al.*, 1987); however, Richieri *et al.* using a more sensitive fluorescence-based assay, reported that LFABP and IFABP have similar affinities for saturated fatty acids but that LFABP has a fivefold greater affinity than IFABP for unsaturated fatty acids (Richieri *et al.*, 1994). In humans, a polymorphism in the IFABP gene results in a substitution of Thr^{54} for Ala^{54} in approximately 15% to 30% of various populations examined (Baier *et al.*, 1995; Rissanen *et al.*, 1997; Sipilainen *et al.*, 1997; Hegele *et al.*, 1996). The Thr^{54}-containing form of human IFABP was shown to have twofold greater affinity for fatty acids than the Ala^{54} form (Baier *et al.*, 1995). The affinities of the two LFABP binding sites differ by up to tenfold, however the differences diminish with temperature, with a unique temperature dependence for each fatty acid, such that the affinities for two bound palmitates are roughly equivalent at physiologic temperature, whereas those of two bound oleates remain divergent (Richieri *et al.*, 1996). Thus it appears that LFABP is able to bind more fatty acids with, on average, a higher affinity relative to IFABP. Based in part on such observations, it has been suggested that LFABP may serve primarily as an intracellular reservoir for fatty acids and, perhaps, other hydrophobic ligands. It is worth noting, however, that the levels and types of ligands that each of these proteins actually binds within the enterocyte milieu has not been determined. This seemingly simple question has proven difficult to address, owing to a combination of factors including the presence of other intracellular proteins that can also bind the various FABP ligands such as fatty acyl CoAs (acyl CoA binding protein), fatty acids (putative membrane fatty acid transporters), and possibly cholesterol (sterol binding proteins); the uncertain amount and affinity of intracellular membranes into which fatty acids and other potential ligands can partition; and the uncertainties regarding intracellular concentrations of the various ligands.

2.3. Regulation of Enterocyte FABP Expression

The developmental regulation of IFABP and LFABP mRNA levels in the rat small intestine appears qualitatively similar. Earliest expression is detected during late gestation, followed by marked upregulation within 24 hours of birth, a decline at the suckling–weaning transition, and a subsequent further increase in expression during adult maturation (Gordon *et al.*, 1985). The decline at weaning is thought to be related to the abrupt switch in diet from high-fat maternal milk to low-fat chow, although this has not been directly evaluated. Though protein expression studies in intestine have not been reported, it was found that in the liver the LFABP levels during development paralleled the described changes in mRNA levels (Sheridan *et al.*, 1987); thus it is likely that intestinal expression of the FABPs reflects their mRNA levels as well.

As noted previously, the expression of LFABP and IFABP in the intestine is restricted to particular regions along the cephalocaudal axis, as well as to specific cells along the villus–crypt axis. Gordon and colleagues have used transgenic mice expressing fusion genes, constructed of various enterocyte FABP promoter sequences linked to the human growth hormone gene, to map the regulatory elements responsible for these complex patterns of expression. They have demonstrated that the gradients of intestinal expression of LFABP are maintained by multiple suppressor elements distributed between nucleotides -4000 and $+21$ of the LFABP 5' flanking region, whereas suppression of expression in the stomach was apparently regulated by sequences outside this region (Simon *et al.*, 1993). A specific 35-nucleotide sequence containing a heptad repeat motif at position -132, shown to be important for the suppression of kidney expression, binds to a 90 kDa nuclear protein in a number of cell types including intestine, and it has been suggested that this protein may represent a novel transcription factor that modulates epithelial-specific gene expression (Simon *et al.*, 1997). A unique 20-base–pair element in the IFABP proximal promoter region functions as a suppressor of expression in the distal small intestine and colon as well as in crypt cells, and it has been shown to bind as yet unidentified small intestinal nuclear proteins (Simon *et al.*, 1995).

Consensus binding sites for the general transcription factor CCAAT/enhancer binding protein (C/EBP) are found in both the LFABP and the IFABP genes, whereas the presence of other regulatory elements may suggest differential regulation and therefore function. In the IFABP gene, for example, are found several copies of a 14-base–pair sequence that is known to bind the steroid hormone receptor family members hepatic nuclear factor-4 and apolipoprotein regulatory protein-1 (Rottman and Gordon, 1993). This consensus sequence is also present in the 5' flanking regions of cellular retinol binding protein II and apolipoprotein AI, which are also highly expressed in the intestinal absorptive cell, suggesting coordinate regulation of several genes believed to be involved in the cellular processing of dietary lipid. The LFABP promoter contains a consensus peroxisome-proliferater response element (PPRE) located between nucleotides -66 and -75 (Simon *et al.*, 1993), similar to those found in a number of genes involved in peroxisomal fatty acid oxidation, suggesting a role for LFABP in this process. As LFABP is not thought to be present within peroxisomes *per se,* its role if any is uncertain but may relate to the maintainance or delivery of substrate supply.

Not surprisingly, potent regulators of LFABP expression in intestine include peroxisome proliferaters such as clofibrate, which caused more than a doubling of intestinal LFABP mRNA and protein levels (Bass *et al.*, 1985). Although a twofold change in expression is modest in a relative sense, the high concentration of the protein in the enterocyte implies an enormous absolute increase in LFABP content. No PPRE has been described in the IFABP gene, and clofibrate treatment caused no change in IFABP mRNA levels and a small increase in IFABP protein levels, which could be a secondary effect due to alterations in other peroxisome-proliferater-responsive genes, perhaps LFABP (Bass *et al.,* 1985). Indeed, in Caco-2 cells transfected with the human IFABP gene, expression of LFABP declined, implying that each enterocyte FABP could be involved in modulating the others' expression (Baier *et al.,* 1996).

LFABP and IFABP levels are also regulated somewhat differently by dietary lipid content. Long-term feeding of a diet containing 20% to 40% fat resulted in increased LFABP expression in the proximal small intestine (Bass, 1988; Lin *et al.,* 1994) and increased IFABP levels in the distal small intestine (Ockner and Manning, 1974). Low-fat/high-carbohydrate diets led to decreased levels of both FABPs (Bass, 1988; Ockner and Manning, 1974). It was recently shown that simply doubling the lipid content in mouse diet from 3% to 6% by weight resulted in two- to threefold increases in LFABP mRNA and protein levels in proximal intestine (Poirier *et al.,* 1997). No alterations were found in IFABP levels, but this may be due to the lower levels of lipid used relative to earlier reports. Furthermore, the various studies have used different dietary oils, and control groups have in some cases been fed chow diets that contain different TG fatty acid compositions than the experimental diets; thus it remains uncertain whether lipid type and/or amount may be responsible for the observed changes in gene expression. Despite very low levels of LFABP in control mouse ileum, ileal infusion of small quantities of various long-chain fatty acids resulted in the appearance of LFABP message in that segment, with arachidonic and docosahexaenoic acids having greatest impact. Interestingly, the most potent modulator of LFABP expression in this system was a nonmetabolizable analogue of palmitic acid, α-bromopalmitate, suggesting that unesterified fatty acids are responsible for the effects of dietary lipid (Poirier *et al.,* 1997).

Intestinal isograft implantation studies have demonstrated that the developmental and proximal–distal gradients of intestinal FABP expression are not dependent on the presence of lumenal lipid or, in fact, luminal contents at all (Rubin *et al.,* 1992), suggesting that the nutritional regulation of FABP expression is apparently limited to modification of programmed patterns of expression, similar to what has been proposed for enzymes of intestinal lipid metabolism (Trotter and Storch, 1993a, 1993b). Nevertheless, the villus-to-crypt gradient of LFABP mRNA expression in neonatal intestinal isografts was substantially disordered, with a more homogeneous distribution throughout the villus than that found for normal intestine (Gutierrez *et al.,* 1995). This pattern is similar to that which has been found following fasting (Iseki and Kondo, 1990). It may be, therefore, that signals from the intestinal lumenal milieu, which could include dietary or biliary lipids or other nutrients, as well as hormonal or nervous stimuli, are important in regulating the levels of enterocyte FABP expression. The recapitulation of the effects of lipids on FABP expression in Caco-2 cells (Poirier *et al.,* 1997) indicates that nutritional rather than hormonal signals or tissue innervation are likely to be the responsible mediators. In fact, there is currently little evidence to support hormonal regulation of LFABP or IFABP expression in intestine. Although the IFABP 5' flanking region contains cAMP response elements (Veerkamp and Maatman, 1995), their functional significance is not known. In addition, the levels of both proteins are essentially equivalent in male and female rat intestine (Bass *et al.,* 1985).

The presence of a PPRE in the LFABP gene is suggestive that, as with other genes found to be regulated by fatty acids and by fibrate drugs, regulation of expression in response to lipid may be mediated by nuclear peroxisome proliferator-activated receptor (PPAR) transcription factors. In the intestine, the PPARα, PPARδ (also termed the fatty-acid-activated receptor or FAAR), and PPARγ subtypes have been identified (Braissant *et al.*, 1996; Poirier *et al.*, 1997). The miminal induction of IFABP by peroxisome proliferaters and by dietary fatty acids may reflect the absence of a PPRE. The functional implications of the differential regulation of LFABP and IFABP have not been investigated as yet. It may be speculated that LFABP, despite its relatively high level of constitutive expression, can be induced to accommodate increased lipid load caused either by dietary fat or lipid metabolic flux, consistent with a possible role as a cellular buffer for fatty acids. If IFABP has a more specialized role to play in fatty acid disposition in the enterocyte, its constitutive expression may already be maximal, and elevated fatty acid levels can be accomodated by its homologue, LFABP.

Taken together, the abundant expression, ligand-binding properties, localization, and regulation of expression of IFABP and LFABP point to a central role for these proteins in intestinal lipid absorption and processing, and afford indirect evidence for a role for FABPs in cellular fatty acid transport. There is also a large body of work demonstrating that addition of LFABP to *in vitro* systems of membrane-bound lipid metabolic enzymes can modulate enzyme activities (reviewed in Bass, 1985; Glatz and van der Vusse, 1996). Not all of these results are consistent, however, most likely reflecting differential substrate partitioning based on the relative amounts and types of subcellular membrane fractions employed and the levels of FABP used. These data, although not as clear-cut as those pertaining to protein expression and ligand binding, also afford indirect support for a role for FABPs in intracellular fatty acid disposition.

3. Functions of Intestinal FABPs in Lipid Transport

This section reviews the experimental approaches that more directly address the transport function of the enterocyte FABPs. In this regard, transport can be defined in a number of ways. It may be taken simply to mean the property of ligand binding based on equilibrium affinity, with consequent effects on ligand partitioning to cell membranes or enzymes, which may in turn influence net uptake and, hence, cellular lipid metabolism. Typically, however, a transporter is considered a protein that moves its ligand from one site to another, as in the case of a transmembrane transporter, for instance. For an intracellular protein to function as a transporter in this latter sense, it must act in a directional manner within the cell, serving to target its ligand in some specific manner based on properties in addition to simple ligand-binding affinity. As is discussed, there is considerable theoretical and experimental evidence indicating that LFABP and IFABP are involved in the overall process of cellular fatty acid transport. Whether they behave directly as true transporters, and/or indirectly as cytoplasmic sites of fatty acid partitioning, is not yet definitively known.

3.1. Studies in Intact Cells

An interesting and potentially revealing approach to the question of FABP function has been a series of studies utilizing transfection of the LFABP or IFABP gene into L-cell fibroblasts or embryonic stem cells, with subsequent assessment of various parameters of lipid transport and metabolism (Jefferson *et al.,* 1990, 1991; Schroeder *et al.,* 1993; Prows *et al.,* 1995; Murphy *et al.,* 1996; Prows and Schroeder, 1997; Atshaves *et al.,* 1998). The levels of expression obtained were somewhat low relative to those found in proximal jejunum, however, they were greater than the levels of endogenous fatty-acid-binding proteins found in the host cells (Jefferson *et al.,* 1991; Prows *et al.,* 1995; Prows and Schroeder, 1997) and thus afford an opportunity to compare cells with and without LFABP or IFABP expression. It was shown that the increase in fluorescence intensity was twofold greater when the fluorescent *cis*-parinaric fatty acid analogue was added to cells expressing LFABP compared to control cells or cells expressing IFABP, and this was interpreted as evidence for an LFABP-mediated transport function (Prows *et al.,* 1995). As the quantum yield for *cis*-parinaric acid binding to LFABP compared with IFABP differs by a similar degree, and as that of the endogenous FABP is not known, however, interpretation of these data in terms of relative transport properties is uncertain. When IFABP was expressed at a twofold higher concentration in L-cells, it was found that *cis*-parinaric acid uptake actually decreased by approximately 25% relative to the cells with lesser IFABP levels (Prows *et al.,* 1997). Conversely, however, the decreased level of IFABP expression in differentiated relative to undifferentiated embryonic stem cells was also correlated with a decrease in uptake (Atshaves *et al.,* 1998). Whether these seemingly opposing results are due to host cell differences, alterations in related aspects of lipid transport in different clonal populations, or other factors, the interpretation of these relatively modest effects of IFABP expression on fatty acid uptake is not straightforward. An absence of dose-dependent effects of IFABP on fatty acid uptake in the embryonic stem cells was also observed (Atshaves *et al.,* 1998), as was an absence of IFABP dose response for the incorporation of radiolabeled fatty acid into various neutral lipid classes (Prows and Schroeder, 1997), again raising questions regarding an interpretation in terms of transport effects. Finally, although some evidence for differential metabolic targeting of oleate by LFABP relative to IFABP was found, the changes were for the most part small and in some cases inconsistent (Murphy *et al.,* 1996; Prows *et al.,* 1995). Interestingly, the L-cells expressing LFABP were found to have an increase in total cellular phospholipids relative to controls; thus a higher membrane content might also contribute to the apparent increases in net fatty acid uptake (Murphy *et al.,* 1996). Whether this greater phospholipid concentration is somehow a direct result of LFABP expression or is secondary to other metabolic changes, to clonal variability, or to other processes is not known.

In Caco-2 intestinal cells transfected with either of the two human forms of IFABP, it was found that the Thr^{54}-IFABP-transfected cells had a twofold increase in net fatty acid uptake and secreted approximately fivefold more triacylglycerol into the basolateral medium relative to Ala^{54}-IFABP-transfected cells (Baier *et al.,* 1996). As mentioned earlier, it was found that in differentiated cells of both lines, the endogenous levels of LFABP were decreased relative to control cells, although LFABP levels nevertheless remained two- to threefold higher than levels of IFABP (Baier *et al.,* 1996). Moreover, it has been subsequently demonstrated that, contrary to earlier indications, Caco-2 cells do in fact express IFABP (Le Beyec *et al.,* 1997). Thus, interpretation of these interesting experiments by direct comparisons to control Caco-2 cells is difficult, and a direct assessment of IFABP function is complex. IFABP or LFABP expression may lead to alterations in cellular fatty acid levels that, in turn, lead to alterations in the expression of other relevant gene products. Overall, therefore, the transfection studies have provided some indications of a transport function for the enterocyte FABPs and remain a promising approach, however, results to date remain suggestive but inconclusive.

The most convincing data that LFABP is involved in intracellular fatty acid transport have come from an elegant series of experiments initiated by Luxon and Weisiger, in which the effective diffusion (D_{eff}) of the fluorescent fatty acid NBD-stearate (NBDS) was evaluated in LFABP-expressing hepatocytes or Hep G2 cells (Luxon and Weisiger, 1993; Luxon, 1996; Luxon and Milliano, 1997). Using the technique of fluorescence recovery after photobleaching (FRAP), they showed that the D_{eff} for NBDS was 65% faster in hepatocytes from female compared to male rats and that this difference correlated closely with the cellular LFABP levels (Luxon and Weisiger, 1993). Analysis of the partitioning of NBDS between cellular membranes compared to the cytosolic compartment showed that twice as much NBDS was present in cytosol from female rat hepatocytes relative to samples from male cells. This was interpreted as underlying the differences in diffusion rate, because membrane-bound fatty acid would be far less mobile than FABP-bound fatty acid (Luxon and Weisiger, 1993). As there are clearly many additional differences between male and female cells in addition to their levels of LFABP, subsequent experiments utilized increasing concentrations of α-bromopalmitate in order to displace NBDS from LFABP. This resulted in a proportional decrease in the D_{eff}, with the displaced NBDS partitioning to a greater extent into relatively immobile subcellular membranes (Luxon, 1996). Streptolysin O treatment, used to promote leakage of cytosolic proteins including LFABP, resulted in a dramatic decrease in the rate of fatty acid movement in Hep G2 cells, consistent with low levels of solubilized versus membrane-bound probe (Luxon and Milliano, 1997). In a series of "model cytoplasm" experiments, samples containing defined levels of liposomes and various FABPs were analyzed using the FRAP method. The results showed that albumin and LFABP caused a concentration-dependent increase in the rate of NBDS diffusion and a decrease in membrane-bound NBDS. Notably, the effect of albumin was 100- to 300-fold greater than that of LFABP, in direct proportion to its greater fatty-acid-binding affinity and capacity. This suggested that the effects of LFABP may be relatively nonspecific, such that any protein that prevents fatty acids from partitioning into "immobile" membranes will increase the D_{eff}. (Luxon and Milliano, 1997). Interestingly, however, when streptolysin O-treated Hep G2 cells were incubated with protein-containing solutions in order to create defined cytoplasmic conditions, albumin was only fourfold more effective than equimolar LFABP in increasing the NBDS diffusion rate (Luxon and Milliano, 1997), implying perhaps that effects of LFABP within the cell may be more specific than those of a generic FABP. In the embryonic stem cells transfected with two different levels of IFABP discussed previously, a dose-dependent increase in NBDS D_{eff} was observed (Atshaves *et al.*, 1998). Collectively, the solution FRAP experiments are quite intriguing in that they demonstrate in intact cells that enterocyte FABPs may indeed play a direct role in intracellular fatty acid transport.

3.2. Modeling of FABP Function in Fatty Acid Transport

In addition to the cellular studies discussed previously, various other approaches have been taken to address the question of an FABP transport function. Although a number of studies in model systems have used other members of the FABP protein family, results may be relevant to the enterocyte FABPs as well, and are in some instances included in the following discussion.

LFABP was found to increase the flux of oleic acid through a lipid–water interface threefold (Weisiger *et al.,* 1989). Similarly, LFABP and heart FABP were reported to increase the amount and rate of movement of radiolabeled fatty acid between two compartments, although a substantial loss of label to the surface of the apparatus was not accounted for (Peeters *et al.,* 1989). The addition of fish muscle FABP to a cytosolic preparation devoid of FABPs under conditions where unbound ligand was in great excess, resulted in a sixfold increase in the apparent diffusion coefficient for oleic acid (Stewart *et al.,* 1991). Theoretical treatments have also concluded that cytoplasmic FABPs could mediate an increase in fatty acid diffusive flux across a cell, largely by virtue of increasing the gradient for unbound fatty acids across the plasma membrane and reducing the membrane-bound fraction of fatty acid (Tipping and Ketterer, 1981).

In a series of *in vitro* experiments using a fluorescence resonance energy transfer assay, a quantitative as well as a mechanistic comparison of the fatty acid transfer properties of LFABP and IFABP was undertaken (Hsu and Storch, 1996). The time scale of fatty acid movement from an FABP to a membrane is on the scale of seconds or less, necessitating the use of fluorescent-tagged fatty acids in order to obviate the need for physical separation of the FABP (fatty acid donor) and the phospholipid membranes (fatty acid acceptor). The studies examined the rate of anthroyloxy-labeled fatty acid (AOFA) movement from the FABP to model acceptor vesicles containing a nonexchangeable quencher of AOFA fluorescence, NBD-phosphatidylcholine. The results showed that fatty acid transfer from IFABP to membranes was faster than from LFABP. For an AOFA analogue of the saturated fatty acid stearate, and for a derivative of the unsaturated oleate, transfer was approximately tenfold faster from IFABP (Hsu and Storch, 1996; Storch *et al.,* 1996).

It was also found that transfer of AOFA from LFABP or IFABP to model membranes was described by entirely different mechanisms. In order to distinguish between transfer that occurred by diffusion of AOFA through the aqueous medium from that which occurred via direct interactions of the FABP with acceptor membranes, the approach taken was to modulate the aqueous phase solubility of the ligand and determine whether corresponding changes in transfer rate from FABP were observed. For AOFA transfer from LFABP, the rate was directly proportional to ligand solubility; increasing fatty acid chain length, acyl chain saturation, and increasing aqueous phase ionic strength all resulted in slower rates of AOFA transfer, consistent with decreased aqueous solubility of the fatty acid (Kim and Storch, 1992; Hsu and Storch, 1996). Moreover, the rate of AOFA transfer from LFABP to membranes was modulated by neither the concentration of acceptor membranes nor their composition, indicating that transfer rate was independent of the membrane acceptor and occurred via dissociation of the AOFA from LFABP followed by aqueous diffusion (Hsu and Storch, 1996; Kim and Storch, 1992). Earlier studies examining the rate of acyl CoA synthesis in microsomes separated from donor vesicles by a dialysis membrane suggested that LFABP increased enzyme activity by enhancing dissociation of fatty acid substrate from the donor membrane into the aqueous phase (McCormack and Brecher, 1987), in agreement with the present hypothesis.

The results obtained for AOFA transfer from IFABP to membranes were markedly different, in that the transfer rates were directly proportional to the concentration of acceptor vesicle phopholipid. Furthermore, the rate of AOFA transfer showed a large degree of regulation by the phospholipid composition of the membranes, with transfer to negatively charged membranes occurring dramatically faster than transfer to zwitterionic membranes (Hsu and Storch, 1996). These results are strongly suggestive of an interaction between

IFABP donor and membrane acceptor, and they imply that fatty acid transfer from IFABP to a phospholipid membrane involves an intermediate in which the protein and the membrane are in physical contact. Based on these differences between IFABP and LFABP, as well as on other indirect evidence as discussed earlier, it was hypothesized that fatty acid transfer from IFABP may involve targeted interactions of the protein with specific membrane lipid and/or protein domains, whereas LFABP may function in the capacity of a cytosolic reservoir for fatty acids (Hsu and Storch, 1996; Storch *et al.*, 1996). Thus, both of the enterocyte FABPs are hypothesized to function in the transport of fatty acid, albeit via distinctly different mechanisms and, presumably, with distinctly different cellular effects.

In a recent report, the transfer of fatty acid between two membranes has been modeled in the absence and presence of a ligand-binding protein under defined conditions, in order to simulate the potential role of FABPs in modulation of fatty acid transfer within a cell (Vork *et al.*, 1997). It was demonstrated that a large range of dissociation and association rates for fatty acid with FABP was without substantial impact on simulated net transfer between the two membrane surfaces unless collisional interaction of the FABP with the membranes was included in the model. Net transfer was most efficient for the case where FABP–membrane interactions occurred at both the donor and acceptor surfaces (Vork *et al.*, 1997). In this regard, we have recently found that AOFA transfer from phospholipid membranes to IFABP, that is, in the opposite direction from that previously examined (Hsu and Storch, 1996), appears to occur during effective collisional interactions between membrane donor and protein acceptor (Thumser and Storch, 2000). These data suggest that fatty acid trafficking by IFABP may utilize membrane–protein interactions for both acquisition and delivery of ligand. Interestingly, results to date indicate that the regulation of AOFA transfer in both directions is not identical, implying that "collisional complexes" between IFABP and membranes may be different for holo- and apoproteins (Thumser and Storch, 2000). Although X ray crystallographic analysis revealed only small differences in conformation between apo- and holo-IFABP's (Scapin *et al.*, 1992; Sacchettini *et al.*, 1992), recent studies by Cistola and colleagues have demonstrated substantial differences in protein backbone disorder, with greater flexibility found in discrete regions of apo-IFABP compared to the holoprotein (Hodsdon and Cistola, 1997a, 1997b). It is possible, therefore, that structural differences between IFABP with or without bound ligand could be differentially recognized at target membranes.

3.3. Structural Mechanism of Fatty Acid Transport from Enterocyte FABPs to Membranes

The hypothesis that IFABP can interact with membranes has now been given experimental support, using two separate approaches. In the first, it has been shown that preincubation of anionic vesicles with IFABP can prevent the subsequent binding of the peripheral membrane protein, cytochrome c (Corsico *et al.*, 1998). In addition, as monitored using infrared spectroscopy, monolayers of dimyristoylphosphatidic acid develop lipid domains on addition of IFABP to the underlying subphase (Wu, *et al.*, 2000).

The large effects of membrane charge on AOFA transfer rate suggest that positively charged amino acid residues on the surface of IFABP are important for the formation of a collisional complex with the membrane. Indeed, it was recently found that neutralization of IFABP surface lysine residues by chemical modification eliminated the effects of membrane surface charge on fatty acid transfer rate (Hsu and Storch, unpublished observations). Studies with heart and adipocyte FABPs showed qualitatively similar results, further implicating cationic surface residues in protein–membrane interactions (Herr *et al.*, 1995, 1996). Site-directed mutagenesis of specific lysine residues in heart FABP (Herr *et al.*, 1996) and adipocyte FABP (Liou and Storch, 1998) has shown that the so-called portal domain of these FABPs, comprising the helix-turn-helix region and opposing β-turns (Fig. 1), is likely to be involved in collisional transfer of fatty acids to membranes. Interestingly, it is this same portal region that shows the greatest degree of conformational difference between holo- and apo-IFABP (Hodsdon and Cistola, 1997a).

In order to test the hypothesis that the helical domain of IFABP is involved in the putative "collisional" mechanism of fatty acid transfer to membranes, a helixless construct of IFABP was employed (Kim *et al.*, 1996). This IFABP variant was found to fold properly and to bind long-chain fatty acids in the interior binding cavity (Cistola *et al.*, 1996; Steele *et al.*, 1998). The results of AOFA transfer experiments clearly indicated that the mechanism of fatty acid transfer from the helixless IFABP was quite different than that of the wild-type protein in that it did not involve a collisional interaction with the membrane. Cytochrome c competition experiments supported this conjecture; little inhibition of cytochrome c–membrane interaction was observed following preincubation with the helixless protein, in contrast to the native IFABP (Corsico *et al.*, 1998). It is hypothesized that the amphipathic α-I helix of IFABP, in particular, is likely to be critical for formation of IFABP–membrane contacts, as ion–pair interactions between amphipathic helices and anionic membranes are well known (Anantharamaiah *et al.*, 1993). Although the α-II helix is not amphipathic in nature, it nevertheless forms long-range interactions with the β-turn between strands C and D and is therefore also likely to play a role in the entrance/exit of fatty acid from the binding pocket of IFABP.

4. Concluding Remarks

The potential for IFABP to acquire and deliver fatty acids via membrane–protein interactions implies that it might participate in the directed movement of fatty acids within the cell. It is worth noting that following subcellular fractionation of small intestinal absorptive cells, 16% of IFABP was found associated with membranes despite repeated washing (Ockner and Manning, 1974). If IFABP is in fact a membrane-interactive protein, is may be associating with membrane lipids, membrane proteins, or both. It is possible that the marked effects of anionic vesicles, in increasing the rate of AOFA transfer from IFABP to membranes (Hsu and Storch, 1996), may be reflecting interactions of the protein with an acidic domain(s) of a specific membrane protein(s). Logical cellular protein targets for FABP interactions might include microsomal and mitochondrial acyl CoA synthetases, for metabolic utilization of fatty acid; intracellular or membrane-bound phospholipases and triglyceride lipases, for acquisition of newly released fatty acid; and putative transmembrane fatty acid transport proteins, for acquisition of fatty acids taken up across the plasma membrane. The similar tissue localization and coordinate regulation of FABPs and the 88 kDa transmembrane fatty acid transporter (FAT) in the intestine provides indirect evidence for complementary functional properties (Poirier *et al.*, 1996). Interestingly, bovine milk fat globule membranes were shown to contain a complex of FAT (also known as CD36) and the heart type FABP (Spitsberg *et al.,* 1995), thereby suggesting that vectorial movement of fatty acids into and within cells is not only logical but also likely.

Acknowledgments

Critical review of the manuscript by Dr. Alfred A. E. Thumser and Ms. Mina Lebitz is gratefully acknowledged.

References

Alpers, D. H., Strauss, A. W., Ockner, R. K., Bass, N. M., and Gordon, J. I., 1984, Cloning of a cDNA encoding rat intestinal fatty acid binding protein, *Proc. Natl. Acad. Sci. USA* **81:**313–317.

Anantharamaiah, G. M., Jones, M. K., and Segrest, J. P., 1993, *The Amphipathic Helix,* CRC Press, Boca Raton, FL, pp. 110–143.

Atshaves, B. P., Foxworth, W. B., Frolov, A., Roths, J., Kier, A., Oetama, B., Piedrahita, J., and Schroeder, F., 1998, Cellular differentiation and I-FABP protein expression modulate fatty acid uptake and diffusion, *Am. J. Physiol.* **274:**C633–C644.

Baier, L. J., Sacchettini, J. C., Knowler, W. C., Eads, J., Paolisso, G., Tataranni, P. A., Mochizuki, H., Bennett, P. H., Bogardus, C., and Prochazka, M., 1995, An amino acid substitution in the human intestinal fatty acid binding protein is associated with increased fatty acid binding, increased fat oxidation, and insulin resistance. *J. Clin. Invest.* **95:**1281–1287.

Baier, L. J., Bogardus, C., and Sacchettini, J. C., 1996, A polymorphism in the human intestinal fatty acid binding protein alters fatty acid transport across Caco-2 cells, *J. Biol. Chem.* **271:**10892–10896.

Banaszak, L., Winter, N., Xu, Z., Bernlohr, D. A., Cowan, S., and Jones, T. A., 1994, Lipid-binding proteins: A family of fatty acid and retinoid transport proteins, *Adv. Prot. Chem.* **45:**89–151.

Bass, N. M., 1985, Function and regulation of hepatic and intestinal fatty acid binding proteins, *Chem. Phys. Lipids* **38:**95–114.

Bass, N. M., 1988, The cellular fatty acid binding proteins: Aspects of structure, regulation, and function, *Intl. Review Cytol.* **3:**143–184.

Bass, N. M., Manning, J. A., Ockner, R. K., Gordon, J. I., Seetharam, S., and Alpers, D. H., 1985, Regulation of the biosynthesis of two distinct fatty acid-binding proteins in rat liver and intestine: Influences of sex difference and clofibrate, *J. Biol. Chem.* **260:**1432–1436.

Braissant, O., Foufelle, F., Scotto, C., Dauca, M., and Wahli, W., 1996, Differential expression of peroxisome proliferator-activated receptors (PPARs): Tissue distribution of PPAR-alpha, -beta, and -gamma in the adult rat, *Endocrinology* **137:**354–366.

Burrier, R. E., and Brecher, P., 1986, Binding of lysophosphatidylcholine to the rat liver fatty acid binding protein, *Biochim. Biophys. Acta* **879:**229–239.

Burrier, R. E., Manson, C. R., and Brecher, P., 1987, Binding of acyl-CoA to liver fatty acid binding protein: Effect on acyl-CoA synthesis, *Biochim. Biophys. Acta* **919:**221–230.

Carey, M. C., and Hernell, O., 1992, Digestion and absorption of fat, *Sem. Gastrointestinal Dis.* **3:**189–208.

Cistola, D. P., Sacchettini, J. C., Banaszak, L. J., Walsh, M. T., and Gordon, J. I., 1989, Fatty acid interactions with rat intestinal and liver fatty acid-binding proteins expressed in *Escherichia coli:* A comparative ^{13}C NMR study, *J. Biol. Chem.* **264:**2700–2710.

Cistola, D. P., Kim, K., Rogl, H., and Frieden, C., 1996, Fatty acid interactions with a helix-less variant of intestinal fatty acid-binding protein, *Biochem.* **35:**7559–7565.

Corsico, B., Cistola, D. P., Frieden, C., and Storch, J., 1998, The helical domain of intestinal fatty acid binding protein is critical for collisional transfer of fatty acids to phospholipid membranes, *Proc. Natl. Acad. Sci.* **95:**12174–12178.

Dutta-Roy, A. K., Gopalswamy, N., and Trulzsch, D. V., 1987, Prostaglandin E_1 binds to Z protein of rat liver, *Eur. J. Biochem.* **162:**615–619.

Eads, J., Sacchetini, J. C., Kromminga, A., and Gordon, J. I., 1993, *Escherichia coli*-derived rat intestinal fatty acid binding protein with bound myristate at 1.5 Å resolution and *I*-FABP$^{Arg\ 106\text{->}Gln}$ with bound oleate at 1.74 Å resolution, *J. Biol. Chem.* **268:**26375–26385.

Glatz, J. F. C., and van der Vusse, G. J., 1996, Cellular fatty acid-binding proteins: Their function and physiological significance, *Prog. Lipid Res.* **35**:243–282.

Gordon, J. I., Smith, D. P., Alpers, D. H., and Strauss, A. W., 1982, Cloning of a complementary deoxyribonucleic acid encoding a portion of rat intestinal preapolipoprotein AIV messenger ribonucleic acid, *Biochem.* **21**:5424–5431.

Gordon, J. I., Elshourbagy, N., Lowe, J. B., Liao, W. S., Alpers, D. H., and Taylor, J. M., 1985, Tissue specific expression and developmental regulation of two genes coding for rat fatty acid binding proteins, *J. Biol. Chem.* **260**:1995–1998.

Gutierrez, E. D., Grapperhaus, K. J., and Rubin, D. C., 1995, Ontogenic regulation of spatial differentiation in the crypt–villus axis of normal and isografted small intestine, *Am. J. Physiol.* **269**:G500–G511.

Hallden, G., and Aponte, G. W., 1997, Evidence for a role of the gut hormone pyy in the regulation of intestinal fatty acid-binding protein transcripts in differentiated subpopulations of intestinal epithelial cell hybrids, *J. Biol. Chem.* **272**:12591–12600.

Haunerland, N., Jagschies, G., Schulenberg, H., and Spener, F., 1984, Fatty-acid-binding proteins: Occurrence of two fatty-acid-binding proteins in bovine liver cytosol and their binding of fatty acids, cholesterol, and other lipophilic ligands, *Hoppe-Seyler's Z. Physiol. Chem.* **365**:365–376.

He, Y., Wang, H., Elias, G., Hsu, K. T., Storch, J., and Stark, R. E., 1999, Titration of 15N LFABP with oleate augments information on protein-fatty acid interactions derived from other physical methods, paper presented at the sixth Keystone Symposium on Frontiers of NMR in Molecular Biology.

Hegele, R. A., Harris, S. B., Hanley, A. J., Sadikian, S., Connelly, P. W., and Zinman, B., 1996, Genetic variation of intestinal fatty acid-binding protein associated with variation in body mass in aboriginal Canadians, *J. Clin. Endocrinol. Metab.* **81**:4334–4337.

Herr, F. M., Wardlaw, S. A., Kakkad, B., Albrecht, A., Quick, T. C., and Ong, D. E., 1993, Intestinal vitamin A metabolism: Coordinate distribution of enzymes and CRBP(II), *J. Lipid Res.* **34**:1545–1554.

Herr, F. M., Matarese, V., Bernlohr, D. A., and Storch, J., 1995, Surface lysine residues modulate the collisional transfer of fatty acid from adipocyte fatty acid binding protein to membranes, *Biochem.* **34**:11840–11845.

Herr, F. M., Aronson, J., and Storch, J., 1996, Role of portal region lysine residues in electrostatic interactions between heart fatty acid binding protein and phospholipid membranes, *Biochem.* **35**:1296–1303.

Hodsdon, M. E., and Cistola, D. P., 1997a, Discrete backbone disorder in the NMR structure of apo intestinal fatty acid-binding protein: Implications for the mechanism of ligand entry, *Biochem,* **36**:1450–1460.

Hodsdon, M. E., and Cistola, D. P., 1997b, Ligand binding alters the backbone mobility of intestinal fatty acid-binding protein as monitored by ^{15}N NMR relaxation and ^{1}H exchange, *Biochem.* **36**:2278–2290.

Hodsdon, M. E., Ponder, J. W., and Cistola, D. P., 1996, The NMR solution structure of intestinal fatty acid-binding protein complexed with palmitate: Application of a novel distance geometry algorithm, *J. Mol. Biol.* **264**:585–602.

Hsu, K. T., and Storch, J., 1996, Fatty acid transfer from liver and intestinal fatty acid binding-proteins to membranes occurs by different mechanisms, *J. Biol. Chem.* **271**:13317–13323.

Hsu, K. T., and Storch, J. Unpublished observations.

Hubbell, T., Behnke, W. D., Woodford, J. K., and Schroeder, F., 1994, Recombinant liver fatty acid binding protein interacts with fatty acyl-coenzyme A, *Biochem.* **33**:3327–3334.

Iseki, S., and Kondo, H., 1990, Light microscopic localization of hepatic fatty acid binding protein mRNA in jejunal epithelia of rats using *in situ* hybridization, immunohistochemical, and autoradiographic techniques, *J. Histochem. Cytochem.* **38**:111–115.

Iseki, S., Hitomi, M., Ono, T., and Kondo, H., 1989, Immunocytochemical localization of hepatic fatty acid binding protein in the rat intestine: Effect of fasting, *Anat. Rec.* **223**:283–291.

Jefferson, J. R., Powell, D. M., Rymaszewski, Z., Kukowska-Latallo, J., Lowe, J. B., and Schroeder, F., 1990, Altered membrane structure in transfected mouse L-cell fibroblasts expressing rat liver fatty acid-binding protein, *J. Biol. Chem.* **265**:11062–11068.

Jefferson, J. R., Slotte, J. P., Nemecz, G., Pastuszyn, A., Scallen, T. J., and Schroeder, F., 1991, Intracellular sterol distribution in transfected mouse L-cell fibroblasts expressing rat liver fatty acid-binding protein, *J. Biol. Chem.* **266**:5486–5496.

Kaikaus, R. M., Bass, N. M., and Ockner, R. K., 1990, Functions of fatty acid binding proteins, *Experientia* **46**:617–630.

Kanda, T., Iseki, S., Hitomi, M., Kimura, H., Odani, S., Kondo, H., Matsubara, Y., Muto, T., and Ono, T., 1989,

Purification and characterization of a fatty-acid-binding protein from the gastric mucosa of rats: Possible identity with heart fatty-acid-binding protein and its parietal cell localization, *Eur. J. Biochem.* **185:**27–33.

Keuper, H. J. K., Klein, R. A., and Spener, F., 1985, Spectroscopic investigations on the binding site of bovine hepatic fatty acid binding protein: Evidence for the existence of a single binding site for two fatty acid molecules, *Chem. Phys. Lipids* **38:**159–173.

Kim, H. K., and Storch, J., 1992, Free fatty acid transfer from rat liver fatty acid-binding protein to phospholipid vesicles: Effect of ligand and solution properties, *J. Biol. Chem.* **267:**77–82.

Kim, K., Cistola, D. P., and Frieden, C., 1996, Intestinal fatty acid-binding protein: The structure and stability of a helix-less variant, *Biochem.* **35:**7553–7558.

Le Beyec, J., Delers, F., Jourdant, F., Schreider, C., Chambaz, J., Cardot, P., and Pincon-Raymond, M., 1997, A complete epithelial organization of Caco-2 cells induces I-FABP and potentializes apolipoprotein gene expression, *Exp. Cell Res.* **236:**311–320.

Lin, M. C., Arbeeny, C., Bergquist, K., Kienzle, B., Gordon, D. A., and Wetterau, J. R., 1994, Cloning and regulation of hamster microsomal triglyceride transfer protein. The regulation is independent from that of other hepatic and intestinal proteins which participate in the transport of fatty acids and triglycerides, *J. Biol. Chem.* **269:**29138–29145.

Liou, H., and Storch, J., 1998, The helical cap domain is important for fatty acid transfer from adipocyte and heart fatty acid-binding proteins to membranes, *FASEB J.* **12:**A514 (Abstract).

Lowe, J. B., Sacchettini, J. C., Laposata, M., McQuillan, J. J., and Gordon, J. I., 1987, Expression of rat intestinal fatty acid-binding protein in *Escherichia coli:* Purification and comparison of ligand binding characteristics with that of *Escherichia coli*-derived rat liver fatty acid-binding protein, *J. Biol. Chem.* **262:**5931–5937.

Lucke, C., Zhang, F., Ruterjans, H., Hamilton, J. A., and Sacchettini, J. C., 1996, Flexibility is a likely determinant of binding specificity in the case of ileal lipid binding protein, *Structure* **4:**785–800.

Luxon, B. A., 1996, Inhibition of binding to fatty acid binding protein reduces the intracellular transport of fatty acids, *Am. J. Physiol.* **271:**G113–G1120.

Luxon, B. A., and Milliano, M. T., 1997, Cytoplasmic codiffusion of fatty acids is not specific for fatty acid binding protein, *Am. J. Physiol.* **273:**C859–C867.

Luxon, B. A., and Weisiger, R. A., 1993, Sex differences in intracellular fatty acid transport: Role of cytoplasmic binding proteins, *Am. J. Physiol.* **265:**G831–G841.

McCormack, M., and Brecher, P., 1987, Effect of liver fatty acid binding protein on fatty acid movement between liposomes and rat liver microsomes, *Biochem. J.* **244:**717–723.

Mishkin, S., Stein, L., Gatmaitan, Z., and Arias, I. M., 1972, The binding of fatty acids to cytoplasmic proteins: Binding to Z-protein in liver and other tissues of the rat, *Biochem. Biophys. Res. Commun.* **47:**997–1003.

Murphy, E. J., Prows, D. R., Jefferson, J. R., and Schroeder, F., 1996, Liver fatty acid-binding protein expression in transfected fibroblasts stimulates fatty acid uptake and metabolism, *Biochim. Biophys. Acta* **1301:**191–198.

Nemecz, G., and Schroeder, F., 1991, Selective binding of cholesterol by recombinant fatty acid binding proteins, *J. Biol. Chem.* **266:**17180–17186.

Nemecz, G., Hubbell, T., Jefferson, J. R., Lowe, J. B., and Schroeder, F., 1991a, Interaction of fatty acids with recombinant rat intestinal and liver fatty acid-binding proteins, *Arch. Biochem. Biophys.* **286:**300–309.

Nemecz, G., Jefferson, J. R., and Schroeder, F., 1991b, Polyene fatty acid interactions with recombinant intestinal and liver fatty acid-binding proteins: Spectroscopic studies, *J. Biol. Chem.* **266:**17112–17123.

Ockner, R. K., and Manning, J. A., 1974, Fatty acid-binding protein in small intestine: Identification, isolation, and evidence for its role in cellular fatty acid transport, *J. Clin. Invest.* **54:**326–338.

Ockner, R. K., and Manning, J. A., 1976, Fatty acid binding protein: Role in esterification of absorbed long chain fatty acid in rat intestine, *J. Clin. Invest.* **58:**632–641.

Ockner, R. K., Manning, J. A., Poppenhausen, R. B., and Ho, W. K. L., 1972, A binding protein for fatty acids in cytosol of intestinal mucosa, liver, myocardium, and other tissues, *Science* **177:**56–58.

Peeters, R. A., Veerkamp, J. H., and Demel, R. A., 1989, Are fatty acid-binding proteins involved in fatty acid transfer? *Biochim. Biophys. Acta* **1002:**8–13.

Poirier, H., Degrace, P., Niot, I., Bernard, A., and Besnard, P., 1996, Localization and regulation of the putative membrane fatty-acid transporter (FAT) in the small intestine. Comparison with fatty acid-binding proteins (FABP), *Eur. J. Biochem.* **238:**368–373.

Poirier, H., Niot, I., Degrace, P., Monnot, M., Bernard, A., and Besnard, P., 1997, Fatty acid regulation of fatty acid-binding protein expression in the small intestine, *Am. J. Physiol.* **273:**G289–G295.

Prows, D. R., and Schroeder, F., 1997, Metallothionein-II_A promoter induction alters rat intestinal fatty acid binding protein expression, fatty acid uptake, and lipid metabolism in transfected L-cells, *Arch. Biochem. Biophys.* **340:**135–143.

Prows, D. R., Murphy, E. J., and Schroeder, F., 1995, Intestinal and liver fatty acid binding proteins differentially affect fatty acid uptake and esterification in L-cells, *Lipids* **30:**907–910.

Richieri, G. V., Ogata, R. T., and Kleinfeld, A. M., 1994, Equilibrium constants for the binding of fatty acids with fatty acid-binding proteins from adipocyte, intestine, heart, and liver measured with the flourescent probe ADIFAB, *J. Biol. Chem.* **269:**23918–23930.

Richieri, G. V., Ogata, R. T., and Kleinfeld, A. M., 1996, Thermodynamic and kinetic properties of fatty acid interactions with rat liver fatty acid-binding protein, *J. Biol. Chem.* **271:**31068–31074.

Riehl, T. E., Bass, N. M., and Stenson, W. F., 1990, Metabolism of 15-hydroxyeicosatetraenoic acid by Caco-2 cells, *J. Lipid Res.* **31:**773–780.

Rissanen, J., Pihlajamaki, J., Heikkinen, S., Kekalainen, P., Kuusisto, J., and Laakso, M., 1997, The Ala$^{54 \rightarrow Thr}$ polymorphism of the fatty acid binding protein 2 gene does not influence insulin sensitivity in Finnish nondiabetic and NIDDM subjects, *Diabetes* **46:**711–712.

Rolf, B., Oudenampsen-Kruger, E., Borchers, T., Faergeman, N. J., Knudsen, J., Lezius, A., and Spener, F., 1995, Analysis of the ligand binding properties of recombinant bovine liver-type fatty acid binding protein, *Biochim. Biophys. Acta* **1259:**245–253.

Rottman, J. N., and Gordon, J. I., 1993, Comparison of the patterns of expression of rat intestinal fatty acid binding protein/human growth hormone fusion genes in cultured intestinal epithelial cell lines and in the gut epithelium of transgenic mice, *J. Biol. Chem.* **268:**11994–12002.

Rubin, D. C., Swietlicki, E., Roth, K. A., and Gordon, J., 1992, Use of fetal intestinal isografts from normal and transgenic mice to study the programming of positional information along the duodenal-to-colonic-axis, *J. Biol. Chem.* **267:**15122–15133.

Sacchettini, J. C., and Gordon, J. I., 1993, Rat intestinal fatty acid binding protein: A model system for analyzing the forces that can bind fatty acids to proteins, *J. Biol. Chem.* **268:**18399–18402.

Sacchettini, J. C., Gordon, J. I., and Banaszak, L. J., 1988, The structure of crystalline *Escherichia coli*-derived rat intestinal fatty acid-binding protein at 2.5 Å resolution, *J. Biol. Chem.* **263:**5815–5819.

Sacchettini, J. C., Gordon, J. I., and Banaszak, L. J., 1989, Crystal structure of rat intestinal fatty-acid-binding protein: Refinement and analysis of the *Escherichia coli*-derived protein with bound palmitate, *J. Mol. Biol.* **208:**327–339.

Sacchettini, J. C., Hauft, S. M., Van Camp, S. L., Cistola, D. P., and Gordon, J. I., 1990, Developmental and structural studies of an intracellular lipid binding protein expressed in the ileal epithelium, *J. Biol. Chem.* **265:**19199–19207.

Sacchettini, J. C., Scapin, G., Gopaul, D., and Gordon, J. I., 1992, Refinement of the structure of *Escherichia coli*-derived rat intestinal fatty acid binding protein with bound oleate to 1.75 Å resolution: Correlation with the structures of the apoprotein and the protein with bound palmitate, *J. Biol. Chem.* **267:**23534–23545.

Scapin, G., Gordon, J. I., and Sacchettini, J. C., 1992, Refinement of the structure of recombinant rat intestinal fatty acid-binding apoprotein at 1.2 Å resolution, *J. Biol. Chem.* **267:**4253–4269.

Schroeder, F., Jefferson, J. R., Powell, D., Incerpi, S., Woodford, J. K., Colles, S., Myers-Payne, S., Emge, T., Hubbell, T., Moncecchi, D., Prows, D., and Heyliger, C. E., 1993, Expression of rat L-FABP in mouse fibroblasts: Role in fat absorption, *Mol. Cell. Biochem.* **123:**73–83.

Schulenberg-Schell, H., Schafer, P., Keuper, H. J. K., Stanislawski, B., Hoffmann, E., Ruterjans, H., and Spener, F., 1988, Interactions of fatty acids with neutral fatty-acid-binding protein from bovine liver, *Eur. J. Biochem.* **170:**565–574.

Sheridan, M., Wilkinson, T. C. I., and Wilton, D. C., 1987, Studies on fatty acid-binding proteins: Changes in the concentration of hepatic fatty acid-binding protein during development in the rat, *Biochem. J.* **242:**919–922.

Shields, H. M., Bates, M. L., Bass, N. M., Best, C. J., Alpers, D. H., and Ockner, R. K., 1986, Light microscopic immunocytochemical localization of hepatic and intestinal types of fatty acid-binding proteins in rat small intestine, *J. Lipid Res.* **27:**549–557.

Simon, T. C., Roth, K. A., and Gordon, J. I., 1993, Use of transgenic mice to map *cis*-acting elements in the liver fatty acid-binding protein gene (FABP1) that regulate its cell lineage-specific, differentiation-dependent, and spatial patterns of expression in the gut epithelium and in the liver acinus, *J. Biol. Chem.* **268:**18345–18358.

Simon, T. C., Roberts, L. J., and Gordon, J. I., 1995, A 20-nucleotide element in the intestinal fatty acid binding protein gene modulates its cell lineage specific, differentiation-dependent, and cephalocaudal patterns of expresion in transgenic mice. *Proc. Natl. Acad. Sci.* **92:**8685–8689.

Simon, T. C., Cho, A., Tso, P., and Gordon, J. I., 1997, Suppressor and activator functions mediated by a repeated heptad sequence in the liver fatty acid-binding protein gene (FABPL), *J. Biol. Chem.* **272**:10652–10663.

Sipilainen, R., Uusitupa, M., Heikkinen, S., Rissanen, A., and Laakso, M., 1997, Variants in the human intestinal fatty acid binding protein 2 gene in obese subjects, *J. Clin. Endocrinol. Metab.* **82**:2629–2632.

Spitsberg, V. L., Matitashvili, E., and Gorewit, R. C., 1995, Association and coexpression of fatty-acid-binding protein and glycoprotein CD36 in the bovine mammary gland, *Eur. J. Biochem.* **230**:872–878.

Steele, R. A., Emmert, D. A., Kao, J., Hodsdon, M. E., Frieden, C., and Cistola, D. P., 1998, The three-dimensional structure of a helix-less variant of intestinal fatty acid-binding protein, *Prot Sci* **7**:1332–1339.

Stewart, J. M., Driedzic, W. R., and Berkelaar, J. A. M., 1991, Fatty-acid-binding protein facilitates the diffusion of oleate in a model cytosol system, *Biochem. J.* **275**:569–573.

Storch, J., 1993, Diversity of fatty acid-binding protein structure and function: Studies with fluorescent ligands, *Mol. Cell. Biochem.* **123**:45–53.

Storch, J., Herr, F. M., Hsu, K. T., Kim, H. K., Liou, H. L., and Smith, E. R., 1996, The role of membranes and intracellular binding proteins in cytoplasmic transport of hydrophobic molecules: Fatty acid-binding proteins, *Comp. Biochem. Physiol.* **115B**:333–339.

Sweetser, D. A., Heuckeroth, R. O., and Gordon, J. I., 1987, The metabolic significance of mammalian fatty-acid-binding proteins: Abundant proteins in search of a function, *Ann. Rev. Nutr.* **7**:337–359.

Sweetser, D. A., Birkenmeier, E. H., Hoppe, P. C., McKeel, D. W., and Gordon, J. I., 1988a, Mechanisms underlying generation of gradients in gene expression within the intestine: An analysis using transgenic mice containing fatty acid binding protein-human growth hormone fusion genes, *Genes Dev.* **2**:1318–1332.

Sweetser, D. A., Hauft, S. M., Hoppe, P. C., Birkenmeier, E. H., and Gordon, J. I., 1988b, Transgenic mice containing intestinal fatty acid-binding protein-human growth hormone fusion genes exhibit correct regional and cell-specific expression of the reporter gene in their small intestine, *Proc. Natl. Acad. Sci. USA* **85**:9611–9615.

Takikawa, H., and Kaplowitz, N., 1986, Binding of bile acids, oleic acid, and organic anions by rat and human hepatic Z protein, *Arch. Biochem. Biophys.* **251**:385–392.

Thompson, J., Winters, N., Terwey D., Bratt, J., and Banaszak, L., 1997, The crystal structure of the liver fatty acid-binding protein, *J. Biol. Chem.* **272**:7140–7150.

Thumser, A. E. A., and Storch, J., 2000, Liver and intestinal fatty acid-binding proteins obtain fatty acids from phospholipid vesicles by different mechanisms, *J. Lipid Res.* **41**: 647–656.

Thumser, A. E. A., and Wilton, D. C., 1995, The binding of natural and fluorescent lysophospholipids to wild-type and mutant rat liver fatty acid-binding protein and albumin, *Biochem. J.* **307**:305–311.

Thumser, A. E. A., and Wilton, D. C., 1996, The binding of cholesterol and bile salts to recombinant rat liver fatty acid-binding protein, *Biochem. J.* **320**:729–733.

Tipping, E., and Ketterer, B., 1981, The influence of soluble binding proteins on lipophile transport and metabolism in hepatocytes, *Biochem. J.* **195**:441–452.

Trotter, P. J., and Storch, J., 1993a, Fatty acid esterification during differentiation of the human intestinal cell line Caco-2, *J. Biol. Chem.* **268**:10017–10023.

Trotter, P. J., and Storch, J., 1993b, Nutritional control of fatty acid esterification in differentiating Caco-2 intestinal cells is mediated by cellular diacylglycerol concentrations, *J. Nutr.* **123**:728–736.

Veerkamp, J. H., and Maatman, R. G. H. J., 1995, Cytoplasmic fatty acid-binding proteins: Their structure and genes, *Prog. Lipid Res.* **34**:17–52.

Vork, M. M., Glatz, J. F. C., and Van Der Vusse, G. J., 1997, Modeling intracellular fatty acid transport: Possible mechanistic role of cytoplasmic fatty acid-binding protein, *Prostagl. Leukotr. Ess. Fatty Acids* **57**:11–16.

Wang, H., He, Y., Hsu, K. T., Magliocca, J. F., Storch, J., and Stark, R. E., 1998, ^{1}H, ^{15}N and ^{13}C resonance assignments and secondary structure of apo liver fatty acid-binding protein, *J. Biomol. NMR* **12**:197–199.

Weisiger, R. A., Pond, S. M., and Bass, L., 1989, Albumin enhances unidirectional fluxes of fatty acid across a lipid-water interface: Theory and experiments, *Am. J. Physiol.* **257**:G904–G916.

Wilkinson, T. C. I., and Wilton, D. C., 1987, Studies on fatty acid-binding proteins: The binding properties of rat liver fatty acid-binding protein, *Biochem. J.* **247**:485–488.

Wu, F., Corsico, B., Flach, C. R., Cistola, D. P., Storch, J., and Mendelsohn, R., 2000, Deletion of the helical motif in the intestinal fatty acid-binding protein reduces its interactions with membrane monolayers: Brewster angle microscopy, IR reflection-absorption spectroscopy and surface pressure studies. Submitted for publication.

CHAPTER 10

Microsomal Triglyceride Transfer Protein

Role in the Assembly of Intestinal Lipoproteins

JOHN R. WETTERAU

The microsomal triglyceride transfer protein (MTP) is an intracellular lipid transfer protein that catalyzes the transport of triglyceride (TG), cholesteryl ester (CE), and phospholipid molecules between membranes. It is an unusual lipid transfer protein in that it is a complex of two proteins of molecular weight 55 kDa and 97 kDa. Intracellular lipid transfer proteins described to date are generally composed of a single polypeptide and have molecular weights in the range of 10,000 to 35,000. High levels of MTP activity are found in the lumen of microsomes isolated from liver and intestine (Wetterau and Zilversmit, 1986). The tissue and subcellular location of MTP, combined with its ability to promote neutral lipid (TG and CE) transfer, suggested that MTP plays a role in the assembly of the triglyceride rich, apolipoprotein B- (apoB) containing lipoproteins, very low density lipoprotein (VLDL), and chylomicrons. This hypothesis has since been confirmed.

1. Characterization of the Microsomal Triglyceride Transfer Protein

MTP was originally isolated from bovine liver. Characterization of the purified protein revealed that it is composed of a small 55 kDa subunit that is the multifunctional protein, protein disulfide isomerase (PDI), and a unique, large, 97 kDa subunit (Wetterau *et al.*, 1990). A variety of approaches have been used to show that the active transfer protein is a complex of two proteins. These include the following:

1. a demonstration that both proteins as well as transfer activity are co-immunoprecipitated by antibodies specific for either PDI or the 97 kDa subunit (Wetterau *et al.*, 1990)
2. using sedimentation equilibrium to show that the molecular weight of the complex is approximately 150,000 (Wetterau *et al.*, 1991b)

JOHN R. WETTERAU • Department of Metabolic Research, Bristol-Myers Squibb, Princeton, New Jersey 08543.
Intestinal Lipid Metabolism, edited by Charles M. Mansbach II *et al.*, Kluwer Academic/Plenum Publishers, 2001.

3. expression studies in a heterologous system which indicated both proteins must be co-expressed to reconstitute an active transfer protein (Ricci *et al.*, 1995).

PDI is best known for its disulfide isomerase and chaperone activities that are used to catalyze the proper folding of newly synthesized proteins within the lumen of the endoplasmic reticulum (ER). However, there is precedent for PDI forming a stable complex with another protein. It is also the β subunit of the $\alpha 2\beta 2$ tetrameric enzyme, prolyl-4-hydroxylase.

The large subunit of MTP is highly conserved between species. There is greater than 85% identity between the human protein (Sharp *et al.*, 1993) and that of bovine, mouse (Nakamuta *et al.*, 1996), and hamster (Lin *et al.*, 1994). The amino acid sequence of the 97 kDa subunit is not highly homologous to any previously characterized protein. However, Shoulders *et al.* (1994) used several criteria including low levels of homology, gene structure, and secondary structural predictions to propose that the MTP large subunit is a member of the vitellogenin gene family. In egg-laying animals, vitellogenin is cleaved to form lipovitellin, which associates with lipid to form a transport and storage lipoprotein complex. Preliminary X ray crystallographic structural analysis of MTP confirmed that the MTP large subunit and lipovitellin share structural folds (Thompson *et al.*, 1998).

Although the role of PDI in the transfer protein is not completely understood, some aspects of its function are emerging. Clearly, both subunits of MTP are required for functional activity. The association of the two proteins into a complex does not appear to be a reversible process (Wetterau *et al.*, 1991a). Dissociation of the complex can be promoted by low concentrations of denaturants, chaotropic agents, or nondenaturing detergents. Once PDI and the large subunit are dissociated, they cannot be reconstituted into an active complex because the large subunit forms an insoluble aggregate. Similar findings have been found when MTP is expressed in Sf9 insect cells using the baculovirus expression system. The production of active transfer protein requires the coexpression of the large subunit and of PDI. In the absence of PDI expression, the large subunit forms an insoluble aggregate (Ricci *et al.*, 1995).

PDI probably plays a role in retaining the MTP complex within the lumen of the ER. Both MTP subunits contain a signal sequence that targets their synthesis to the rough ER and translocation into the ER lumen. ER proteins that contain a Lys-Asp-Glu-Leu (KDEL) amino terminal sequence are, in part, retained within the lumen of the ER by a receptor that recognizes the KDEL sequence. Whereas PDI contains an amino terminal KDEL sequence, the MTP large subunit does not. The association of the large subunit with PDI probably plays a role in maintaining its subcellular location. Although this has not been demonstrated experimentally for MTP, it has been shown that the KDEL sequence on PDI is required for proper ER retention of the α subunit of prolyl-4-hydroxylase that, like the MTP large subunit, does not contain an ER retention sequence (Vuori *et al.*, 1992).

The disulfide isomerase activity of PDI is inhibited when PDI forms a complex with the 97 kDa subunit of MTP. Whether this is due to the two homologous active sites of PDI (amino acid sequence -CGHC-) being buried or to changes in the conformation of PDI is not known. By coexpressing the MTP large subunit with wild-type PDI or PDI in which the two active sites have been mutated to create PDI without disulfide isomerase activity, Lamberg *et al.* (1996) have shown that the disulfide isomerase activity of PDI is not necessary for MTP to express its transfer activity.

The mechanism of MTP-mediated lipid transport has been investigated by examining the kinetics of the lipid transport process and the lipid-binding properties of the protein. When TG or CE transport was measured as a function of substrate concentration, the observed transfer rates were consistent with ping-pong bi-bi kinetics, indicative of a shuttle transport mechanism (Atzel and Wetterau, 1993). In this process, MTP binds to a membrane and extracts individual lipid molecules, it dissociates from the membrane, MTP with bound lipid then diffuses to an acceptor membrane, and it binds to the acceptor membrane and deposits its bound lipid molecules. The shuttle mechanism predicts that MTP with bound lipid is an intermediate in the transfer reaction. If MTP is incubated with donor membranes and then is re-isolated, it is found to bind up to one neutral lipid molecule and two to three phospholipid molecules (Atzel and Wetterau, 1994).

Further support for a shuttle mechanism has been provided by directly measuring the two half reactions of the transfer reaction (Atzel and Wetterau, 1994). The transfer of neutral lipid or phospholipid from a membrane to MTP, as well as from MTP to a membrane, have both been shown to occur. These studies further revealed that there are two types of lipid molecule binding sites on MTP. The neutral lipid and a portion of the phospholipid transfer between MTP and a membrane is rapid, whereas a portion of the phospholipid transfer is slow. This suggests that all the lipid molecule binding sites in MTP do not contribute equally to rapid lipid transport. The role of the slow sites is not known.

MTP activity is routinely monitored by measuring MTP-mediated transfer of radiolabeled lipid molecules from donor to acceptor synthetic small unilamellar phosphatidylcholine (PC) vesicles (Wetterau and Zilversmit, 1985). When TG or CE transport is being measured, the substrates for the assay usually contain about 200 phospholipid molecules for every one molecule of TG or CE. Neutral lipid has poor solubility in a phospholipid membrane, thus limiting the amount that can be incorporated into a PC vesicle to a few mole percent. When MTP activity is expressed as the percent of donor lipid transported per time, it shows a clear preference for neutral lipid transfer relative to phospholipid transfer. It transports TG about 25 times faster than PC. However, if MTP activity is expressed as the number of molecules transferred per time, then, because the donor membranes contain so many more phospholipid than neutral lipid molecules, MTP is actually transporting more phospholipid molecules than TG molecules. If MTP had no preference, it would transport 200 PC molecules for every one TG in a membrane containing 200 PC and one TG molecules. However, it actually transports about ten PC for every TG molecule, clearly demonstrating that when MTP binds to a membrane, it preferentially extracts and transports neural lipid, compared to phospholipid.

2. MTP Expression

Intestinal MTP expression and its regulation in response to various diets have been investigated in hamsters (Lin *et al.,* 1994). MTP large subunit mRNA levels are highest in the combined duodenum plus jejunum. The ileum, the liver, and the colon have about 40, 10, and 5% of the levels found in the duodenum plus jejunum, respectively. Similar trends are found for MTP activity and for large subunit protein.

MTP gene expression is differentially regulated in the liver and the intestine. Although a 31-day high-fat diet increases MTP large subunit mRNA levels in both the liver and the intestine, only intestinal message levels are increased by an acute (24 hr) high-fat diet. In contrast, in streptozotocin-treated diabetic rats, hepatic message levels are decreased with insulin treatment, whereas intestinal levels are unchanged (Wetterau *et al.,* 1997). In all cases, the regulation is modest, with maximal observed changes being only about twofold.

3. Studies of MTP in Abetalipoproteinemic Subjects

Early characterization of MTP suggested that it may play a role in the synthesis of the apoB-containing lipoproteins, VLDL, and chylomicrons. These large lipid–protein complexes consist of a triglyceride droplet surrounded by phospholipid, free cholesterol, and protein. For VLDL that are made in the liver, the primary structural protein is apoB-100, a single polypeptide with a molecular weight of about 550 kDa. Chylomicrons that are made in the intestine from dietary lipid are structurally similar to VLDL except that they are larger and their primary structural protein is apoB-48, which has a molecular weight of about 260 kDa. Early studies of MTP led to the hypothesis that in VLDL and chylomicron assembly, MTP shuttles lipid molecules from the ER membrane where lipid synthesis occurs to developing apoB particles within the ER lumen.

Studies of MTP in abetalipoproteinemic subjects provided the first direct evidence that MTP is involved in the assembly of plasma lipoproteins. Abetalipoproteinemia is a rare genetic disease in which affected individuals are unable to produce apoB-containing lipoproteins (Kane and Havel, 1995). As a result of the defect, subjects have plasma cholesterol levels of around 40 mg/dl and TG levels less than 10 mg/dl. Their apoB gene is normal and their ability to synthesize cholesterol, triglyceride, and phospholipid is not impaired. Due to the block in lipoprotein production, subjects have fat-filled enterocytes, have fat malabsorption, and may develop fatty liver. The primary pathologies associated with the disease are secondary to a deficiency of fat-soluble vitamins, particularly vitamin E.

The initial examination of MTP in four unrelated abetalipoproteinemic subjects revealed that neither MTP activity nor protein could be detected in intestinal biopsies obtained from the four subjects (Wetterau *et al.,* 1992). In contrast, MTP activity and protein were readily detected in intestinal biopsies from control subjects. In addition, MTP activity was also detectable in an individual with homozygous hypobetalipoproteinemia and in another with Anderson's disease. These latter two subjects have an inability to produce chylomicrons due to a genetic defect in the gene encoding apoB in the case of hypobetalipoproteinemia, or an unknown genetic defect in the case of Anderson's disease. Thus, the observed absence of MTP in the abetalipoproteinemic subjects appeared to be a primary defect in the disease and not secondary to an inability to produce chylomicrons or to detect MTP under conditions where chylomicrons are not made.

Characterization of the gene encoding MTP in two of the four abetalipoproteinemic individuals revealed mutations that readily explained the absence of MTP (Sharp *et al.,* 1993). One subject was homozygous for a nonsense mutation whereas the second was homozygous for a frameshift mutation. Both mutations predict a severely truncated MTP large subunit. In the two remaining subjects, disease-causing mutations that predict more subtle changes in the protein were identified. One subject was homozygous for a nonsense mutation that predicts that the large subunit is truncated by only 30 amino acids (Ricci *et al.,* 1995). The fourth subject was a compound heterozygote, with one allele containing a mutation that results in a deletion of the 10th of 18 exons encoding MTP, whereas the other allele had a missense mutation that results in Arg540 being changed to a His (Rehberg *et al.*, 1996). In addition to these four subjects, the gene encoding the large subunit of MTP has been characterized in eight additional abetalipoproteinemic subjects by Shoulders *et al.* (1993) and by Narcisi *et al.* (1995). In all cases, mutations that readily predict defective MTP were identified.

To understand how the subtle mutations in the MTP large subunit identified by Ricci *et al.* (1995) and by Rehberg *et al.* (1996) disrupt MTP to the extent that MTP protein and activity cannot be detected in intestinal biopsies, wild-type and the two mutant MTP large subunits were coexpressed with PDI in Sf9 insect cells. Although wild-type protein produced an active MTP protein complex, both mutant large subunits were unable to associate with PDI to form an active complex. The mutant proteins formed insoluble aggregates. The results of these studies suggest that *in vivo,* the mutations prevent normal protein–protein interactions with PDI, resulting in a misfolded protein that is rapidly degraded.

4. Models for Studying the Assembly of apoB-Containing Lipoproteins

The assembly of VLDL and chylomicrons occurs by a complex sequence of steps that are just beginning to be unraveled. The elucidation of this process has been aided by the use of two liver-derived cell lines, Hep G2 and McA-RH7777. Hep G2 cells, like human liver from which they are derived, produce lipoproteins containing apoB-100. In contrast, McA-RH7777 cells, which are derived from rat hepatocytes, produce both apoB-48- and apoB-100-containing lipoproteins. Rat liver also produces both apoB-100 and apoB-48 lipoproteins. In addition, a human colon-derived cell line, Caco-2, has been used to study lipoprotein assembly. This cell line also secretes both apoB-100 and apoB-48. Although apoB-48 is the dominant form of apoB produced in human intestine, the intestine appears to also produce trace levels of apoB-100 lipoproteins (Hoeg *et al.,* 1990).

More recently, nonhepatic, nonintestinal cell lines in which apoB and MTP have been co-expressed have been used to investigate lipoprotein assembly. These studies allow one to separate those processes that are general to cells co-expressing MTP and apoB from those particular to hepatocytes or to enterocytes. Although cell culture models have been useful for studying lipoprotein assembly, the different cell lines and culture conditions used by investigators, as well as the recent use of cells transfected with plasmids overexpressing apoB, have produced varying results that complicate efforts to identify the sequence of events that best represent what may be occurring when lipoproteins are produced in animals or in humans.

The assembly of lipoproteins has been extensively studied in liver and liver-derived cell lines. In contrast, intestinal lipoprotein assembly has not been studied in detail. To develop a model for the role of MTP in intestinal lipid metabolism and lipoprotein assembly, it will be assumed that a model for lipoprotein assembly that is emerging, based primarily on studies conducted with liver-derived cells, is applicable to the intestine. Current circumstantial evidence suggests the two processes share many similarities; however, there are also some clear differences.

Newly synthesized apoB may follow a pathway leading to its secretion as a lipoprotein particle or alternately, to its intracellular degradation. Based on studies performed in cell culture, apoB may be degraded either co-translationally, following its association with

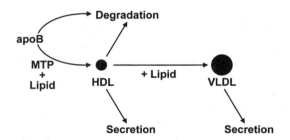

Figure 1. Hepatic apoB metabolism. Following its synthesis, apoB may follow a pathway leading to its degradation, the formation of a small lipoprotein particle with the density of HDL, or a mature, lipid-rich VLDL particle.

the ER membrane, or after it has formed a small, HDL-density lipoprotein particle (for a review, see Yao *et al.,* 1997). A similarly complex picture has been proposed for the synthesis of secretion-competent lipoprotein particles. ApoB co-translationally associates with lipid (Borén *et al.,* 1992) and forms a small, high-density, primordial lipoprotein particle. This lipid-poor lipoprotein may be secreted directly or may be converted to a lipid-rich VLDL that is subsequently secreted. This pathway (Fig. 1) has become known as the two-step model for lipoprotein assembly (for a review, see Innerarity *et al.,* 1996). More recently, it was shown that membrane-associated apoB can serve as an intermediate in the synthesis of a lipoprotein particle (Rustaeus *et al.,* 1998). Alternatively, based on studies performed in rats, it has been proposed that a full complement of lipid is added to a lipoprotein particle at an early stage of its production (Rusiñol *et al.,* 1993). It is not clear whether this latter scenario represents an alternate pathway to the two-step model for lipoprotein assembly or if the difference is simply kinetic in nature. Conditions that favor rapid lipoprotein assembly, such as those found in the rat, may preclude the identification of an intermediate in the process.

A clear example of the two-step model for lipoprotein assembly is found in McA-RH7777 cells. Borén *et al.* (1994) showed that when cells are cultured in fatty-acid-deficient media, apoB-48 associates with lipid to make a primordial HDL-density lipoprotein particle that is retained within the cell and is only slowly secreted. When the cells are cultured in the presence of a lipid-rich media, VLDL are made and are rapidly secreted from the cell. To demonstrate a precursor–product relationship between the high-density particles and the lipid-rich VLDL, they performed a pulse chase study in which apoB was labeled and was chased in the absence of fatty acids. Under these conditions, all the intracellular labeled apoB-48 was in HDL. The culture media was then switched to a fatty-acid-rich media. The labeled intracellular HDL particles were converted to lipid-rich VLDL that were rapidly secreted, confirming that small, dense apoB-48 lipoproteins convert to large lipidated particles. Recent studies suggest that the synthesis of an apoB-100 VLDL also occurs through an HDL intermediate in McA-RH7777 cells (Rustaeus *et al.,* 1998).

Although the precise pathway leading to apoB degradation or lipoprotein production may vary depending on the system studied, it is clear that in cell culture models, lipid availability regulates lipoprotein production (for reviews, see Dixon and Ginsberg, 1993; Sniderman and Cianflone, 1993). An example of this is the regulation of apoB-48 production in McA-RH7777 cells by fatty acids (see the previous paragraphs). There have been similar findings in Hep G2 cells where addition of oleic acid to the culture media has been reported to increase the number of apoB particles produced (Dixon *et al.,* 1991) or to induce the production of more lipid-rich particles (Ellsworth *et al.,* 1986). The regulation of apoB secretion occurs co- or post-translationally. ApoB mRNA levels and apoB synthesis are not affected by lipid addition to the growth media. Thus within the cell, there are competitive processes occurring—apoB degradation and apoB conversion to a secretion-competent lipoprotein particle. Lipid addition to apoB stabilizes it and helps it progress toward the production of a secretion-competent lipoprotein. However, the applicability of the regulation observed in cell culture to lipoprotein production *in vivo* is unclear.

Some components of the assembly process do not appear to be unique to an hepatocyte or an enterocyte and can be reproduced in nonhepatic, nonintestinal cell lines cotransfected with plasmids expressing apoB and MTP. When apoB fragments larger than about 30% of the size of apoB-100 are expressed in HeLa or in COS cells, only trace quantities of apoB are secreted into the media. In contrast, the co-expression of MTP results in the production of small, dense lipoprotein particles (Gordon *et al.,* 1994; Leiper *et al.,* 1994; Patel and Grundy, 1996; Wang *et al.,* 1996). However, these systems fail to produce lipid-rich VLDL particles. In HeLa cells, HDL-density particles are produced when apoB-53 (representing the amino-terminal 53% of apoB-100) is co-expressed with MTP (Gordon *et al.,* 1994). Analogous to what is observed with liver-derived cells, the addition of cholesterol and fatty acids to the growth media of HeLa cells expressing MTP and apoB regulates the number and nature of the lipoproteins produced. Lipid supplements result in a doubling of the number of particles produced and a general decrease in their density. However, even when using a lipid-rich media, only 15% of the secreted particles have a density less than 1.063 g/ml, the density of a low-density lipoprotein (LDL) particle. Thus the expression of MTP and apoB in a nonhepatic, nonintestinal cell line appears to reconstitute some aspects, but not all, of native lipoprotein assembly.

5. Intestinal Assembly of apoB-Containing Lipoproteins

5.1. Comparison to Hepatic VLDL Production

There are clear differences in the synthetic pathway for a hepatic VLDL particle and an intestinal chylomicron particle. This is most obvious when one considers that a newly synthesized chylomicron particle is 75 to 600 nm in diameter compared to a newly synthesized VLDL particle, which has a diameter in the range of 30 to 80 nm. Whether chylomicrons are made by an extension of the two-step model proposed for hepatic VLDL synthesis has not been directly addressed experimentally, but circumstantial evidence suggests that they may. Intracellular apoB particles of HDL, LDL, VLDL, and chylomicron density have been identified in rat intestine (Magun *et al.,* 1988). The smaller particles may serve as intermediates in the formation of larger ones, although this has not been shown experimentally. A large pool of apoB that was not associated with a lipoprotein particle was also found in these studies.

There is a pool of intestinal apoB present in the nonabsorptive state that can be rapidly mobilized upon fat absorption. Glickman *et al.* (1978) administered a protein synthesis inhibitor to isolated rat jejunal segments. This was followed by lipid administration one hour later. Isolated cells were examined 5 and 60 minutes following lipid addition under conditions that favor lipid absorption and lipoprotein production. Intracellular apoB was observed immediately after lipid addition (even though protein synthesis had been inhibited for over an hour); however, by 60 min post administration of lipid, it was depleted. Apparently preformed apoB was being used to produce lipoproteins, similar to what was observed for apoB-48 VLDL production in McA-RH7777 cells by Borén *et al.* (1994).

As has been observed in liver-derived cell lines, lipid availability regulates intestinal lipoprotein production both in cell culture and *in vivo*. The increased lipid available following a fat meal results in the production of larger, more lipid-rich lipoproteins. Hayashi *et al.* (1990) found that in rats, this can occur with no change in the total number of particles secreted. In contrast, in Caco-2 cells, as has been found in Hep G2 cells, oleic acid stimulates apoB secretion (Moberly *et al.*, 1990).

Why chylomicron-size particles are made in the intestine and not in the liver is not understood. Both VLDL and chylomicron-size apoB-48 particles are made in the intestine, possibly utilizing diverging or separate assembly pathways. The larger size of hepatic apoB-100, compared to intestinal apoB-48, does not appear to limit the size of the particle formed. Human enterocytes produce trace quantities of apoB-100 lipoproteins, so apoB-100 is functional in the context of the intestine. This has been confirmed in transgenic mice expressing exclusively apoB-100 in the intestine (Farese *et al.*, 1996). Furthermore, preliminary analysis suggested the presence of chylomicron-size particles in the enterocytes of these animals (communicated by R. Hamilton and J. S. Wong within Farese *et al.*, 1996).

There are distinguishing features of hepatic VLDL and intestinal chylomicron synthesis. At times of active fat absorption in the intestine, triglyceride is synthesized through the monoacylglycerol pathway (reacylation of *sn*-2 monoacylglycerol). In contrast, in the liver and the intestine in a postabsorptive state, the glycerol-3-phosphate pathway (phosphatidic acid intermediate) is dominant for triglyceride synthesis. Whether these alternate lipid synthetic pathways, or perhaps utilization of different compartments within the ER, control the nature of the particles produced is not known.

A genetic disease affecting lipoprotein assembly provides the most compelling evidence that there is at least one distinct step involved in intestinal chylomicron production that is not shared with hepatic VLDL production. Subjects with Anderson's disease (also called chylomicron retention disease) have a defect in the production of apoB-containing lipoproteins which is restricted to the intestine (Kane and Havel, 1995). Although it is believed the missing factor in subjects with Anderson's disease is required for the synthesis of a chylomicron particle, it has not been conclusively determined whether the factor is directly involved in particle assembly or secretion.

5.2. Comparison of Intestinal VLDL Production to Chylomicron Production

There is evidence that different pathways are used in the production of intestinal VLDL and chylomicrons. Hayashi *et al.* (1990) found that Pluronic L-81 (L81), a hydrophobic detergent, inhibits the secretion of large chylomicrons in fat-fed rats, yet is does not effect VLDL secretion. There was no effect of L81 on total apoB secreted, only on the size of the particles produced, suggesting that L81 selectively affects the chylomicron synthetic pathway. Additional evidence for divergent pathways comes from the work of Mahley *et al.* (1971), who found that within the Golgi of epithelial cells of rat small intestine, VLDL and chylomicron-size lipoproteins tend to segregate into separate vesicles. Furthermore, in

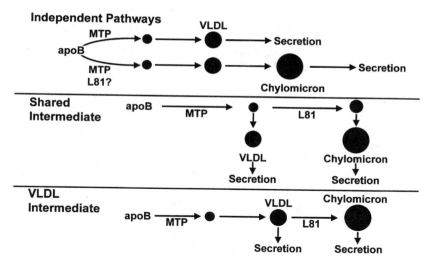

Figure 2. Intestinal VLDL and chylomicron assembly. Based on studies of intestinal lipoprotein production as well as on studies of lipoprotein assembly performed with liver-derived cell lines, several models for the assembly of intestinal VLDL and of chylomicrons may be proposed. The different models vary in the extent to which the pathways leading to VLDL and to chylomicrons overlap. However, all three models have independent steps that may explain the known differences in intestinal VLDL and chylomicron assembly. Possible MTP-dependent, or L81-sensitive steps are shown.

some cases, a biphasic distribution of particles are secreted into lymph, suggesting distinct production pathways for VLDL and chylomicrons, rather than a continuum of the same pathway (Hayashi *et al.,* 1990).

Ockner *et al.* (1969) found that intraduodenal infusion of different fatty acids to rats resulted in differential stimulation of VLDL or chylomicron secretion. For example, linoleic acid increased lymph triglyceride levels in the chylomicron fraction, whereas palmitic acid increased lymph triglyceride in both the VLDL and chylomicron fractions. In addition, the VLDL and chylomicrons produced had different fatty acid compositions. This supports a model in which distinct steps and lipid pools are involved in the production of intestinal VLDL and chylomicrons.

Collectively, these findings suggest that the formation of intestinal VLDL and chylomicron particles involve distinct steps. However, the data does not exclude that the pathway for the formation of a chylomicron passes through intermediates that are shared by the VLDL pathway. Chylomicron assembly may be an extension of the two-step model used to describe hepatic lipoprotein assembly. Minimally, a third step would be required to explain chylomicron assembly, one in which a VLDL-size particle is converted to a large chylomicron. This could be an L81-sensitive step not utilized in the formation of a VLDL particle. Several models for the divergence of VLDL and chylomicron assembly that are consistent with the available data are presented in Fig. 2. The extent to which intestinal VLDL and chylomicron assembly pathways share common steps needs to be further clarified.

6. Role of MTP in Lipoprotein Assembly

In *in vitro* lipid transfer assays, MTP has the ability to promote the transport of TG, CE, and a variety of phospholipid molecules between synthetic membranes. This raises a question as to which transfer activities of MTP are important for lipoprotein assembly. This issue was partially addressed by studies performed with an MTP inhibitor, BMS-200150 (Jamil *et al.,* 1996). This compound binds to MTP and inhibits TG and CE transfer between small unilamellar vesicles with an IC_{50} of around 2 µM. It also inhibits apoB secretion from Hep G2 cells with a similar IC_{50}. Interestingly, the compound has limited ability to inhibit phosphatidylcholine transfer. Although virtually all neutral lipid transfer can be inhibited by BMS-200150, the maximum inhibition of phosphatidylcholine transfer is only 30%. Thus a neutral lipid selective MTP inhibitor blocks the production of apoB lipoproteins. This study does not address the role of phosphatidylcholine transfer. A phospholipid-selective inhibitor would be required to address the role of phospholipid transfer in lipoprotein production.

Several lines of evidence indicate that MTP is involved at an early stage in lipoprotein assembly, or the first step in the two-step model for lipoprotein assembly (see Fig. 1). First, cell culture models have demonstrated that small lipoproteins of HDL or LDL size (products of the first step) are secretion competent. However, these lipoproteins do not appear in abetalipoproteinemic subjects, indicating that the block in lipoprotein secretion that occurs in subjects lacking MTP occurs at a step before the formation of these smaller apoB lipoproteins. This is further supported by heterologous expression of MTP and apoB in various nonhepatic, nonintestinal cell lines. When apoB alone is expressed in these cells, little apoB is secreted into the media. However, when MTP is co-expressed with apoB, small HDL- or LDL-density particles are produced, showing that MTP expression reconstitutes the first step in lipoprotein assembly in these cells.

Studies performed by Benoist and Grand-Perret (1997) further elucidate the critical role that MTP plays early in the assembly pathway. When Hep G2 cells were pulsed for 10 min with radiolabeled amino acids and then were chased, apoB-100 secretion was inhibited if an MTP inhibitor was added at the beginning of the chase period. However, if the inhibitor was added 10 min into the chase, apoB secretion was not affected, indicating that when the assembly process had progressed to a certain stage, it was no longer sensitive to MTP inhibition. When lipoprotein secretion was blocked, apoB failed to accumulate within the cell, suggesting that it was being rapidly degraded. This was confirmed by showing that the addition of protease inhibitors to MTP-inhibitor-treated Hep G2 cells resulted in the accumulation of intracellular apoB. Utilizing puromycin to synchronize apoB-100 synthesis in a pulse–chase protocol, they further demonstrated that in the absence of MTP activity, co-translational degradation of apoB begins after about 70% of the protein has been translated. Thus in Hep G2 cells, MTP is involved early in the assembly process, apparently mediating the addition of lipid to apoB, which stabilizes it from co- and post-translational degradation, and directing it toward a pathway that results in the formation of a mature lipoprotein particle.

Utilizing the protocol developed by Borén *et al.* (1994) to separate the two steps of apoB-48 lipoprotein assembly in McA-RH7777 cells, Gordon *et al.* (1996) showed that MTP is required for the production of the primordial HDL lipoproteins by showing that their production is blocked by an MTP inhibitor. However, they found that normal levels of MTP activity were not required for the conversion of the HDL-density particles to VLDL. Using a pulse–chase protocol, they produced labeled intracellular HDL by culturing the cells in a fatty-acid-deficient media. They then inhibited the MTP and found that the conversion of these particles to lipid-rich VLDL particles was not blocked when a fatty-acid-rich media was used to drive the production of VLDL. However, using the same MTP inhibitor but a different experimental protocol, Wang *et al.* (1997) found that the HDL-to-VLDL conversion is sensitive to a loss of MTP activity in McA-RH7777 cells.

Although there is no information currently available that directly addresses the precise role of MTP in intestinal apoB lipoprotein production, the effect of MTP inhibitors on the production of lipoproteins by differentiated Caco-2 cells has been studied. Using two different MTP inhibitors, apoB-100 secretion from these cells was found to be more sensitive to MTP inhibition than apoB-48, which was relatively insensitive (Haghpassand *et al.*, 1996; Greevenbroek *et al.*, 1998). The largest effect of the MTP inhibitors was on the inhibition of triglyceride secretion. With both inhibitors, the net effect of MTP inhibition in Caco-2 cells was the production of fewer, more dense particles. Although these results may be interpreted to suggest that MTP inhibition affects the lipidation of a primordial lipoprotein, an alternate explanation would be that more than one pathway for lipoprotein assembly occurs in Caco-2 cells and that the pathway leading to lipid-rich lipoproteins is more sensitive to MTP inhibition, possibly at an early step in the pathway. In contrast to these findings, Luchoomun *et al.* (1997) found that apoB-48 secretion from nondifferentiated Caco-2 cells stably transfected with a vector expressing apoB-48 could be readily inhibited. The expression of both apoB-48 and apoB-100 as well as the possibility of multiple assembly pathways will make it difficult to elucidate the precise role of MTP in lipoprotein assembly in Caco-2 cells.

Extending the findings from studies performed in liver-derived cell lines, one would predict that MTP plays a crucial role in the early stages of intestinal VLDL and chylomicron assembly. This is also directly supported by the observation that apoB-48 HDL or LDL are not produced in abetalipoproteinemic subjects, even though these particles are secretion competent. Thus the block in intestinal lipoprotein assembly that occurs in the absence of MTP occurs prior to the formation of small apoB-48 particles. MTP inhibitor studies indicate that the neutral lipid transfer activity of MTP is important in these early steps, with the role of phospholipid transfer still undefined. It also remains to be determined what role MTP plays in the latter stages of assembly, in the lipidation of primordial high-density particles to form VLDL, or in the further lipidation to form chylomicrons.

References

Atzel, A., and Wetterau, J. R., 1993, Mechanism of microsomal triglyceride transfer protein catalyzed lipid transport, *Biochem.* **32**:10444–10450.

Atzel, A., and Wetterau, J. R., 1994, Identification of two classes of lipid molecule binding sites on the microsomal triglyceride transfer protein, *Biochem.* **33**:15382–15388.

Benoist, F., and Grand-Perret, T., 1997, Co-translational degradation of apolipoprotein B100 by the proteasome is prevented by microsomal triglyceride transfer protein, *J. Biol. Chem.* **272**:20435–20442.

Borén, J., Graham, L., Wettesten, M., Scott, J., White, A., and Olofsson, S.-O., 1992, The assembly and secretion of ApoB 100-containing lipoproteins in Hep G2 cells. ApoB 100 is cotranslationally integrated into lipoprotein, *J. Biol. Chem.* **267**:9858–9867.

Borén, J., Rustaeus, S., and Olofsson, S.-O., 1994, Studies on the assembly of apolipoprotein B-100- and B-48-containing very low density lipoprotein in McA-RH7777 cells, *J. Biol. Chem.* **269**:25879–25888.

Dixon, J. L., and Ginsberg, H. N., 1993, Regulation of hepatic secretion of apolipoprotein B-containing lipoproteins: Information obtained from cultured liver cells, *J. Lipid Res.* **34:**167–179.

Dixon, J. L., Furukawa, S., and Ginsberg, H. N., 1991, Oleate stimulates secretion of apolipoprotein B-containing lipoproteins from Hep G2 cells by inhibiting early intracellular degradation of apolipoprotein B, *J. Biol. Chem.* **266:**5080–5086.

Ellsworth, J. L., Erickson, S. K., and Cooper, A. D., 1986, Very low and low density lipoprotein synthesis and secretion by the human hepatoma cell line Hep-G2: Effects of free fatty acid, *J. Lipid Res.* **27:**858–874.

Farese, R. V., Jr., Véniant, M. M., Cham, C. M., Flynn, L. M., Pierotti, V., Loring, J. F., Traber, M., Ruland, S., Stokowski, R. S., Huszar, D., and Young, S. G., 1996, Phenotypic analysis of mice expressing exclusively apolipoprotein B-48 or apolipoprotein B-100, *Proc. Natl. Acad. Sci. USA* **93:**6393–6398.

Glickman, R. M., Kilgore, A., and Khorana, J., 1978, Chylomicron apoprotein localization within rat intestinal epithelium: Studies of normal and impaired lipid absorption, *J. Lipid Res.* **19:**260–268.

Gordon, D. A., Jamil, H., Sharp, D., Mullaney, D., Yao, Z., Gregg, R. E., and Wetterau, J., 1994, Secretion of apolipoprotein B-containing lipoproteins from HeLa cells is dependent on expression of the microsomal triglyceride transfer protein and is regulated by lipid availability, *Proc. Natl. Acad. Sci. USA* **91:**7628–7632.

Gordon, D. A., Jamil, H., Gregg, R. E., Olofsson, S.-O., and Borén, J., 1996, Inhibition of the microsomal triglyceride transfer protein blocks the first step of apolipoprotein B lipoprotein assembly but not the addition of bulk core lipids in the second step, *J. Biol. Chem.* **271:**33047–33053.

Greevenbroek, M. M. J.v., Robertus-Teunissen, M. G., Erkelens, D. W., and de Bruin, T. W. A., 1998, Participation of the microsomal triglyceride transfer protein in lipoprotein assembly in Caco-2 cells: Interaction with saturated and unsaturated dietary fatty acids, *J. Lipid Res.* **39:**173–185.

Haghpassand, M., Wilder, D., and Moberly, J. B., 1996, Inhibition of apolipoprotein B and triglyceride secretion in human hepatoma cells (Hep G2), *J. Lipid Res.* **37:**1468–1480.

Hayashi, H., Fujimoto, K., Cardelli, J. A., Nutting, D. F., Bergstedt, S., and Tso, P., 1990, Fat feeding increases size, but not number, of chylomicrons produced by small intestine, *Am. J. Physiol.* **259:**G709–G719.

Hoeg, J. M., Sviridov, D. D., Tennyson, G. E., Demosky, S. J., Jr., Meng, M. S., Bojanovski, D., Safonova, I. G., Repin, V. S., Kuberger, M. B., Smirnov, V. N., Higuchi, K., Gregg, R. E., and Brewer, H. B., Jr., 1990, Both apolipoproteins B-48 and B-100 are synthesized and secreted by the human intestine, *J. Lipid Res.* **31:** 1761–1769.

Innerarity, T. L., Borén, J., Yamanaka, S., and Olofsson, S.-O., 1996, Biosynthesis of apolipoprotein B48-containing lipoproteins, *J. Biol. Chem.* **271:**2353–2356.

Jamil, H., Gordon, D. A., Eustice, D. C., Brooks, C. M., Dickson, J. K., Jr., Chen, Y., Ricci, B., Chu, C.-H., Harrity, T. W., Ciosek, C. P., Jr., Biller, S. A., Gregg, R. E., and Wetterau, J. R., 1996, An inhibitor of the microsomal triglyceride transfer protein inhibits apoB secretion from Hep G2 cells, *Proc. Natl. Acad. Sci. USA* **93:**11991–11995.

Kane, J. P., and Havel, R. J., 1995, Disorders of the biogenesis and secretion of lipoproteins containing the B apolipoproteins, in: *The Metabolic and Molecular Bases of Inherited Disease,* 7th edition. (C. R. Scriver, A. L. Beaudet, W. S. Sly, and D. Valle, eds.), McGraw-Hill, New York, pp.1853–1885.

Lamberg, A., Jauhiainen, M., Metso, J., Ehnholm, C., Shoulders, C., Scott, J., Pihlajaniemi, T., and Kivirikko, K. I., 1996, The role of protein disulphide isomerase in the microsomal triacylglycerol transfer protein does not reside in its isomerase activity, *Biochem. J.* **315:**533–536.

Leiper, J. M., Bayliss, J. D., Pease, R. J., Brett, D. J., Scott, J., and Shoulders, C. C., 1994, Microsomal triglyceride transfer protein, the abetalipoproteinemia gene product, mediates the secretion of apolipoprotein B-containing lipoproteins from heterologous cells, *J. Biol. Chem.* **269:**21951–21954.

Lin, M. C. M., Arbeeny, C., Bergquist, K., Kienzle, B., Gordon, D. A., and Wetterau, J. R., 1994, Cloning and regulation of hamster microsomal triglyceride transfer protein. The regulation is independent from that of other hepatic and intestinal proteins which participate in the transport of fatty acids and triglycerides, *J. Biol. Chem.* **269:**29138–29145.

Luchoomun, J., Zhou, Z., Bakillah, A., Jamil, H., and Hussain, M. M., 1997, Assembly and secretion of VLDL in nondifferentiated Caco-2 cells stably transfected with human recombinant apoB48 cDNA, *Arterioscler. Thromb. Vasc. Biol.* **17:**2955–2963.

Magun, A. M., Mish, B., and Glickman, R. M., 1988, Intracellular apoA-I and apoB distribution in rat intestine is altered by lipid feeding, *J. Lipid Res.* **29:**1107–1116.

Mahley, R. W., Bennett, B. D., Morré, D. J., Gray, M. E., Thistlethwaite, W., and LeQuire, V. S., 1971, Lipoproteins associated with the Golgi apparatus isolated from epithelial cells of rat small intestine, *Lab. Invest.* **25:**435–444.

Moberly, J. B., Cole, T. G., Alpers, D. H., and Schonfeld, G., 1990, Oleic acid stimulation of apolipoprotein B secretion from Hep G2 and Caco-2 cells occurs post-transcriptionally, *Biochim. Biophys. Acta* **1042:**70–80.

Nakamuta, M., Chang, B. H.-J., Hoogeveen, R., Li, W.-H., and Chan, L., 1996, Mouse microsomal triglyceride transfer protein large subunit: cDNA cloning, tissue-specific expression and chromosomal localization, *Genomics* **33:**313–316.

Narcisi, T. M. E., Shoulders, C. C., Chester, S. A., Read, J., Brett, D. J., Harrison, G. B., Grantham, T. T., Fox, M. F., Povey, S., de Bruin, T. W. A., Erkelens, D. W., Muller, D. P. R., Lloyd, J. K., and Scott, J., 1995, Mutations of the microsomal triglyceride transfer-protein gene in abetalipoproteinemia, *Am. J. Hum. Genet.* **57:**1298–1310.

Ockner, R. K., Hughes, F. B., and Isselbacher, K. J., 1969, Very low density lipoproteins in intestinal lymph: Role in triglyceride and cholesterol transport during fat absorption, *J. Clin. Invest.* **48:**2367–2373.

Patel, S. B., and Grundy, S. M., 1996, Interactions between microsomal triglyceride transfer protein and apolipoprotein B within the endoplasmic reticulum in a heterologous expression system, *J. Biol. Chem.* **271:**18686–18694.

Rehberg, E. F., Samson-Bouma, M.-E., Kienzle, B., Blinderman, L., Jamil, H., Wetterau, J. R., Aggerbeck, L. P., and Gordon D. A., 1996, A novel abetalipoproteinemia genotype. Identification of a missense mutation in the 97-kDa subunit of the microsomal triglyceride transfer protein that prevents complex formation with protein disulfide isomerase, *J. Biol. Chem.* **271:**29945–29952.

Ricci, B., Sharp, D., O'Rourke, E., Kienzle, B., Blinderman, L., Gordon, D., Smith-Monroy, C., Robinson, G., Gregg, R. E., Rader, D. J., and Wetterau, J. R., 1995, A 30-amino acid truncation of the microsomal triglyceride transfer protein large subunit disrupts its interaction with protein disulfide-isomerase and causes abetalipoproteinemia, *J. Biol. Chem.* **270:**14281–14285.

Rusiñol, A., Verkade, H., and Vance, J. E., 1993, Assembly of rat hepatic very low density lipoproteins in the endoplasmic reticulum, *J. Biol. Chem.* **268:**3555–3562.

Rustaeus, S., Stillemark, P., Lindberg, K., Gordon, D., and Olofsson, S.-O., 1998, The microsomal triglyceride transfer protein catalyzes the post-translational assembly of apolipoprotein B-100 very low density lipoprotein in McA-RH7777 cells, *J. Biol. Chem.* **273:**5196–5203.

Sharp, D., Blinderman, L., Combs, K. A., Kienzle, B., Ricci, B., Wager-Smith, K., Gil, C. M., Turck, C. W., Bouma, M.-E., Rader, D. J., Aggerbeck, L. P., Gregg, R. E., Gordon, D. A., and Wetterau, J. R., 1993, Cloning and gene defects in microsomal triglyceride transfer protein associated with abetalipoproteinemia, *Nature* **365:**65–69.

Shoulders, C. C., Brett, D. J., Bayliss, J. D., Narcisi, T. M. E., Jarmuz, A., Grantham, T. T., Leoni, P. R. D., Bhattacharya, S., Pease, R. J., Cullen, P. M., Levi, S., Byfield, P. G. H., Purkiss, P., and Scott, J., 1993, Abetalipoproteinemia is caused by defects of the gene encoding the 97 kDa subunit of a microsomal triglyceride transfer protein, *Hum. Mol. Genetics* **2:**2109–2116.

Shoulders, C. C., Narcisi, T. M. E., Read, J., Chester, S. A., Brett, D. J., Scott, J., Anderson, T. A., Levitt, D. G., and Banaszak, L. J., 1994, The abetalipoproteinemia gene is a member of the vitellogenin family and encodes an alpha-helical domain. *Nat. Structural Biol.* **1:**285–286.

Sniderman, A. D., and Cianflone, K., 1993, Substrate delivery as a determinant of hepatic apoB secretion, *Arterioscler. Thromb.* **13:**629–636.

Thompson, J., Sack, J., Jamil, H., Wetterau, J., Einspahr, H., and Banaszak, L., 1998, Structure of the microsomal triglyceride transfer protein in complex with protein disulfide isomerase, *Biophys. J.* **74:**A138.

Vuori, K., Pihlajaniemi, T., Myllylä, R., and Kivirikko, K. I., 1992, Site-directed mutagenesis of human protein disulphide isomerase: Effect on the assembly, activity and endoplasmic reticulum retention of human prolyl-4-hydroxylase in *Spodoptera frugiperda* insect cells, *EMBO J.* **11:**4213–4217.

Wang, S., McLeod, R. S., Gordon, D. A., and Yao, Z., 1996, The microsomal triglyceride transfer protein facilitates assembly and secretion of apolipoprotein B-containing lipoproteins and decreases cotranslational degradation of apolipoprotein B in transfected COS-7 cells, *J. Biol. Chem.* **271:**14124–14133.

Wang, Y., McLeod, R. S., and Yao, Z., 1997, Normal activity of microsomal triglyceride transfer protein is required for the oleate-induced secretion of very low density lipoproteins containing apolipoprotein B from McA-RH7777 cells, *J. Biol. Chem.* **272:**12272–12278.

Wetterau, J. R., and Zilversmit, D. B., 1985, Purification and characterization of microsomal triglyceride and cholesteryl ester transfer protein from bovine liver microsomes, *Chem. Phys. Lipids* **38:**205–222.

Wetterau, J. R., and Zilversmit, D. B., 1986, Localization of intracellular triacylgylcerol and cholesteryl ester transfer activity in rat tissues, *Biochim. Biophys. Acta* **875:**610–617.

Wetterau, J. R., Combs, K. A., Spinner, S. N., and Joiner, B. J., 1990, Protein disulfide isomerase is a component of the microsomal triglyceride transfer protein complex, *J. Biol. Chem.* **265:**9800–9807.

Wetterau, J. R., Combs, K. A., McLean, L. R., Spinner, S. N., and Aggerbeck, L. P., 1991a, Protein disulfide isomerase appears necessary to maintain the catalytically active structure of the microsomal triglyceride transfer protein, *Biochem.* **30:**9728–9735.

Wetterau, J. R., Aggerbeck, L. P., Laplaud, P. M., and McLean, L. R., 1991b, Structural properties of the microsomal triglyceride transfer protein complex, *Biochem.* **30:**4406–4412.

Wetterau, J. R., Aggerbeck, L. P., Bouma, M.-E., Eisenberg, C., Munck, A., Hermier, M., Schmitz, J., Gay, G., Rader, D. J., and Gregg, R. E., 1992, Absence of microsomal triglyceride transfer protein in individuals with abetalipoproteinemia, *Science* **258:**999–1001.

Wetterau, J. R., Lin, M. C. M., and Jamil, H., 1997, Microsomal triglyceride transfer protein, *Biochim. Biophys. Acta* **1345:**136–150.

Yao, Z., Tran, K., and McLeod, R. S., 1997, Intracellular degradation of newly synthesized apolipoprotein B, *J. Lipid Res.* **38:**1937–1953.

CHAPTER 11

Intestinal Synthesis of Triacylglycerols

ARNIS KUKSIS and RICHARD LEHNER

1. Introduction

During intestinal fat absorption, luminal lipolytic products are absorbed across the microvillus border of the enterocyte. Inside the cell, lipolytic products are reesterified and assembled with apolipoproteins to yield chylomicrons, which are released at the basolateral membrane and enter the blood stream via the lymph (Black, 1995; Davidson, 1994). The major routes for fatty acid esterification in the enterocytes are the monoacylglycerol and phosphatidic acid pathways, both of which ultimately yield triacylglycerols (Johnston, 1977). In the monoacylglycerol pathway, which is characteristic of enterocytes and is of primary importance following lipid ingestion, triacylglycerols are the major products. In the phosphatidic acid pathway, fatty acids are esterified to glycerol-3-phosphate, which is derived from glucose or glycerol. The phosphatidic acid is converted to diacylglycerol and subsequently to triacylglycerols as well as glycerophospholipids. Although the two pathways have been generally considered to function separately (Johnston, 1977), evidence for their convergence at the level of 2-monoacylglycerols via a deacylation/reacylation cycle has been presented (Yang and Kuksis, 1991; Lehner and Kuksis, 1992). The isolation of pancreatic lipase from rat enterocytes (Tsujita *et al.*, 1996) provides potential explanations for past controversies while creating new ones.

   This chapter updates earlier reviews of various aspects of intestinal triacylglycerol biosynthesis including enzyme isolation (Johnston, 1977; Kuksis and Manganaro, 1986), physiology (Desnuelle, 1986; Tso, 1994), physical chemistry (Thompson and Dietschy, 1981; Carey *et al.*, 1983) and stereochemistry (Kuksis and Manganaro, 1986) as well as a more general recent review of triacylglycerol biosynthesis (Lehner and Kuksis, 1996). Other pertinent reports deal with triacylglycerol synthesis and secretion by the human colon adenocarcinoma cell line Caco-2 (Van Greevenbroek *et al.*, 1995). A specific characteristic of the Caco-2 cell line is that it uses the phosphatidic acid pathway for the synthesis of triacylglycerols (Trotter and Storch, 1993).

ARNIS KUKSIS and RICHARD LEHNER • Banting and Best Department of Medical Research, University of Toronto, Toronto M5G 1L6 Canada.
Intestinal Lipid Metabolism, edited by Charles M. Mansbach II *et al.,* Kluwer Academic/Plenum Publishers, 2001.

2. Sources of Fatty Acids and Glycerol

Diet and *de novo* synthesis are the two primary sources of fatty acids and glycerol for triacylglycerol synthesis, and the fatty acids released from body stores and those generated by chain elongation and desaturation of exogenous and endogenous fatty acids are secondary sources. The fatty acids from both primary and secondary sources may enter the pathway of triacylglycerol biosynthesis in the form of acyl-CoA esters and as partial acylglycerols. The glycerol for the phosphatidic acid pathway is derived mainly in the form of dihydroxyacetone and glycerol-3-phosphate from glycolysis because the enterocyte possesses minimal glycerol kinase activity (Thompson and Dietschy, 1981; Mansbach and Parthasarathy, 1982).

2.1. Lipolysis of Lumenal Acylglycerols

After ingestion, triacylglycerols are first hydrolyzed by a gastric acid lipase, which preferentially attacks the short- and medium-chain fatty acids in the *sn*-3 position of the triacylglycerol molecule (Bezard and Bugaut, 1986). Emulsified acylglycerols enter the duodenum where they are mixed with bile and pancreatic juice. Pancreatic lipase acts on triacylglycerols at the oil–water interface yielding small amounts of *sn*-1,2- and *sn*-2,3-diacylglycerols as intermediates and 2-monoacylglycerols and free fatty acids as final products. Over short periods of time, the enzyme discriminates against long-chain saturated and long-chain polyunsaturated fatty acids (Bottino *et al.*, 1967; Yang *et al.*, 1990). As a result, the long-chain fatty acids are initially released more slowly than the shorter chain fatty acids. Prolonged digestion, however, leads to hydrolysis of all fatty acids in the primary positions and to formation of 2-monoacylglycerols characteristic of the secondary position (Yang *et al.*, 1990; Ikeda *et al.*, 1993). Some 15% to 20% of the total monoacylglycerol product is in the *sn*-1(3) isomer form arising from 2-monoacylglycerols by acyl migration. The *sn*-1(3)-monoacylglycerols are subject to further hydrolysis by pancreatic lipase (Wang, 1986).

Lumenal lipolysis also results in a release of fatty acids from glycerophospholipids. Thus, the Ca^{++}-dependent pancreatic phospholipase A_2 gives rise to the *sn*-1-monoacyl (lyso) derivatives, which are absorbed intact and serve as acceptors of acyl groups in the microvillus cells resulting in resynthesis of glycerophospholipids (Wang, 1986). The brush border membrane of mature enterocytes from rats (Pind and Kuksis, 1987, 1988, 1989), guinea pigs (Diagne *et al.*, 1987; Gassama-Diagne *et al.*, 1989) and rabbits (Pind and Kuksis, 1991; Boll *et al.*, 1993) has been shown to contain a phospholipase B, which attacks both *sn*-1- and *sn*-2 positions of glycerophospholipids. Later, the enzyme was observed to hydrolyze also mono- and diacylglycerols (Gassama-Diagne *et al.*, 1992) and retinyl esters (Rigtrup *et al.*, 1994). Takemori *et al.* (1998) and Tojo *et al.* (1998) have now identified and cloned the functional domain of this enzyme and have determined its tissue distribution.

2.2. *De novo* Synthesis of Fatty Acids

The other primary source of fatty acids for triacylglycerol genesis is *de novo* synthesis. The liver has the highest capacity for fatty acid synthesis, but intestinal mucosa also possesses significant activity. The rate of *de novo* formation of long-chain acids is rapid in fed animals, especially when the diet has little or no fat, but slow in starved animals (Goodridge, 1991). The fatty acids from diet and from *de novo* synthesis are subject to chain elongation and desaturation. The first double bond introduced into a saturated acyl chain is generally in the Δ^{-9} position, so that substrates for further desaturation contain either a Δ^{-9} double bond or one derived from the Δ^{-9} position by chain elongation. The other double bonds are introduced by further oxidative desaturation. Animal systems cannot introduce double bonds beyond the Δ^{-9} position (Cinti *et al.*, 1992).

2.3. Lipolysis of Endogenous Triacylglycerols

A secondary source of fatty acids for intestinal triacylglycerol biosynthesis is the lipolysis of adipose tissue triacylglycerols; as a result of this process, free fatty acids and small amounts of monoacylglycerols are released into blood for transport to other tissues as albumin-bound complexes. Recent studies have shown that the molecular structure of fatty acids influences their mobilization from fat cells (Raclot and Groscolas, 1993). The *in vitro* mobilization of 52 fatty acids ranging in chain length from C_{12} to C_{24} and having zero to 6 double bonds, including 23 pairs of positional isomers, showed that under conditions of simulated lipolysis, individual fatty acids are more readily mobilized from fat cells when they are short and unsaturated and when their double bonds are closer to the methyl end of the chain.

Lipoprotein lipase releases free fatty acids and 2-monoacylglycerols into plasma from both chylomicron and VLDL triacylglycerols (Santamarina-Tojo and Dugi, 1994). The released hydrolysis products are rapidly cleared by tissues or are bound by albumin. Lipoprotein lipase preferentially attacks the *sn*-1-acyl group of triacylglycerols (Akesson *et al.*, 1983) and the 1-acyl group of glycerophospholipids (Zandonella *et al.*, 1995). The hepatic lipase also attacks preferentially the *sn*-1-acyl group of triacylglycerols and partial acylglycerols to yield 2-monoacylglycerols as final products (Akesson *et al.*, 1983). Unlike lipoprotein lipase, hepatic triacylglycerol lipase is also effective in the transacylation of lipids (Wang, 1986). It is not known whether or not any of the 2-monoacylglycerols released into plasma by the endogenous lipases are cleared and re-esterified by the intestine along with the free fatty acids.

3. Mucosal and Intracellular Transport of Fatty Acids and Monoacylglycerols

Intestinal triacylglycerol synthesis depends on the availability of free fatty acids and 2-monoacylglycerols or glycerol. These substrates are delivered to the triacylglycerol synthetase complex by the various transport systems, which in view of the separate coverage in the book (chapters 9, 10, and 12) are considered here only briefly.

3.1. Mucosal Transport

Lumenal monoacylglycerols and free fatty acids, solubilized as bile salt micelles, are transferred to the mucosal cell. It is still unclear how these hydrophobic compounds move through the aqueous phase and across the hydrophilic outer leaflet of the brush border membrane after their egress from micellar structures (Thompson *et al.,* 1993). Binding of monoacylglycerols to the putative fatty acid transporters has not been reported. Bosner *et al.* (1988, 1989) obtained evidence for heparin-modulated reversible binding of pancreatic lipase to the brush border membranes. Immobilization of the lipase on the membrane would allow the intercalation of fatty acids and 2-monoacylglycerols into the membrane in their monomeric (submicellar) form, thus obviating the necessity for solubilization with bile salts and the formation of large mixed micelles. It is not known if and how the pancreatic-lipase-facilitated uptake of fatty acids and monoacylglyceerols may be related to the intestinal reacylation of monoacylglycerols due to pancreatic lipase functioning as a monoacylglycerol acyltransferase (Tsujita *et al,* 1996). In a subsequent study, Lopez-Candales *et al.* (1993) have shown a 39-fold increase in cholesteryl esters when Caco-2 cells were incubated with cholesterol in the presence of exogenously added cholesterol esterase. Basal uptake, which occurs without the assistance of cholesterol esterase, is associated with low levels of esterification and may direct cholesterol to outer monolayer locations. In contrast, a cholesterol-esterase-mediated pathway not only is much more efficient as a cholesterol transport system but also leads to high levels of ester formation within the cell. The authors have postulated that neutral lipids can, in part, move across biological membranes via a mechanism involving enzymes anchored to membrane proteoglycans such as those found in the brush border of the enterocyte.

3.2. Intracellular Transport

Having crossed the brush border membrane, free fatty acids and monoacylglycerols must be delivered to their site of processing, the endoplasmic reticulum. Although no definite transport system has been demonstrated to mediate such vectorial delivery, the primary candidates for this function appear to be the intracellular fatty-acid-binding proteins (FABPs; chapter 9; Poirier *et al.,* 1997). It has also been suggested, but not demonstrated (Minich *et al.,* 1997), that *l*-FABP also binds monoacylglycerols, lysophospholipids, and fatty acyl-CoA esters. Transfection of 3T3 fibroblasts with FABppm cDNA lead to cell surface expression of FABPpm and an increase in saturable fatty acid uptake (Isola *et al.,* 1995). A fatty acid translocase (FAT) with high levels of expression in muscle, intestine, and adipose tissue has been identified (Harmon and Abumrod, 1993) and its mRNA expression is increased by dexamethasone, fatty acids, and other lipid molecules (Amri *et al.,* 1995). Expression of FAT cDNA in fibroblasts leads to increased fatty acid uptake (Ibrahimi *et al.,* 1996). FAT binds various long-chain fatty acids with similar affinities (Baillie *et al.,* 1996).

3.3. Location of Biosynthetic Pathways

In the intestine, both pathways of triacylglycerol formation were first localized to microsomes. Later, cytochemical experiments involving the determination of free CoA distribution suggested that the monoacylglycerol pathway was primarily associated with smooth endoplasmic reticulum, whereas the glycerol-3-phosphate pathway was largely confined to the rough endoplasmic reticulum membranes (Cartwright *et al.,* 1997). There exists ample experimental evidence indicating that both pathways are several times more active in villus cells than in crypt cells (Shiau, 1987). The proposed topology of triacylglycerol biosynthesis is consistent with observations of high concentration of smooth endoplasmic reticulum just beneath the terminal web of the apical membrane of the enterocyte (Friedman and Cardell, 1977).

The subcellular localization of the enzyme in the microsomal fractions was investigated in both the liver and the intestine (Coleman and Bell, 1983). Using proteases and non-permeable inhibitors, they found that the diacylglycerol acyltransferase (DGAT) active site was located on the cytoplasmic side of the endoplasmic reticulum. The rat intestinal microsomal monoacylglycerol acyltransferase (MGAT) activity was found to be resistant to inactivation by subtilisin, also suggesting lumenal localization of the active site (Hulsmann and Kurpershoek-Davidov, 1976). Owen *et al.* (1997) have recently challanged this view and suggested that in rat liver microsomes two DGAT isozymes may exist, one overt (cytosol facing) and one with latent (lumenal) orientation. The rationale for two forms of the enzyme may be that cytosol-facing DGAT may be involved in synthesis of triacylglycerol destined for intracellular storage and the lumenal DGAT may catalyze re-esterification of diacylglycerols mobilized from the storage pool by lipolysis for secretion as lipoproteins.

The enzyme 1-acyl-*sn*-glycerol-3-phosphate acyltransferase (AGPAT) has been localized to the endoplasmic reticulum by confocal immunofluorescence (Aguado and Campbell, 1998). The dihydroxyacetone phosphate acyltransferase of intestinal mucosa has been shown to be associated with the peroxisome fraction (Ruyter *et al.*, 1992) as already established for the liver (Hajra, 1995).

4. Biosynthesis of Triacylglycerols

The intestine synthesizes triacylglycerols by two basic pathways—the phosphatidic acid pathway and the monoacylglycerol pathway. The phosphatidic acid pathway is responsible for all synthesized triacylglycerol when 2-monoacylglycerol is not available (i.e., during fasting or feeding of fatty acid alkyl esters). During fat absorption in monogastric animals, the monoacylglycerol pathway is the major route of triacylglycerol formation.

4.1. Phosphatidic Acid Pathway

Triacylglycerol biosynthesis from glycerol-3-phosphate in intestinal cells was first documented in the early 1960s (for a review of early work, see Johnston, 1977). However, when large amounts of long-chain 2-monoacylglycerols are absorbed by the enterocyte, the glycerol-3-phosphate pathway accounts for only about 20% to 30% of formed triacylglycerol (Yang *et al.*, 1995). Phosphatidic acid can also be synthesized via ATP-dependent phosphorylation of *sn*-1,2-diacylglycerol by diacylglycerol kinase. Various forms of mammalian diacylglycerol kinase have been reported (Sakane *et al.*, 1989; Yada *et al.*, 1990) and some have been cloned (Gotto and Kondo, 1996; Houssa *et al.*, 1997). The contribution of this enzyme to the net phosphatidate biosynthesis in the intestine has not been determined. Figure 1 outlines the phosphatidic acid pathway of triacylglycerol biosynthesis.

4.1.1. Glycerophosphate Acyltransferase (GPAT)

GPAT has not been investigated in the intestine beyond the initial observations made by the early workers (see Johnston, 1977). The work on the liver enzyme has been reviewed

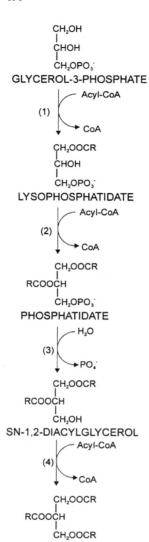

CH_2OH
|
$CHOH$
|
$CH_2OPO_3^-$
GLYCEROL-3-PHOSPHATE
(1) Acyl-CoA → CoA

CH_2OOCR
|
$CHOH$
|
$CH_2OPO_3^-$
LYSOPHOSPHATIDATE
(2) Acyl-CoA → CoA

CH_2OOCR
|
$RCOOCH$
|
$CH_2OPO_3^-$
PHOSPHATIDATE
(3) H_2O → PO_4^-

CH_2OOCR
|
$RCOOCH$
|
CH_2OH
SN-1,2-DIACYLGLYCEROL
(4) Acyl-CoA → CoA

CH_2OOCR
|
$RCOOCH$
|
CH_2OOCR
TRIACYLGLYCEROL

Figure 1. Biosynthesis of triacylglycerols via the phosphatidic acid pathway: (1) glycerol-3-phosphate acyltransferase; (2) 1-acylglycerol-3-phosphate acyltransferase; (3) phosphatidic acid phosphohydrolase; (4) diacylglycerol acyltransferase.

(Yamashita *et al.*, 1973). Mammalian tissues are now recognized as containing two forms of GPAT, one present in the mitochondrial membranes and the other in the microsomal membranes with the former showing preference for saturated fatty acyl-CoA (Bell and Coleman, 1980, 1983). The first structural information on the enzyme arrived with the purification of an 83 kDa GPAT from *Escherichia coli* membranes (Green *et al.*, 1981). Later, purified liver enzyme was found to be inactive unless reconstituted with phospholipid (Scheideler and Bell, 1986), notably, cardiolipin and phosphatidylglycerol (Scheideler and Bell, 1989). Active and latent forms of GPAT from *E. coli* have also been characterized (Scheideler and Bell, 1991). Shin *et al.* (1991) cloned a murine p90 protein with 30% identity with the *E. coli* GPAT based on the amino acid sequence deduced from cDNA. The p90 mRNA was dramatically induced in livers of fasted mice fed a high-carbohydrate diet. The p90 mRNA was expressed in high levels in the liver, in muscle, and in the kidney with lower levels in the brain. The protein was also detected in fully differentiated adipocytes but not in preadipocytes (Yet *et al.,* 1993). Mitochondrial GPAT mRNA levels are induced during differentiation of 3T3-L1 preadipocytes to adipocytes, and this regulated expression appears to require both nuclear factor Y (NF-Y) and adipocyte determination and differentiation-dependent factor 1/sterol regulatory element-binding proteins (ADD1/SREBPs), indicating a link between the regulation of genes involved in cholesterol homeostasis, fatty acid synthesis, and glycerolipid synthesis (Ericsson *et al.,* 1997). An 85 kDa GPAT was also recently solubilized and purified from rat liver mitochondria (Vancura and Haldar, 1994).

4.1.2. Dihydroxyacetonephosphate Acyltransferase (DHAPAT)

DHAPAT catalyzes the acylation of dihydroxyacetone phosphate to form 1-acyl-dihydroxyacetone phosphate. This reaction represents the first step in synthesis of ether-linked glycerolipids, but it can also lead to glycerol ester lipids following the enzymatic reduction of the reaction product by NADPH to 1-acyl-*sn*-glycero-3-phosphate (Hajra, 1995). DHAPAT was solubilized with zwitterionic detergent (3–[3–cholamidopropyl)dimethylammoniol]–1–propane-sulfonate (CHAPS) from guinea pig liver peroxisomal membranes. The enzyme activity co-purified with a protein of an apparent molecular mass of 69 kDa (Webber and Hajra, 1992). The purified enzyme had different kinetic parameters when compared to the membrane-bound form.

4.1.3. Acylglycerophosphate Acyltransferase (AGPAT)

The 1-acyl-*sn*-glycerol-3-phosphate acyltransferase has also not been specifically assayed in the intestinal mucosa. The enzyme was purified 7.5-fold from rat liver microsomes by solubilization of the membranes with Triton X-100 (Yamashita and Numa, 1972). A human-expressed sequence tag (EST) was identified by homology with coconut AGPAT and was used to isolate a full-length human cDNA from a heart muscle library (Eberhardt *et al.,* 1997). COS 7 cells expressing the cDNA exhibited AGPAT activity with a preference for arachidonoyl-CoA as the acyl donor. Northern blot analysis indicated expression in most tissues including intestine, with liver and pancreas being the highest expressors. Another human cDNA clone that was 48% identical to the previously mentioned AGPAT at the amino acid level was isolated and designated as hLPAAT alpha (lysophosphatidic acid acyltransferase; Aguado and Campbell, 1998).

4.1.4. Phosphatidic Acid Phosphohydrolase (PAPH)

PAPH converts phosphatidic acid to *sn*-1,2-diacylglycerol. The enzyme is present in both the soluble and the particulate cellular fractions and the diacylglycerol production has been proposed to be regulated by the fatty-acid-induced translocation of the enzyme from the cytosol to the endoplasmic reticulum (Martin *et al.,* 1987; Gomez-Munoz *et al.,* 1992). PAPH appears to consist of several isoforms that can be distinguished from each other with respect to subcellular localization, cation dependences, and sensitivity to various inhibitors (Gomez-Munoz *et al.,* 1992; Jamal *et al.,* 1991). One isoform of the enzyme, a plasma-membrane-bound type 2 PAP, was recently purified, cDNA isolated, and expressed (Kai *et al.,* 1996; Hooks *et al.,* 1998).

4.1.5. *Diacylglycerol Acyltransferase (DGATpa)*

A diacylglycerol acyltransferase specifically associated with the phosphatidic acid pathway (DGATpa) has not been isolated. The *sn*-1,2-diacylglycerols released by PAPH are acylated to triacylglycerols via an acyl-CoA-dependent DGATpa, which may not be the same as that associated with the triacylglycerol synthetase complex (DGATmg). Owen *et al.* (1997) have reported the existence of a cytosolic form and a lumenal form of DGAT on the basis of two distinct activities participating in the biosynthesis of VLDL triacylglycerols by the liver. Neither of the DGATs has been purified and it is not known whether either of them is related to that associated with the monoacylglycerol pathway of triacylglycerol biosynthesis in the intestine.

4.2. Monoacylglycerol Pathway

The enzymatic reactions utilizing monoacylglycerols for di- and for triacylglycerol biosynthesis have been investigated in the intestinal mucosa of various animal species, and evidence has been obtained for the existence of a triacylglycerol synthetase complex (Johnston, 1977) as well as for a dissociation and reconstitution of the enzyme complex (Kuksis and Manganaro, 1986; Lehner and Kuksis, 1996). Figure 2 outlines the monoacylglycerol pathway of triacylglycerol biosynthesis.

4.2.1. *Triacylglycerol Synthetase Complex*

The isolation of the triacylglycerol synthetase as a detergent-solubilized complex does not distinguish between a naturally occurring triacylglycerol synthetase assembly and a mixture of membrane-bound enzymes pooled on the basis of their common solubility in a detergent. A characteristic ratio of enzyme activities and a co-precipitation of all component enzymes by antibodies prepared to individual enzymes suggest the existence of a true enzyme complex in the intestine.

4.2.1a Purification. Rao and Johnston (1966) reported a 16-fold purification of MGAT along with DGAT and acyl-CoA ligase activities from hamster intestinal microsomes. The purification of the synthetase complex involved extraction of the enzyme activities from the membranes by sonication in the absence of a detergent and by ammonium sulfate precipitation of the extract. The enzyme present in the precipitate was unstable and 50% of the activity was lost upon storage for 12 hours at 0°C. The complex obtained from hamster microsomes eluted in the void volume on Sephadex G-200, indicating either a large molecular mass or the formation of protein aggregates (Rao and Johnston, 1966). Four major polypeptides of unspecified molecular mass were observed upon analysis of the purified fraction by nondenaturing sodium dodecyl sulfate-polyacrylamide gel electrophoresis (SDS-PAGE). Subsequently, Manganaro and Kuksis (1985b) obtained a 10-fold purification of

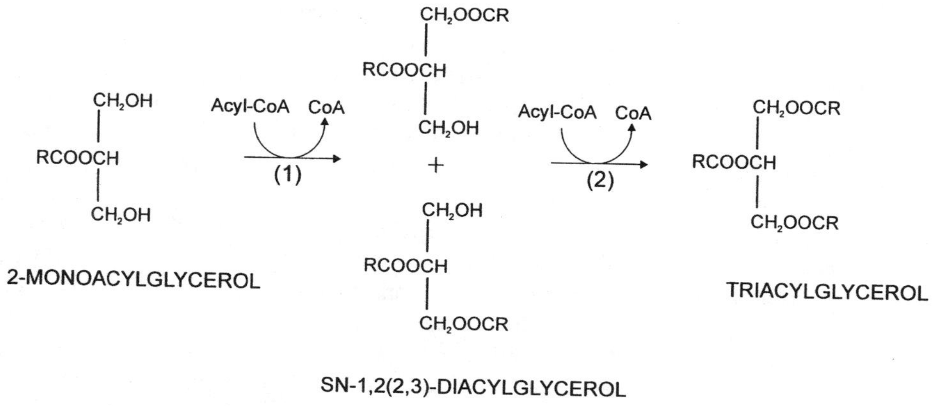

Figure 2. Biosynthesis of triacylglycerols via the 2-monoacylglycerol pathway: (1) monoacylglycerol acyltransferase; (2) diacylglycerol acyltransferase.

the triacylglycerol synthetase complex from rat intestinal mucosa by solubilization of microsomal membranes with an ionic detergent (sodium taurocholate) and chromatography of the solubilized extract on phenyl Sepharose. The estimated molecular mass of the complex purified from rat microsomes was in excess of 350 kDa (Manganaro and Kuksis, 1985a). Further chromatography of the complex on size exclusion columns was accompanied by loss of the DGAT activity.

4.2.1b Characterization. More recently, a dye-affinity chromatography of CHAPS-solubilized rat intestinal microsomes yielded four active components for the triacylglycerol synthetase complex: MGAT, DGAT, acyl-CoA ligase (FACS), and acyl-CoA acyltransferase (AAT; Lehner and Kuksis, 1995). The purified fraction contained four major polypeptides of apparent molecular masses of 52, 54, 58, and 68 to 70 kDa. The 54 kDa protein was associated with the AAT activity (Lehner and Kuksis, 1993a, 1993b) that was proposed to act as an acyl-CoA binding subunit of the hetero-oligomeric complex (Lehner and Kuksis, 1995). The 54 kDa protein was shown to bind covalently acyl groups from acyl-CoA via a thiol ester linkage (Lehner and Kuksis, 1993a). The presence of an enzyme complex rather than a mixture was indicated by the insolubilization of the MGAT and DGAT activities during immunoprecipitation of the AAT (Lehner and Kuksis, 1995).

4.2.2. 2-Monoacylglycerol acyltransferase (MGAT)

The biochemical and stereochemical aspects of MGAT have been extensively studied in the microsomes of rat and hamster intestine (Johnston, 1977), in those of neonatal livers of rats (Coleman and Haynes, 1986; Coleman *et al.,* 1986), guinea pigs (Coleman *et al.,* 1987) and in those of adipose tissue of migrating birds (Mostafa *et al.,* 1994). Attempts to purify MGAT have met with limited success only (Kuksis and Manganaro, 1986; Lehner and Kuksis, 1996).

4.2.2a Purification. Manganaro and Kuksis (1985a) reported the isolation of MGAT from the triacylglycerol synthetase complex by gel filtration in the presence of guanidine chloride. The purified enzyme possessed only the MGAT activity. It required acyl-CoA for acylation as it no longer was able to synthesize it from free fatty acids, ATP, and CoA. The purified MGAT migrated as a 37 kDa polypeptide band upon SDS-PAGE. It is not known if this 37 kDa protein is a subunit of a polymeric enzyme or a proteolytic breakdown product, although a 15 kDa peptide frequently accompanied the appearance of the 37 kDa polypeptide. On the basis of the specific activity of the 37 kDa polypeptide, it was estimated that the MGAT activity had been purified 600-fold (Manganaro and Kuksis, 1985b). The enzyme appears to have an overall acidic nature as it was bound to an anion exchanger from which it could be eluted by a sodium chloride gradient (Manganaro and Kuksis, 1985a, 1985b).

Bhat *et al.* (1993) were able to solubilize 56% of the MGAT activity from neonatal rat liver microsomes by the non-ionic detergent Triton X-100. They reported a 40-fold purification of the activity from solubilized microsomes. Further attempts to purify the enzyme using hydrophobic, dye, or CoA matrices were not successful. The solubilized and partially purified liver MGAT showed a preference for monoacylglycerols containing linoleic and linolenic acids (Bhat *et al.,* 1993) and was less thermolabile and less sensitive to manganese inhibition than its intestinal counterpart. The liver MGAT was regulated by *sn*-1,2-diacylglycerols and by other specific lipids in Triton X-100/phospholipid micelles (Bhat *et al.,* 1994). Sphingosine, but not ceramide, was found to inhibit the enzyme in the detergent/phospholipid micelles (Bhat *et al.,* 1995).

Recently, Tsujita *et al.* (1996) have reported that rat intestinal mucosa contains high diacylglycerol synthesizing activity, which converts 2-monoacylglycerols into diacylglycerols in the presence of free fatty acids independent of coenzyme A and ATP. The MGAT activity was purified from rat intestinal mucosa by successive chromatographic separations on diethylaminoethyl (DEAE) cellulose, CM- Sephadex, and anti-IgG-Sepharose against pancreatic lipase. The enzyme was electrophoretically homogeneous, and its molecular weight was 49 kDa, which is identical with that of rat pancreatic lipase. This enzyme was free of DGAT activity and was not precipitated with other components of the triacylglycerol synthetase complex. The activity of the purified enzyme was inhibited by addition of the antibody. Using immunocytochemical techniques, it was found that immunoreactive protein against rat pancreatic lipase was uniformly distributed within the absorptive cells of the intestine but was absent from the microvillar membrane. These results suggest that pancreatic lipase is present in intestinal absorptive cells and that it may contribute to the resynthesis of diacylglycerol from monoacylglycerol and fatty acids in these cells. Triacylglycerol- and diacylglycerol-hydrolyzing activities of the purified enzyme and pancreatic lipase were inhibited by addition of intestinal mucosa extract. The mechanism of the inhibition was not discussed. Although MGAT activity of intestinal mucosal homogenate was inhibited by about 65% by addition of antibody against rat pancreatic lipase, it was not established whether or not the remaining activity was due to the microsomal MGAT, which functions as part of the triacylglycerol synthetase complex. Pancreatic lipase is not known to be expressed in the intestinal mucosa or to be absorbed across the epithelial wall. On the other hand, it has been found using immunocytometry that cholesterol esterase actually crosses the brush border membrane (Gallo *et al.,* 1980).

4.2.2b Stereospecificity. The carbon atom at the 2-position of 2-monoacylglycerol constitutes a prochiral center. The immediate acylation product may therefore be either *sn*-1,2-diacylglycerol, sn-2,3-diacylglycerol, or both depending on the stereospecificity of the enzyme. In evert sacs of rat intestinal mucosa the main products of 2-monoacylglycerol acylation were sn-1,2-diacylglycerol (53%–63%) and sn-2,3-diacylglycerols (37%–47%; Breckenridge and Kuksis, 1972). These ratios were subsequently confirmed in other preparations of intestinal segments, tissue homogenates, and isolated cells (Bugaut *et al.*, 1984; Kuksis and Manganaro, 1986). These results differ from those obtained with isolated microsomes, as first noted by Johnston *et al.* (1970). Paltauf and Johnston (1971) compared the esterification of the sn-1-, sn-2-, and sn-3-monoalkylglycerols, which cannot be isomerized, and the corresponding monoacylglycerol derivatives as fatty acid acceptors in rat and hamster microsomes. Generally, the MGAT showed marked preference for the acylation of the sn-1- position, resulting in the synthesis of 80% to 100% of sn-1,2-diacylglycerol. Similar specificity was obtained by Coleman *et al.* (1986) who employed the diacylglycerol kinase of *Escherichia coli* to identify the sn-1,2-diacylglycerol products of the microsomal MGAT. More recently, the ratios of enantiomeric diacylglycerols resulting from acylation of 2-monoacylglycerols by microsomal and partially purified MGAT were determined by chiral phase high performance liquid chromatography (HPLC) (Lehner *et al.*, 1993), a direct and unambiguous method for separation, identification and quantitation of the sn-1,2-, sn-2,3- and X-1,3-diacylglycerols. Figure 3 shows the ultraviolet (UV) absorption profile of the dinitrophenylurethane (DNPU) derivatives of standard diacylglycerols (A) and of the diacylglycerols (B) recovered from incubation of the microsomes with radiolabeled 2-oleoylglycerol and oleoyl-CoA. The incubation products are seen to consist largely of the 1,2-diacyl-sn-glycerols with smaller, but clearly recognizable anmounts, of the sn-2,3-enantiomers. The incubation blank (C) contains UV-absorbing peaks that coincide with the 1,2-diacyl-sn-glycerol standards. Collection and rechromatography of the peaks corresponding to the 1,2- and 2,3-diacyl-sn-glycerols (B) showed the absence of cross-contamination among the enantiomers (D and E). The sn-1,2-diacylglycerols made up 80% to 90% of the diacylglycerol products. Clearly, a difference existed between the results obtained with intact cells and with microsomes, which could not be explained on the basis of fatty acid specificity. The possibility that endogenous pancreatic lipase contributes various amounts of rac-1,2-diacylglycerols to the diacylglycerol pool of the intestinal mucosa (Tsujita *et al.*, 1996) must now be considered as a possible explanation for the discrepancy between the results obtained with microsomes and with whole cells.

4.2.2c Chain-length Specificity. The chain-length specificity of acyl donors and acceptors for the reaction has been studied *in vitro* and the results may not reflect the true specificity of the enzyme *in vivo*. Comparisons of utilization of long-chain versus short-chain substrates are compromised by differences in their solubility in aqueous media. The use of organic solvents or detergents as solubilizers *in vitro* interferes with membrane integrity and leads to extraction of the enzyme from its natural environment. Nevertheless, results from *in vitro* assays carried out in several laboratories (for a review, see Kuksis and Manganaro, 1986; Lehner and Kuksis, 1996) indicate that saturated and unsaturated 2-monoacylglycerols and acyl-CoA are utilized for diacylglycerol synthesis at different rates. The highest esterification rates were obtained with monodecanoylglycerol and monoundecanoylglycerol. Monooleoylglycerol was utilized at twice the rate of monostearoylglycerol. Employing the *in situ* generating system of acyl-CoA, myristoyl-CoA, palmitoyl-CoA, stearoyl-CoA, oleoyl-CoA, linoleoyl-CoA, and arachidonoyl-CoA were utilized with similar efficiency regardless of the chain length and unsaturation of the 2-monoacylglycerol acceptor. Coleman

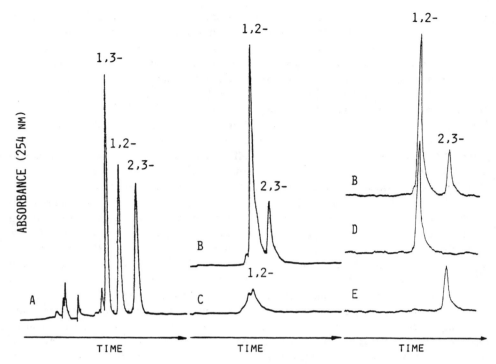

Figure 3. Ultraviolet absorption profile of the 3,5-dinitrophenylurethane derivatives of standard diacylglycerols (A), of diacylglycerols recovered from incubation of intestinal microsomes with radiolabeled 2-oleoylglycerol and oleoyl-CoA (B), of endogenous diacylglycerols (C), and of the recovered and rechromatographed 1,2-diacylglycerols (D) and 2,3-diacylglycerols (E). 1,3-, X-I,3-diacylglycerols; 1,2-, *sn*-1,2-diacylglycerols; 2,3-, *sn*-2,3-diacylglycerols. Chiral phase HPLC conditions as described elsewhere (Lehner, Kuksis, and Itabashi, 1993) .

and Haynes (1984) found that neonatal rat liver microsomal MGAT exhibited a marked preference for 2-oleoylglycerol. Lower activities were obtained with short-chain and with saturated 2-monoacylglycerols. Neonatal liver enzyme also exhibited specificity for acyl-CoA chain length. Activity observed with octanoyl-CoA was only about 8% of that obtained with palmitoyl-CoA, whereas acetyl-CoA was not esterified. More recently, Xia *et al.* (1993) and Bhat *et al.* (1993) have reported an increased affinity of hepatic MGAT for 2-monoacylglycerols that contain polyunsaturated acyl chains.

4.2.3. Diacylglycerol Acyltransferase (DGATmg)

The DGATmg reaction involves acyl-CoA-dependent acylation of diacylglycerols derived from esterification of 2-monoacylglycerol (monoacylglycerol pathway) or from hydrolysis of endogenous triacylglycerols during the deacylation/reacylation cycle (see the following section).

4.2.3a Purification. A partial purification of intestinal DGAT was first reported by Rao and Johnston (1966). Sonication of hamster intestinal microsomes and ammonium sulphate precipitation resulted in a 16-fold increase of specific activity of the enzyme in the pellet that also included other components of the putative triacylglycerol synthetase complex—MGAT and acyl-CoA ligase.

Manganaro and Kuksis (1985b) obtained a 10-fold purification of DGAT from rat intestinal microsomes using 2% sodium taurocholate (detergent/protein ratio of approximately 1:3). Phenyl Sepharose chromatography of a taurocholate-solubilized extract permitted the recovery of the DGAT activity together with the other activities of the triacylglycerol synthetase complex—MGAT and acyl-CoA ligase (Manganaro and Kuksis, 1985b). This active fraction did not contain any detectable GPAT activity, which was either lost or inactivated upon solubilization or during chromatography (Manganaro and Kuksis, 1985b). About 50% of the solubilized and partially purified DGAT activity of the complex was lost upon storage at $-20°C$ for 2 weeks even in the presence of glycerol and a protease inhibitor. Gel filtration of the partially purified enzyme complex resulted in complete loss of the DGAT activity, while retaining some MGAT activity. Denaturing SDS-PAGE of the purified enzyme extract indicated protein bands in the 52 to 56 kDa and 70 to 90 kDa range, but it was not determined which one of the many polypeptide bands present in this preparation was associated with DGAT activity (Manganaro and Kuksis, 1985b).

Solubilization of rat liver microsomes with 8 mM Triton X-100 followed by gel filtration chromatography of the solubilized extract led to the recovery of DGAT, *sn*-1-acyl-3-glycerophosphorylcholine acyltransferase, GPAT, and AGPAT activities in the turbid, void volume fractions (Hosaka *et al.,* 1977). DGAT was subsequently resolved from AGPAT and to a lesser extent from GPAT by sucrose density gradient centrifugation. More detailed solubilization characteristics of the DGAT were obtained by Polokoff and Bell (1980) for the rat liver microsomal enzyme. The best solubilization results were obtained with bile salt detergents, sodium deoxycholate, and sodium cholate. The cholate-solubilized activity was purified three- to fourfold by Sepharose 4B chromatography and ninefold by centrifugation in 10% to 20% sucrose gradient with recovery of 80%. The purified enzyme was strongly dependent on added magnesium and phospholipids for activity. This active fraction also contained enriched PAPH activity as well as detectable activities of other enzymes involved in the biosynthesis of triacylglycerols and neutral glycerophospholipids via the glycerol-3-phosphate pathway.

Subsequently, Andersson *et al.* (1994) raised monoclonal antibodies to rat liver microsomal proteins that eluted at the 50 to 80 kDa range upon gel filtration of sodium carbonate-solubilized microsomes. An antibody reacting with a 60 kDa protein on western blots was capable of immunoprecipitating the DGAT activity. Immunoaffinity chromatography resulted in 415-fold purification of the DGAT activity coinciding with the elution of a 60 kDa protein.

An independent indication of the mass of rat liver diacylglycerol acyltransferase has been obtained by the radiation inactivation procedure (Ozasa *et al.,* 1989). Exposure of frozen microsomal preparations to high-energy electrons resulted in a dose-dependent loss of the activity. By measuring the incorporation of radiolabeled acyl-CoA and endogenous diacylglycerol into triacylglycerols, a target size of 72 ± 4 kDa was obtained.

DGAT has also been purified 20-fold from solubilized microsomes of soybean cotyledons (Kwanyuen and Wilson, 1986). The enzyme activity was recovered in the void volume upon gel filtration together with about 10% of the total applied protein. No other acyltransferase or lipase activities were detected in this fraction. Delipidation of the active eluate by acetone precipitation resulted in a loss of the enzyme activity (Kwanyuen and Wilson, 1990). Denaturing SDS-PAGE revealed several polypeptide bands, with the major components migrating at apparent molecular masses of 84, 66, 59, 41, 29, and 25 kDa (Kwanyuen and Wilson, 1990). Little *et al.* (1994) have reported the dispersion of DGAT from a particulate fraction of microspore-derived embryos of oilseed rape (*B. napus* L cv Reston). The particulate fraction was dispersed with octanoyl *N*-methylglucamide (MEGA-8) followed by partial purification and fractionation of DGAT using Mono Q anion exchange chromatography. Purifications of up to 150-fold over the specific activity in the 1,500 to 100,000 × g particulate fraction were obtained. Particulate and dispersed DGAT preparations lost 25% to 30% of their initial activity when stored for 3 days at 4°C.

Kamisaka *et al.* (1997) have reported purification of DGAT to apparent homogeneity from the lipid body fraction of an oleaginous fungus, *Mortierella ramanniana* var. *angulispora*. They solubilized the lipid body fraction with 0.1% Triton X-100 and fractionated the detergent-soluble extract on Yellow 86 agarose, Heparin-Sepharose, and Superdex-200. The final fraction was enriched almost 5000-fold for DGAT activity and was devoid of other acyltransferase or esterase activities. Analysis of this fraction by SDS-PAGE yielded a single polypeptide band of apparent molecular mass of 53 kDa. The purified enzyme was activated by inclusion of anionic phospholipids such as phosphatidic acid and was inhibited by sphingosine.

Recently, Cases *et al.* (1998), using coding sequences from acyl-CoA:cholesterol acyltransferase (ACAT) homology search of the EST databases, identified a clone with a homology to ACAT. Expression of the full length cDNA in the baculovirus expression system resulted in fivefold increase in DGAT activity indicating the cDNA codes for DGAT. The DGAT gene expression and DGAT activity coincided with accumulation of triacylglycerols during differentiation of 3T3-L1 cells into adipocytes. No other acyltransferase activity was detected when a variety of substrates, including cholesterol, were used as acyl acceptors. Unfortunately, these substrates did not include 2-monoacylglycerols. The DGAT mRNA was detected in all human tissues tested with the highest expression being observed in the small intestine. Interestingly, the level of DGAT mRNA in the liver was relatively low and the authors suggested that the liver may express another DGAT isoform. The DGAT cDNA coded for a protein of 488 amino acids with a relative electrophoretic mobility of 50 kDa. Proteins of an apparently similar molecular mass are present in the purified triacylglycerol synthetase (TGS). It remains to be determined whether the same DGAT isoform functions in both monoacylglycerol (MG) and phosphatidic acid (PA) pathways in the enterocyte.

4.2.3b Stereospecificity. Microsomal preparations from hamster intestine were reported to convert 1,3-diacylglycerol and 1-alkyl-3-acyl-*sn*-glycerol to triacylglycerols, whereas 1-acyl-2-alkylglycerol was not acylated by either hamster or rat enzyme preparations (Paltauf and Johnston, 1971). Likewise, O'Doherty *et al.* (1972) reported preferential acylation of *sn*-1-stearoyl-2-linoleoylglycerol with several saturated and unsaturated long-chain fatty acids, except for stearate, which reacted more readily with *sn*-2-palmitoyl-3-oleoylglycerol. Excluding the latter, enantiomeric utilization ratios of *sn*-1,2- to *sn*-2,3-diacylglycerols were in the range of 60:40 for most fatty acid donors, whereas linoleate gave an 85:15 ratio. Chiral column HPLC analysis of diacylglycerols remaining after reaction of rat intestinal DGAT with 1,2-dioleoyl-*rac*-glycerol, however, showed that the 2,3-dioleoyl-*sn*-glycerols were utilized at the same rate as the *sn*-1,2-enantiomer (Lehner *et al.*, 1993).

4.2.3c Chain-length Specificity. O'Doherty *et al.* (1972) obtained evidence that different molecular species of diacylglycerols are utilized at different rates. When an equimolar mixture of enantiomeric diacylglycerols were incubated with radioactive fatty acids in the presence of rat intestinal microsomes, the *sn*-1-stearoyl-2-linoleoylglycerol reacted more readily than the *sn*-2-palmitoyl-3-oleoylglycerol with all common fatty acids, except stearate, which was taken up more readily by the palmitoyloleoylglycerol. There was no difference in the yields of triacylglycerols formed from palmitic and oleic acids when incubated with three different *sn*-2,3-diacylglycerols. The utilization of various molecular species of *sn*-1,2- and *sn*-2,3-diacylglycerols by the microsomes of rat liver and Ehrlich ascites cells gave comparable yields from palmitic and oleic acids (O'Doherty and Kuksis, 1974). Hosaka *et al.* (1977) also concluded that the acyl donor specificity of the liver DGAT is very broad. Myristoyl-, palmitoyl-, oleoyl-, linoleoyl-, and arachidonoyl-CoA all were found to be effective acyl donors. Stearoyl CoA was utilized fairly efficiently when present at high concentrations.

4.2.4. Acyl-CoA Ligase (Synthetase; FACS)

The intestinal FACS activity has been solubilized from microsomes and has been partially purified by hydrophobic (Manganaro and Kuksis, 1985a, 1985b) or affinity chromatography (Lehner and Kuksis, 1995), where the enzyme co-eluted with MGAT and DGAT. The rat liver microsomal long-chain FACS activity has been solubilized with 5 mM Triton X-100 and has been purified 100-fold by sequential chromatography on blue-Sepharose, hydroxylapatite and phosphocellulose (Tanaka *et al.*, 1979). The purified enzyme migrated on denaturing polyacrylamide gel electrophoresis with an estimated molecular mass of 76 kDa. Both the microsomal and the mitochondrial enzymes showed little preference for either saturated or unsaturated long-chain fatty acids (Tanaka *et al.*, 1979). A complementary DNA encoding this enzyme was isolated and sequenced (Suzuki *et al.*, 1990). The enzyme was predicted to contain 699 amino acid residues corresponding to a calculated molecular mass of 78,177. Distribution of the enzyme in rat tissues was also examined (Suzuki *et al.*, 1990). Using Northern blot hybridization analysis with a 520-base–pair fragment of the cDNA, a high level of expression was observed in liver, heart, and adipose tissues. Surprisingly, the signal for small intestine was only 10% of that observed in liver. It is possible that intestinal mucosal cells express a tissue specific isoenzyme. Existence of other than microsomal FACS (FACS 1) within a cell has been demonstrated by the presence of enzyme activity in mitochondria (Tanaka *et al.*, 1979), in peroxisomes (Uchiyama *et al.*, 1996), and in the plasma membrane of hepatocytes (Davidson and Cantrill, 1986). The plasma membrane enzyme showed slight preference for polyenoic fatty acids, although both saturated and unsaturated fatty acids of 12 to 24 carbons were suitable substrates. The peroxisomal FACS had a high sequence similarity to the plasma membrane FACS but no homology to the microsomal enzyme. Recently, a brain-specific microsomal FACS (FACS 3; Fujino *et al.*, 1997) and a homologous enzyme expressed in several tissues except for liver and lung (FACS 4; Piccini *et al.*, 1998) have been characterized with respect to chromosomal localization but not with respect to their function. Complementary DNAs encoding the human (Abe *et al.*, 1992) and rat brain (Fujino and Yamamoto, 1992) FACS were isolated. The deduced amino acid sequences were 85% and 65% identical with the rat acyl-CoA ligase sequence (Schaffer and Lodish, 1994). It is not currently known how many different FACS isoenzymes function in mammalian cells. *Saccharomyces cerevisie* contains at least three FACS genes, each coding for an enzyme with different chain-length specificity (Knoll *et al.*, 1994). Three forms of rat FACS mRNA with 5'-untranslated region heterogeneity were isolated (Suzuki *et al.*, 1995). The different mRNAs were found to be generated by alternative transcription from three different promoters in the FACS gene (Suzuki *et al.*, 1995).

4.2.5. Acyl CoA Acyltransferase (Hydrolase; ACH)

Acyl-CoA hydrolase (ACH) or acyl-CoA acyltransferase (AAT) activity has been observed in intestinal cell homogenates of various animal species and in man. Palmitoyl-CoA hydrolase activity was determined in hamster (Rao and Johnston, 1966) and in rat (Rodgers, 1969) intestinal microsomes. The enzyme has been partially purified from rat intestinal microsomes by ammonium sulphate precipitation, gel filtration, and anion exchange chromatography (De Jong *et al.*, 1978). Sodium SDS-PAGE showed several polypeptide bands between 60 and 75 kDa. The active fraction also exhibited hydrolytic activity toward long-chain (16 and 18 acyl carbons) and medium-chain (8 acyl carbons) monoacylglycerols and toward short-chain triacylglycerol (tributyroylglycerol). A 54 kDa AAT from rat intestinal mucosa has been characterized (Lehner and Kuksis, 1993a, 1993b). The enzyme was found to be specific for long-chain acyl-CoAs and did not hydrolyze other glycerolipids. It was postulated to act as an acyl-CoA binding subunit of the proposed triacylglycerol synthetase complex (Lehner and Kuksis, 1995). Immunoblot analysis of AAT showed crossreacting proteins in other rat and human tissues (Lehner and Kuksis, 1993a, 1993b).

It was found that incubations of isolated rat intestinal microsomes with monoacylglycerols and acyl-CoA favor hydrolysis to acylation (Hulsmann and Kurpershoek-Davidov, 1976; Ailhaud *et al.*, 1964). Clearly, the intact cell contains some mechanism that down-regulates the hydrolytic activity because most of the absorbed monoacylglycerol is acylated. The hydrolytic activity could be inhibited *in vitro* by preincubation of microsomes with 10 μM diethyl *p*-nitrophenyl phosphate (De Jong *et al.*, 1978), which is known to be a serine esterase inhibitor. Similarly, 10 to 50 mM sodium or potassium fluoride has been reported to inhibit the intestinal hydrolase (Johnston, 1977).

4.3. Deacylation/Reacylation Cycle

The possibility of triacylglycerol hydrolysis in the intestine prior to transport out was suggested by Mansbach *et al.* (1987) because the triacylglycerol particles accumulated in the intestine during fat absorption appeared to be too large for secretion as chylomicrons. The authors speculated that the endogenous triacylglycerols were hydrolyzed to free fatty acids and were transported via the portal vein. Although no proof was presented for this possibility, evidence was obtained for structural differences between the triacylglycerols accumulating in the enterocyte and those appearing in the lymph.

4.3.1. Characterization

Yang and Kuksis (1991) determined the composition of the sn-1- and sn-3-positions of lymph chylomicron triacylglycerols obtained after feeding fish oil triacylglycerols (monoacylglycerol pathway) and the corresponding fatty acid methyl and ethyl esters (glycerol-3-phosphate pathway). The chylomicron triacylglycerols arising via the 2-monoacylglycerol and the glycerol-3-phosphate pathways differed mainly in the composition of the fatty acids in the sn-2- position, but showed remarkable similarity in the fatty acid composition in the sn-1- and sn-3- positions. This similarity between the two types of triacylglycerol structures is consistent with a lipase hydrolysis of di- or triacylglycerols, formed via the phosphatidic acid pathway, to monoacylglycerols prior to reacylation, to triacylglycerols and secretion as chylomicrons in the lymph. It has been since reported (Wiggins and Gibbons, 1992) that about 70% of the newly synthesized liver triacylglycerol may undergo a deacylation/reacylation before secretion as VLDL. It is not known which route the resynthesis of these lipolytic products (2-monoacylglycerols or diacylglycerols) takes as the adult liver is not believed to contain MGAT activity (Coleman and Haynes, 1984), which would be necessary for an efficient VLDL secretion. Figure 4 outlines a hypothetical scheme for the assembly of VLDL and chylomicron triacylgycerols based on the previously mentioned considerations. The triacylglycerols derived via the phosphatidic acid pathway are shown to go into cytoplasmic storage from which they are retrieved via a triacylglycerol hydrolase and one or more acyltransferases for subsequent secretion. The operation of such a scheme depends on the demonstration of endogenous lipases and appropriate acyltransferases in both liver and intestine.

4.3.2. Isolation of Endogenous Lipases

Both acid (Rao and Mansbach, 1990) and alkaline (Rao and Mansbach, 1993) triacylglycerol lipase activities have been described in intestinal cytosol. To these must now be added the pancreatic lipase activity recently demonstrated in the enterocytes (Tsujita *et al.*, 1996). Triacylglycerol- and diacylglycerol-hydrolyzing activities of the purified enzyme,

Figure 4. A hypothetical scheme of assembly of endogenous triacylglycerols for secretion as VLDL and chylomicrons. "Fat droplet" represents cytoplasmic storage triacylglycerols, which must undergo partial lipolysis before reacylation and incorporation into the lipoproteins. G-3-P, glycerol-3-phosphate; DG, diacylglycerol; TG, triacylglycerol; VLDL, very low density lipoprotein or chylomicron.

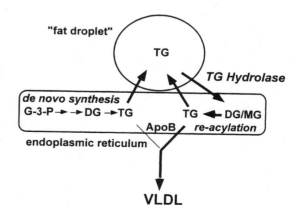

however, were inhibited by addition of intestinal mucosa extract. It is not known to what extent, if any, these enzymes might participate in the deacylation/reacylation cycle of the intestine. A candidate enzyme (triacylglycerol hydrolase) for such a cycle, however, appears to have been recently purified from porcine liver microsomes and has been characterized with respect to acyl chain specificity and tissue distribution (Lehner and Verger, 1997). The hydrolase co-sediments with subcellular organelles containing enzymes that are required for synthesis and assembly of lipoproteins (Lehner *et al.*, 1999). In addition, the hydrolase associates with isolated liver fat droplets. The developmental expression of the hydrolase mRNA and consequently the protein coincides with ontogeny of VLDL secretion by the liver (during weaning). Moreover, the enzyme is absent from liver-derived cell lines known to be impaired in VLDL assembly and secretion. The rat triacylglycerol hydrolase cDNA was isolated and was found to belong to a family of microsomal carboxylesterases (Lehner and Vance, 1999). Expression of the cDNA in the rat hepatoma cell line McArdle RH-7777 resulted in increased depletion of prelabeled triacylglycerol stores and increased utilization of the stored triacylglycerol for lipoprotein secretion, which coincided with increased apoB-100 secretion in the VLDL density range. The lysomal acid lipase initially thought to participate in the deacylation/reacylation process (Francone *et al.*, 1989) apparently is not involved (Hilaire *et al.*, 1993, 1994).

4.3.3. Diacylglycerol Transacylase (DGTA)

A long-chain DGAT activity present in rat intestinal and liver microsomes has also been characterized and the 52 kDa enzyme was purified to homogeneity (Lehner and Kuksis, 1993b). The transacylase catalyzes a unique reaction whereby triacylglycerols are synthesized from diacylglycerols in the absence of acyl-CoA by an intermolecular fatty acyl transfer between two diacylglycerol molecules (Lehner and Kuksis, 1993b). The microsomal activity of the DGAT was low (0.8 nmol/mg protein/min) compared with the acyl-CoA-dependent reaction catalyzed by DGAT (5.5 nmol/mg protein/min; Lehner and Kuksis, 1992). Although the enzyme utilized the same substrate as DGATmg, it exhibited very different chromatographic properties. *In vitro* assays show that the 2-monoacylglycerol, a by-product of the transacylation reaction, is readily utilized for triacylglycerol synthesis in the usual acyl-CoA-dependent manner. The transacylase is not stereospecific because both sn-1,2- and sn-2,3-diacylglycerols are suitable substrates. Figure 5 illustrates the biosynthesis of triacylglycerols from X-1,2-diacylglycerols by DGTA. The reaction proceeds independent of acyl-CoA. The contribution of diacylglycerol transacylase to the overall triacylglycerol synthesis in the intestine is presently not known.

Triacylglycerol formation via an acyl-CoA-independent transacylation has been observed also in microsomal preparations of developing safflower (Stobart *et al.*, 1997) and castor bean (Mancha and Stymne, 1997) seeds. Lin *et al.* (1998) have also observed that the incorporation of ricinoleate from 2-ricinoylglycerophosphocholine into triacylglycerols did not require CoA, whereas incorrporation of free ricinoleate required CoA. The major route to triacylglycerols containing ricinoleate in castor microsomes, however, proceeded via ricinoyl-CoA (Lin *et al.*, 1998). Yamashita *et al.* (1997) have recently reviewed the acyltransferases and the transacylases involved in fatty acid remodeling of glycerophospholipids and metabolites of bioactive lipids in mammalian cells.

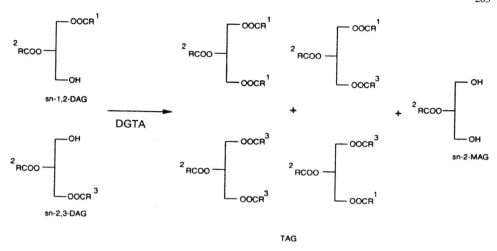

Figure 5. Biosynthesis of triacylglycerols from X-1,2-diacylglycerols via diacylglycerol transacylase (DGTA). The reaction proceeds independent of acyl-CoA (Lehner and Kuksis, 1993b). MAG, monoacylglycerols; DAG, diacylglycerols.

5. Regulation of Triacylglycerol Biosynthesis

The presence of both biosynthetic and hydrolytic activities in the enterocytes complicates *in vivo* investigations, and the absence of purified enzymes represents a major obstacle to advancement of knowledge in this area of research. Multiple points may exist at which regulation of triacylglycerol synthesis occurs. Uptake of substrates at the apical surface, transport to the intracellular membranes, and acylation into triacylglycerols are all potential control points.

5.1. Metabolic Interactions

Both monoacylglycerol and phosphatidic acid pathways of triacylglycerol synthesis presumably compete for the same pool of fatty acids and both produce diacylglycerol intermediates. Because the phosphatidic acid pathway is believed to be the only route for glycerophospholipid biosynthesis, apart from reacylation of any absorbed dietary lysophospholipids, the cell would be expected to continue its operation to maintain the synthesis of phospholipids, which are required for efficient assembly and secretion of chylomicrons (O'Doherty *et al.*, 1973; Kuksis *et al.*, 1979).

The major difference between *in vivo* and *in vitro* acylation of 2-monoacylglycerols to triacylglycerols in the intestinal mucosa is that only a small quantity of diacylglycerol intermediate accumulates *in vivo*, whereas incubations with isolated subcellular fractions yield diacylglycerol as the predominant product. This indicates that some cellular interaction may be necessary for reproducing the *in vivo* conditions *in vitro*. Either some factors are lost or their function is interfered with during cell disruption and fractionation. None of these factors have been identified.

Modulation of acyltransferase activities by low-molecular mass fatty-acid-binding proteins has been already noted. Other proteins may also be present and may play some role in the acylation process. Knudsen *et al.* (1989) have isolated and characterized a 10 kDa cytosolic acyl-CoA-binding protein from rat liver and Rosendal *et al.* (1993) and Mandrup *et al.* (1993) have suggested that it may function in the cell as an acyl-CoA pool generator, transporter, or a protector against acyl-CoA hydrolysis by long-chain ACH. The prevention of acyl-CoA hydrolysis affected only the conversion of diacylglycerols to triacylglycerols but not the acylation of *sn*-glycerol-3-phosphate to phosphatidic acid. It was suggested that phospholipid synthesis is directly linked to the synthesis of long-chain acyl-CoA without involving any auxiliary transport proteins, whereas acyl-CoA synthesized for triacylglycerol formation is bound by the acyl-CoA-binding protein before being delivered to the triacylglycerol-synthesizing enzymes (Rasmussen *et al.*, 1993).

Wetterau and Zilversmit (1986) reported a triacylglycerol-transfer protein (MTP) in rat intestinal mucosa. The heterodimeric protein was shown to play a key role in the assembly and secretion of apoB-containing lipoprotein particles. MTP apparently functions by removing the triacylglycerols from their site of synthesis and transferring them to the site of prechylomicron/VLDL assembly (Wetterau *et al.*, 1991). Co-expression of MTP with apoB in heterologous cells results in the assembly of lipoproteins (Leiper *et al.*, 1994; Gordon *et al.*, 1994). Transfection of L-cell fibroblasts with I-FABP cDNA resulted in increased cellular mass of triacylglycerols and cholesteryl esters, suggesting a role for I-FABP in neutral lipid synthesis (Prows *et al.*, 1996). The mechanism of stimulation of fatty acid esterification and/or activation by FABP is unclear. On the other hand, FABP preparations have been shown to stimulate activities of several other enzymes that utilize long-chain fatty acids or fatty acyl-CoA as substrates, including long-chain FACS, *sn*-glycerol-3-phosphate acyltransferase, DGATpa, and ACAT (Jamdar and Cao, 1995; Lehner and Kuksis, 1996).

5.2. Activation and Inhibition

Ahdieh *et al.* (1998) have described the effects of intestinal ischemia on epithelial DGAT and MGAT activities and their recovery in response to luminal *l*-glutamine, the primary intestinal fuel, and transforming growth factor-α, a mitogenic hormone similar to epidermal growth factor present in breast milk. Ischemic damage and recovery were analyzed in mucosa from Thiry-Vella loops in the mid-ileum of 7-week-old pigs.

Early *in vitro* studies demonstrated that diacylglycerols synthesized by acylation of monoacylglycerols were not converted to phosphatidylcholine (Johnston *et al.*, 1970, 1967). These reactions, however, were carried out under conditions that are inhibitory to cholinephosphotransferase, the enzyme catalyzing transfer of the choline moiety to diacylglycerols (Lehner and Kuksis, 1992). In addition, monoacylglycerols and their ether analogues have been claimed to inhibit GPAT activity in both intestine (Polheim *et al.*, 1973) and liver (Coleman, 1988) *in vitro*. More recent experiments indicated that, under noninhibitory conditions to cholinephosphotransferase (in the absence of detergents), diacylglycerols formed via the monoacylglycerol pathway are readily converted to phosphatidylcholine (Lehner and Kuksis, 1992).

Furthermore, inhibition studies (Grigor and Bell, 1982) have suggested that only one diacylglycerol acyltransferase functions for both phosphatidic acid and monoacylglycerol pathways. There is now evidence that liver contains at least two DGAT activities (Owen *et al.,* 1997). It is not known whether or not both DGAT activities are inhibited under the conditions of Grigor and Bell (1982). Various other steps in the two pathways of triacylglycerol biosynthesis are inhibited by the common metabolic inhibitors as already noted.

5.3. Regulation of DGAT Activity by Phosphorylation

It has been suggested that hepatic and adipose tissue DGAT may be subject to regulation by phosphorylation /dephosphorylation mechanisms (Haagsman *et al.,* 1981, 1982; Maziere *et al.,* 1986; Rodriguez *et al.,* 1992). Incubation of isolated rat hepatocytes with glucagon decreased DGAT activity 53%, whereas cholinephosphotransferase activity did not change (Haagsman *et al.,* 1981). Cyclic-AMP mimicked the effect of glucagon. The DGAT activity was inactivated by 46% upon incubation of rat liver microsomes with cytosol in the presence of $MgCl_2$ and ATP. Of the nucleotides tested, ATP was most effective and could not be replaced by β,γ-Methylene ATP (Haagsman *et al.,* 1982). The presence of fluoride (50 mM), a phosphoprotein phosphatase inhibitor, in the homogenization medium resulted in lower microsomal DGAT activity. Fluoride inhibited this activation, suggesting that a phosphatase is the activating factor (Haagsman *et al.,* 1982). These results indicate that the liver DGAT may be active in its dephosphorylated state. Using an experimental design similar to that mentioned previously, Rodriguez *et al.* (1992) demonstrated inactivation of rat adipose tissue DGAT. They also showed that the ATP + cytosol-inactivated microsomal activity could be reactivated by incubation with partially purified rat liver phosphoprotein phosphatase. The DGAT inactivation was insensitive to Ser/Thr kinase inhibitor H-7 but was sensitive to Tyr kinase inhibitors genistein and tyrphostin-25, suggesting that adipose tissue DGAT may be regulated by a tyrosine kinase/phosphatase action (Lau and Rodriguez, 1996). The opposite result was obtained for DGAT in hamster fibroblasts (Maziere *et al.,* 1986). Here, incorporation of labeled exogenous fatty acids into triacylglycerols was enhanced two- to threefold in the presence of dibutyryl (db) cAMP, whereas no stimulation of phospholipid synthesis was observed under the same conditions. In hamster fibroblasts, DGAT-specific activity increased threefold in dbcAMP treated cells in the presence but not in the absence of 50 mM fluoride, suggesting that the active form of the enzyme is phosphorylated. A similar stimulation of triacylglycerol synthesis by cAMP was observed in Balb/c macrophages and in bovine endothelial cells, in contrast to a similar treatment in rat hepatocytes where decreased incorporation of fatty acids into triacylglycerols was observed (Haagsman *et al.,* 1981; Maziere *et al.,* 1986). These apparent discrepancies need to be reconciled. One of the criticisms that may be raised against the hypothesis that microsomal DGAT is regulated by phosphorylation/dephosphorylation is the use of a small endogenous diacylglycerol pool for the enzymic assays. The modulation by cAMP was not observed when ethanol-dispersed exogenous diacylglycerols were employed (Haagsman *et al.,* 1981). An ATP-dependent phosphorylation of the endogenous diacylglycerol into phosphatidic acid by diacylglycerol kinase would diminish the endogenous diacylglycerol pool and consequently the yield of the triacylglycerol product. In the intestine, inclusion of cofactors that would favor phosphorylation (ATP, divalent cations) resulted in the increase of specific activities of both MGAT and DGAT without influencing the overall ratios of synthesized di-and triacylglycerols.

6. Summary and Conclusions

Early work established the monoacylglycerol and the phosphatidic acid pathways of acyl-CoA-dependent reesterification of dietary fatty acids by the intestine. A sequential acylation of absorbed 2-monoacylglycerols to triacylglycerols occurs via MGAT and DGAT, which are isolated together with acyl-CoA synthetase and AAT as a triacylglycerol synthetase complex. The complex has been resolved into the acyltransferase components, which have been extensively purified. In several instances antibodies have been prepared to the purified proteins. The acyltransferases associated with the phosphatidic acid pathway of triacylglycerol biosynthesis have not been isolated or extensively investigated in the intestine. Recent experiments have suggested that the triacylglycerols generated via the phosphatidic acid pathway are degraded to 2-monoacylglycerols before reacylation via the monoacylglycerol pathway and secretion as chylomicrons.

Other recent studies demonstrate that pancreatic lipase is present in intestinal absorptive cells and that it contributes to the resynthesis of diacylglycerols from 2-monoacylglycerols. The formation of *rac*-1,2-diacylglycerols by the lipase could account for the near random generation of the *sn*-1,2(2,3)-diacylglycerols by intact enterocytes and by cell homogenates and the apparent stereospecific formation of *sn*-1,2-diacylglycerols by purified microsomes. Furthermore, the isolation of a microsomal DGTA, which yields triacylglycerols and 2-monoacylglycerols, demonstrates that the resynthesis of dietary fat in the intestine is much more complicated than suggested by earlier work.

Other questions remain regarding the subcellular segregation of the metabolic pathways for triacylglycerol and glycerophospholipid biosynthesis and for chylomicron secretion. Studies performed with the liver suggest that DGAT activity occurs on both sides of the microsomal membrane. Both overt and latent activities of DGAT appear to be involved in VLDL triacylglycerol secretion by the liver, and two DGATs may be involved in chylomicron triacylglycerol synthesis and secretion by the intestine.

In conclusion, substantial progress has been obtained in the field of triacylglycerol synthesis in mammals. Several of the key enzymes involved in triacylglycerol biosynthesis have been solubilized and purified to homogeneity. Some of them (mitochondrial GPAT, FACS; mouse DGAT) have also been cloned and tissue distribution has been determined. More effort should now be directed to obtaining the complete amino acid sequence of MGAT, DGAT, AAT, and DGTA. Knowledge at the molecular level of the enzymes involved in such important metabolic processes as fat absorption and triacylglycerol biosynthesis and would be a step to a better understanding of the structure–function relationship and the regulation of glycerolipid biosynthesis in general.

Acknowledgment

The studies of the authors and their collaborators referred to in this review were supported by grants from the Medical Research Council of Canada, Ottawa, Canada and the Heart and Stroke Foundation of Ontario, Toronto, Canada.

References

Abe, T., Fujino, T., Fukuyama, R., Minoshima, S., Shimizu, N., Toh, H., Suzuki, H., and Yamamoto, T., 1992, Human long chain acyl-CoA synthetase: Structure and chromosomal location, *J.Biochem.* **111:**123–128.

Aguado, B., and Campbell, R. D., 1998, Characterization of a human lysophosphatidic acid acyltransferase that is encoded by a gene located in the Class III region of the human major histocompatibility complex, *J. Biol. Chem.* **273**:4096–4105.

Ahdieh, N., Blikslager, A. T., Bhat, B. G., Coleman, R. A., Argenzio, R. A., and Rhoads, J. M., 1998, L-glutamine and transforming growth factor-α enhance recovery of monoacylglycerol acyltransferase and diacylglycerol acyltransferase activity in porcine postischemic ileum, *Pediatr. Res.* **43**:227–233.

Ailhaud, G., Samuel, D., Lazduwski, M., and Desnuelle, P., 1964, Quelques observations sur le mode d'action de la monoglyceride transacylase et de la diglyceride transacylase de la muqueuse intestinale, *Biochim. Biophys. Acta* **84**:643–664.

Akesson, B., Gronowitz, S., Herslof, B., Michelsen, P., and Olivecrona, T., 1983, Stereospecicity of different lipases, *Lipids* **18**:313–318.

Amri, E-Z., Bonino, F., Ailhaud, G., Abumrad, N. A., and Grimaldi, P. A., 1995, Cloning of a protein that mediates transcriptional effects of fatty acids in preadipocytes. Homology to peroxisome proliferator-activated receptors, *J. Biol. Chem.* **270**:2367–2371.

Andersson, M., Wettesten, M., Borén, J., Magnusson, A., Sjöberg, A., Rustaeus, S., and Olofsson, S-O. , 1994, Purification of diacylglycerol: Acyltransferase from rat liver to near homogeneity, *J. Lipid Res.* **35**:535–545.

Baillie, A. G. S., Coburn, C. T., and Abumrad, N. A., 1996, Reversible binding of long-chain fatty acids to purified FAT, the adipose CD36 homolog, *J. Membr. Biol.* **153**:75–81.

Bell, R. M., and Coleman, R. A., 1980, Enzymes of glycerolipid synthesis in eukaryotes, *Annu. Rev. Biochem.* **43**:243–277.

Bell, R. M., and Coleman, R. A., 1983, Enzymes of triacylglycerol formation in mammals, in: *The Enzymes,* 3rd ed., Volume 16, (P. D. Boyer, ed.), Academic Press, New York, pp. 87–111.

Bezard, J., and Bugaut, M., 1986, Absorption of glycerides containing short, medium, and long chain fatty acids, in: *Fat Absorption,* Volume I (A. Kuksis, ed.), CRC Press, Boca Raton, FL, pp. 119–158.

Bhat, B. G., Bardes, E. S-G., and Coleman, R. A., 1993, Solubilization and partial purification of neonatally expressed rat hepatic microsomal monoacylglycerol acyltransferase, *Arch. Biochem. Biophys.* **300**:663–669.

Bhat, B. G., Wang, P., and Coleman, R. A., 1994, Hepatic monoacylglycerol acyltransferase is regulated by *sn*-1,2-diacylglycerol and by specific lipids in Triton X-100/phospholipid-mixed micelles, *J. Biol. Chem.* **269**:13172–13178.

Bhat, B. G., Wang, P., and Coleman, R. A., 1995, Sphingosine inhibits rat hepatic monoacylglycerol acyltransferase in Triton X-100 mixed micelles and isolated hepatocytes, *Biochem.* **34**:11237–11244.

Black, D. D., 1995, Intestinal lipoprotein metabolism, *J. Pediatr. Gastroenterol. Nutr.* **20**:125–147.

Boll, W., Schmid-Chanda, T., Semenza, G., and Mantei, N., 1993, Messenger RNAs expressed in intestine of adult but not baby rabbits. Isolation of cognate cDNAs and characterization of a novel brush border protein with esterase and phospholipase activity, *J. Biol. Chem.* **268**:12901–12911.

Bosner, M. S., Gulick, T., Riley, D. J. S., Spilburg, C. A., and Lange, L. G., 1988, Receptor-like function of heparin in the binding and uptake of neutral lipids, *Proc. Natl. Acad. Sci. U.S.A.* **85**:7438–7442.

Bosner, M. S., Gulick, T., Riley, D. J. S., Spilburg, C. A., and Lange, L. G., 1989, Heparin-modulated binding of pancreatic lipase and uptake of hydrolyzed triglycerides in the intestine, *J. Biol. Chem.* **264**:20261–20264.

Bottino, N., Vandenburg, G. A., and Reiser, R., 1967, Resistance of certain long-chain polyunsaturated fatty acids of amrine oils to pancreatic lipase hydrolysis, *Lipids* **2**:489–493.

Breckenridge, W. C., and Kuksis, A., 1972, Stereochemical course of diacylglycerol formation in rat intestine, *Lipids* **7**:256–259.

Bugaut, M., Myher, J. J., Kuksis, A., and Hoffman, A. G. D., 1984, An examination of the stereochemical course of acylation of 2-monoacylglycerols by rat intestinal villus cells using [^{2}H$_3$]palmitic acid, *Biochim. Biophys. Acta* **792**:254–269.

Carey, M. C., Small, D. M., and Bliss, C. M., 1983, Lipid digestion and absorption, *Annu. Rev. Physiol.* **45**:651–677.

Cartwright, I. J., Higgins, J. A., Wilkinson, J., Bellevia, S., Kendrick, J. S., and Graham, J. M., 1997, Investigation of the role of lipids in the assembly of very low density lipoproteins in rabbit hepatocytes, *J. Lipid Res.* **38**:531–545.

Cases, S., Smith, S. J., Zheng, Y-W., Myers, H. M., Lear, S. R., Sande, E., Novak, S., Collins, C., Welch, C. B., Lusis, A. J., Erickson, S., and Farese, R. V., Jr., 1998, Identification of a gene encoding an acyl CoA:diacylglycerol acyltransferase, a key enzyme in triacylglycerol synthesis. *Proc. Natl. Acad. Sci. USA* **95**, 13018–13023.

Cinti, D. L., Cook, L., Nagi, M. N., and Suneja, S. K., 1992, The fatty acid chain elongation system of mammalian endoplasmic reticulum, *Prog. Lipid Res.* **31**:1–51.

Coleman, R. A., 1988, Hepatic *sn*-alkyl-3-phosphate acyltransferase: Effect of monoacylglycerol analogs on mitochondrial and microsomal activities, *Biochim. Biophys. Acta* **963**, 367–374.

Coleman, R. A., and Bell, R. M., 1983, Topography of membrane-bound enzymes that metabolize complex lipids, in: *The Enzymes,* 3rd ed., Volume 16 (P. D. Boyer, ed.), Academic Press, New York, pp. 605–625.

Coleman, R. A., and Haynes, E. B., 1984, Hepatic monoacylglycerol acyltransferase, *J. Biol. Chem.* **259**:8934–8938.

Coleman, R. A., and Haynes, E. B., 1986, Monoacylglycerol acyltransferase. Evidence that the activities from rat intestine and suckling liver are tissue-specific isozymes, *J. Biol. Chem.* **261**:224–228.

Coleman, R. A., Walsh, J. P., Millington, D. S., and Maltby, D. A., 1986, Stereospecificity of monoacylglycerol acyltransferase activity from rat intestine and suckling rat liver, *J. Lipid Res.* **27**:158–165.

Coleman, R. A., Haynes, E. B., and Coats, C. D., 1987, Ontogeny of microsomal activities of triacylglycerol synthesis in guinea pig liver, *J. Lipid Res.* **28**:320–325.

Davidson, B. C., and Cantrill, R. C., 1986, Rat hepatocyte plasma membrane acyl-CoA synthetase activity, *Lipids* **21**:571–574.

Davidson, N. O., 1994, Cellular and molecular mechanisms of small intestinal lipid lipid transport, in: *Physiology of the Gastrointestinal Tract,* (L. R. Johnson, ed.), Raven Press, New York, pp. 1909–1934.

De Jong, B. J. P., Kalkman, C., and Hulsmann, W. C., 1978, Partial purification and properties of monoacylglycerol lipase and two esterases from isolated rat small intestinal epithelial cells, *Biochim. Biophys. Acta* **530**: 56–66.

Desnuelle, P., 1986, Pancreatic lipase and phospholipase, in: *Molecular and Cellular Basis of Digestion* (P. Desnuelle, H. Sjostrom, and O. Noren, eds.), Elsevier Science Publishers B. V., Amsterdam, pp. 275–296.

Diagne, A., Mitjaville, S., Fauvel, J., Chap, H., and Douste-Blazy, L., 1987, Intestinal absorption of ester and ether glycerophospholipids in guinea pig. Role of a phospholipase A_2 from brush border membrane, *Lipids* **22**:33–40.

Eberhardt, C., Gray, P. W., and Tjoelker, L. W., 1997, Human lysophosphatidic acid acyltransferase, *J. Biol. Chem.* **272**:20299–20305.

Ericsson, J., Jackson, S. M., Kim, J. B., Spiegelman, B. M., and Edwards, P. A., 1997, Identification of glycerol-3-phosphate acyltransferaser as an adipocyte determination and differentiation factor 1- and sterol regulatory element-binding protein-responsive gene, *J. Biol. Chem.* **272**:7298–7305.

Francone, O. L., Kalopissis, A-D., and Griffaton, G., 1989, Contribution of cytoplasmic storage triacylglycerol to VLDL-triacylglycerol in isolated rat hepatocytes, *Biochim. Biophys. Acta* **1002**:28–36.

Friedman, H. I., and Cardell, R. R., Jr., 1977, Alterations in the endoplasmic reticulum and Golgi complex of intestinal epithelial cells during fat absorption and after termination of this process: A morphological and morphometric study, *Anat. Rec.* **188**:77–102.

Fujino, T., and Yamamoto, T., 1992, Cloning and functional expression of a novel long-chain acyl-CoA synthetase expressed in brain, *J. Biochem.* (Tokyo) **111**:197–203.

Fujino, T., Man-Jong, K., Minekura, H., Suzuki, H., and Yamamoto, T. T., 1997, Alternative translation initiation generates acyl-CoA synthetase 3 isoforms with heterogeneous amino termini, *J. Biochem.* (Tokyo) **122**:212–216.

Gallo, L. L., Chiang, Y., Vahouny, G. V., and Treadwell, C. R., 1980, Localization and origin of rat intestinal cholesterol esterase determined by immunocytochemistry, *J. Lipid. Res.* **21**:537–545.

Gassama-Diagne, A., Fauvel, J., and Chap, H., 1989, Purification of a new, calcium-independent, high molecular weight phospholipase A_2/lysophospholipase (phospholipase B) from guinea pig intestinal brush border membrane, *J. Biol. Chem.* **264**:9470–9475.

Gassama-Diagne, A., Rogalle, P., Fauvel, J., Wilson, M., Klaebe, A., and Chap, H., 1992, Substrate specificity of phospholipase B from guinea pig intestine. A glycerol ester lipase with broad specificity, *J. Biol. Chem.* **267**:13418–13424.

Gomez-Munoz, A., Hatch, G. M., Martin, A., Jamal, Z., Vance, D. E., and Brindley, D. N., 1992, Effects of okadaic acid on the activities of two distinct phosphatidate phosphohydrolases in rat hepatocytes, *FEBS Lett.* **301**:103–106.

Goodridge, A. G., 1991, Fatty acid synthesis in eucaryotes, in: *Biochemistry of Lipids, Lipoproteins and Membranes* (D. E. Vance and J. Vance, eds.), Elsevier, Amsterdam, pp. 111–139.

Gordon, D. A., Jamil, H., Sharp, D., Mullaney, D., Yao, Z., Gregg, R. E., and Wetterau, J., 1994, Secretion of apolipoprotein B-containing lipoproteins from He La cells is dependent on expression of the microsomal triglyceride transfer protein and is regulated by lipid availability, *Proc. Natl. Acad. Sci. U.S.A.* **91**:7626–7632.

Gotto, K., and Kondo, H., 1996, A 104-kDa diacylglycerol kinase containing ankyrin-like repeats localizes in the nucleus, *Proc. Natl. Acad. Sci. USA* **93**:11196–11201.

Green, P. R., Merrill, A, H., Jr., and Bell, R. M., 1981, Membrane phospholipid synthesis in *Escherichia coli:* Purification, reconstitution and characterization of sn-glycerol-3-phosphate acyltransferase, *J. Biol. Chem.* **256**:11151–11159.

Grigor, M. R., and Bell, R. M., 1982, Separate monoacylglycerol and diacylglycerol acyltransferase functions in intestinal triglyceride synthesis, *Biochim. Biophys. Acta* **712**:464–472.

Haagsman, H. P., de Haas, C. G. M., Geelen, M. J. H., and van Golde, L. M. G., 1981, Regulation of triacylglycerol synthesis in the liver. A decreased diacylglycerol acyltransferase activity after treatment of isolated rat hepatocytes with glucagon, *Biochim. Biophys. Acta* **664**:74–81.

Haagsman, H. P., de Haas, C. G. M., Geelen, M. J. H., and van Golde, L. M. G., 1982, Regulation of triacylglycerol synthesis in the liver, *J. Biol. Chem.* **257**:10593–10598.

Hajra, A. K., 1995, Glycerolipid biosynthesis in peroxisomes (microbodies), *Progr. Lipid Res.* **34**:343–364.

Harmon, C. M., and Abumrad, N. A., 1993, Binding of sulfosuccinimidyl fatty acids to adipocyte membrane proteins: Isolation and amino terminal sequence of an 88-kD protein implicated in transport of long-chain fatty acids, *J. Membrane Biol.* **133**:43–49.

Hilaire, N., Négre-Salvayre, A., and Salvayre, R., 1993, Cytoplasmic triacylglycerols and cholesteryl esters are degraded in two separate catabolic pools in cultured human fibroblasts, *FEBS Lett.* **328**:230–234.

Hilaire, N., Négre-Salvayre, A., and Salvayre, R., 1994, Cellular uptake and catabolism of high-density-lipoprotein triacylglycerols in human cultured fibroblasts: Degradation block in Neutral Lipid Storage Disease, *Biochem. J.* **297**:467–473.

Hooks, S. B., Ragan, S. P., and Lynch, K. R., 1998, Identification of a novel human phosphatidic acid phosphatase type 2 isoform, *FEBS Lett.* **427**:188–192.

Hosaka, K., Schiele, U., and Numa, S., 1977, Diacylglycerol acyltransferase from rat liver microsomes. Separation and acyl-donor specificity, *Eur. J. Biochem.* **76**:113–118.

Houssa, B., Schaap, D., Van der Wal, J., Goto, K., Kondo, H., Yamakawa, A., Shibata, M., Takenawa, T., and Van Blitterswijk, W. J., 1997, Cloning of a novel human diacylglycerol kinase (DGKθ) containing three cysteine-rich domains, a proline-rich region, and a pleckstrin homology domain with an overlapping ras-associating domain, *J. Biol. Chem.* **272**:10422–10428.

Hulsmann, W. C., and Kurpershoek-Davidov, R., 1976, Topographic distribution of enzymes involved in glycerolipid synthesis in rat small intestinal epithelium, *Biochim. Biophys. Acta* **450**:288–295.

Ibrahimi, A., Sfeir, Z., Magharaie, H., Amri, E-Z., Grimaldi, P., and Abumrad, N. A., 1996, Expression of the CD36 homolog (FAT) in fibroblast cells: Effects on fatty acid transport, *Proc. Natl. Acad. Sci. USA* **93**:2646–2651.

Ikeda, I., Imasato, Y., Nagao, H., Sasaki, E., Sugano, M., Imaizumi, K., and Yazawa, K., 1993, Lymphatic transport of eicosapentaenoic and docosahexaenoic acids as triglyceride, ethyl ester and free acid, and their effect on cholesterol transport in rats, *Life Sci.* **52**:1371–1379.

Isola, L. M., Zhou, S-L., Kiang, C-L., Stump, D. D., Bradbury, M. W., and Berk, P. D., 1995, 3T3 fibroblasts transfected with cDNA for mitochondrial aspartate aminotransferase express plasma membrane fatty acid-binding protein and saturable fatty acid uptake, *Proc. Natl. Acad. Sci. USA* **92**:9866–9870.

Jamal, Z., Martin, A., Gomez-Munoz, A., and Brindley, D. N., 1991, Plasma membrane fractions from rat liver contain phosphatidic acid phosphohydrolase distinct from that in the endoplasmic reticulum and cytosol, *J. Biol. Chem.* **266**:2988–2996.

Jamdar, S. C., and Cao, W. F., 1995, Triacylglycerol biosynthesis enzymes in lean and obese Zucker rats, *Biochim. Biophys. Acta* **1255**:237–243.

Johnston, J. M., 1977, Gastrointestinal tissue, in: *Lipid Metabolism in Mammals* (F. Snyder, ed.), Plenum Press, New York, pp. 151–187.

Johnston, J. M., Rao, G. A., and Lowe, P. A., 1967, The separation of the α-glycerophosphate and monoglyceride pathways in the intestinal biosynthesis of triglycerides, *Biochim. Biophys. Acta* **137**:578–580.

Johnston, J. M., Paltauf, F., Schiller, C. M., and Schultz, L. D., 1970, The utilization of the (-glycerophosphate and monoglyceride pathways for phosphatidylcholine biosynthesis in the intestine, *Biochim. Biophys. Acta* **218**:124–133.

Kai, M., Wada, I., Imai, S-I., Sakane, F., and Kanoh, H., 1996, Identification and cDNA cloning of 35-kDA phosphatidic acid phosphatase (Type 2) bound to plasma membranes, *J. Biol. Chem.* **271**:18931–18938.

Kamisaka, Y., Mishra, S., and Nakahara, T., 1997, Purification and characterization of diacylglycerol acyltransferase from the lipid body fraction of an oleaginous fungus, *J. Biochem.* **121**:1107–1114.

Knoll, L. J., Johnson, D. R., and Gordon, J. I., 1994, Biochemical studies of three *Saccharomyces cerevisiae* acyl-CoA synthetases, Faa1p, Faa2p, and Faa3p, *J. Biol. Chem.* **269**:16348–16356.

Knudsen, J., Hojrup, P., Hansen, H. O., Hansen, H. F., and Roepstorff, P., 1989, Acyl-CoA-binding protein in the rat. Preparation, binding characteristics, tissue concentrations and amino acid sequence, *Biochem. J.* **262**: 513–519.

Kuksis, A., and Manganaro, F., 1986, Biochemical characterization and purification of intestinal acylglycerol acyltransferases, in *Fat Absorption*, Volume I (A. Kuksis, ed.), CRC Press, Boca Raton, FL, pp. 223–259.

Kuksis, A., Shaikh, N. A., and Hoffman, A. G. D., 1979, Lipid absorption and metabolism, *Environ. Health Perspect.* **33**:45–55.

Kwanyuen, P., and Wilson, R. F., 1986, Isolation and purification of diacylglycerol acyltransferase from germinating soyabean cotyledons, *Biochim. Biophys. Acta* **877**:238–245.

Kwanyuen, P., and Wilson, R. F., 1990, Subunit and amino acid composition of diacylglycerol acyltransferase from germinating soybean cotyledons, *Biochim. Biophys. Acta* **1039**:67–72.

Lau, T. E., and Rodriguez, M. A., 1996, A protein tyrosine kinase associated with the ATP-dependent inactivation of adipose diacylglycerol acyltransferase, *Lipids* **31**:277–283.

Lehner, R., and Kuksis, A., 1992, Utilization of 2-monoacylglycerols for phosphatidylcholine biosynthesis in the intestine, *Biochim. Biophys. Acta* **1125**:171–179.

Lehner, R., and Kuksis, A., 1993a, Purification of an acyl-CoA hydrolase from rat intestinal microsomes, *J. Biol. Chem.* **268**:24726–24733.

Lehner, R., and Kuksis, A., 1993b, Triacylglycerol synthesis by an *sn*-1,2(2,3)-diacylglycerol transacylase from rat intestinal microsomes, *J. Biol. Chem.* **268**:8781–8786.

Lehner, R., and Kuksis, A., 1995, Triacylglycerol synthesis by purified triacylglycerol synthetase of rat intestinal mucosa, *J. Biol. Chem.* **270**:13630–13636.

Lehner, R., and Kuksis, A., 1996, Biosynthesis of triacylglycerols, *Progr. Lipid Res.* **35**:169–201.

Lehner, R., and Vance, D. E., 1999, Cloning and expression of a cDNA encoding a hepatic microsomal lipase that mobilizes stored triacylglycerol, *Biochem. J.* **343**:1–10.

Lehner, R., and Verger, R., 1997, Purification and characterization of a porcine microsomal triacylglycerol hydrolase, *Biochem.* **36**:1861–1868.

Lehner, R., Cui, Z., and Vance, D. E., 1999, Subcellular localization, developmental expression and characterization of a liver triacylglycerol hydrolase, *Biochem. J.* **338**:761–768.

Lehner, R., Kuksis, A., and Itabashi, Y., 1993, Stereospecificity of monoacylglycerol and diacylglycerol acyltransferases from rat intestine as determined by chiral phase high-performance liquid chromatography, *Lipids* **28**:29–34.

Leiper, J. M., Bayliss, J. D., Pease, R. J., Brett, D. J., Scott, J., and Shoulders, C. C., 1994, Microsomal triglyceride transfer protein, the abetalipoproteinemia gene product, mediates secretion of apolipoprotein B-containing lipoproteins from heterologous cells, *J. Biol. Chem.* **269**:21951–21954.

Lin, J-T., Woodruff, C. L., Lagouche, O. J., Mckeon, T. A., Stafford, A. E., Goodrich-Tanrikulu, M., Singleton, J. A., and Haney C. A., 1998, Biosynthesis of triacylglycerols containing ricinoleate in castor microsomes using 1-acyl-2-oleoyl-*sn*-glycero-3-phosphocholine as the substrate of oleoyl-12-hydroxylase, *Lipids* **33**:59–69.

Little, D., Weselake, R., Pomeroy, K., Furukawa-Soffer, T., and Bagu, J., 1994, Solubilization and characterization of diacylglycerol acyltransferase from microspore-derived cultures of oilseed rape, *Biochem. J.* **304**: 951–958.

Lopez-Candales, A., Bosner, M. S., Spilburg, C. A., and Lange, L. G., 1993, Cholesterol transport function of pancreatic cholesterol esterase: directed sterol uptake and esterification in enterocytes, *Biochem.* **32**:12085–12089.

Mancha, M., and Stymne, S., 1997, Remodeling of triacylglycerols in microsomal preparations from developing castor bean (*Ricinus communis* L.) endosperm, *Planta* **203**:51–57.

Mandrup, S., Jepsen, R., Skott, H., Rosendal, J., Hojrup, P., Kristiansen, K., and Knudsen, J., 1993, Effect of heterologous expression on acyl-CoA-level and composition in yeast, *Biochem. J.* **290**:369–374.

Manganaro, F., and Kuksis, A., 1985a, Purification and preliminary characterization of 2-monoacylglycerol acyltransferase from rat intestinal villus cells, *Can. J. Cell Biol.* **63**:341–347.

Manganaro, F., and Kuksis, A., 1985b, Rapid isolation of a triacylglycerol synthetase from intestinal mucosa, *Can. J. Cell Biol.* **63**, 107–114.

Mansbach, C. M., and Parthasarathy, S., 1982, A re-examination of the fate of glyceride glycerol in neutral lipid absorption and transport, *J. Lipid Res.* **23**:1009–1019.

Mansbach, C. M., Arnold, A., and Garret, M., 1987, Effect of chloroquine on intestinal lipid metabolism, *Am. J. Physiol.* **253**:G673–G678.

Martin, A., Hales, P., and Brindley, D. N., 1987, A rapid assay for assessing the activity and the Mg^{++} and Ca^{++} requirements of phosphatidate phosphohydrolase in cytosolic and microsomal fractions of rat liver, *Biochem. J.* **245**:347–355.

Maziere, C., Maziere, J-C., Mora, L., Auclair, M., and Polonovski, J., 1986, Cyclic AMP increases incorporation of exogenous fatty acids into triacylglycerols in hamster fibroblasts, *Lipids* **21**:525–528.

Minich, D. M., Vonk, R. J., and Verkade, H. J., 1997, Intestinal absorption of essential fatty acids under physiological and essential fatty acid-deficient conditions, *J. Lipid Res.* **38**:1709–1721.

Mostafa, N., Bhat, B. G., and Coleman, R. A., 1994, Adipose monoacylglycerol:acyl-coenzyme A acyltransferase activity in the white-throated sparrow (*Zonotrichia albicollis*): Characterization and function in a migratory bird, *Lipids* **29**, 785–791.

O'Doherty, P. J. A., and Kuksis, A., 1974, Microsmal synthesis of fi- and triacylglycerols in rat liver and Ehrlich ascites cells, *Can. J. Biochem.* **52**:514–524.

O'Doherty, P. J. A., Kuksis, A., and Buchnea, D., 1972, Enantiomeric diglycerides as stereospecific probes in triglyceride synthesis *in vitro, Can. J. Biochem.* **50**:881–887.

O'Doherty, P. J. A., Kakis, G., and Kuksis, A., 1973, Role of lecithin in intestinal fat absorption, *Lipids* **8**:249–255.

Owen, M. R., Corstorphine, C. C., and Zammit, V. A., 1997, Overt and latent activities of diacylglycerol acyltransferase in rat liver microsomes: Possible roles in very low density lipoprotein triacylglycerol secretion, *Biochem. J.* **323**:17–21.

Ozasa, S., Kempner, E. S., and Erickson, S. K., 1989, Functional size of acyl-CoA:diacylglycerol acyltransferase by radiation inactivation, *J. Lipid Res.* **30**:1759–1762.

Paltauf, F., and Johnston, J. M., 1971, The metabolism *in vitro* of enantiomeric 1-O-alkylglycerols and 1,2- and 1,3-alkylacylglycerols in the intestinal mucosa, *Biochim. Biophys. Acta* **239**:47–56.

Piccini, M., Vitelli, F., Bruttini, M., Pober, B. R., Jonsson, J. J., Villanova, M., Zollo, M., Borsani, G., Ballabio, A., and Renieri, A., 1998, FACL4, a new gene encoding long-chain acyl CoA synthase 4, is deleted in a family with Alport syndrome, elliptocytosis, and mental retardation, *Genomics* **47**:350–358.

Pind, S., and Kuksis, A., 1987, Isolation of purified brush border membranes from rat jejunum containing a Ca^{2+}-independent phospholipase A_2 activity, *Biochim. Biophys. Acta* **901**:78–87.

Pind, S., and Kuksis, A., 1988, Solubilization and assay of phospholipase A_2 activity from rat jejunal brush border membranes, *Biochim. Biophys. Acta* **938**:211–221.

Pind, S., and Kuksis, A., 1989, Association of the intestinal brush-border membrane phospholipase A_2 and lysophospholipase activities (phospholipase B) with a stalked membrane protein, *Lipids* **24**:357–362.

Pind, S., and Kuksis, A., 1991, Further characterization of a novel phospholipase B (phospholipase A_2/lysophospholipase) from intestinal brush border membranes, *Biochem. Cell Biol.* **69**:346–357.

Poirier, H., Niot, I., Degrace, P., Monnot, M-C., Bernard, A., and Besnard, P., 1997, Fatty acid regulation of fatty acid-binding protein expression in the small intestine, *Am. J. Physiol.* **273**:G289–G295.

Polheim, D., David, J. S. K., Schultz, F. M., Wylie, M. B., and Johnston, J. M., 1973, Regulation of triglyceride biosynthesis in adipose and intestinal tissue, *J. Lipid Res.* **14**:415–421.

Polokoff, M. A., and Bell, R. M., 1980, Solubilization, partial purification and characterization of rat liver microsomal diacylglycerol acyltransferase, *Biochim. Biophys. Acta* **618**:129–142.

Prows, D. R., Murphy, E. J., Moncecchi, D., and Schroeder, F., 1996, Intestinal fatty acid-binding protein expression stimulates fibroblast fatty acid esterification, *Chem. Phys. Lipids* **84**:47–56.

Raclot, T., and Groscolas, R., 1993, Differential mobilization of white adipose tissue fatty acids according to chain length, unsaturation, and positional isomerism, *J. Lipid Res.* **34**:1515–1526.

Rao, G. A., and Johnston, J. M., 1966, Purification and properties of triglyceride synthetase from the intestinal mucosa, *Biochim. Biophys. Acta* **125**:465–473.

Rao, R. H., and Mansbach, C. M., 1990, Acid lipase in rat intestinal mucosa: Physiological parameters, *Biochim. Biophys. Acta* **1043**:273–280.

Rao, R. H., and Mansbach, C. M., 1993, Alkaline lipase in rat intestinal mucosa: Physiological parameters, *Arch. Biochem. Biophys.* **304**:483–489.

Rasmussen, J. T., Rosendal, J., and Knudsen, J., 1993, Interactions of acyl-CoA-binding protein (ACBP) on processes for which acyl-CoA is a substrate, product or inhibitor, *Biochem. J.* **292**:907–913.

Rigtrup, K. M., Kakkad, B., and Ong, D. E., 1994, Purification and partial characterization of a retinyl ester hydrolase from the brush border of rat small intestinal mucosa: Probable identity with brush border phospholipase B, *Biochem.* **33**:2661–2666.

Rodgers, J. B., Jr., 1969, Assay of acyl-CoA:monoglyceride acyltransferase from rat small intestine using continuous recording spectrophotometry, *J. Lipid Res.* **10**:427–432.

Rodriguez, M. A., Dias, C., and Lau, T. E., 1992, Reversible ATP-dependent inactivation of adipose diacylglycerol acyltransferase, *Lipids* **27**:577–581.

Rosendal, J., Ertbjerg, P., and Knudsen, J., 1993, Characterization of ligand binding to acyl-CoA-binding protein, *Biochem. J.* **290**:321–326.

Ruyter, B., Lund, J. S., Thomassen, M., and Christiansen, E. N., 1992, Studies of dihydroxyacetone phosphate acyltransferase in rat small intestine. Subcellular localization and effect of partially hydrogenated fish oil and clofibrate. *Biochem. J.* **282**:565–570.

Sakane, F., Yamada, K., and Kanoh, H., 1989, Different effects of sphingosine, R59022 and anionic amphiphiles on two diacylglycerol kinase isozymes purified from porcine thymus cytosol, *FEBS Lett.* **255**:409–413.

Santamarina-Tojo, S., and Dugi, K., 1994, Structure, function, and role of lipoprotein lipase in lipoprotein metabolism, *Curr. Opin. Lipidol.* **5**:117–125.

Schaffer, J. E., and Lodisch, H. F., 1994, Expression cloning and characterization of a novel adipocyte long chain fatty acid transport protein, *Cell* **79**:427–436.

Scheideler, M. A., and Bell, R. M., 1986, Efficiency of reconstitution of the membrane-associated *sn*-glycerol-3-phosphate acyltransferase of *Escherichia coli, J. Biol. Chem.* **261**:10990–10995.

Scheideler, M. A., and Bell, R. M., 1989, Phospholipid dependence of homogeneous, reconstituted *sn*-glycerol-3-phosphate acyltransferase of *Escherichia coli, J. Biol. Chem.* **264**:12455–12461.

Scheideler, M. A., and Bell, R. M., 1991, Characterization of active and latent forms of the membrane-associated *sn*-glycerol-3-phosphate acyltransferase, *J. Biol. Chem.* **266**:14321–14327.

Shiau, Y-F., 1987, Lipid digestion and absorption, in: *Physiology of the Gastrointestinal Tract,* 2nd Edition (L. R. Johnson, ed.), Raven Press, New York, pp.1527–1556.

Shin, D-H., Paulauskis, J. D., Moustaid, N., and Sul, H. S., 1991, Transcriptional regulation of p90 with sequence homology to *Escherichia coli* glycerol-3-phosphate acyltransferase, *J. Biol. Chem.* **266**:23834–23839.

Stobart, K., Mancha, M., Lenman, M., Dahlqvist, A., and Stymne, S., 1997, Triacylglycerols are synthesized and utilized by transacylation reaction in microsomal preparations of developing safflower (*Carthamus tinctorius* L.) seeds, *Planta* **203**:58–66.

Suzuki, H., Kawarabayasi, Y., Kondo, J., Abe, T., Nishikawa, K., Kimura, S., Hashimoto, T., and Yamamoto, T., 1990, Structure and regulation of rat long-chain acyl-CoA synthetase, *J. Biol. Chem.* **265**:8681–8685.

Suzuki, H., Watanabe, M., Fujino, T., and Yamamoto, T., 1995, Multiple promoters in rat acyl-CoA synthetase gene mediate differential expression of multiple transcripts with 5′-end heterogeneity, *J. Biol. Chem.* **270**: 9676–9682.

Takemori, H., Zolotaryov, F. N., Ting, L., Urbain, T., Komatsubara, Hatano, O., Okamoto, M., and Tojo, H., 1998, Identification of functional domains of rat intestinal phospholipase B/lipase. Its cDNA cloning, expression, and tissue distribution, *J. Biol. Chem.* **273**:2222–2231.

Tanaka, T., Hosaka, K., Hoshimaru, M., and Numa, S., 1979, Purification and properties of long-chain acyl-Coenzyme A synthetase from rat liver, *Eur. J. Biochem.* **98**:165–172.

Thompson, A. B. R., and Dietschy, J. M., 1981, Intestinal lipid absorption: Major extracellular and intracellular events, in: *Physiology of the Gastrointestinal Tract* (L. R. Johnson, ed.), Raven Press, New York, pp. 1147–1220.

Thompson, A. B. R., Schoeller, C., Keelan, M., Smith, L., and Clandinin, M. T., 1993, Lipid absorption: Passing through the unstirred layers, brush border membrane, and beyond, *Can. J. Physiol. Pharmacol.* **71**:531–555.

Tojo, H., Ichida, T., and Okamoto, M., 1998, Purification and characterization of a catalytic domain of rat intestinal phospholipase B/lipase associated with brush border membranes, *J. Biol. Chem.* **273**:2214–2221.

Trotter, P. J., and Storch, J., 1993, Fatty acid esterification during differentiation of the human intestinal cell line Caco-2, *J. Biol. Chem.* **268**:10017–10023.

Tso, P., 1994, Intestinal lipid absorption, in: *Physiology of the Gastrointestinal Tract* (L. R. Johnson, ed.), Raven Press, New York, pp. 1867–1907.

Tsujita, T., Miyazaki, T., Tabei, R., and Okuda, H., 1996, Coenzyme A-independent monoacylglycerol acyltransferase from rat intestinal mucosa, *J. Biol. Chem.* **271**:2156–2161.

Uchiyama, A., Aoyama, T., Kamijo, K., Uchida, Y., Kondo, N., Orii, T., and Hashimoto, T., 1996, Molecular cloning of cDNA encoding rat very long-chain acyl-CoA synthetase, *J. Biol. Chem.* **271**:30360–30365.

Vancura, A., and Haldar, D., 1994, Purification and characterization of glycerophosphate acyltransferase from rat liver mitochondria, *J. Biol. Chem.* **269**:27209–27215.

Van Greevenbroek, M. M. J., Voorhout, W. F., Erkelens, D. W., Van Meer, G., and de Brui, T. W. A., 1995, Palmitic acid and linoleic acid metabolism in Caco-2 cells: Different triglyceride synthesis and lipoprotein secretion, *J. Lipid Res.* **36**:13–24.

Wang, C-S., 1986, Hydrolysis of dietary glycerides and phosphoglycerides: Fatty acid and positional specificity of lipases and phospholipase, in: *Fat Absorption,* Volume I (A. Kuksis, ed.), CRC Press, Boca Raton, FL, pp. 83–117.

Webber, K. O., and Hajra, A. K., 1992, Purification of dihydroxyacetone phosphate acyltransferase from guinea pig liver proxisomes, *Arch. Biochem. Biophys.* **300**:88–97.

Wetterau, J. R., and Zilversmit, D. B., 1986, Localization of intracellular triacylglycerol and cholesterol ester transfer activity in rat tissues, *Biochim. Biophys. Acta* **875**:610–617.

Wetterau, J. R., Aggerbeck, L. P., Lapland, P. M., and McClean, L. R., 1991, Structural properties of the microsomal triglyceride-transfer protein complex, *Biochem.* **30**:4406–4412.

Wiggins, D., and Gibbons, G. F., 1992, The lipolysis/esterification cycle of hepatic triacylglycerol, its role in the secretion of very low density lipoprotein and its response to hormone and sulphonylureas, *Biochem. J.* **284**:457–462.

Xia, T., Mostafa, N., Bhata, B. G., Florant, G. L., and Coleman, R. A., 1993, Selective retention of essential fatty acids: The role of hepatic monoacylglycerol acyltransferase, *Am. J. Physiol.* **265**:R414-R419.

Yada, Y, Ozeki, T., Kanoh, H., and Nozawa, 1990, Purification and characterization of cytosolic diacylglycerol kinases of human platelets, *J. Biol. Chem.* **265**:19237–19243.

Yamashita, S., and Numa, S., 1972, Partial purification and properties of glycerophosphate acyltransferase from rat liver. Formation of 1-acylglycerol-3-phosphate and palmityl co-enzyme A, *Eur. J. Biochem.* **31**:565–573.

Yamashita, S., Hosaka, K., and Numa, S., 1973, Acyl-donor specificities of partially purified 1-acylglycerophosphate acyltransferase, 2-acylglycerophosphate acyltransferase and 1-acylglycerophosphocholine acyltransferase from rat liver microsomes, *Eur. J. Biochem.* **38**:25–31.

Yamashita, A., Sugiura, T., and Waku, K., 1997, Acyltransferases and transacylases involved in fatty acid remodeling of phospholipids and metabolism of bioactive lipids in mammalian cells, *J. Biochem.* **122**:1–16.

Yang, L-Y., and Kuksis, A., 1991, Apparent convergence (at 2-monoacylglycerol level) of phosphatidic acid and 2-monoacylglycerol pathways of synthesis of chylomicron triacylglycerols, *J. Lipid Res.* **32**:1173–1186.

Yang, L- Y., Kuksis, A., and Myher, J. J., 1990, Lipolysis of menhaden oil triacylglycerols and corresponding fatty acid alkyl esters by pancreatic lipase *in vitro:* A reexamination, *J. Lipid Res.* **31**:137–148.

Yang, L-Y., Kuksis, A., and Myher, J. J., 1995, Biosynthesis of chylomicron triacylglycerols by rats fed glyceryl or alkyl esters of menhaden oil fatty acids, *J. Lipid Res.* **36**:1046–1057.

Yet, S-F, Lee, S, Hahm, Y. T., and Sul, H. S., 1993, Expression and identification of p90 as the murine mitochondrial glycerol-3-phosphate acyltransferase, *Biochem.* **32**:9486–9491.

Zandonella, G., Haalck, L., Spener, F., Faber, K., Paltauf, F., and Hermetter, A., 1995, Inversion of lipase stereospecificity for fluorogenic alkyldiacylglycerols. Effect of substrate solubilization, *Eur. J. Biochem.* **231**:50–55.

CHAPTER 12

Triacylglycerol Movement in Enterocytes

CHARLES M. MANSBACH II

1. Introduction

This chapter covers the movement of triacylglycerols (TGs) within enterocytes. Relatively little is known about how TG vectorially transits the intestinal absorptive cell compared to other aspects of intestinal cellular function. Intracellular TG movement is important because in humans, the intestine must absorb 100 to 150 gm of fat per day on a typical Western diet, which, if inefficiently done, may result in steatorrhea, the excessive excretion of fat in the stool, and may also lead to fat-soluble vitamin deficiencies. Normally, the end result of the complex process of lipid absorption is the export of TG containing long-chain fatty acids (FA) into the lymph in a unique, specialized lipoprotein particle, the chylomicron (Green and Glickman, 1981) and a slightly more dense particle, the very low density lipoprotein (VLDL, Ockner *et al.*, 1969b; Tso *et al.*, 1987). More recent studies have begun to shed some light on how TG moves from its site of synthesis in the endoplasmic reticulum (ER) to its exit site at the basolateral membrane of the enterocyte as shown schematically in Fig. 1.

2. The Movement of Triacylglycerol from its Site of Synthesis to its Exit from the Enterocyte as a Chylomicron

2.1. The Topography of Enzymes that Synthesize Triacylglycerol

Intracellular TG movement begins with its synthesis, which is reviewed by Kuksis and Lehner (Chapter 11). The enzymes responsible for this are located in the ER in the intestine (Lehner and Kuksis, 1995). Their transverse location has not yet been examined but if the intestine is analogous to the liver in this regard, these enzymes are located on the cytoplasmic face of the ER (Coleman and Bell, 1978). ER enzymes are known to express their

CHARLES M. MANSBACH II • Department of Medicine, Division of Gastroenterology, The University of Tennessee, Memphis, Tennessee 38163, *and* The Office of Research and Development Medical Research Service, Department of Veterans Affairs, Memphis, Tennessee 38104.
Intestinal Lipid Metabolism, edited by Charles M. Mansbach II *et al.*, Kluwer Academic/Plenum Publishers, 2001.

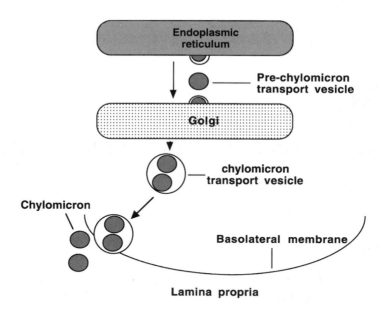

Figure 1. The movement of triacylglycerol from the endoplasmic reticulum (ER) to the Golgi via the pre-chylomicron transport vesicles. Each of these vesicles transports one chylomicron. On the *trans* face of the Golgi, chylomicron transport vesicles are formed that can move multiple chylomicrons to the basolateral membrane. The chylomicrons are released from the enterocytes via a reverse pinocytotic mechanism.

activity either overtly (enzyme present on the cytosolic hemileaflet and activity expressed directly) or latently (enzyme present on the lumenal hemileaflet and maximal activity expressed only in the presence of detergent, which enables the substrate to traverse the ER membrane to gain access to the enzyme). The preponderance of evidence would suggest that all physiologically important diacylglycerol acyltransferase (DGAT), the terminal enzyme in TG synthesis, as well as monoacylglycerol acyltransferase (MGAT), are on the cytosolic hemi-leaflet of the ER (Hulsmann and Kupershook-Davidov, 1976; Coleman and Bell, 1978; Coleman and Haynes, 1985), although there has been a question raised as to the existence of a latent form of DGAT (Owen *et al.*, 1997). The original data by Coleman and Bell (1978) were based on the susceptibility of both MGAT and DGAT to proteolytic attack in which the vast majority of total enzyme activity was destroyed in the absence of detergent with only a small additional reduction in activity when detergent was added.

2.1.1. The Physiological Importance of the Topographical Site of Triacylglycerol Synthesis

The specific location of the TG-synthetic enzymes is important because the site of synthesis of TG plays a role in considering how the TG is integrated into the developing chylomicron. If the TG is synthesized on the cytosolic side of the ER hemileaflet, then the TG must traverse the ER membrane to gain access to its lumen where the developing chylomicron resides. This could occur through dissolution of the TG in the ER membrane, which can occur at levels of up to 3% (Hamilton and Small, 1981). Although the molar percentage composition of TG in the membrane is small, the rate at which TG can translocate across the ER membrane may be very fast. Unfortunately there is no specific information on TG, but diacylglycerol (DG), which is physicochemically similar to TG, rapidly moves across phospholipid monolayers (Walsh and Bell, 1986) so that TG movement to the ER lumen may not be a limiting factor in chylomicron–TG accretion. Under conditions where the ER membrane composition is altered, however, TG translocation may be partially restricted just as ER membrane compositional changes in the intestine can alter the activity of complex lipid synthetic enzymes (Mansbach and Parthasarathy, 1979). Should TG translocation occur slower than the TG synthetic rate, then TG would accumulate on the cytosolic face of the ER. This is an untested hypothesis with unknown consequences but, if true, the cytosolic-face-TG could be a substrate for either the cytosolic (Rao and Mansbach, 1993) or microsomal lipase (Coleman and Haynes, 1983). Alternately, if TG were synthesized in significant amounts on the lumenal side of the ER membrane, it would have immediate access to the developing chylomicron.

2.1.2. Triacylglycerol Is Rapidly Synthesized and Forms Chylomicrons in a Two-Step Process

Presuming that the TG synthetic enzymes are localized to the cytosolic side of the ER-membrane, FAs and MG arrive from the cytosol likely via the intestinal fatty-acid-binding protein (I-FABP) as described by Storch in chapter 9 (Trotter and Storch, 1991; Hsu and Storch, 1996). Despite the multiple enzymatic steps involved (see chapter 11), TG synthesis occurs rapidly. A morphological time course study by Jersild (1966) showed that one minute after placing chyme in rat *in situ* intestinal sacs, small lipid droplets were visible in the ER and the Golgi at electron microscopy (EM). At 2 min, the amount of fat droplets increased greatly in both organelles. In 5 min, some fat droplets were seen intercellularly, suggesting that they are ready to exit the intestine into the lymph. This describes morphologically a very rapid process that is substantiated by physiological studies that showed that 30 sec after the intraduodenal injection of [^{3}H]oleate in a rat that was actively absorbing lipid, 79 % of the absorbed [^{3}H]oleate was in TG with 25 % of the [^{3}H]oleate absorbed (Mansbach and Nevin, 1998).

Recently, an apolipoprotein B (apoB) knock-out mouse model has been used to investigate the assembly of the chylomicron in the ER (Hamilton *et al.,* 1998). In this model, the mice are transgenic for human apoB in which case they express apoB in the liver but not the intestine. EM studies show two types of particles in the ER that are consistent with a model in which apoB, phospholipid, and a small amount of TG form in the ER lumen as one type of particle. The second particle forms more distally in the ER as a large TG-rich droplet without apoB being present. These two particles are presumed to fuse to form the final chylomicron. Interestingly, in the knock-out model, little TG is transported to the Golgi, suggesting that apoB might play a role in signaling that the nascent chylomicron is available for transport to the Golgi.

2.2. The Movement of Triacylglycerol from the ER to the Golgi

The addition of TG to the developing chylomicron in the ER is carefully described in chapter 10. Once partially matured in the ER, the chylomicron is ready to move to the Golgi apparatus. How this journey through the cytosol is made is a subject of controversy. One possibility is that microtubules deliver the developing chylomicron to the Golgi (Moreau *et al.*, 1991). This potential is made less likely by findings in our laboratory that have shown that intestinal ER, labeled with [^{3}H]-TG, can transport the TG to Golgi from donor rats (Kumar and Mansbach, 1997). Further, the addition of the microtubular polymerization inhibitor, colchicine, did not impede TG delivery to the Golgi. The second possibility is that a vesicle is utilized to deliver the developing chylomicrons to the Golgi. This thesis is supported by studies that showed that the addition of an excess of phosphatidylcholine (PC) or PC and cholesterol as small unilamellar vesicles, chylomicrons, or chylomicrons isolated from the plasma, did not quench the movement of [^{3}H]-TG to the Golgi. In sum, these data suggest that a vesicle, enclosing the developing chylomicron, was involved in the transport process. Of particular interest was the finding that intestinal ER delivered [^{3}H]-TG only to intestinal Golgi but not to liver or kidney Golgi (Kumar and Mansbach, 1997). These data lead to the hypothesis that there is a specific protein(s) that is expressed on the surface of the putative vesicle, which we have named the pre-chylomicron transport vesicle (PCTV), whose function is to target the vesicle to the Golgi (Kumar and Mansbach, 1999). Further, our data showing no vesicular TG transport to liver Golgi supports the prior work of Moreau *et al.* (1991) in liver, which showed that phospholipids but not TG were transported from the ER to the Golgi by a vesicular mechanism.

2.2.1. *The Movement Of Triacylglycerol from the ER to the Golgi Is Rate Limiting*

In order to support the thesis that the PCTV was physiologically important, the rate of movement of [^{3}H]-TG from the ER to the Golgi was correlated in intestinal cells isolated from rats infused intraduodenally with TO that had either an intact or a diverted bile duct and from rats infused with TO supplemented with PC (Kumar and Mansbach, 1997). The results obtained from these *in vitro* experiments were parallel to those obtained *in vivo* using similarly treated rats that had a main mesenteric lymph duct fistula. In both cases, the rate of TG movement into the lymph or to the Golgi was fastest in the rats supplemented with PC and was slowest in the rats that had been bile fistulated.

These data support previous work that suggested that the movement of TG from the ER to the Golgi is rate limiting under four different circumstances.

1. During active fat absorption, TG appears to be trapped in the ER awaiting transport to the Golgi (Friedman and Cardell, 1977).
2. In the clinical condition, the bacterial overgrowth syndrome, in which bacteria proliferate in the lumen of the intestine, lipid can be seen at electron microscopy accumulating in the ER (Ament *et al.*, 1972).
3. When chylomicron output into the lymph is severely restricted by the non-ionic detergent, Pluronic L-81 (L-81), the block in chylomicron output appears to be at the level of the ER (Tso *et al.*, 1981a).
4. In the unusual genetic condition, chylomicron retention disease, in which chylomicrons are formed but their secretion is blocked, developing chylomicrons are identified in the ER but not the Golgi (Nemeth *et al.*, 1995).

For these reasons the ER to Golgi transport step appears to be crucial to the output rate of chylomicrons into the lymph.

The ability of the ER to transport TG to the Golgi was more directly tested in recently reported experiments (Mansbach and Dowell, 2000) in which rats were infused intraduodenally with increasing amounts of non-radiolabeled glyceryltrioleate (TO), from 22.5 to 135 μmol/h, to mass steady state and then observed for the rapidity with which they obtained radiolabeled steady state in the ER and Golgi on the introduction of ^{3}H-TO into the infusion while maintaining the mass at the same rate. As the amount of infused TO increased, so did the length of time it took to obtain radiolabel steady state conditions in the ER, from 10 to 60 min. Further, at the 135 μmol/h level, not all the TG was able to cross the ER membrane as shown by studies in which this TG, but not the TG within the ER lumen, was hydrolyzed by lipase added *in vitro*. When the ER membrane was permeabilized by taurocholate, then lipase gained access to the sequestered TG in the ER lumen and hydrolyzed it. In sum, these data show the progressive decrease in the proportion of ER-TG that can be transported to the Golgi as the lipid input load increases until at the largest lipid input rate, more TG is synthesized than can cross the ER lumen; i.e. maximum rates of TG transport to the Golgi have been exceeded.

2.2.2. The Mechanism by Which the Developing Chylomicron Moves from the ER to the Golgi

The mechanism by which the vesicle buds from the ER, traverses the intervening space in the cytosol, and is targeted specifically to the intestinal Golgi is unknown. One possibility is the same mechanism employed by secretory proteins as elucidated by Rothman and Orci (1992). Further credence to this postulate was given by the finding that the movement of TG from the ER to the Golgi was N-ethylmaleimide-1 (NEM) sensitive (Kumar and Mansbach, 1997). However, it was found that TG transport was GTP independent, the activity was not inhibited by GTPγS, a non-hydrolyzable GTP analogue, and antibodies to rab's 1 and 2 did not reduce the rate of TG movement to the Golgi. These data essentially rule out the secretory protein pathway as a mechanism (Rothman and Orci, 1992). Further studies showed that TG transport was cytosol and ATP dependent and that the cytosolic component included a protein (Kumar and Mansbach, 1997). In sum, the data support a vesicle, PCTV, that encloses the developing chylomicron in its transit to the Golgi. We propose that the PCTV has on its surface a specific targeting protein enabling it to dock specifically on the *cis* face of the intestinal Golgi. It then fuses with the Golgi in an ATP-dependent step and delivers its cargo into the Golgi lumen.

In recent studies we have isolated the PCTV as a 200 nm, lipid-containing vesicle (Kumar and Mansbach, 1999). The PCTV was found to have unique proteins as compared to the ER and the Golgi and to express only small amounts of enzymes that are associated with the ER and the Golgi. The PCTV delivered its TG cargo solely to the Golgi and not to the ER, demonstrating the vectorial nature of the transport. Of particular interest is that immunoblot studies showed that apoA-I was present in both the ER and the Golgi but only in small amounts in the PCTV. Because mature chylomicrons have apoA-I on their surface, these data strongly suggest that the PCTV is a post-ER, pre-Golgi particle. Further, because it would appear that apoA-I is carried to the Golgi by a route different than the PCTV, the data offer a mechanism by which apoA-I secretion could be uncoupled from apoB-48 secretion. Indeed, apoA-I secretion has been found to be continuously secreted even when apoB secretion is suppressed (Hamilton *et al.,* 1986; Gordon *et al.,* 1996; Haghpassand *et al.,* 1996).

2.2.3. The Physiological Importance of a Pre-Chylomicron Transport Vesicle

The presence of a vesicle is important not only to transport the chylomicron to its intended target but also to isolate the vesicle from a cytosolic lipase (Rao and Mansbach, 1991, 1993) that would otherwise hydrolyze the chylomicron-TG intracellularly. This lipase has been cloned and sequenced in unreported studies (Mahan, Heda, Rao, and Mansbach, 2000) that have shown that the lipase is in fact pancreatic, colipase-dependent lipase except that it is expressed in the intestine. The activity of the intestinal lipase is comparable to that expressed in the pancreas and is very active when purified (12 mmol/mg prot/min using TO as substrate; Rao and Mansbach, 1991). This would result in the rapid hydrolysis of any TG accessible to the enzyme.

2.3. The Movement of the Pre-Chylomicron from the Golgi to the Lymph

Once the TG, now in the center of the developing chylomicron (Green and Glickman, 1981), is in the Golgi, its further processing is not evident. It is possible, however, that additional TG is added to the chylomicron during its transit through the Golgi because both MGAT and DGAT are expressed in the Golgi, although at a reduced specific activity compared to the ER (Kumar and Mansbach, 1997). After the Golgi, a different and larger vesicular structure is used to transport multiple chylomicrons to the basolateral surface of the enterocyte where it is targeted to clathrin-coated pits (Sabesin and Frase, 1977). These vesicles are likely to be tethered to microtubules as the mechanism by which they are moved to the basolateral membrane as suggested by the finding that colchicine greatly reduces the ability of the intestine to export chylomicrons (Glickman *et al.*, 1976). The movement of these vesicles from the Golgi to the basolateral membrane is the only step in the process in which microtubules could be significantly involved. The vesicles then fuse with the basolateral membrane and the chylomicrons are exocytosed into the laminal propria by reverse pinocytosis (Sabesin and Frase, 1977; Tso *et al.*, 1985).

The rate of movement of chylomicrons across the lamina propria to the lymph ductules is governed by the state of hydration of the tissue (Tso *et al.*, 1985). Using the appearance time of [³H]oleate in the lymph, Tso *et al.* (1985) were able to show that the less well hydrated the interstitial matrix as measured by lymph flow, the longer the appearance time of the [³H]oleate in the lymph. The shortest appearance time (14 min) was with a lymph flow of 4 μl/min or greater in the rat, suggesting that this flow rate indicates adequate tissue hydration.

2.4. Determinants of the Rate of Export of Chylomicrons by the Intestine

2.4.1. Phosphatidylcholine Availability is Important for Chylomicron Transport

The ability of the intestine to export chylomicrons into the lymph is determined by at least two other factors—the availability of PC and the location along the length of the intestine where lipid absorption is occurring. Abundant evidence shows that the lumenal availability of PC is important with respect to chylomicron output rates (Bennett-Clark, 1978; Tso *et al.*, 1978, 1981b; Mansbach and Arnold, 1986b). These studies have all shown that in the absence of lumenal PC as caused by bile diversion, the output rate of chylomicrons is reduced by comparison to bile-duct-intact rats. By contrast, when additional PC is added to large lipid infusions, chylomicron output is increased as compared to unsupplemented controls (Bennett-Clark, 1978; Mansbach and Arnold, 1986b). This increase could either be due to the direct absorption of the extra PC as *sn*-1-acyl PC (lyso-PC) from the intestine and its incorporation into chylomicron PC (Scow *et al.,* 1967; Mansbach, 1977) or from the increase in biliary PC output that is seen under these conditions. Indeed, bile PC is preferred as a chylomicron PC precursor as compared to other forms of dietary PC (Mansbach, 1977). Because bile PC output is increased on feeding rats a fat supplemented diet (Knox *et al.,* 1991), the increase in TG output into the lymph seen under these conditions (Mansbach and Arnold, 1986b) could be due to a similar mechanism.

An unanswered question is that of the necessity for dietary (biliary) PC when the intestine can synthesize PC *de novo* using the Kennedy pathway (Mansbach, 1973; Mansbach and Partharsarathy, 1979; Mansbach and Arnold, 1986a) as a supplementary pathway. One possibility is that the *de novo* synthesis of PC is not able to keep up with the requirements for the PC needed for chylomicron formation. This is suggested by the finding that in a bile-fistula rat, whose lymph lipid output is reduced as compared to a bile-intact animal, the inclusion of choline in the intraduodenal lipid infusion only partially restores lymph TG output, whereas the addition of lyso-PC to the intraduodenal infusion fully restores lymph TG output to the same level seen in the bile-duct-intact condition (Tso *et al.,* 1978). Even under physiological conditions, where rat bile PC output is 4 to 5 μmol/h (Turley and Dietschy, 1979; Knox *et al.,* 1991), not enough PC is available to support maximal chylomicron output even though only 2 to 4 μmol/h of chylomicron-PC is put out into the lymph during fat absorption (Mansbach, 1977; Bennett-Clark, 1978; Tso *et al.,* 1981b). This is shown by the increase in chylomicron-TG export on the addition of PC to the intraduodenal lipid infusion (Bennett-Clark, 1978; Mansbach and Arnold, 1986b). In sum, these data would suggest that a significant amount of absorbed lyso-PC is not effectively utilized for chylomicron-PC formation.

It is not clear as to why enough PC is not available from the intestinal lumen. PC itself is not absorbed intact (Scow *et al.,* 1967) but must first be hydrolyzed by pancreatic phospholipase A_2 to the absorbable *sn*-1-lyso-PC. The re-acylation of lyso-PC to PC is mediated by lyso-PC acyltransferase (Mansbach, 1972) that is expressed in the intestine. The activity of this enzyme is greatest in the villus tips of the proximal intestine, the site of maximal lipid absorption, suggesting that it is involved in the lipid absorption process (Mansbach, 1972). That this is the case is shown by the incorporation of *sn*-1-lyso PC into chylomicron-PC (Scow *et al.,* 1967). Nevertheless, it is clear that to a significant extent lumenal PC is stripped of its acyl groups entirely and thus is lost to the chylomicron-PC precursor pool. In this way, ultimately not enough PC is available to support maximal chylomicron-TG output unless PC input into the intestine is increased by dietary supplementation.

2.4.2. The Location of Lipid Absorption May Influence the Rate of Chylomicron Output

The other determinant for the rate of chylomicron export from the enterocytes is the location along the intestine that is involved in the absorptive process. It was first suggested in studies in rats that examined the amount of lipid remaining in various portions of the intestine after lipid infusion. These studies showed that the distal intestine was unable to clear lipid as quickly as the proximal gut (Bennett-Clark *et al.,* 1973). This was followed by further studies in which lipid, bile, and pancreatic juice were infused into the proximal or distal intestine for 2 days and the rate of output of TG into the lymph was observed (Wu *et al.,* 1975). These studies continued to show reduced rates (40 %) of TG output into the lymph when the distal intestine was perfused as compared to the proximal one. The question was next asked if the ileal defect could be due to the lack of perfusion of the distal intestine with lipid for longer time periods. When FA and MG were infused into the distal intestine for 7 days prior to study, the distal intestine still accumulated lipid to a greater extent than did the proximal intestine, indicating a defect in chylomicron output (Wu and Clark, 1976). Finally, electron microscopic evidence showed that the proximal intestine had cleared absorbed lipid but the distal intestine had not 6 hr after a proximal or distal lipid infusion, respectively (Sabesin *et al.*, 1977). These results should be contrasted with those more recently obtained by Tso's group (Tsuchiya *et al.*, 1996). In these studies, the ileum was transposed to be the proximal portion of the bowel and lipid was infused into the duodenum. The result was that no defect in the processing of lipid by the neoproximal, "distal," intestine was observed. Differences between these studies may be explained in part by the amount of lipid administered, which was more physiological in Tsuchiya *et al.*'s (1996) studies and was designed to be twice the maximal absorption rate for rats in the studies reported by Clark's (Sabesin *et al.*, 1977) group. However, in support of a defect in the ileum are the studies by Jersild and Clayton (1971) in which chyme was placed in the intestinal lumen with morphological assessment of lipid absorption performed by timed tissue samples examined by EM. When the chyme was placed in the ileum, large amounts of lipid were present as droplets in the cytoplasm but a relative paucity of lipid was found in both the ER and the Golgi lumens as compared to when chyme was placed in the jejunum. Only a few chylomicrons were found in the lamina propria. These data raise the question of a factor, present in jejunum but absent in ileum, that promotes the movement of TG across the ER membrane. In sum the data support the thesis that there is a defect in chylomicron formation in the *in situ* ileum but not when it is transposed to the proximal intestine.

2.4.3. Hormones and Cytokines Influence the Rate of Chylomicron Output in Cell Culture Systems

Studies have been performed in cell culture systems to define whether cytokines or hormones affect lipid secretion by the various cell types studied. The intestinal cryptal cell line, IEC-6 cells, has been used to examine the effect of transforming growth factor-β (TGF-β) on the ability of these cells to increase the very small amount of TG that they normally secrete into the medium. No effect was noted (Mehran *et al.*, 1994). In Caco-2 cells, a human colon cancer cell line that differentiates into intestinal cells with the maturity of 12-day-old rat jejunal cells, tumor necrosis factor-α (TNF-α) was shown to inhibit lipid secretion (Mehran *et al.*, 1995). Similarly, human fetal intestinal explants responded to exposure to hydrocortisone with a reduced ability to secrete lipids (Loirdighi *et al.*, 1997). In sum, the hormones and cytokines studied have resulted in little progress being made to get the cell lines or the explants to secrete a reasonable proportion of the intracellular TG available to be secreted.

3. Endogenous Acyl Groups are Recruited by the Enterocyte and Exit the Intestine via the Portal Vein

3.1. Large Amounts of Endogenous Acyl Groups are Present in the Intestine during Lipid Absorption

In addition to the TG that moves through the intestinal cell and exits in chylomicrons, a considerable amount of TG is present in the intestinal cell whose acyl groups are not derived from the diet. This is best illustrated in experiments in which large amounts of TO (135 μmol/h) are infused intraduodenally until a mass steady state is reached and the TG-acyl group composition of the mucosa is determined. These studies have consistently shown that 40% to 50 % of mucosal TG-FAs are not derived from the intestinal lumen (Mansbach and Partharsarathy, 1982; Mansbach and Dowell, 1992; Mansbach and Nevin, 1994). These are entirely unexpected findings because it would be thought that with the large amount of oleate from TG-oleate fluxing through the cell, the vast majority of the TG-FA would be oleate. For example, when a total of 945 μmol of TO was infused intraduodenally over 7 hr into rats with 100 to 150 μmol of TG expected to be present in the mucosa at any one time under these conditions (Mansbach and Dowell, 1992), only 60 % of the intestinal mucosal TG was from the infusate. This indicates that large quantities of endogenous acyl groups flux into the intestinal mucosa during lipid absorption and/or are derived from enterocyte phospholipid acyl groups. The proportion of nondietary lipid in the mucosa may be increased in part on the basis that the endogenous TG-acyl groups may have a turnover rate that is slower than that of the TG in the pool that subserves chylomicron formation (Mansbach and Arnold, 1986b) and thus, over time, it becomes a larger proportion of the total TG-FA than would be present acutely. An overall schema of the fate of TG-FA coming from endogenous sources is shown in Fig. 2.

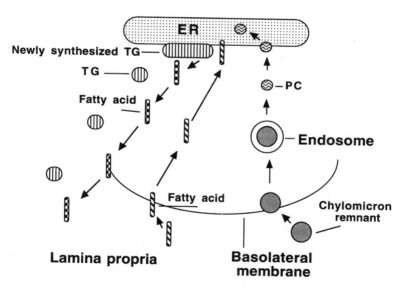

Figure 2. The flow of endogenous lipids in the enterocyte. Chylomicron remnants enter the cell from the laminal propria by a receptor-dependent mechanism. Inside the enterocyte, the remnants are proposed to enter the endosomal system where phosphatidylcholine (PC) is stripped away. The PC goes to the ER where it may be utilized for chylomicron formation or for membrane synthesis. Fatty acids enter the enterocyte from the lamina propria and go to the ER where they are synthesized to triacylglycerol (TG). This TG may be hydrolyzed to form fatty acids that then exit the cell via the basolateral membrane or it may be exocytosed as TG in an unknown lipoprotein particle.

3.2. The Origin of Endogenous Acyl Groups

3.2.1. Circulating Fatty Acids Enter the Intestinal Cells

Where could these endogenous FA come from? Circulating FA is one potential. When radiolabeled palmitate was injected as a bolus IV, only a minority of the FA was present as mucosal TG; the majority was in water soluble oxidation products in both rats and man (Gangl and Ockner, 1975; Gangl and Renner, 1978). By contrast, if oleate is infused IV bound to albumin until steady state conditions are reached, then 86% of the oleate radioactivity is found to be in mucosal TG (Mansbach and Dowell, 1992). Interestingly, at steady state, the specific activity of the mucosal TG was significantly higher than that of chylomicrons that were collected from the lymph in the same animals. This indicates that the mucosal TG pool labeled by the incoming, endogenous FA-oleate was a poor precursor of chylomicron-TG as compared to the TG pool labeled by dietary FAs. Unfortunately, the amount of mucosal TG that could be ascribed to the circulating oleate could not be calculated due to uncertainties concerning the rate of turnover of the mucosal endogenous TG-FA pool and whether all the $[^3H]$-dpm in the mucosa were $[^3H]$oleate.

3.2.2. Chylomicron Remnants Are Internalized by the Enterocytes

Another potential source of endogenous TG-FA is from chylomicron remnants. Chylomicron remnants are the resulting lipoprotein particles from the metabolism of chylomicrons predominately by adipose tissue and muscle in the periphery. During metabolism of the chylomicrons by lipoprotein lipase, multiple changes occur: TG is removed from its core, phospholipid buds off of its surface that is now too large for the reduced mass of TG and cholesteryl-ester remaining (Redgrave and Small, 1979), C-lipoproteins elute from its surface, and apolipoprotein E accumulates (Mjos *et al.,* 1975). Of interest with respect to the potential delivery of nondietary TG-FA to the mucosa is the change in the TG-FA composition of the resulting remnant particle as compared to the parent chylomicron. For example, the remnants have proportionally more arachidonate as compared to the parent chylomicron (Nilsson and Landin, 1988; Mansbach and Dowell, 1995) as well as more stearate and less oleate. Thus chylomicron remnants could deliver TG-FAs to the mucosa that are different from those of the diet if the remnants were internalized by the enterocytes. When this question was first studied (Redgrave, 1970), the vast majority of the chylomicron remnants were thought to go to the liver. This was certainly true for cholesteryl-ester but not necessarily for remnant-TG. Indeed, in the fasting state the intestine was found to effectively compete with the liver for $[^3H$-TG$]$-remnant uptake when remnants were constantly infused IV until a steady state TG-specific activity was reached (Mansbach and Dowell, 1995). By contrast, remnant-cholesterol was mostly taken up by the liver. When TO was given intraduodenally and the study was repeated, the quantity of remnant-TG taken up by the mucosa was greater but the proportion of the total $[^3H]$-TG taken up by the liver was even greater than during fasting conditions. It was postulated that during fat absorption, the remnant receptors on the intestinal basolateral membranes become saturated but those of the liver do not, resulting in the liver being able to take up a greater proportional share. In sum, the intestinal mucosa can receive nondietary FA from circulating FA or from specific FAs from chylomicron remnants.

3.3. The ER Is the Site of the Split in the Triacylglycerol Stream Going either to Chylomicrons or the Triacylglycerol Storage Pool

Because FAs entering from the circulation and acyl groups entering from chylomicron remnants both result in TG-FAs that are different than dietary TG-FAs and these nondietary TGs are not primarily selected as chylomicron-TG precursors, at some level in the enterocyte they must split from the stream of TG entering the chylomicron precursor pool. Recent work has suggested that this occurs on the cytoplasmic surface of the ER (Mansbach and Nevin, 1998). In these studies, a mass steady state was developed in rat intestinal mucosa by the constant infusion of TO intraduodenally. [^{3}H]-TO was then introduced into the infusion while keeping the mass the same. The increase in [^{3}H]-TG-specific activity was followed over time in the ER and the Golgi (Fig.3). Of interest was that the specific activity of the ER-TG was always lower than the Golgi-TG. At the radioactive steady state, the Golgi-TG had a specific activity that approached that expected for chylomicron-TG under these infusion conditions whereas the specific activity of the ER-TG was significantly less. Further, the rate of increase in TG-specific activity was greater in the Golgi-TG than the ER-TG, suggesting that only a portion of the ER-TG pool was transferred to the Golgi and that the TG in this pool was more representative of dietary TG than the TG remaining with the ER. These data imply that once TG translocates to the lumen of the ER, it eventually exits the enterocyte as chylomicron-TG. By contrast, the TG that does not translocate across the ER membrane does not enter the chylomicron-TG precursor pool. This leads to the thesis that the two TG streams split at the level of the ER as shown in Fig. 4.

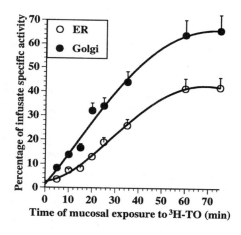

Figure 3. The percentage of infusate-specific activity in the ER and the Golgi as a function of the time of ^{3}[H]-TO infusion. Rats were infused intraduodenally with TO, 135 μmol/hr, for 6 hr which was supplemented at t_0 with ^{3}[H]-TO. At the indicated times, the mucosa was harvested and the ER and Golgi fractions were isolated. The lipids from the organelles were extracted and the TG-specific activity (dpm/μmol) was determined. The organelle-specific activity was compared to that of the infusate. The data are the mean ± SEM ($N = 5$). The Golgi had a greater specific activity than the ER at all time points ($p < 0.05$). At mass steady state, after 6 hr of TO infusion, the ER contained 29 ± 3 nmol TG and the Golgi containd 63 ± 8 nmol TG. The specific activity of the infusate was 27,450 ± 2000 dpm/μmol TO. The equations for the linear regression lines of the data from t_0 to $t = 35$ min are: ER $y = 0.766x - 1.21$, $r^2 = 0.9762$; Golgi $y = 1.26x + 1.75$, $r^2 = 0.9521$. Chylomicron TG-specific activity as a percentage of infusate specific activity under similar ^{3}[H]-TO infusion conditions have been shown to be 75 % (Mansbach *et al.,* 1987) and 77 % (Mansbach and Partharsarathy, 1982). (From *J. Lipid Res.* **39,** 1998. Used with permission.).

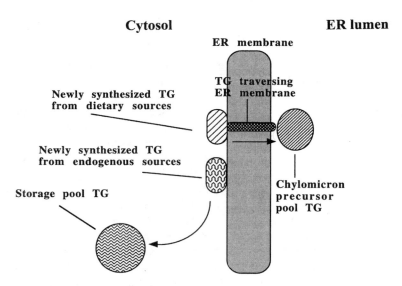

Figure 4. The intracellular movement of TGs in the enterocyte. Newly synthesized TG is shown on the cytosolic face of the ER membrane as two separate pools. The TG can go into either a storage pool or it can traverse the ER membrane into the ER lumen where it contributes to the developing chylomicron. The TG that enters the ER lumen is mainly from dietary derived sources such as fatty acid and monoacylglycerols (MG) and its glyceride glycerol is from diet-derived MG. The TG entering the storage pool is mainly from endogenous sources. Note that the two TG pools are suggested to be separated on the ER membrane enabling the identification of TG from each source so that it can be targeted appropriately.

4. How Triacylglycerol is Recruited for Triacylglycerol-Rich Lipoprotein Formation

The intestine appears to be different than the liver with respect to how it accesses TG for TG-rich lipoprotein particle formation. In the liver, TG enters a storage pool as a first step prior to its incorporation into VLDL. When more TG is required for VLDL formation, the TG storage pool is accessed and the TG in the pool is partially hydrolyzed (Mooney and Lane, 1981; Yang *et al.,* 1995b), predominantly to DG. To determine whether the intestine employed a similar strategy, experiments were performed by infusing ([^{3}H]glyceryl, [^{14}C]oleoyl) TO intraduodenally (40 μmol/h) while TG exit from the intestine was blocked by the co-infusion of the non-ionic detergent, Pluronic L-81 (Tso *et al.,* 1980). After 8 hr of lipid infusion, the L-81 was removed from the infusate and the inhibition of TG export was relieved. During the unblocking phase, the intraduodenal infusate was changed to linoleate (120 μmol/h). Lymph was constantly collected and chylomicron-TG was analyzed for its TG-FA composition. When the inhibition was relieved, the chylomicron-TG appearing in the lymph was predominantly TG-oleate with only small amounts of TG-linoleate (Halpern *et al.,* 1988). It would be assumed that if significant amounts of TG hydrolysis occurred prior to chylomicron-TG formation, this would be maximized during the unblocking stage.

Because the intestine was flooded with linoleate during this time, linoleate would have been used as the preferred FA for partial glyceride acylation formed as a result of the proposed hydrolysis. Because it was not, it is likely that in the intestine, unlike the liver, TG is not partially hydrolyzed prior to its final assembly in chylomicrons. This view is supported by studies in which either menhaden oil or FA-esters derived from the oil were given by gavage to rats and thoracic duct chylomicron-TG was evaluated from the point of view of the specific esterification site of the various FAs to glycerol (Yang *et al.*, 1995a).

An alternate view has been proposed by Yang *et al.* (1995a, 1995b) that states that TG formed by the *de novo* TG synthetic pathway, in which phosphatidic acid is an intermediate, first enters a storage pool and subsequently is partially hydrolyzed to MG with re-acylation to TG prior to its exit from the intestinal cell as chylomicron-TG. This route might be particularly pertinent in conditions in which dietary FA and MG-driven TG synthesis occurs more rapidly on the cytoplasmic face of the ER than the TG can be transported across the ER membrane to participate in chylomicron formation.

5. The Intestine Produces Very Low Density Lipoproteins

Besides chylomicrons, the intestine also exports the heavier, smaller, TG-rich particle, VLDLs (Ockner *et al.*, 1969a). These differ from chylomicrons (Sf > 400) by their greater density (Sf = 102) and their greater ratio of PC to TG. It is clear that the VLDL in the lymph come from the intestine and not from the plasma because under fasting conditions, bile diversion greatly reduces the output of VLDL, as well as chylomicrons, into the lymph. Of further interest is the finding that the specific FA administered intraduodenally influences the distribution of TG into chylomicrons or VLDL. For example, when palmitate and monooleate were infused intraduodenally into a lymph-fistula rat, 31 % of the TG in the lymph was carried by VLDL as compared to 17 % when oleate-monooleate were infused (Ockner *et al.*, 1969a). In general, investigators have found that the amount of TG transported by VLDL is much less than by chylomicrons (Mansbach and Arnold, 1986b; Tso *et al.*, 1987). This is best illustrated by a study in which the amount of lipid infused intraduodenally into rats was progressively increased. The amount of TG appearing in the lymph as chylomicron-TG increased in parallel with the lipid infusion, whereas the amount of VLDL-TG also increased initially but rapidly reached a plateau so that progressively more of the infused TG was carried by the chylomicrons (Shiau *et al.*, 1985; Tso *et al.*, 1987). This maximizes the efficient use of the surface monolayer surrounding the TG-rich lipoproteins. Because VLDLs are smaller than chylomicrons, it takes more surface area per TG using VLDL as the transport vehicle than when chylomicrons are used. Similarly, when an increasing load of TG is given intraduodenally, the intestinal response is to produce larger chylomicrons, not to increase their number (Hayashi *et al.*, 1990).

5.1. The Intestine Produces Very Low Density Lipoproteins and Chylomicrons by Differing Pathways

L-81 has been utilized to reveal another facet of intestinal lipid absorption. When TG is infused intraduodenally in the face of L-81 co-infusion, chylomicron output into the lymph is drastically reduced (Tso *et al.*, 1980). The question asked was whether a similar effect would occur if the same amount of PC were infused rather than TG. The interesting and unexpected finding was that TG output into the lymph was maintained but the TG, instead of being in chylomicrons, was exported into the lymph as VLDL (Tso *et al.,* 1984). These data support the thesis that the chylomicron and VLDL TG-secretory pathways are separate. This separation had been previously suggested by EM findings in which it was shown that the Golgi contained either chylomicron- or VLDL-sized particles, but not both together (Mahley *et al.,* 1971), by the suggestion that L-81 does not interfere with VLDL excretion (Tso *et al.,* 1981a), and by the finding that VLDL contains significantly more endogenous TG-FA than chylomicrons (Shiau *et al.,* 1985).

6. The Fate of Triacylglycerol that Does Not Enter the Chylomicron Precursor Pool

What happens to the TG that does not enter the chylomicron precursor pool? Clearly it leaves the intestine because even after many hours of lymph collection, the TG that does not exit the intestine in the lymph is nonetheless cleared from the mucosa (Mansbach and Dowell, 1995). There are two possibilities as to what happens with this TG. The first is that it leaves the intestine as TG but in a form different than chylomicrons. This TG would be proposed to exocytose from the basolateral surface of the enterocyte, traverse the lamina propria, and enter the portal circulation. Evidence for this pathway comes from data in rats in which cannulas in both the carotid artery (superior mesenteric artery surrogate) and the portal vein enabled the establishment of an arterial–portal venous gradient. An ultrasonic sensor measured flow in the portal vein (Mansbach *et al.,* 1991) so that the mass of TG (and/ or FA) that was delivered to the portal vein by the intestine each hour could be calculated under steady state conditions. The resulting data suggested that up to 39 % of the radiolabeled, intraduodenally infused TO was transported via the portal vein presumably to the liver. Much of this was TG. What lipoprotein or other form the TG was in is unknown. In snakes, portomicrons carry TG to the liver from the intestine, and it is possible that some variant of this occurs in this rat model as well. Further evaluation of this problem is made difficult by the relatively small venous–aterial gradients that translate into large amounts of mass fluxed due to the high flow state in the portal vein.

The second possibility to rid the intestine of TG that is not exported in chylomicrons is for the mucosal TG-endogenous-FA to be hydrolyzed to fatty acids and for these to exit the intestine via the portal vein route. Indeed, a striking finding in the portal vein studies was the finding that a considerable amount of endogenous FAs were transported by the portal vein to the liver during fat absorption (Mansbach *et al.,* 1991). In considering potential lipases that could hydrolyze the ER bound TG, two are possible. The first is a microsomal lipase first described by Coleman and Haynes (1983) and subsequently purified as a 60 kDa protein by Lehner and Verger (1997). The enzyme is quite active at 240 nmol/min/mg prot. but it is not nearly as active as colipase stimulated, pancreatic lipase, 4,500 μmol/min/mg prot. (Verger *et al.,* 1969). Although the microsomal lipase was purified from liver, it is known to be present in the intestine as well (Coleman and Haynes, 1983) with a specific activity that is greater than that of liver in the unpurified state. This enzyme is displayed on the cytosolic hemileaflet (Coleman and Haynes, 1983) and thus is geographically well localized to hydrolyze TG that remains on the cytosolic face of the ER.

An additional cytosolic lipase described previously (Rao and Mansbach, 1993) could also perform the hydrolytic function. In either case, FAs would be produced whose composition would be different than that of the diet and these would be expected to exit the cell quickly so that their intracellular concentration does not build up.

7. The Intestine Has Two Pools of Triacylglycerol

In the intestinal cell there are two pools of TG, one of which subserves chylomicron formation (pool A) and the other of which has been termed the storage pool (pool B). It is TG from the storage pool that provides the TG and FA that exit the intestine in the portal vein. These pools turn over at different rates and are compositionally different with respect to their TG-FAs. Pool A turns over at a rate of 0.6/hr; pool B is slower (Mansbach and Arnold, 1986b). When lymph TG output is increased either by fat prefeeding rats or by including PC in an intraduodenal TO infusion, the turnover rate of pool A does not change. In order to increase the output rate therefore, it is evident that the pool size of pool A must increase. When tested, this occurs. For example, rats infused intraduodenally with TO plus PC have a mucosal TG-specific activity that is 78 % of the [^{3}H]-TO infused as compared to those infused with TO alone in which mucosal TG-specific activity is 46 % of the infusate specific activity (Parlier *et al.,* 1989) despite the TG content of the mucosa remaining the same under the two infusion conditions (Mansbach and Arnold, 1986b). This means that the majority of the mucosal TG under the first infusion condition is similar to that expected for chylomicron formation, whereas the TG in the second is distributed more evenly between that which would be expected to exit the cell in chylomicrons and that which would not. If the TG in pool A is increased in size when PC is included in the intraduodenal TO infusion, and the total amount of mucosal TG remains the same (Mansbach and Arnold, 1986b), then the amount of TG in pool B should be reduced as compared to when only TO is infused. This is what has been found (Mansbach and Nevin, 1994). Furthermore, the end result of a smaller pool B when PC and TO are infused should be a reduced flux of FAs into the portal vein (Mansbach and Dowell, 1993) as a result of the hydrolysis of the pool B TG (Tipton *et al.,* 1989). As expected, the amount of FA appearing in the portal vein approached zero under these infusion conditions. In sum, these data make a compelling case for the presence of two intestinal pools of TG that reciprocally increase or decrease in size depending on the amount of TG that is exported in chylomicrons. If chylomicron-TG output increases, then pool A increases in size and pool B decreases and vice versa.

8. Physiological Results of the Two Triacylglycerol Transport Pathways

The data described previously indicate that the intestine can regulate the extent to which it exports mucosal TG as chylomicron-TG or as TG or FA that enter the portal system. Further, it seems likely that the regulatory element is the availability of PC for chylomicron formation. The question arises as to the physiological consequences of TG exported as chylomicrons versus the increase in VLDL output by the liver that results from the TG and/or FA that arrive there from the intestine. Recent data from transgenic mice have enabled at least a partial answer (Young *et al.,* 1995). In these genetically engineered animals, the mouse apoB gene was altered such that mouse apoB is not expressed. These mice died *in utero*. When the human apoB gene was inserted in these mice as a transgene, apoB was expressed in the liver but not the intestine and thus the mice could not form chylomicrons. One of the interesting results of this model is that the circulating high-density lipoprotein (HDL) levels were only 36 % of normal values. Because one of the metabolic products of chylomicron metabolism is HDL (Redgrave and Small, 1979), one interpretation of the transgenic data is that the HDL concentrations are reduced as a consequence of the lack of chylomicron production. This would argue that chylomicron metabolism is a major contributor to plasma HDL. HDL is the reverse cholesterol transporter in the plasma and is associated with a reduction in coronary disease. By contrast, the end product of VLDL metabolism is low-density lipoprotein (LDL), the major cholesterol-carrying vehicle in the plasma and the one associated with the development of atherosclerotic disease. Therefore, whether or not the absorbed lipid is transported from the intestine as chylomicrons may have consequences for HDL production.

References

Ament, M. E., Shimoda, S. S., Saunders, D. R., and Rubin, C. E., 1972, Pathogenesis of steatorrhea in three cases of small intestinal stasis syndrome, *Gastroenterol.* **63**:728–747.

Bennett-Clark, S., 1978, Chylomicron composition during duodenal triglyceride and lecithin infusion. *Am. J. Physiol.* **235**:E183–E190.

Bennett-Clark, S., Lawergren, B., and Martin, J., 1973, Regional intestinal absorptive capacities for triolein: An alternative to markers. *Am. J. Physiol.* **225**:574–585.

Coleman, R., and Bell, R. M., 1978, Evidence that biosynthesis of phosphatidylethanolamine, phosphatidylcholine and triacylglycerol occurs on the cytoplasmic side of microsomal vesicles, *J. Cell. Biol.* **76**:245–253.

Coleman, R. A., and Haynes, E. B., 1983, Differentiation of microsomal from lysosomal triacylglycerol lipase activities in rat liver, *Biochim. Biophys. Acta.* **751**:230–240.

Coleman, R. A., and Haynes, E. B., 1985, Subcellular location and topography of rat hepatic monoacylglycerol acyltransferase activity, *Biochim. Biophys. Acta* **834**:180–187.

Friedman, H. I., and Cardell, R. R., 1977, Alterations in the endoplasmic reticulum and Golgi complex of intestinal epithelial cells during fat absorption and after termination of this process: A morphological and morphometric study, *Anat. Rec.* **188**:77–101.

Gangl, A., and Ockner, R. K., 1975, Intestinal metabolism of plasma free fatty acids: Intracellular compartmentation and mechanisms of control, *J. Clin. Invest.* **55**:803–813.

Gangl, A., and Renner, F., 1978, *In vivo* metabolism of plasma free fatty acids by intestinal mucosa of man, *Gastroenterol.* **74**:847–850.

Glickman, R. M., Perrotto, J., and Kirsch, K., 1976, Intestinal lipoprotein formation: Effect of colchicine, *Gastroenterol.* **70**:347–352.

Gordon, D. A., Jamil, H., Gregg, R. E., Olofsson, S.-O., and Borén, J., 1996, Inhibition of the microsomal triglyceride transfer protein blocks the step of apolipoprotein B lipoprotein assembly but not the addition of bulk core lipids in the second step, *J. Biol. Chem.* **271**:33047–33053.

Green, P. H. R., and Glickman, R. M., 1981, Intestinal lipoprotein metabolism, *J. Lipid. Res.* **22**:1153–1173.

Haghpassand, M., Wilder, D., and Moberly, J. B., 1996, Inhibition of apolipoprotein B and triglyceride secretion in human hepatoma cells (HepG2), *J. Lipid. Res.* **37**:1468–1480.

Halpern, J., Tso, P., and Mansbach, C. M., II, 1988, The mechanism of lipid mobilization by the small intestine after transport blockade, *J. Clin. Invest.* **82**:74–81.

Hamilton, J. A., and Small, D. M., 1981, Solubilization and localization of triolein in phosphatidylcholine bilayers: A ^{13}C NMR study, *Proc. Natl. Acad. Sci.* **78(11)**:6878–6882.

Hamilton, R. L., Guo, L. S., Felker, T. E., Chao, Y. S., and Havel, R. J., 1986, Nascent high density lipoproteins from liver perfusates of orotic acid-fed rats, *J. Lipid Res.* **27**:967–978.

Hamilton, R. L., Wong, J. S., Cham, C. M., Nielsen, L. B., and Young, S. B., 1998, Chylomicron-sized lipid particles are formed in the setting of apolipoprotein B deficiency, *J. Lipid Res.* **39**:1543–1557.

Hayashi, H., Nutting, D. F., Fujimoto, K., Cardelli, J. A., Black, D., and Tso, P., 1990, Transport of lipid and apolipoproteins A-I and A-IV in intestinal lymph of the rat, *J. Lipid Res.* **31:**1613–1625.

Hsu, K.-T., and Storch, J., 1996, Fatty acid transfer from liver and intestinal fatty acid-binding proteins to membranes occurs by different mechanisms, *J. Biol. Chem.* **271:**13317–13323.

Hulsmann, W. C., and Kupershook-Davidov, R., 1976, Topographic distribution of enzymes involved in glycerolipid synthesis in rat small intestinal epithelium, *Biochim. Biophys. Acta* **450:**288–300.

Jersild, R. A., Jr., 1966, A time sequence study of fat absorption in the rat jejunum, *Am. J. Anat.* **118:**135–161.

Jersild, R. A., and Clayton, R. T., 1971, A comparison of the morphology of lipid absorption in the jejunum and ileum of the adult rat, *Am. J. Anat.* **131:**481–504.

Knox, R., Stein, I., Levinson, D., Tso, P., and Mansbach, C. M., II, 1991, Effect of fat prefeeding on bile flow and composition in the rat, *Biochim. Biophys. Acta* **1083:**65–70.

Kumar, N. S., and Mansbach, C. M., II, 1997, Determinants of triacylglycerol transport from the endoplasmic reticulum to the Golgi in intestine, *Am. J. Physiol.* **273:**G18–G30.

Kumar, N. S., and Mansbach, C. M., II, 1999, Prechylomicron transport vesicle: Isolation and characterization, *Am. J. Physiol.* **276:**G378–G386.

Lehner, R., and Kuksis, A., 1995, Triacylglycerol synthesis by purified triacylglycerol synthetase of rat intestinal mucosa, *J. Biol. Chem.* **270:**13630–13636.

Lehner, R., and Verger, R., 1997, Purification and characterization of a porcine liver microsomal triacylglycerol hydrolase, *Biochem.* **36:**1861–1868.

Loirdighi, N., Delvin, M. D., Delvin, D., and Levy, E., 1997, Selective effects of hydrocortisone on intestinal lipoprotein and apolipoprotein synthesis in the human fetus, *J. Cell. Biochem.* **66:**65–76.

Mahan, J., Heda, G. D., Rao, R. H., and Mansbach, C. M., 2000, The intestine expresses pancreatic triacylglycerol lipase: regulation by dietary lipid. Submitted.

Mahley, R. W., Bennett, B. D., Morre, D. J., Gray, M. E., Thistlethwaite, W., and LeQuire, V. S., 1971, Lipoproteins associated with Golgi apparatus isolated from epithelial cells of rat small intestine, *Lab, Invest.* **25:**435–444.

Mansbach, C. M., II, 1972, Lysolecithin acyltransferase in hamster intestinal mucosa, *Lipids* **7:**593–595.

Mansbach, C. M., II, 1973, Complex lipid synthesis in hamster intestine, *Biochim. Biophys. Acta* **296:**386–400.

Mansbach, C. M., II, 1977, The origin of chylomicron phosphatidylcholine in the rat, *J. Clin. Invest.* **60:**411–420.

Mansbach, C. M., II, and Arnold, A., 1986a, CTP: Phosphocholine cytidyltransferase in intestinal mucosa, *Biochim. Biophys. Acta* **875:**516–524.

Mansbach, C. M., II, and Arnold, A., 1986b, Steady-state kinetic analysis of triacylglycerol delivery into mesenteric lymph, *Am. J. Physiol.* **251:**G263–G269.

Mansbach, C. M., II, and Dowell, R. F., 1992, Uptake and metabolism of circulating fatty acids by rat intestine, *Am. J. Physiol.* **261:**G927–G933.

Mansbach, C. M., II, and Dowell, R. F., 1993, Portal transport of long acyl chain lipids: Effect of phosphatidylcholine and low infusion rates, *Am. J. Physiol.* **264:**G1082–G1089.

Mansbach, C. M., II, and Dowell, R. F., 1995, The role of the intestine in chylomicron remnant clearance, *Am. J. Physiol.* **269:**G144–G152.

Mansbach, C. M., II, and Dowell, R. F., 2000, Effect of increasing lipid loads on the ability of the endoplasmic reticulum to transport lipid to the Golgi, *J. Lipid Res.* **41:**605–612.

Mansbach, C. M., II, and Nevin, P., 1994, Effect of Brefeldin A on lymphatic triacylglycerol transport in the rat, *Am. J. Physiol.* **266:**G292–G302.

Mansbach, C. M., II, and Nevin, P., 1998, Intracellular movement of triacylglycerols in the intestine, *J. Lipid Res.* **39:**963–968.

Mansbach, C. M., II, and Partharsarathy, S., 1979, Regulation of *de novo* PC synthesis in rat intestine, *J. Biol. Chem.* **254:**9688–9694.

Mansbach, C. M., II, and Partharsarathy, S., 1982, A re-examination of the fate of glyceride glycerol in neutral lipid absorption and transport, *J. Lipid Res.* **23:**1009–1019.

Mansbach, C. M., II, and Parthasarathy, S., 1979, Regulation of *de novo* phosphatidylcholine synthesis in rat intestine, *J. Biol. Chem.* **254:**9688–9694.

Mansbach, C. M., II, Arnold, A., and Garrett, M., 1987, Effect of chloroquine on intestinal lipid metabolism, *Am. J. Physiol* **253:**G673–G678.

Mansbach, C. M., II, Dowell, R. F., and Pritchett, D., 1991, Portal transport of absorbed lipids in the rat, *Am. J. Physiol.* **261:**G530–G538.

Mehran, M., Seidman, E., Gurbindo, C., and Levy, E., 1994, Lipid esterification and synthesis by the IEC-6 intestinal epithelial crypt cell line: Effect of transforming growth factor beta, *Can. J. Physiol. Pharmacol.* **72:**1272–1276.

Mehran, M., Seidman, E., Marchand, R., Gurbindo, C., and Levy, E., 1995, Tumor necrosis factor-a inhibits lipid and lipoprotein transport by Caco-2 cells, *Am. J. Physiol.* **269:**G953–G960.

Mjos, O. D., Faergeman, O., Hamilton, R. L., and Havel, R. J., 1975, Characterization of remnants produced during the metabolism of triglyceride-rich lipoproteins of blood plasma and intestinal lymph in the rat, *J. Clin. Invest.* **56:**603–615.

Mooney, R. A., and Lane, M. D., 1981, Formation and turnover of triglyceride-rich vesicles in the chick liver cell, *J. Biol. Chem.* **256:**11724–11733.

Moreau, P., Rodriguez, M., Cassagne, C., Morre, D. M., and Morre, D. J., 1991, Trafficking of lipids from the endoplasmic reticulum to the Golgi apparatus in a cell-free system from rat liver, *J. Biol. Chem.* **266:**4322–4328.

Nemeth, A., Myrdal, U., Veress, B., Bergland, L., and Angelin, B., 1995, Studies on lipoprotein metabolism in a family with jejunal chylomicron retention, *Eur. J. Clin. Invest.* **25:**271–280.

Nilsson, A., and Landin, B., 1988, Metabolism of chylomicron arachidonic and linoleic acid in the rat, *Biochim. Biophys. Acta* **959:**288–295.

Ockner, R. K., Hughes, F. B., and Isselbacher, K. J., 1969a, Very low density lipoproteins in intestinal lymph: Origin, composition, and role in lipid transport in the fasting state, *J. Clin. Invest.* **48:**2079–2088.

Ockner, R. K., Hughes, F. B., and Isselbacher, K. J., 1969b, Very low density lipoproteins in intestinal lymph: Role in triglyceride and cholesterol transport during fat absorption, *J. Clin. Invest.* **48:**2367–2373.

Owen, M. R., Cortsorphine, C. G., and Zamimit, V. A., 1997, Overt and latent activities of diacylglycerol acyltransferase, *Biochem. J.* **323:**17–21.

Parlier, R. D., Frase, S., and Mansbach, C. M., II, 1989, The intraenterocyte distribution of absorbed lipid and effects of phosphatidylcholine, *Am. J. Physiol.* **256:**G349–G355.

Rao, R. H., and Mansbach, C. M., II,1991, Intestinal alkaline lipase: Purification and properties, *FASEB J.* **5:**A1466.

Rao, R. H., and Mansbach, C. M., II, 1993, Alkaline lipase in rat intestinal mucosa: Physiological parameters, *Arch. Biochem. Biophys.* **304:**483–489.

Redgrave, T. G.,1970, Formation of cholesterol ester-rich particulate lipid during metabolism of chylomicrons, *J. Clin. Invest.* **49:**465–471.

Redgrave, T. G., and Small, D. M., 1979, Quantitation of the transfer of surface phospholipid of chylomicrons in the rat, *J. Clin. Invest.* **64:**162–171.

Rothman, J. E., and Orci, L., 1992, Molecular dissection of the secretory pathway, *Nature* **355:**409–415.

Sabesin, S. M., and Frase, S., 1977, Electron microscopic studies of the assembly, intracellular transport, and secretion of chylomicrons by rat intestine, *J. Lipid Res.* **18:**456–511.

Sabesin, S. M., Clark, S. B., and Holt, P. R., 1977, Ultrastructural features of regional differences in chylomicron secretion by rat intestine, *Exper. Mol. Pathol.* **26:**277–289.

Scow, R., Stein, Y., and Stein, O., 1967, Incorporation of dietary lecithin and lysolecithin into lymph chylomicrons in the rat, *J. Biol. Chem.* **242:**4919–4924.

Shiau, Y.-F., Popper, D. A., Reed, M., Capuzzi, D., and Levine, G. M., 1985, Intestinal triglycerides are derived from both endogenous and exogenous sources, *Am. J. Physiol.* **248:**G164–G169.

Tipton, A. D., Frase, S., and Mansbach, C. M., II, 1989, The isolation and characterization of a mucosal triacylglycerol pool undergoing hydrolysis, *Am. J. Physiol* **257:**G871–G878.

Trotter, P. J., and Storch, J., 1991, Fatty acid uptake and metabolism in a human intestinal cell line (Caco-2): Comparison of apical and basolateral incubation, *J. Lipid Res.* **32:**293–304.

Tso, P., Lam, J., and Simmonds, W. J., 1978, The importance of the lysophosphatidylcholine in lymphatic transport of fat, *Biochim. Biophys. Acta* **528:**364–372.

Tso, P., Balint, J. A., and Rodgers, J. B., 1980, Effect of hydrophobic surfactant (Pluronic L-81) on lymphatic lipid transport in the rat, *Am. J. Physiol.* **239:**G348–G353.

Tso, P., Balint, J. A., Bishop, M. B., and Rodgers, J. B., 1981a, Acute inhibition of intestinal lipid transport by Pluronic L-81 in the rat, *Am. J. Physiol.* **241:**G487–G497.

Tso, P., Kendrick, M., Balint, J. A., and Simmonds, W. J., 1981b, Role of biliary phosphatidylcholine in the absorption and transport of dietary triolein in the rat, *Gastroenterol.* **80:**60–65.

Tso, P., Drake, D. S., Black, D. D., and Sabesin, S. M., 1984, Evidence for separate pathways of chylomicron and very low density lipoprotein assembly and secretion by rat small intestine, *Am. J. Physiol.* **247**:G599–G610.

Tso, P., Pitts, V., and Granger, N., 1985, Role of lymph flow in intestinal chylomicron transport, *Am. J. Physiol.* **249**:G21–G28.

Tso, P., Lindström, M. B., and Borgström, B., 1987, Factors regulating the formation of chylomicrons and very low density lipoproteins by the rat small intestine, *Biochim. Biophys. Acta* **922**:304–313.

Tsuchiya, T., Kalogeris, T. J., and Tso, P., 1996, Ileal transportation into the upper jejunum affects lipid and bile salt absorption in rats, *Am. J. Physiol.* **271**:G681–G691.

Turley, S. D., and Dietschy, J. M., 1979, Regulation of biliary cholesterol output in the rat: Dissociation from the rate of hepatic cholesterol synthesis, the size of the hepatic cholesteryl ester pool, and the hepatic uptake of chylomicron cholesterol, *J. Lipid Res.* **20**:923–934.

Verger, R., DeHaas, G. H., Sarda, L., and Desnuelle, P., 1969, Purification from porcine pancreas of 2 molecular species with lipase activity, *Biochim. Biophys. Acta* **188**:272–282.

Walsh, J. P., and Bell, R. M., 1986, Sn-1,2-diacylglycerol kinase of *Escherichia coli*, *J. Biol. Chem.* **261**:6239–6247.

Wu, A.-L., and Clark, S. B., 1976, Resistance of intestinal triglyceride transport capacity in the rat to adaptation to altered luminal environment, *Am. J. Clin. Nutr.* **29**:157–168.

Wu, A. L., Bennett-Clark, S., and Holt, P. R., 1975, Transmucosal triglyceride transport rates in proximal and distal rat intestine *in vivo, J. Lipid Res.* **16**:251–257.

Yang, L. Y., Kuksis, A., and Myher, J. J., 1995a, Biosynthesis of chylomicron triacylglycerols by rats fed glyceryl or alkyl esters of menhaden oil fatty acids, *J. Lipid Res.* **36**:1046–1057.

Yang, L.-Y., Kuksis, A., Myher, J. J., and Steiner, G., 1995b, Origin of triacylglycerol moiety of plasma very low density lipoproteins in the rat: Structural studies, *J. Lipid Res.* **36**:125–136.

Young, S. G., Cham, C. M., Pitas, R. E., Burri, B. J., Connolly, A., Glynn, S., Pappu, A. S., Wong, J. S., Hamilton, R. L., and Farese, R. W., Jr., 1995, A genetic model for absent chylomicron formation: Mice producing apolipoprotein B in the liver, but not the intestine, *J. Clin. Invest.* **96**:2932–2946.

CHAPTER 13

Regulation of Intestinal Cholesterol Metabolism

F. JEFFREY FIELD

1. Introduction

The amount of unesterified cholesterol within a mammalian cell is tightly controlled. Cells lack a mechanism to degrade cholesterol and excess accumulation of this sterol will cause cell death. The means by which the cell procures cholesterol, therefore, must be closely regulated. Importantly, however, cholesterol is also required for a multitude of critical cellular functions. Thus, it is vital that the cell maintain an ample and continuous supply of this sterol. For example, cholesterol is necessary for normal membrane structure and function; it is required for cell growth; cholesterol regulates the activities of key enzymes and membrane transport proteins; it is a precursor for vitamins, bile acids, and steroid hormones; and in liver and intestine, it is important for normal lipoprotein synthesis and secretion. This critical balancing act by the cell to have sufficient sterol to function normally but not too much to cause cholesterol excess has implications for the development of atherosclerotic heart and peripheral vascular disease, stroke, and cholelithiasis. Because cholesterol is central to these disease states and, therefore, has an important impact on the general health of a population, the regulation of cholesterol metabolism has been extensively studied.

Cholesterol requirements of most cells are met by two separate but interrelated processes. One process is the endogenous synthesis of cholesterol. This synthetic pathway, which involves over 20 separate reactions originating from one molecule of acetyl-CoA, is regulated primarily by the activity of 3-hydroxy-3-methylglutaryl coenzyme A (HMG-CoA) reductase, which catalyzes the formation of mevalonate from HMG-CoA (Bloch, 1965; Brown and Goldstein, 1980). The second process involves the utilization of circulating lipoprotein cholesterol. Lipoproteins bound to a specific cell surface receptor are internalized and degraded, releasing cholesterol that can then be used by the cell. Under most circumstances, the endogenous synthesis of cholesterol is required only if lipoprotein cho-

F. JEFFREY FIELD • Department of Internal Medicine, University of Iowa College of Medicine, Iowa City, Iowa 52242.

Intestinal Lipid Metabolism, edited by Charles M. Mansbach II *et al.,* Kluwer Academic/Plenum Publishers, 2001.

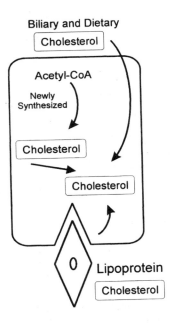

Figure 1. Access to cholesterol by intestinal absorptive cells.

lesterol is insufficient to meet the sterol requirements of the cell (Brown and Goldstein, 1986). Another critical protein for maintaining normal cellular cholesterol homeostasis is acylcoenzyme A: cholesterol acyltransferase (ACAT). This enzyme catalyzes the formation of cholesteryl esters from unesterified cholesterol and CoA-derivatives of fatty acids, thereby allowing the cell to store potentially toxic unesterified cholesterol as harmless cytoplasmic cholesteryl ester lipid droplets (harmless, that is, unless the cholesteryl esters are located in subendothelial macrophages in your left anterior descending coronary artery).

The small intestine is very important in maintaining total body cholesterol homeostasis. Rates of intestinal cholesterol synthesis and the total net amount of cholesterol synthesized there are second only to the liver. Moreover, under conditions of increased absorption of cholesterol, the small intestine may actually contribute more newly synthesized cholesterol to the body than does the liver (Dietschy and Siperstein, 1967; Spady and Dietschy, 1983). Newly synthesized cholesterol from the small intestine contributes to the plasma pool of cholesterol, indicating that changes in intestinal cholesterol production will lead to changes in plasma cholesterol levels (Lindsey and Wilson, 1965). The intestine is also responsible for the absorption of dietary and biliary-derived cholesterol. No other organ or cell is exposed to cholesterol in this way. Thus, the intestinal cell is unique in that, in addition to acquiring cholesterol like other cells by synthesis of cholesterol and uptake of lipoproteins, the absorptive cell can also utilize dietary or biliary cholesterol to maintain a sterol balance (Fig. 1). In response to an influx of fatty acids, the intestine assembles and secretes triacylglycerol-rich lipoprotein particles. Because unesterified and esterified cholesterol are used for surface and core components of the lipoprotein particle, respectively, cholesterol requirements of the intestinal cell are challenged following the ingestion of a fatty meal to ensure normal synthesis of lipoproteins. In addition to the functions of cholesterol synthesis, absorption, and lipoprotein production, the intestine is also a site for the uptake and degradation of plasma lipoproteins. In rodents, for example, the intestine is second to the liver in the clearance of circulating low-density-lipoproteins (LDL; Spady *et al.,* 1983)

In this chapter, I review what is currently known about the regulation of cholesterol metabolism in the intestine. Although significant progress has been made regarding the regulation of cholesterol synthesis and esterification at the cellular and the molecular levels, very few of these studies have been performed in intestine. It is not possible to apply information obtained from other cells to an intestinal cell as the enterocyte is unique, particularly in the way it obtains cholesterol. Moreover, factors that have been clearly shown to regulate cholesterol metabolism in hepatocytes and in other cell types are not as clearly defined as modulators in enterocytes. Results of studies performed in intestine are often conflicting and results from different laboratories are often difficult to compare. This problem is not one of investigators in the field, but rather of the complexity of the organ being studied and of species differences. The rather marked degree of variability of basal rates of cholesterol synthesis and regulation of cholesterol metabolism among species have been well documented (Turley and Dietschy, 1988). Unlike the cellular makeup of the liver, which is quite homogenous, the cellular makeup of the small intestine is very complex. Moreover, the function of the intestine differs along its length. For example, cholesterol and fat-soluble vitamins are absorbed in the proximal intestine, whereas similarly structural molecules, such as bile acids, are taken up in the distal small bowel (Sylvin and Nordstrom, 1970; Dietschy *et al.,* 1966). Functional differences also exist along the vertical axis of intestinal mucosa. Cells within the crypt are actively dividing and are destined to replenish the soon-to-be-lost senescent cells of the upper villus. Cells within crypts do not have access to lumenal contents and do not take part in the absorption of nutrients. In contrast, cells located in the middle to upper regions of the villus are not dividing and are exposed to lumenal contents. What is quite clear from all of this is that the intestine is not a simple organ to study. When the regulation of intestinal cholesterol metabolism is investigated, species differences and the regional and cellular differences within the intestine must be considered.

There are other inherent problems that must be addressed before interpreting results from studies performed in intestine. In many studies, subcellular fractionation is performed to isolate microsomal fractions to determine activities of HMG-CoA reductase and ACAT. Standard techniques for preparing hepatic microsomes are often used to prepare intestinal microsomes. Because of lumenal debris and thick tenacious mucus, intestinal microsomes tend to be heavily contaminated with other unwanted organelles (Field *et al.,* 1982a). Purity from preparation to preparation becomes a significant problem. The confusion that still exists as to the cellular localization of intestinal HMG-CoA reductase and ACAT can be partially explained by this technical problem.

The preparation of intestinal cells to perform metabolic studies is also not an easy task. The age-old technique of "gently scraping" the intestinal mucosa from the underlying muscular layer with a microscope slide is neither gentle nor precise. The scrapings are a mixture of intestinal, mononuclear, endothelial, and muscular cells, and if this technique is used to prepare cells from different segments of the intestine, it can lead to variability in the amount of intestinal cells recovered. Isolated cells can be prepared using buffers containing ethylenediaminetetraacetic acid (Weiser, 1973; Raul *et al.,* 1977). With this technique, different cell populations along the vertical axis of the villus can be recovered with fewer unwanted cells. The major problem with this technique is that soon after the cells are removed from the villus, they begin to die, making metabolic studies very difficult to perform (Shakir *et al.*, 1978; Field and Mathur, 1983b). In contrast, intestinal explants or strips remain viable for longer periods of time and metabolic studies have been performed for up to 24 hours (Gebhard and Cooper, 1978; Field *et al.*, 1982b; Herold *et al.*, 1984). These preparations, however, do not circumvent the problem of having other contaminating cells within the preparation. These unwanted cells could respond to modulators of cholesterol metabolism quite unlike intestinal cells. Until recently, intestinal cell lines have never gained much popularity for the study of cholesterol metabolism. Earlier available cell lines represented cells of the crypt region rather than absorptive cells of the upper villus. These cells also lacked polarity, did not have brush borders, and had very little capacity for absorption (Quaroni and May, 1980; Quaroni *et al.*, 1979). Since the late 1980s, a human intestinal cell line called Caco-2 has been used extensively for studies examining a variety of intestinal functions, including studies of cholesterol absorption, cellular cholesterol trafficking, cholesterol metabolism, and lipoprotein assembly and secretion (Field *et al.*, 1987b, 1995b, 1991; Hughes *et al.*, 1987; Field and Mathur, 1995). Because my laboratory has used the Caco-2 cell line to investigate the regulation of intestinal cholesterol metabolism at the cellular and the molecular levels, this chapter includes a section that details the information generated using this cell as a model for the human enterocyte.

2. Localization of Cholesterol Synthesis and Esterification

In localizing cholesterol synthesis along the length of the small intestine, there are apparent species differences. In the rat, cholesterol synthesis was highest in the duodenum, proximal to the entry of bile, and in the distal segment of the ileum (Dietschy and Siperstein, 1965; Stange and Dietschy, 1983a). Immunocytochemical staining of HMG-CoA reductase found maximal staining in the ileum followed by the jejunum and then the duodenum (Li *et al.*, 1988). It was speculated that in the rat, cholesterol synthesizing activity was highest in areas of the small intestine that were not actively involved in cholesterol absorption, suggesting a possible feedback regulation of cholesterol synthesis by lumenal cholesterol, a somewhat controversial concept I will return to later. In contrast, in the rabbit, HMG-CoA reductase activity in the jejunum was higher or equal to that observed in the ileum (Stange *et al.*, 1981a; Field *et al.*, 1987a). The results of two studies performed in humans are conflicting. Gebhard *et al.* (1985) found similar activites of HMG-CoA reductase in the ileum and the jejunum. Dietschy and Gamel (1971), however, observed higher rates of cholesterol synthesis in the ileum compared to either the duodenum or the jejunum.

The data regarding the location of cholesterol synthesis along the vertical axis of the villus are equally conflicting. Although very early reports had suggested that cells located in the upper regions of the villus were almost completely devoid of cholesterol synthesizing activity (Dietschy and Siperstein, 1965), subsequent results have clearly shown substantial amounts of cholesterol being synthesized in cells of the upper villus (Murthy *et al.*, 1988; Stange and Dietschy, 1983a). Stange and Dietschy (1983b) have calculated that in the rat, 70% to 84% of the total cholesterol synthesized by the intestinal epithelium occurred in cells of the lower villus or crypts. Whether this is totally or partially correct, the explanation for why this should be so makes good sense. To provide sufficient sterol for new membranes, the highest demand for new cholesterol should be in cells that are rapidly dividing. Moreover, because cells of the crypt cannot utilize lumenal cholesterol, they must rely on cholesterol biosynthesis and/or lipoprotein cholesterol to meet their needs. In addition, if lumenal cholesterol does inhibit cholesterol synthesis, as is discussed later, cells of the crypts would not be under such regulation. It would be expected, therefore, that HMG-CoA reductase activity would be higher in cells of the crypts compared to cells of the upper villus. There are studies, however, that would argue that cholesterol synthesizing activity is as high or higher in cells of the upper villus. Compared to crypt cells, Merchent and Heller (1977) found twice as much reductase activity and cholesterol synthetic rates in cells of the upper villus. Muroya *et al.* (1977) observed that cholesterol synthesis was similar in cells prepared from crypts and from the upper to middle villus. In rabbit intestine, reductase activities in crypt and in upper villus cells were found to be similar (Stange *et al.*, 1981a).

The conflicting results are likely secondary to the difficulty of preparing pure populations of cells along the villus gradient. To circumvent this problem, Li *et al.* (1988) and Singer *et al.* (1987) used immunohistochemistry to localize HMG-CoA reductase protein along the villus gradient. Both studies demonstrated that cells of the upper villus contained the greatest amount of reductase protein. Although the amount of reductase protein may not reflect cholesterol synthesizing activity, the results suggest that cells of the upper villus have significant cholesterol synthesizing capacity and likely contribute substantially to the amount of newly synthesized cholesterol produced by the intestine.

Previously, when discussing cholesterol esterification within the small intestine, it was necessary to implicate two proteins that were shown to catalyze the esterification of absorbed cholesterol, the endoplasmic reticulum enzyme, ACAT, and the pancreatic-derived, cholesterol esterase. Due to the efforts of David Hui and his group, however, this is no longer required. By developing a cholesterol esterase knockout mouse to investigate the physiological role for cholesterol esterase in cholesterol absorption, these investigators found that compared to wild-type mice, null mice, lacking the cholesterol esterase gene, absorbed similar amounts of unesterified cholesterol (Howles *et al.*, 1996). Thus cholesterol esterase has no role in the absorption or esterification of lumenal cholesterol. In support of previous *in vitro* cell culture data, cholesterol esterase was found to have a primary role in the absorption of cholesteryl esters by catalyzing their hydrolysis to fatty acids and to unesterified cholesterol in the lumen (Huang and Hui, 1990). Although this appears to be all well and good, and the issue looks to be settled, as I discuss later, there still remains controversy as to the identity of intestinal ACAT (Meiner *et al.*, 1996; Meiner *et al.*, 1997).

Like the location of HMG-CoA reductase, the location of ACAT along the length of the intestine is also species dependent. In the rat, the proximal one third of the small intestine had the highest activity with the lowest activity being found in the distal third (Haugen and Norum, 1976). In contrast, in rabbit intestine, ACAT activity was highest in the midgut with lower activities in the proximal and distal segments (Field *et al.*, 1982b; Field and Salome, 1982). The pattern in human intestine is similar to that observed in the rabbit with fourfold higher ACAT activity in the jejunum compared to proximal or distal segments (Helgerud *et al.*, 1981). Unlike HMG-CoA reductase, the distribution of ACAT activity along the villus axis is fairly uniform (Field *et al.*, 1982b; Stange *et al.*, 1983).

As mentioned, the story of intestinal ACAT is not complete. Farese and his group recently generated mice lacking ACAT (Meiner *et al.*, 1996). Unexpectedly, ACAT gene disruption had little effect on hepatic cholesterol ester metabolism and in addition, ACAT-deficient mice absorbed cholesterol normally. It appeared, therefore, that the liver and the intestine likely contained a cholesterol esterifying enzyme or enzymes that differed from the enzyme that was inactivated in the ACAT "knockout" mice. Using *in situ* hybridization and immunoblotting techniques to investigate tissue distribution of ACAT, no expression was observed in the liver or in the intestine (Meiner *et al.*, 1997). The results suggest that multiple cholesterol-esterifying enzymes likely exist in mammals and another distinct ACAT-like enzyme is responsible for cholesterol esterification in the liver and in the intestine. An ACAT homologue with close sequence homology to "original" ACAT, and which is predominately expressed in the intestine and the liver, is presently being investigated as the candidate ACAT gene in these two organs (personal communication, Robert Farese and Lawrence Rudel).

3. Regulation of Cholesterol Metabolism

3.1. Regulation of Cholesterol Synthesis by Lumenal Cholesterol

Unlike cholesterol synthesis in the liver, which promptly and substantially decreases in response to dietary cholesterol, most of the earlier studies performed in rats failed to demonstrate an inhibitory effect of dietary cholesterol on intestinal cholesterol synthesis (Dietschy and Siperstein, 1967; Stange *et al.*, 1983). From the results of these early studies, it was argued that cholesterol in the intestinal lumen did not regulate cholesterol synthesis in the intestinal cell. The inability of dietary cholesterol to inhibit cholesterol synthesis was also demonstrated in intestines of squirrel monkeys and of humans (Dietschy and Wilson, 1968; Dietschy and Gamel, 1971). It was later observed, however, that if the amount of cholesterol that fluxed through the intestine could be maximized by feeding rats cholesterol combined with a bile acid, inhibition of intestinal cholesterol synthesis could be readily demonstrated (Shefer *et al.*, 1973). This suggested that similar to other cells, intestinal cells will respond to an excess of cholesterol influx by decreasing the rate of cholesterol synthesis. In contrast to the results observed in rats, results in guinea pigs, in hamsters, and in rabbits have demonstrated an unequivocal decrease in intestinal cholesterol synthesis in response to dietary cholesterol, suggesting that species differences exist (Stange *et al.*, 1981a; Swann *et al.*, 1975; Ho, 1975; Andersen *et al.*, 1982).

Additional support for the idea that lumenal cholesterol regulated intestinal cholesterol synthesis came from studies in which rats were fed cholestyramine or *a*-olefin-maleic acid copolymer, agents that inhibit the absorption of bile acids and cholesterol, respectively. Under these conditions of reduced cholesterol flux across the intestine, the rates of intestinal cholesterol synthesis were markedly increased in these animals (Stange *et al.*, 1983).

In an attempt to investigate the role of lumenal cholesterol, independent of other lumenal factors, Gebhard and Cooper (1978) added cholesterol solubilized in ethanol to canine intestinal explants. They found a 23% decrease in the activity of HMG-CoA reductase. The direct effect of cholesterol on intestinal cholesterol synthesis was further confirmed in two other studies using intestinal explants from rabbits and from humans (Herold *et al.*, 1984; Gebhard *et al.*, 1985). Later in this chapter, the regulation of cholesterol synthesis at the cellular and molecular level by the influx of micellar cholesterol is discussed in detail. Suffice it to say, results from the later dietary experiments and from experiments performed in explants support a role for lumenal cholesterol in the regulation of intestinal cholesterol synthesis.

3.2. Regulation of Cholesterol Esterification by Lumenal Cholesterol

Data generated on the effect of absorbed cholesterol on intestinal ACAT are much clearer than the data just described for HMG-CoA reductase. Dietary cholesterol has been shown to consistently increase the activity of ACAT within the intestine (Field *et al.*, 1982b; Stange *et al.*, 1983; Field and Salome, 1982; Norum *et al.*, 1977; Drevon *et al.*, 1979). In contrast, decreasing the flux of cholesterol across the intestine by feeding cholestyramine has resulted in a decrease in ACAT activity (Field and Salome, 1982). Most evidence to date suggests that incoming dietary cholesterol expands a particular intracellular pool of cholesterol that is utilized by intestinal ACAT as substrate.

The activity of ACAT, as measured *in vitro* in microsomes prepared from intestinal mucosa, is not at maximal velocity. The amount of cholesterol is limiting. The measurement of ACAT activity under these conditions essentially reflects the amount of cholesterol available to the enzyme as substrate. When exogenous cholesterol is added to intestinal microsomes *in vitro*, the activity of ACAT is increased (Field and Mathur, 1983b; Suckling and Stange, 1985). Changes in ACAT activity under conditions of cholesterol saturation (added exogenous cholesterol) are more likely to reflect regulation of the enzyme and not just changes in substrate supply. There is still some uncertainty, however, whether cholesterol can also activate ACAT independent of its effect on substrate supply. In cells with a limited amount of cholesterol, cholesterol itself has been shown to serve as a physiological activator of the enzyme, suggesting that ACAT is an allosteric enzyme (Cheng *et al.*, 1995). It is speculated that following the binding of cholesterol to ACAT, the protein undergoes a change in configuration to convert the enzyme into a more active form. Confirmation of this speculation must await purification of the enzyme, a feat that has yet to be accomplished.

At the molecular level, ACAT activity does not appear to be regulated at the level of the protein or mRNA. In rabbits, Pape and his colleagues (1995) found that intestinal ACAT mRNA levels were not altered by a diet enriched in cholesterol and in fat. Likewise, Uelmen *et al.* (1995) observed that compared to chow-fed controls, ACAT mRNA levels in intestine of mice fed an atherogenic diet were either unaltered or decreased. The available data clearly establish a role of dietary cholesterol in the regulation of intestinal cholesterol esterification and ACAT activity. The predominant mechanism for this regulation, however, most likely occurs at the level of substrate supply for the enzyme.

3.3. Regulation of Cholesterol Synthesis by Fatty Acids

Most information regarding the regulation of intestinal cholesterol synthesis by fatty acids is derived from dietary studies in which different types of fat are fed to animals. Bochenek and Rogers (1979) investigated the regulation of intestinal HMG-CoA reductase by feeding rats safflower or tripalmitin oil. After 4 weeks on the respective diets, HMG-CoA reductase activities were not altered when compared to animals ingesting a control chow. In contrast, in another study performed in rats, after only 3 days of a 10% corn oil diet, HMG-CoA reductase activities were increased in the jejunum of these animals (Stange *et al.*, 1983). In rabbits, the regulation of intestinal cholesterol synthesis by dietary fat was also dependent upon the type of fat ingested. In animals fed 2% animal fat with small amounts of cholesterol, intestinal cholesterol synthesis was similar to that observed in animals receiving a control diet (Andersen *et al.*, 1982). However, in another study HMG-CoA reductase activities were decreased throughout the small intestine of rabbits ingesting a diet enriched in n-3 fatty acids (menhaden oil) but decreased only in the midgut of animals ingesting cocoa butter, an oil composed mostly of short-chained saturated fatty acids (Field *et al.*, 1987a). Using a different approach to study the effects of individual fatty acids on cholesterol synthesis, Gebhard and Prigge (1981) infused fatty acids into Thiry-Vella fistulae of dogs and measured HMG-CoA reductase activities before and after the perfusion. Surprisingly, they found that except for hexanoic acid (6:0), all fatty acids infused, 8:0, 12:0, 16:0, 18:0, 18:1, and 18:2, increased reductase activity.

3.4. Regulation of Cholesterol Ester Synthesis by Fatty Acids

Bennett-Clark (1979) infused triolein into the duodenums of rats with lymph fistulae and found intestinal ACAT activity to be inhibited. Moreover, chylomicrons prepared from the mesenteric lymph of these animals were deficient in cholesteryl esters. Bennett-Clark postulated that the promotion of triacylglycerol-rich lipoprotein secretion by a large influx of fatty acids led to a decrease in ACAT activity to provide the intestinal cell with more unesterified cholesterol to assemble the lipoprotein particle. A similar observation was made in intestinal explants. Compared to explants cultured in the absence of fatty acids, explants exposed to high concentrations of oleic acid contained significantly less ACAT activity (Field *et al.*, 1982b).

In dietary studies, the regulation of intestinal ACAT is dependent on the type of fat ingested. In two separate studies, the activity of intestinal ACAT was unchanged by diets enriched in corn oil (Field *et al.*, 1982b; Stange *et al.*, 1983). Similarly, diets enriched in more saturated fats such as coconut oil or cocoa butter had little effect on intestinal ACAT (Field *et al.*, 1987a; Field and Salome, 1982). In contrast, diets containing polyunsaturated fatty acids, safflower, and menhaden oils, which caused a marked increase in microsomal membrane linoleic and n-3 fatty acid content, respectively, resulted in increased activities of intestinal ACAT (Field *et al.*, 1987a; Field and Salome, 1982). Independent of microsomal cholesterol content, membranes enriched in polyunsaturated fatty acids increase the activity of intestinal ACAT.

Information about the influence of individual fatty acids on cholesterol metabolism will not be forthcoming from dietary studies, nor will mechanistic answers regarding how fat or fatty acids regulate intestinal cholesterol metabolism. Questions such as these will need to be addressed at the cellular and the molecular levels under rigidly controlled conditions likely afforded only by cell culture systems.

3.5. Phosphorylation State of HMG-CoA Reductase

It has been well documented that HMG-CoA reductase is acutely regulated by reversible phosphorylation of the protein with phosphorylation inactivating the enzyme (Field *et al.*, 1982a; Nordstrom *et al.*, 1977; Beg *et al.*, 1987). The phosphorylation site on HMG-CoA reductase has been identified as a serine residue located right at the end of the conserved catalytic domain close to the C-terminus (Clarke and Hardie, 1990). Phosphorylation, therefore, likely results in a change in the catalytic activity of the enzyme. Although it had been suggested that phosphorylation of HMG-CoA reductase leads to enhanced degradation of the protein, evidence would now suggest that not to be the case (Zammit and Caldwell, 1992). During conventional preparation of microsomes, HMG-CoA reductase is activated by dephosphorylation through a cytosolic phosphoprotein phosphatase. Including a nonspecific phosphatase inhibitor, such as NaF or KF, in buffers used to prepare microsomes, will maintain reductase in its phosphorylated or catalytically inactive state. It is possible, therefore, to estimate the amount of HMG-CoA reductase that exists in its active form *in vivo* by measuring the activity of reductase in microsomes prepared in the presence (latent) and absence (total) of fluoride (Brown *et al.,* 1979).

In rat liver, for example, it has been estimated that only 10% to 20% of HMG-CoA reductase is in its active form *in vivo* (Brown *et al.,* 1979). In contrast, in rat intestine, 50% to 60% of HMG-CoA reductase is thought to be in its active form *in vivo* (Field *et al.,* 1982a; Oku *et al.,* 1984). In human intestine, Gebhard *et al.,* (1985) estimated that 85% of total reductase was in its active form. The data would suggest, therefore, that compared to hepatic HMG-CoA reductase, in which most of the enzyme is its latent form *in vivo,* intestinal reductase is in its active form. This makes good sense in a tissue that is rapidly dividing and requires cholesterol for the immediate transport of lipids.

Although phosphorylation/dephosphorylation of HMG-CoA reductase may be an efficient and rapid way to regulate cholesterol synthesis within cells, there are no data to substantiate that this is an important regulatory mechanism of cholesterol synthesis *in vivo.*

3.6. Phosphorylation State of ACAT

Early reports suggested that like HMG-CoA reductase, ACAT activity could be regulated acutely by phosphorylation/dephosphorylation of the enzyme (Gavey *et al.,* 1983; Suckling *et al.,* 1983). In contrast to reductase, however, phosphorylation was said to activate ACAT, whereas dephosphorylation resulted in inactivation. This was a very attractive hypothesis. It implied that a cell could rapidly regulate the amount of unesterified cholesterol to meet immediate demands. Under conditions of cellular dephosphorylation, HMG-CoA reductase would be activated and at the same time ACAT would be inactivated. This would result in an increase in cellular availability of unesterified cholesterol by stimulating synthesis and preventing esterification. Although an attractive hypothesis, it has been challenged (Field *et al.,* 1984; Corton and Hardie, 1992). Experiments performed in rabbit microsomes failed to demonstrate evidence for the regulation of ACAT activity by phosphorylation–dephosphorylation (Field *et al.,* 1984). Moreover, the predicted amino acid sequence for ACAT derived from its cDNA did not identify potential phosphorylation sites that would be recognized by classical protein kinases (Chang *et al.,* 1993, 1995). Thus, arguments for the phosphorylation–dephosphorylation of ACAT as a mechanism for regulating cholesterol esterification must be suspect.

3.7. Regulation of Cholesterol Metabolism by Lipoproteins

The question of whether the intestinal cell can utilize lipoprotein cholesterol has been of interest to investigators in this field for a long time. Similar to other cell types that rely on lipoprotein cholesterol to meet their sterol requirements, it would not seem unreasonable to assume that the absorptive cell would also have access to this form of cholesterol. The challenge, however, is to design an experiment that would control for the amount of lumenal cholesterol flux while simultaneously altering lipoprotein levels to investigate the regulation of intestinal cholesterol metabolism by circulating plasma lipoproteins.

Anderson and Dietschy (1976) and Panini *et al.* (1979) induced extreme hypolipidemia in rats by injecting the animals with the drug 4-aminopyrazolo(3,4-*d*)pyrimidine. Following the injection, plasma cholesterol levels decreased to approximately 5 mg/dL. In several organs, including the small intestine, rates of cholesterol synthesis were induced. In addition, if lipoproteins were infused into animals receiving the drug, the increase in cholesterol synthesis in intestine could be partially blunted. These results demonstrated that, similar to other cells, intestinal cells also respond to the deprivation of lipoprotein cholesterol by increasing their synthesis of cholesterol.

In isolated Thiry-Vella fistula, Gebhard and Prigge (1981) showed that HMG-CoA reductase activity was suppressed in fistulae of dogs fed a high-cholesterol diet, suggesting that cells within the isolated segment were taking up increased amounts of lipoprotein cholesterol causing a decrease in cholesterol synthesis. In another study, a parabiont rat model was used to investigate whether lipoprotein cholesterol would regulate intestinal cholesterol synthesis (Purdy and Field, 1984). Paired rats were surgically joined to establish a common circulation. One of the paired animals received an atherogenic diet to induce hyperlipidemia. In the intestine of the paired animals ingesting a control diet but having significant hypercholesterolemia, HMG-CoA reductase activities were decreased and ACAT activities increased.

The combined results of these four studies strongly suggest that cells of the small intestine can take up lipoproteins and respond appropriately to lipoprotein cholesterol by altering cholesterol synthetic rates. What these studies do not do, however, is provide information as to which lipoproteins are regulatory or even if classical cellular lipoprotein receptors are involved in this process. Studies have been done to address this issue. In canine intestinal explants used immediately after preparation, Gebhard and Cooper (1978) were unable to show that either LDL or HDL regulated HMG-CoA reductase activity. In contrast, if intestinal explants were first preincubated in medium deficient in lipoproteins prior to adding LDL or HDL, then HMG-CoA reductase activity could be suppressed. The results would suggest that intestinal lipoprotein receptors are likely low in number and cholesterol deprivation is required to induce the receptor in order to demonstrate a functional response to added LDL or HDL.

Using isolated cells prepared from rat intestine, Suzuki *et al.* (1983) demonstrated saturable binding, internalization, and degradation of labeled LDL or HDL. The receptors for LDL were not the same as for HDL as LDL bound to an apolipoprotein B, E receptor, whereas HDL shared two binding sites that interacted with apo B and apo E or apo A. Similarly, Sviridov *et al.* (1986) demonstrated that HDL-3 was taken up by human isolated intestinal cells by an high-affinity, saturable receptor distinct from the LDL receptor. It was suggested that apo A-I and/or apo A-II were responsible for the binding of HDL-3 to the intestinal cell surface (Sviridov *et al.,* 1992). Moreover, in another study, specific and saturable binding of HDL-3 to basolateral membranes prepared from rat intestinal cells supported results demonstrating that the HDL-3 receptor was distinct from the LDL apo B, E receptor (Kagami *et al.,* 1984). It has also been proposed that specific binding of HDL-3 to its receptor is a means to promote intestinal cholesterol efflux via a novel retroendocytosis pathway (Rogler *et al.,* 1991).

Further support for the notion that small intestinal cells have LDL receptors comes from the work of Fong *et al.* (1989) who demonstrated LDL receptor protein in rat intestinal cells using immunohistochemistry techniques. They found that LDL receptors were present throughout the small intestine both in villus crypts and in upper regions of the villus. In agreement with these results, mRNA for the LDL receptor has also been demonstrated in rat small intestine (Fong *et al.,* 1995). Taken together, the results would suggest that small intestinal cells do have specific receptors for LDL and for HDL and by taking up lipoprotein cholesterol, cholesterol synthetic rates can be regulated. Whether this is an important pathway for the intestinal cell to obtain cholesterol remained in question. To address this, Stange and Dietschy (1983b) studied the clearance rates of labeled sucrose-LDL by rat intestine under a variety of conditions that would drastically alter the availability of lumenal cholesterol. The results of this study demonstrated that the intestinal cell responded to changes in lumenal cholesterol flux by altering its rate of cholesterol synthesis. Despite marked changes in cholesterol flux across the intestinal cell from the lumen, LDL uptake rates by the intestine were unchanged. The authors argued that the contribution of lipoprotein cholesterol to intestinal cell cholesterol metabolism was minimal at best.

3.8. Regulation of Cholesterol Metabolism by Hormones

Little information exists regarding the regulation of intestinal cholesterol metabolism by hormones. What information does exist is conflicting. Feingold (1989) has reviewed the regulation of intestinal cholesterol metabolism in diabetes and the effects of hyper- and hypothyroidism on cholesterol synthesis, and esterification in intestine have also been studied (Field *et al.,* 1986). Although these metabolic disease states do cause changes in intestinal cholesterol metabolism, the results of these studies do not particularly help us understand whether specific hormones regulate cholesterol metabolism within the intestine or whether the ensuing metabolic derangements are responsible. Thus, these studies are not discussed.

Goodman *et al.* (1981) studied the regulation of intestinal cholesterol synthesis by glucagon and insulin in canine intestinal explants. Glucagon, at a concentration of $10-8$ M, significantly decreased HMG-CoA reductase activity and cholesterol synthesis, whereas insulin ($10-6$ M) had the opposite effect, increasing reductase activity and cholesterol synthesis. In *in vitro* assays of HMG-CoA reductase, however, neither hormone demonstrated a direct effect on the enzyme. In biopsies taken from Thiry-Vella fistulae of dogs that were infused intravenously with glucagon or insulin, HMG-CoA reductase activities were decreased following glucagon and were increased following insulin, supporting results from cultured explants (Goodman *et al.,* 1981). In another study, performed in rabbit intestinal explants, however, neither glucagon nor insulin altered the activity of HMG-CoA reductase. Two other hormones were studied. Triamcinolone decreased HMG-CoA reductase activity and triiodothyronine increased activity (Stange *et al.,* 1981b). Again, there may be a problem with species differences but, needless to say, more data will be necessary before conclusions can be drawn regarding the regulation of intestinal cholesterol synthesis by hormones.

4. Regulation of Cholesterol Metabolism in Caco-2 Cells

4.1. Regulation of HMG-CoA Reductase Activity

4.1.1. Sterol Flux

Because it is difficult to control for the amount of cholesterol flux through the small intestine, results of dietary studies investigating the regulation of cholesterol synthesis by

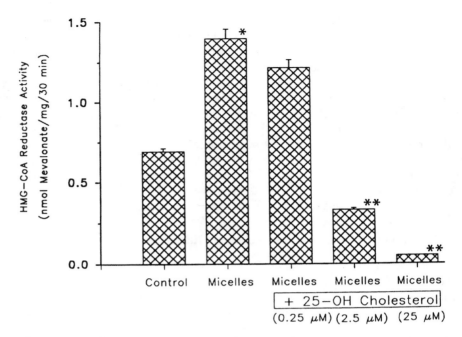

Figure 2. Regulation of HMG-CoA reductase activity by micellar 25-hydroxycholesterol. Reprinted by permission from "Regulation of gene expression and synthesis and degradation of 3-hydroxy-3-methylglutaryl coenzyme A reductase by micellar cholesterol in Caco-2 cells," *J. Lipid Res.* 32, pp. 1811–1821, 1991.

absorbed cholesterol have been conflicting. In cell culture, however, both cholesterol influx and cholesterol efflux can be rigidly controlled (Field *et al.*, 1991). In Caco-2 cells, cholesterol influx from taurocholate micelles decreases the activity of HMG-CoA reductase. A more potent regulatory sterol than cholesterol and one that has been used to study molecular mechanisms of HMG-CoA reductase regulation, 25-hydroxycholesterol, solubilized in micelles, also causes a marked decrease in the activity of HMG-CoA reductase (Fig. 2). It was also found, somewhat unexpectedly, that due to an efflux of cellular cholesterol, reductase activity in cells incubated with taurocholate micelles alone is increased.

The activity of HMG-CoA reductase can be regulated by transcriptional, by translational, and by posttranslational control (Nakanishi *et al.*, 1988; Goldstein and Brown, 1990). In Caco-2 cells incubated with taurocholate micelles alone, HMG-CoA reductase mass is increased (Fig. 3B). In contrast, the influx of 25-hydroxycholesterol substantially reduces HMG-CoA reductase mass (Fig. 3D). Thus, in intestinal cells, the regulation of reductase activity by changes in cholesterol flux are paralleled by changes in HMG-CoA reductase mass.

One possible mechanism for regulating HMG-CoA reductase mass by sterol flux is at the level of gene expression. Cholesterol efflux (cells incubated with micelles alone) increases HMG-CoA reductase mRNA levels (Fig. 4). In contrast, 2.5 μM of 25-hydroxy-

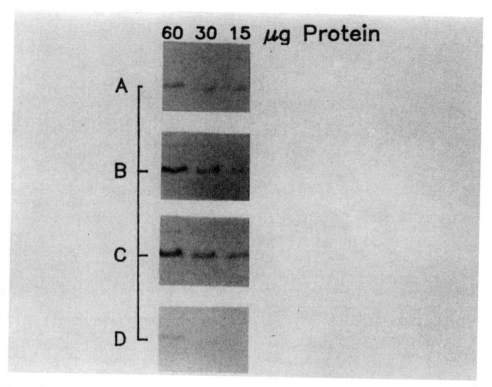

Figure 3. Regulation of HMG-CoA reductase mass by micellar 25-hydroxycholesterol. A. Control B. Micelles C. Micelles plus 150 uM cholesterol D. Micelles plus 2.5 uM 25-hydroxycholesterol. Reprinted by permission from "Regulation of gene expression and synthesis and degradation of 3-hydroxy-3-methylglutaryl coenzyme A reductase by micellar cholesterol in Caco-2 cells," *J. Lipid Res.* 32, pp. 1811–1821, 1991.

cholesterol, a concentration that causes profound inhibition of reductase activity, does not alter mRNA levels of reductase. Even at 25 μM, reductase mRNA levels are decreased by no more than 50%. These results suggested to us that in intestinal cells, cholesterol efflux increases HMG-CoA reductase activity by increasing gene transcription leading to an increase in the amount of reductase mass. In contrast, following sterol influx, a decrease in reductase activity and mass cannot be explained solely at the level of gene expression.

The half-life of reductase in Caco-2 cells is approximately 2.1 hr, a value similar to the half-life of reductase in other cell types (Edwards *et al.*, 1983; Faust *et al.*, 1982). Cholesterol efflux causes a decrease in reductase degradation ($t1/2 = 4.6$ hr), whereas the rate of synthesis of HMG-CoA reductase is not altered. In contrast, sterol influx modestly increases degradation to a $t1/2$ of 1.4 hr; but probably more significant, the rate of synthesis of reductase is profoundly inhibited.

The results of the regulation of HMG-CoA reductase activity by sterol flux in Caco-2 cells is summarized in Table 1. Cholesterol efflux increases the activity of HMG-CoA re-

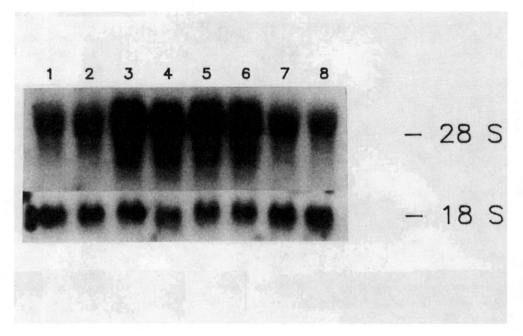

Figure 4. Regulation of HMG-CoA reductase levels by micellar 25-hydroxycholesterol. Lanes 1,2: Control medium; lanes 3,4: Micelles; lanes 5,6: Micelles plus 2.5 uM 25-hydroxycholesterol; lanes 7,8: Micelles plus 25 uM 25-hydroxycholesterol. Top blot: HMG-CoA reductase mRNA; Bottom blot: Actin mRNA. Reprinted by permission from "Regulation of gene expression and synthesis and degradation of 3-hydroxy-3-methylglutaryl coenzyme A reductase by micellar cholesterol in Caco-2 cells," *J. Lipid Res.* 32, pp. 1811–1821, 1991.

ductase by increasing the amount of reductase mass that, in turn, is likely caused by an increase in HMG-CoA reductase gene expression (transcriptional) and a decrease in the rate of degradation of the enzyme (posttranslational). Under conditions of sterol influx, HMG-CoA reductase activity appears to be regulated at three different levels. Sterol influx inhibits reductase activity by decreasing the amount of HMG-CoA reductase mass. This occurs with a modest decrease in mRNA abundance (transcriptional), an increase in the rate of degradation of the enzyme (posttranslational), and a profound decrease in its rate of synthesis (translational). Thus, under controlled cell culture conditions, the regulation of HMG-CoA reductase activity in Caco-2 cells is not dissimilar to other cell types studied. It can be stated with assurance then, that changes in apical cholesterol flux do indeed regulate cholesterol synthesis and HMG-CoA reductase activity in intestinal cells.

4.1.2. β-sitosterol

A related sterol to cholesterol, β-sitosterol, a plant sterol that differs from cholesterol only by an ethyl group on carbon-24, is not absorbed by human intestine to any appreciable extent (Gould *et al.,* 1969; Salen *et al.,* 1970). Exactly where in the pathway of sterol

Table 1. Regulation of HMG-CoA Reductase and LDL Expression by Sterol Flux in CaCo-2 Cells

	HMG-CoA Reductase				LDL Receptor			
	Activity	Mass	mRNA	Synthesis	Degradation	Activity	Mass	mRNA	Synthesis	Degradation
Efflux	$++^a$	$++$	$++$	$\pm^c$	$--$	nd^d	$+$	$++$	$++$	$\pm$
Influx	$---^b$	$---$	$-$	$---$	$+$	$-$	$--$	$--$	$--$	$\pm$

$^a+$, increased; $^b-$, decreased; $^c\pm$, unchanged; ^{d}nd, not determined.

absorption the intestinal cell discriminates between cholesterol and β-sitosterol remains controversial (Kuksis and Huang, 1962; Sylvin and Borgström, 1969; Bhattacharyya, 1981; Field and Mathur, 1983a; Child and Kuksis, 1983; Heinemann *et al.*, 1991). Somewhat unexpectedly, when cholesterol and β-sitosterol were being compared, we found that β-sitosterol could regulate cholesterol synthesis in Caco-2 cells (Field *et al.*, 1997). As demonstrated previously, HMG-CoA reductase activity and cholesterol synthesis are decreased following cholesterol influx (Fig. 5). With oleic acid in the micelle to promote triacylglycerol-rich lipoprotein assembly and secretion, cholesterol influx decreases HMG-CoA reductase mass (Fig. 6, lane 2) and mRNA abundance (Fig. 7, lane 2). These results provide further evidence that the absorption of cholesterol regulates HMG-CoA reductase activity, and it does so at the levels of enzyme mass and gene expression. When β-sitosterol is substituted for cholesterol in the micelle, cholesterol synthesis and HMG-CoA reductase activity are also inhibited (Fig. 5). Similarly, although not as potent as cholesterol, the influx of β-sitosterol decreases both reductase mass (Fig. 6, lane 3) and mRNA levels (Fig. 7, lane 3). As is discussed later, this suggests that the plant sterol can substitute for cholesterol in activating the sterol-regulatory-element-binding protein pathway to cause a decrease in transcription of the HMG-CoA reductase gene.

4.1.3. Lysophosphatidylcholine

In the lumen of the gut, significant amounts of lysophosphatidylcholine are generated by the hydrolysis of biliary and exogenous phosphatidylcholine. An influx of lysophosphatidylcholine into intestinal cells causes an increase in the synthesis and secretion of triacylglycerol-rich lipoproteins. These secreted lipoproteins, however, are markedly depleted in cholesteryl esters, suggesting that during the influx of lysophosphatidylcholine, cholesterol metabolism is being altered in the small intestinal cell (Field *et al.*, 1994). More-

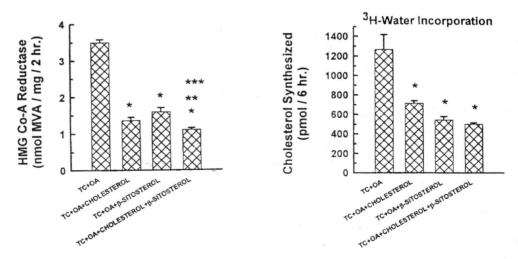

Figure 5. Regulation of HMG-CoA reductase activity and cholesterol synthesis by cholesterol and/or β-sitosterol. Reprinted by permission from *J. Lipid Res.* 38, 1997.

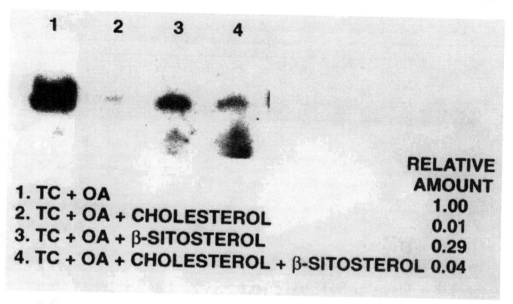

1. TC + OA — 1.00
2. TC + OA + CHOLESTEROL — 0.01
3. TC + OA + β-SITOSTEROL — 0.29
4. TC + OA + CHOLESTEROL + β-SITOSTEROL — 0.04

RELATIVE AMOUNT

Figure 6. Regulation of HMG-CoA reductase mass by cholesterol and/or β-sitosterol. Reprinted by permission from "Effect of micellar β-sitosterol on cholesterol metabolism in Caco-2 cells," *J. Lipid Res.* 38, pp. 348–360, 1997.

over, results of studies performed in humans ingesting lecithin (phosphatidylcholine) have demonstrated a lowering of serum cholesterol (Beil and Grundy, 1980). Because the mechanism or mechanisms for these observations may be occurring at the level of the intestine, the regulation of cholesterol metabolism following the influx of lysophosphatidylcholine has been studied (Muir *et al.,* 1996).

Rates of cholesterol synthesis and HMG-CoA reductase activities are increased in cells incubated with micelles containing lysophosphatidylcholine. Similar to results with taurocholate micelles alone, micelles containing lysophosphatidylcholine also cause a marked efflux of cholesterol from cells. With cholesterol efflux, HMG-CoA reductase mass is increased as is mRNA levels, suggesting (as expected from previous results) that intestinal cells respond to cholesterol depletion by increasing HMG-CoA reductase activity at the levels of enzyme mass and gene expression. It was also found that if cholesterol is included in the micelle-containing lysophosphatidylcholine, the increases in cholesterol synthesis, HMG-CoA reductase activity, and enzyme mass are completely prevented. These results indicate that in intestinal cells, gene transcription for reductase is activated by cholesterol efflux. By including cholesterol in the micelle, cellular cholesterol depletion is prevented, thereby, negating "the signal" to stimulate transcription generated by cholesterol efflux. These observations are compatible with what is known about the sterol-regulatory-element-binding protein pathway in other cells and suggests that this pathway exists in intestinal cells (see the following section).

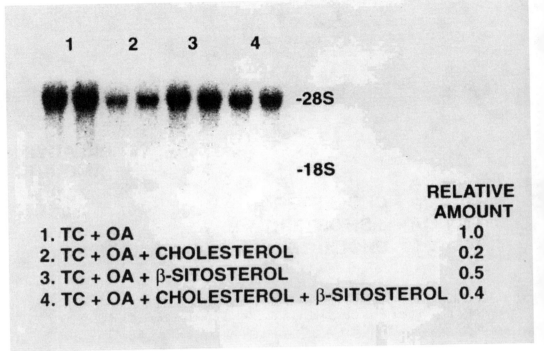

1 2 3 4

-28S

-18S

	RELATIVE AMOUNT
1. TC + OA	1.0
2. TC + OA + CHOLESTEROL	0.2
3. TC + OA + β-SITOSTEROL	0.5
4. TC + OA + CHOLESTEROL + β-SITOSTEROL	0.4

Figure 7. Regulation of HMG-CoA reductase mRNA levels by cholesterol and/or β-sitosterol. Reprinted by permission from "Effect of micellar β-sitosterol on cholesterol metabolism in Caco-2 cells," *J. Lipid Res.* 38, pp. 348–360, 1997.

4.2. Regulation of ACAT Activity

4.2.1. Sterol Flux

Unfortunately, little information exists on the regulation of ACAT activity in Caco-2 cells and what data do exist do not address the cellular or molecular mechanisms for the regulation. This paucity of information is not from neglect, because the topic is an important one, but the probes to study ACAT at the protein and RNA level have only recently become available. Moreover, as discussed previously, there remains controversy as to the identification of intestinal ACAT.

In an early study, it was observed that the rate of cholesterol esterification was decreased in cells incubated with micelles devoid of cholesterol (Field *et al.*, 1987b). At the time, the mechanism for this inhibition of ACAT activity by micelles was uncertain. It is now clear that loss of plasma membrane cholesterol is responsible for this inhibition. Moreover, when Caco-2 cells are incubated with micelles containing lysophosphatidylcholine, the activity of ACAT measured *in vitro* is also markedly decreased. If exogenous cholesterol is added to the *in vitro* assays, however, ACAT activity is unaltered. It is clear that lysophosphatidylcholine is depleting a pool of cholesterol used by ACAT as substrate.

Following the influx of cholesterol, cholesterol esterifying activity, that is, ACAT activity, is increased. The increase in ACAT activity is directly proportional to the amount of cholesterol solubilized in the micelle, and new enzyme synthesis is not required. Thus, in Caco-2 cells, changes in apical cholesterol flux regulate ACAT activity by expanding or depleting a pool of cholesterol used by ACAT as substrate.

There is a growing body of evidence to suggest that the predominant substrate for ACAT is cholesterol derived from the plasma membrane (Lange, 1994; Lange *et al.*, 1993; Tabas *et al.*, 1988; Nagy and Freeman, 1990; Liscum and Dahl, 1992). This appears to hold true for intestinal cells as well (Field *et al.*, 1995a, 1995b). When Caco-2 cells are driven to produce triacylglycerol-rich lipoproteins, it is cholesterol derived from the plasma membrane that is preferentially esterified by ACAT and is used for the transport of triacylglycerols. The esterification of newly synthesized cholesterol or cholesterol derived from absorption is negligible compared to the amount of plasma membrane cholesterol that is esterified. Moreover, inhibitors of the membrate transporter, *p*-glycoprotein, interfere with the trafficking of cholesterol from the plasma membrane to the endoplasmic reticulum, the site of ACAT. It has been suggested that *p*-glycoprotein or another member of this membrane transporter family may have a role in intracellular cholesterol transport and, thus, the regulation of ACAT activity (Field *et al.*, 1995a; Debry *et al.*, 1997).

4.3. Regulation of the LDL Receptor

Humans spend a significant portion of their time in the postabsorptive state. If dietary and/or biliary cholesterol decrease the expression of the LDL receptor at the basolateral membrane of the enterocyte, the contribution of the small intestine to the clearance of plasma LDL would decrease. As already alluded to, this may have important implications as to the effect of diet on LDL levels, and hence atherosclerosis, in humans. The effect of micellar cholesterol influx on the expression of the LDL receptor has been studied in Caco-2 cells grown on semipermeable filters separating an upper (apical surface) and lower (basolateral surface) well (Field *et al.*, 1993).

It was found that conditions of cholesterol efflux increase the amount of LDL receptor mass (Fig. 8, micelles) and mRNA levels (Fig. 9, lanes 3,4). In contrast, micellar cholesterol or 25-hydroxycholesterol influx decreases LDL receptor protein (Fig. 8) and mRNA levels (Fig. 9, lanes 5–8) with the polar sterol being more potent. Thus, it is clear that changes in cholesterol flux at the apical membrane regulate the expression of the intestinal LDL receptor. In Caco-2 cells, degradation of LDL receptor protein plays no role in regulating receptor expression in response to changes in sterol flux. Synthesis of the LDL receptor, however, parallels the changes in mRNA levels. Cholesterol efflux increases the rate of receptor synthesis and cholesterol influx causes a decrease. The apical influx of sterol also decreases the functional expression of the receptor. Less LDL is bound to the basolateral membrane of Caco-2 cells and less is degraded in response to cholesterol influx at the apical membrane. These results taken together indicate that the intestinal LDL receptor responds to changes in cholesterol flux occurring at the absorptive surface (results summarized in Table 1). It is likely, therefore, that under normal conditions, the expression of the intestinal LDL receptor is continuously being down-regulated by the presence of cholesterol in the lumen. We would postulate that the intestine would only play a minor role in the catabolism of plasma LDL in humans.

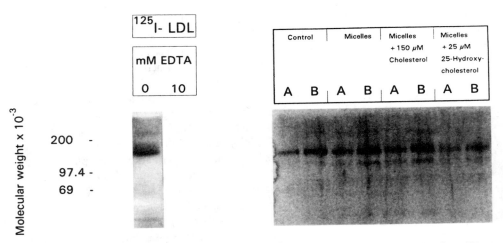

Figure 8. Regulation of LDL receptor mass by micellar sterols. Reprinted by permission from "Regulation of LDL receptor expression by luminal sterol flux in Caco-2 cells," *Arteriosc. Throm. Vasc. Biol.* 13, pp. 729–737, 1993.

5. Sterol-Regulatory-Element-Binding Protein Pathway

The manner in which cells respond to accumulating membrane cholesterol by repressing transcription of genes for HMG-CoA reductase and for the LDL receptor has been ascribed to a family of membrane-bound transcription factors called sterol-regulatory-ele-

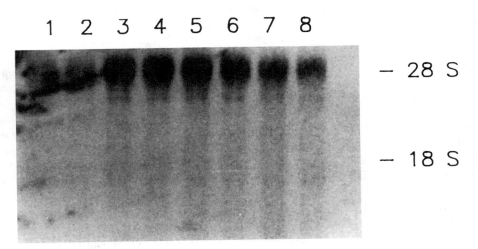

Figure 9. Regulation of LDL receptor mRNA levels by micellar 25-hydroxycholesterol. Lanes 1,2: Control medium; lanes 3,4: Micelles; lanes 5,6: Micelles plus 2.5 uM 25-hydroxycholesterol; lanes 7,8: Micelles plus 25 uM 25-hydroxycholesterol. Reprinted by permission from "Regulation of LDL receptor expression by luminal sterol flux in Caco-2 cells," *Arteriosc. Thromb. Vasc. Biol.* 13, pp. 729–737, 1993.

ment-binding proteins 1 and 2 (SREBPs; Wang *et al.,* 1994; Yokoyama *et al.,* 1993; Hua *et al.,* 1993; Brown and Goldstein, 1997). In addition to genes involved in cholesterol metabolism, genes governing the regulation of fatty acid synthesis are also activated by SREBPs. In a two-step proteolytic process, cholesterol depletion causes the release of a mature form of SREBP from the endoplasmic reticulum membrane that then enters the nucleus and activates transcription. When cholesterol accumulates in membranes, the first proteolytic cleavage of SREBP is repressed; SREBP remains membrane-bound and transcription of the target genes is decreased. At the present time, there is little information about the SREBP pathway in the intestine, except that SREBPs are expressed at the RNA level in intestine (Jay Horton, personal communication). In mice lacking the SREBP-1 gene, intestinal cholesterol synthetic rates are similar to controls, suggesting that SREBP-1 is not required for cholesterol synthesis in the intestine. Intestinal fatty acid synthesis, however, is decreased in SREBP-1 knockout mice, suggesting that SREBP-1 may be important for supporting fatty acid synthesis in the gut (Shimano *et al.,* 1996, 1997; Jay Horton, personal communication). Moreover, most tissues studied to date have predominantly one particular isoform of SREBP-1 (1c). In the intestine, in contrast, there seems to be more of the 1a isoform. Compared to the 1c isoform of SREBP-1, the 1a isoform appears to be the more potent in affecting genes regulating cholesterol and fatty acid synthesis. This might suggest that in tissues that have high cell turnover and require more cholesterol for membrane synthesis, or in cells that transport large amounts of lipids and thus require more fatty acids and cholesterol, the more active form of SREBP-1 is expressed. This new area of research remains open for future investigation in the small intestine.

6. Summary

From the information that is presently available on the regulation of cholesterol metabolism in the small intestine and in intestinal cells in culture, I think it is reasonable to state that similar to other cells that strive to maintain cholesterol balance by regulating the influx of cholesterol, intestinal cells also respond appropriately to changes in cholesterol flux. For example, in animal models, under conditions of increased intestinal cholesterol influx secondary to increased ingestion of cholesterol, of certain fats, or by elevated levels of lipoprotein cholesterol, intestinal cholesterol synthesis decreases and cholesterol esterification increases. In contrast, with a decrease in cholesterol influx (feeding cholestyramine or *a*-olefin-maleic acid copolymer, or by inducing severe hypocholesterolemia), the reverse occurs—intestinal cholesterol synthesis increases and cholesterol esterification decreases. Table 2 summarizes the results obtained in several different animal species and experimental models.

The mechanisms that are involved in the regulation of intestinal cholesterol metabolism by changes in sterol flux are summarized in Fig. 10 and in Table 1 from data obtained in Caco-2 cells. Cholesterol efflux increases the rate of cholesterol synthesis and the expression of the LDL receptor. Depletion of cholesterol from cells increases the mass of HMG-CoA reductase by decreasing the rate of degradation of the enzyme and by increasing its gene expression. LDL receptor mass is also increased, and it would appear that this is due to an increased amount of mRNA and increased synthesis of the receptor. Cholesterol influx decreases the rate of cholesterol synthesis and the expression of the LDL re-

Table 2. Regulation of Intestinal Cholesterol Metabolism

Apical	Cholesterol Synthesis and/or HMG-CoA Reductase Activity	Cholesterol Esterification and/or ACAT Activity
Cholesterol	$-^a, \pm^c$	$++^b$
Fat		
Safflower	$\pm$	$+$
Corn	$+$	$\pm$
Menhaden	$-$	$+$
Cocoa	$+, \pm$	$\pm$
Triolein	nd^d	$-$
Coconut	nd	6
Intracellular		
Phosphorylation/ Dephosphorylation	$+$	$+, -$
Basolateral		
Lipoproteins	$-$	$+$
Hormones		
Insulin	$+, \pm$	$-$
Glucagon	$-, \pm$	nd
Progesterone	$+$	$-$
Triamcinolone	$-$	nd
Triiodothyronine	$+$	nd

$^a+$, increased; $^b-$, decreased; $^c\pm$, unchanged; ^{d}nd, not determined.

ceptor. HMG-CoA reductase mass is markedly decreased under these conditions, and decreased translational efficiency of the message and increased degradation of the protein appear to be playing the predominant roles in this regulation. In contrast, a decrease in LDL receptor mass can be explained solely at the level of gene expression for the receptor.

The proximal small intestine is constantly being bathed with either dietary or biliary cholesterol. Under normal conditions, then, it is doubtful that the absorptive cell would be deficient in cholesterol or would be losing cellular cholesterol into the lumen (efflux). This influx of lumenal cholesterol would "replace" the plasma membrane pool of cholesterol that is being utilized for the transport of lipids and thus, there would be little need for the absorptive cell to alter either cholesterol synthesis or LDL cholesterol uptake. In fact, it is likely that both HMG-CoA reductase and LDL receptor expression would be chronically suppressed in the upper small intestine. This will differ in the distal small intestine that encounters much less lumenal cholesterol. Changes in cholesterol synthesis and in LDL receptor expression may play more important roles in regulating the amount of cholesterol available to an enterocyte of the distal small bowel. Despite the regional and local differences of cholesterol metabolism within the small intestine, changes in cholesterol flux appropriately regulate cholesterol metabolism in the intestinal cell.

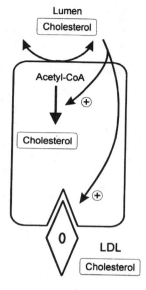
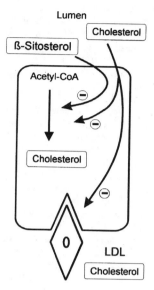

Figure 10. Regulation of intestinal cholesterol metabolism by changes in cholesterol flux.

References

Andersen, J. M., and Dietschy, J. M., 1976, Cholesterogenesis: Derepression in extrahepatic tissues with 4-aminopyrazolo [3,4-d] pyrimidine, *Science* **193:**903–905.

Andersen, J. M., Turley, S. D., and Dietschy, J. M., 1982, Relative rates of sterol synthesis in the liver and various extrahepatic tissues of normal and cholesterol-fed rabbits. Relationship to plasma lipoprotein and tissue cholesterol levels, *Biochim. Biophys. Acta* **711:**421–430.

Beg, Z. H., Stonik, J. A, and Brewer, J. B., Jr., 1987, Modulation of the enzymic activity of 3-hydroxy-3-methyl-glutaryl coenzyme A reductase by multiple kinase systems involving reversible phosphorylation: A review, *Metabolism* **36:**900–917.

Beil, F. U., and Grundy, S. M., 1980, Studies on plasma lipoproteins during absorption of exogenous lecithin in man, *J. Lipid Res.* **21:**525–536.

Bennett-Clark, S., 1979, Mucosal coenzyme A-dependent cholesterol esterification after intestinal perfusion of lipids in rats, *J. Biol. Chem.* **254:**1534–1536.

Bhattacharyya, A. K., 1981, Uptake and esterification of plant sterols by rat small intestine, *Am. J. Physiol.* **240:**G50–55.

Bloch, K., 1965, The biological synthesis of cholesterol, *Science* **150:**19–28.

Bochenek, W. J., and Rogers J. B., 1979, Dietary regulation of 3-hydroxy-3-methylgutaryl-CoA reductase from rate intestine, *Biochem. Biophys. Acta* **575:**57–62.

Brown, M. S., and Goldstein, J. L., 1980, Multivalent feedback regulation of HMG-CoA reductase, a control mechanism coordinating isoprenoid synthesis and cell growth, *J. Lipid Res.* **21:**505–517.

Brown, M. S., and Goldstein, J. L., 1986, A receptor-mediated pathway for cholesterol homeostasis, *Science* **232:**34–47.

Brown, M. S., and Goldstein, J. L., 1997, The SREBP pathway: Regulation of cholesterol metabolism by proteolysis of a membrane-bound transcription factor, *Cell* **89:**331–340.

Brown, M. S., Goldstein, J. L., and Dietschy, J. M., 1979, Active and inactive forms of 3-hydroxy-3-methylglutaryl coenzyme A reductase in the liver of the rat, *J. Biol. Chem.* **254:**5144–5149.

Chang, C. C. Y., Huh, H. Y., Cadigan, K. M., and Chang, T. Y., 1993, Molecular cloning and functional expression of human acyl-coenzyme A:cholesterol acyltransferase cDNA in mutant Chinese hamster ovary cells, *J. Biol. Chem.* **268:**20747–20755.

Chang, T. Y., Chang, C. C. Y., and Cheng, D., 1995, Acyl-coenzyme A:cholesterol acyltransferanse, *J. Biol. Chem.* **270:**29532–29540.

Cheng, D., Chang, C. C. Y., Qu, X., and Chang, T. Y., 1995, Activation of acyl-coenzyme A: Cholesterol acyltransferase (ACAT) by cholesterol or by oxysterol in a cell-free system, *J. Biol. Chem.* **270:**685–695.

Child, P., and Kuksis, A., 1983, Critical role of ring structure in the differential uptake of cholesterol and plant sterols by membrane preparations *in vitro, J. Lipid Res.* **24:**1196–1209.

Clarke, P. R., and Hardie, D. G., 1990, Regulation of HMG-CoA reductase: Identification of the site phosphorylated by the AMP-activated protein kinase *in vitro* and in intact rat liver, *EMBO J.* **9:**2439–2446.

Corton, J. M., and Hardie, D. G., 1992, Evidence against a role for phosphorylation/dephosphorylation in the regulation of acyl-CoA:cholesterol acyltransferase, *Eur. J. Biochem.* **204:**203–208.

Debry, P., Nash, E. A., Neklason, D. W., and Metherall, J. E., 1997, Role of multidrug resistance *P*-glycoproteins in cholesterol esterification. *J. Biol. Chem.* **272:**1026–1031.

Dietschy, J. M., and Gamel, W. G., 1971, Cholesterol synthesis in the intestine of man: Regional differences and control mechanisms, *J. Clin. Invest.* **50:**872–880.

Dietschy, J. M., and Siperstein, M. D., 1965, Cholesterol synthesis by the gastrointestinal tract: Localization and mechanisms of control. *J. Clin. Invest.* **44:**1311–1327.

Dietschy, J. M., and Siperstein, M. D., 1967, Effect of cholesterol feeding and fasting on sterol synthesis in seventeen tissues of the rat, *J. Lipid Res.* **8:**97–104.

Dietschy, J. M., and Wilson, J. D., 1968, Cholesterol synthesis in the squirrel monkey: Relative rates of synthesis in various tissues and mechanisms of control, *J. Clin. Invest.* **47:**166–174.

Dietschy, J. M., Salomon, H. S., and Siperstein, M. D., 1966, Bile acid metabolism. I. Studies on the mechanisms of intestinal transport, *J. Clin. Invest.* **45:**832–846.

Drevon, C. A., Lilljeqvist, A-C., Schreiner, B., and Norum, K. R., 1979, Influence of cholesterol/fat feeding on cholesterol esterification and morphological structures in intestinal mucosa from guinea pigs, *Atherosclerosis* **34:**207–219.

Edwards, P. A., Lan, S.-F., Tanaka, R. D., and Fogelman, A. M., 1983, Mevalonolactone inhibits the rate of synthesis and enhances the rat of degradation of 3-hydroxy-3-methylglutaryl coenzyme A reductase in rat hepatocytes, *J. Biol. Chem.* **258:**7272–7275.

Faust, J. R., Luskey, R. K., Chin, J., Goldstein, J. L., and Brown, M. S., 1982, Regulation of synthesis and degradation of 3-hydroxy-3-methylglutaryl-coenzyme A reductase by low density lipoprotein and 25-hydroxycholesterol in UT-1 cells, *Proc. Natl. Acad. Sci. USA* **79:**5205–5209.

Feingold, K. R., 1989, Importance of small intestine in diabetic hypercholesterolemia, *Diabetes* **38:**141–145.

Field, F. J., and Mathur, S. N., 1983a, β-sitosterol: Esterification by intestinal acylcoenzyme A:cholesterol acyltransferase (ACAT) and its effect on cholesterol esterification, *J. Lipid Res.* **24:**409–417.

Field, F. J., and Mathur, S. N., 1983b, Regulation of acyl CoA: cholesterol acyltransferase by 25-hydroxycholesterol in rabbit intestinal microsomes and absorptive cells, *J. Lipid Res.* **24:**1049–1059.

Field, F. J., and Mathur, S. N., 1995, Intestinal lipoprotein synthesis and secretion, *Prog. Lipid Res.* **34:**185–198.

Field, F. J., and Salome, R. G., 1982, Effect of dietary fat saturation, cholesterol and cholestyramine on acyl-CoA: cholesterol acyltransferase activity in rabbit intestinal microsomes, *Biochim. Biophys.* Acta **12:**557–570.

Field, F. J., Erickson, S. K., Shrewsbury, M. A., and Cooper, A. D., 1982a, 3-Hydroxy-3-methylglutaryl coenzyme A reductase from rat intestine: Subcellular localization and *in vitro* regulation, *J. Lipid Res.* **23:**105–113.

Field, F. J., Cooper, A. D., and Erickson, S. K., 1982b, Regulation of rabbit intestinal acyl coenzyme A-cholesterol acyltransferase *in vivo* and *in vitro, Gastroenterol.* **83:**873–880.

Field, F. J., Hennig, B, and Mathur, S. N., 1984, *In vitro* regulation of 3-hydroxy-3-methylglutarylcoenzyme A reductase and acylcoenzyme A: cholesterol acyltransferase activities by phosphorylation-dephosphorylation in rabbit intestine, *Biochim. Biophys.* Acta **802:** 9–16.

Field, F. J., Albright, E., and Mathur, S. N., 1986, The effect of hypothyroidism and thyroxine replacement on hepatic and intestinal HMG-CoA reductase and ACAT activities and biliary lipids in the rat, *Metabolism* 35:1085–1089.

Field, F. J., Albright, E., and Mathur, S. N., 1987a, Effect of dietary *n*-3 fatty acids on HMG-CoA reductase and ACAT activities in liver and intestine of the rabbit, *J. Lipid Res.* 28:50–58.

Field, F. J., Albright, E., and Mathur, S. N., 1987b, Regulation of cholesterol esterification by micellar cholesterol in Caco-2 cells, *J. Lipid Res.* 28:1057–1066.

Field, F. J., Shreves, T., Fujiwara, D., Murthy, S., Albright, E., and Mathur, S. N., 1991, Regulation of gene expression and synthesis and degradation of 3-hydroxy-3-methylglutaryl coenzyme A reductase by micellar cholesterol in Caco-2 cells, *J. Lipid Res.* 32:1811–1821.

Field, F. J., Fujiwara, D., Born, E., Chappell, D. A., and Mathur, S. N., 1993, Regulation of LDL receptor expression by luminal sterol flux in Caco-2 cells, *Arterioscler. Thromb.* 13:729–737.

Field, F. J., Born, E., Chen, H., Murthy, S., and Mathur, S. N., 1994, Lysophosphatidylcholine increases the secretion of cholesteryl ester-poor triacylglycerol-rich lipoproteins by Caco-2 cells, *Biochem. J.* 304:35–42.

Field, F. J., Born, E., Chen, H., Murthy, S., and Mathur, S. N., 1995a, Esterification of plasma membrane cholesterol and triacylglycerol-rich lipoprotein secretion in CaCo-2 cells: Possible role of *p*-glycoprotein, *J. Lipid Res.* 36:1533–1543.

Field, F. J., Born, E., and Mathur, S. N., 1995b, Triacylglycerol-rich lipoprotein cholesterol is derived from the plasma membrane in Caco-2 cells, *J. Lipid Res.* 36:2651–2660.

Field, F. J., Born, E., and Mathur, S. N., 1997, Effect of micellar β-sitosterol on cholesterol metabolism in CaCo-2 cells, *J. Lipid Res.* 38:348–360.

Fong, L. G., Bonney, E, Kosek, J. C., and Cooper, A. D., 1989, Immunohistochemical localization of low density lipoprotein receptors in adrenal gland, liver, and intestine, *J. Clin. Invest.* 84:847–856.

Fong, L. G., Fujishima, S. E., Komaromy, M. C., Pak, Y. K., Ellsworth, J. L., and Cooper, A. D., 1995, Location and regulation of low-density lipoprotein receptors in intestinal epithelium, *Am. J. Physiol.* 269: G60–72.

Gavey, K. L., Trujillo, D. L., and Scallen, T. J., 1983, Evidence for phosphorylation/deposphorylation of rat liver acyl-CoA: cholesterol acyltransferase, *Proc. Natl. Acad. Sci. USA* 80:2171–2174.

Gebhard, R. L., and Cooper, A. D., 1978, Regulation of cholesterol synthesis in cultured canine intestinal mucosa, *J. Biol. Chem.* 8:2790–2796.

Gebhard, R. L., and Prigge, W. F., 1981, *In vivo* regulation of canine intestinal 3-hydroxy-3-methylglutaryl coenzyme A reductase by cholesterol, lipoprotein, and fatty acids, *J. Lipid Res.* 22:1111–1118.

Gebhard, R. L., Stone, B. G., and Prigge, W. F., 1985, 3-Hydroxy-3-methylglutaryl coenzyme A reductase activity in the human gastrointestinal tract, *J. Lipid Res.* 26:47–53.

Goldstein, J. L., and Brown, M. S., 1990, Regulation of the mevalonate pathway, *Nature* 343:425–430.

Goodman, MW., Prigge, W. F., Gebhard, R. L., 1981, Hormonal regulation of canine intestinal cholesterol synthesis, *Am. J. Physiol.* 240:G274–G280.

Gould, R. G., Jones, R. J., LeRoy, G. V., Wissler, R. W., and Taylor, G. B., 1969, Absorbability of β-sitosterol in humans, *Metabolism* 18:652–662.

Haugen, R., and Norum, K. R., 1976, Coenzyme-A-dependent esterification of cholesterol in rat intestinal mucosa, *Scand. J. Gastroenterol.* 11:615–621.

Heinemann, T., Kullak-Ublick G. A., Pietruck, B., and vonBergmann, K., 1991, Mechanisms of action of plant sterols on inhibition of cholesterol absorption. Comparison of sitosterol and sitostanol, *Eur. J. Clin. Pharmacol.* 40:S59–73.

Helgerud P., Saarem, K., and Norum, K. R., 1981, Acyl-CoA: cholesterol acyl-transferase in human small intestine: Its activity and some properties of the enzymic reaction, *J. Lipid Res.* 22:271–277.

Herold, G., Schneider, A., Ditschuneit, H., and Stange, E. F., 1984, Cholesterol synthesis and esterification in cultured intestinal mucosa. Evidence for compartmentation, *Biochim. Biophys. Acta* 796:27–33.

Ho, K-J., 1975, Effect of cholesterol feeding on circadian rhythm of hepatic and intestinal cholesterol biosynthesis in hamsters, *Proc. Soc. Exp. Biol. Med.* 150:271–277.

Howles, P. N., Carter, C. P., and Hui, D. Y., 1996, Dietary free and esterified cholesterol absorption in cholesterol esterase (bile salt-stimulated lipase) gene-targeted mice, *J. Biol. Chem.* 271:7196–7202.

Hua, X., Yokoyama, C., Wu, J., Briggs, M. R., Brown, M. S., Goldstein, J. L., and Wang, X., 1993, SREBP-2, a second basic-helix-loop-helix-leucine zipper protein that stimulates transcription by binding to a sterol regulatory element, *Proc. Natl. Acad. Sci. USA* 90:11603–11607.

Huang, Y., and Hui, D. Y., 1990, Metabolic fate of pancreas-derived cholesterol esterase in intestine: An *in vitro* study using Caco-2 cells, *J. Lipid Res.* **31:**2029–2037.

Hughes, T. E., Sasak, W. V., Ordovas, J. M., Forte, T. M., Lamon-Fava, S., and Schaefer, E. J., 1987, A novel cell line (Caco-2) for the study of intestinal lipoprotein synthesis, *J. Biol. Chem.* **262:**3762–3767.

Kagami, A., Fidge, N., Suzuki, N., and Nestel, P., 1984, Characteristics of the binding of high-density lipoprotein, by intact cells and membrane preparations of rat intestinal mucosa, *Biochim. Biophys. Acta* **795:**179–190.

Kuksis, A., and Huang, T. C., 1962, Differential absorption of plant sterols in the dog, *Can. J. Biochem. Physiol.* **40:**1493–1504.

Lange, Y., 1994, Cholesterol movement from plasma membrane to rough endoplasmic reticulum, *J. Biol. Chem.* **269:**1–4.

Lange, Y., Strebel, F., and Steck, T. L., 1993, Role of the plasma membrane in cholesterol esterification in rat hepatoma cells, *J. Biol. Chem.* **268:**13838–13843.

Li, A. C., Tanaka, R. D., Callaway, K., Fogelman, A. M., and Edwards, P. A., 1988, Localization of 3-hydroxy-3-methylglutaryl CoA reductase and 3-hydroxy-3-methylglutaryl CoA synthase in the rat liver and intestine is affected by cholestyramine and mevinolin, *J. Lipid Res.* **29:**781–796.

Lindsey, C. A., and Wilson, J. D., 1965, Evidence for a contribution by the intestinal wall to the serum cholesterol of the rat, *J. Lipid Res.* **6:**173–181.

Liscum, L., and Dahl, N. K., 1992, Intracellular cholesterol transport, *J. Lipid Res.* **33:**1239–1253.

Meiner, V., Tam, C., Gunn, M. D., Dong, L-M., Weisgraber, K. H., Novak, S., Myers, H. M., Erickson, S. K., and Farese R. V., Jr., 1997, Tissue expression studies on the mouse acyl-CoA: cholesterol acyltransferase gene (Acact): Findings supporting the existence of multiple cholesterol esterification enzymes in mice, *J. Lipid Res.* **38:**1928–1933.

Meiner, V. L., Cases, S., Myers, H. M., Sande, E. R., Bellosta, S., Schambelan, M., Pitas, R. E., McGuire, J., Herz, J., and Farese, R. V., Jr., 1996, Disruption of the acyl-CoA: cholesterol acyltransferase gene in mice: Evidence suggesting multiple cholesterol esterification enzymes in mammals, *Proc. Natl. Acad. Sci. USA* **93:**14041–14046.

Merchant, J. L., and Heller, R. A., 1977, 3-Hydroxy-3-methylglutaryl coenzyme A reductase in isolated villus and crypt cells of the rat ileum, *J. Lipid Res.* **18:**722–732.

Muir, L. V., Born, E., Mathur, S. N., and Field, J. F., 1996, Lysophosphatidylcholine increases 3-hydroxy-3-methylglutaryl-coenzyme A reductase gene expression in Caco-2 cells, *Gastroenterol.* **110:**1068–1076.

Muroya, H., Sodhi, H. S., and Gold, R. G., 1977, Sterol synthesis in intestinal villi and crypt cells of rats and guinea pigs, *J. Lipid Res.* **18:**301–308.

Murthy, S., Albright, E., Mathur, S. N., and Field, F. J., 1988, Modification of Caco-2 cell membrane fatty acid composition by eicosapentaenoic acid and palmitic acid: Effect on cholesterol metabolism, *J. Lipid Res.* **29:**773–780.

Nagy, L., and Freeman, D. A., 1990, Effect of cholesterol transport inhibitors on steroidogenesis and plasma membrane cholesterol transport in cultured MA-10 Leydig tumor cells, *Endocrinology* **126:**2267–2276.

Nakanishi, M., Goldstein, J. L., and Brown, M. S., 1988, Multivalent control of 3-hydroxy-3-methylglutaryl coenzyme A reductase, *J. Biol. Chem.* **263:** 8929–8937.

Nordstrom, J. L., Rodwell, V. W., and Mitschelen, J. J., 1977, Interconversion of active and inactive forms of rat liver hydroxymethylglutaryl-CoA reductase, *J. Biol. Chem.* **252:**8924–8934.

Norum, K. R, Lilljeqvist, A-C., and Drevon, C. A., 1977, Coenzyme-A-dependent esterification of cholesterol in intestinal mucosa from guinea-pig. Influence of diet on the enzyme activity, *Scan. J. Gastroenterol.* **12:**281–288.

Oku, H., Ide, T., and Sugano, M., 1984, Reversible inactivation-reactivation of 3-hydroxy-3-methylglutaryl coenzyme A reductase of rat intestine, *J. Lipid Res.* **25:**254–261.

Panini, S. R., Lehrer, G., Rogers, D. H., and Rudney, H., 1979, Distribution of 3-hydroxy-3-methylglutaryl coenzyme A reductase and alkaline phosphatase activities in isolated ileal epithelial cells of fed, fasted, cholestyranine-fed, and 4-aminopyrazolo [3,4-d] pyrimidine-treated rats, *J. Lipid Res.* **20:**879–889.

Pape, M. E., Schultz, P. A., Rea, T. J., DeMattos, R. B., Kieft, K., Bisgaier, C. L., Newton, R. S., and Krause, B. R., 1995, Tissue specific changes in acyl-CoA: cholesterol acyltransferase (ACAT) mRNA levels in rabbits, *J. Lipid Res.* **36:**823–838.

Purdy, B. H., and Field, F. J., 1984, Regulation of acylcoenzyme A: cholesterol acyltransferase and 3-hydroxy-3-methylglutaryl coenzyme A reductase activity by lipoproteins in the intestine of parabiont rats, *J. Clin. Invest.* **74:**351–357.

Quaroni, A., and May, R. J., 1980, Establishment and characterization of intestinal epithelial cell cultures, *Methods Cell. Biol.* **21**:403–427.

Quaroni, A., Wands, J. Trestad, R. L., and Isselbacher, K. J., 1979, Epithelioid cell cultures from rat small intestine. Characterization by morphologic and immunologic criteria, *J. Cell Biol.* **80**:248–265.

Raul, F., Simon, P., Kendinger, M., and Haffen, K., 1977, Intestinal enzymes activities in isolated villus and crypt cells during post-natal development of the rat, *Cell Tissue Res.* **176**:167–178.

Rogler, G., Herold G., and Stange, E. F., 1991, HDL3-retroendocytosis in cultured small intestinal crypt cells: A novel mechanism of cholesterol efflux, *Biochim. Biophys. Acta* **1095**:30–38.

Salen, G., Ahrens, E. H., Jr., and Grundy, S. M., 1970, Metabolism of β-sitosterol in man, *J. Clin. Invest.* **49**:952–967.

Shakir, K. M. M., Sundaram, S. G., and Margolis, S., 1978, Lipid synthesis in isolated intestinal cells, *J. Lipid Res.* **19**:433–442.

Shefer, S., Hauser, S. Lapar V., and Mosbach E. H., 1973, Regulatory effects of dietary sterols and bile acids on rat intestinal HMG CoA reductase, *J. Lipid Res.***14**:400–405.

Shimano, H., Horton, J. D., Hammer, R. E., Shimomura, I., Brown, M. S., and Goldstein, J. L., 1996, Overproduction of cholesterol and fatty acids causes massive liver enlargement in transgenic mice expressing truncated SREBP-1a, *J. Clin. Invest.* **98**:1575–1584.

Shimano, H., Shimomura, I., Hammer, R. E., Herz, J., Goldstein, J. L., Brown, M. S., and Horton, J. D., 1997, Elevated levels of SREBP-2 and cholesterol synthesis in livers of mice homozygous for a targeted disruption of the SREBP-1 gene, *J. Clin. Invest.* **100**:2115–2124.

Singer, I. I., Kawka, D. W., McNally, S. E., Scott, S., Alberts, A. W., Chen, J. S., and Huff, J. W., 1987, Hydroxymethylglutaryl-coenzyme A reductase exhibits graded distribution in normal and mevionolin-treated ileum, *Arteriosclerosis* **7**:144–151.

Spady, D. K., and Dietschy, J. M., 1983, Sterol synthesis *in vivo* in 18 tissues of the squirrel monkey, guinea pig, rabbit, hamster, and rat, *J. Lipid Res.* **24**:303–315.

Spady, D. K., Bilheimer, D. W., and Dietschy, J. M., 1983, Rates of receptor-dependent and independent low density lipoprotein uptake in the hamster, *Proc. Natl. Acad. Sci . USA* **80**:3499–3503.

Stange, E. F., and Dietschy, J. M., 1983a, Absolute rates of cholesterol synthesis in rat intestine *in vitro* and *in vivo:* A comparison of different substrates in slices and isolated cells, *J. Lipid Res.* **24**:72–82.

Stange, E. F., and Dietschy, J. M., 1983b, Cholesterol synthesis and low density lipoprotein uptake are regulated independently in rat small intestinal epithelium, *Proc. Natl. Acad. Sci. USA* **80**:5739–5743.

Stange, E. F., Preclik, G., Schneider, A., Seiffer, E., and Ditschuneit, H., 1981a, Hormonal regulation of 3-hydoxy-3-methylglutaryl coenzyme A reductase and alkaline phosphatase in cultured intestinal mucosa, *Biochim. Biophys. Acta* **678**:202–206.

Stange, E. F., Alavi, M., Schneider, A., Ditschuneit, H., and Poley, J. R., 1981b, Influence of dietary cholesterol, saturated and unsaturated lipid on 3-hydroxy-3-methylglutaryl CoA reductase activity in rabbit intestine and liver, *J. Lipid Res.* **22**:47–56.

Stange, E. F., Suckling, K. E., and Dietschy, J. M., 1983, Synthesis and coenzyme A-dependent esterification of cholesterol in rat intestinal epithelium, *J. Biol. Chem.* **258**:12868–12875.

Suckling, K. E., and Stange, E. F., 1985, Role of acyl-CoA:cholesterol acyltransferase in cellular cholesterol metabolism, *J. Lipid Res.* **26**:647–671.

Suckling, K. E., Stange, E. F., and Dietschy, J. M., 1983, Dual modulation of hepatic and intestinal acyl-CoA: cholesterol acyltransferase activity by (de-)phosphorylation and substrate supply *in vitro, FEBS Let* **151**:111–116.

Suzuki, N., Fidge, N., Nestel, P., and Yin, J., 1983, Interaction of serum lipoproteins with the intestine. Evidence for specific high density lipoprotein-binding sites on isolated rat intestinal mucosal cells, *J. Lipid Res.* **24**:253–264.

Sviridov, D. D., Safonova, I. G., Gusev, V. A., Talalaev, A. G., Tsibulsky, V. P., Ivanov, V. O., Preobrazensky, S. N., Repin, V. S., and Smirnov, V. N., 1986, Specific high affinity binding and degradation of high density lipoproteins by isolated epithelial cells of human small intestine, *Metabolism* **35**:588–595.

Sviridov, D. D., Ehnholm C., Tenkanen H., Pavlov, M-Yu., Safonova, I. G., and Repin, V. S., 1992, Studies on the proteins involved in the interaction of high-density lipoprotein with isolated human small intestine epithelial cells, *FEBS Let* **303**:202–204.

Swann, A., Wiley, M. H., and Siperstein, M. D., 1975, Tissue distribution of cholesterol feedback control in the guinea pig, *J. Lipid Res.* **16**:360–366.

Sylvin, C., and Borgström, B., 1969, Absorption and lymphatic transport of cholesterol and sitosterol in the rat, *J. Lipid Res.* **10:**179–182.

Sylvin, C., and Nordstrom, C., 1970, The site of absorption of cholesterol and sitosterol in the rat small intestine, *Scand. J. Gastroenterol.* **5:**57–63.

Tabas, I., Rosoff, W. J., and Boykow, G. C., 1988, Acylcoenzyme A: cholesterol acyltransferase in macrophages utilizes a cellular pool of cholesterol oxidase-accessible cholesterol as substrate, *J. Biol. Chem.* **263:**1266–1272.

Turley, S. D., and Dietschy, J. M., 1988, The metabolism and excretion of cholesterol by the liver, in: *The Liver: Biology and Pathobiology,* 2nd ed. (I. M. Arias, W. B. Jakoby, H. Popper, D. Schachter, and D. A. Shafritz, eds.), Raven Press, New York, pp. 617–641.

Uelmen, P. J., Oka, K., Sullivan, M., Chang, C. C. Y., Chang, T. Y., and Chan, L., 1995, Tissue-specific expression and cholesterol regulation of acylcoenzyme A:cholesterol acyltransferase (ACAT) in mice, *J. Biol. Chem.* **270:**26192–26201.

Wang, X., Sato, R., Brown, M. S., Hua, X., and Goldstein, J. L., 1994, SREBP-1, a membrane-bound transcription factor released by sterol-regulated proteolysis, *Cell* **77:**53–62.

Weiser, M. M., 1973, Intestinal epithelial cell surface membrane glycoprotein synthesis, *J. Biol. Chem.* **248:**2536–2541.

Yokoyama, C., Wang, X., Briggs, M. R., Admon, A., Wu, J., Hua, X., Goldstein, J. L., and Brown, M. S., 1993, SREBP-1, a basic helix-loop-helix-leucine zipper protein that controls transcription of the LDL receptor gene, *Cell* **75:**187–197.

Zammit, V. A., and Caldwell A. M., 1992, Direct demonstration that increased phosphorylation of 3-hydroxy-3-methylglutaryl-CoA reductase does not increase its rate of degradation in isolated rat hepatocytes, *Biochem. J.* **284:**901–904.

CHAPTER 14

Regulation of Intestinal Apolipoprotein Gene Expression

DENNIS D. BLACK

1. Introduction

Apolipoproteins are lipid-binding peptides that are surface components of lipoprotein particles. They play essential roles in the assembly, secretion, and peripheral metabolism of lipoprotein particles. The liver and the small intestine are the two major organs responsible for the synthesis and secretion of lipoproteins containing lipid of endogenous and exogenous origin, respectively. Although many details of the uptake, processing, and secretion of the lipid components of intestinal lipoproteins have been elucidated, the mechanisms of regulation of intestinal apolipoprotein expression by lipid absorption and by other factors remain relatively poorly understood.

The predominant apolipoproteins expressed in the small intestine are apo B, A-I, A-IV, and C-III (Table 1). Therefore, this chapter focuses on the regulation of the expression of these apolipoproteins in the small intestine. Apo B exists as two distinct forms in the human, the rat, and the swine (Black, 1995; Hamilton, 1994; Kane, 1983; Olofsson et al., 1987). Apo B-100 is the larger species and is a component of liver-derived very-low-density lipoprotein (VLDL) and low-density lipoprotein (LDL) and contains the LDL receptor-binding domain (Hamilton, 1994; Hospattankar et al., 1986; Hui et al., 1984; Olofsson et al., 1987). Apo B-48 is the smaller form and is found in intestinal chylomicrons and VLDL and does not contain the LDL receptor-binding sequence (Olofsson et al., 1987). In both the liver and the intestine, the incorporation of apo B is essential for the secretion of triacylglycerol-rich lipoproteins (Hamilton, 1994; Innerarity et al., 1996; Olofsson et al., 1987).

ApoA-I is the major apolipoprotein of plasma high-density lipoproteins (HDL), as well as a major protein component of intestinal chylomicrons and VLDL, and it is produced by both the liver and the intestine in the human, the rat, and the swine (Black and Davidson,

DENNIS D. BLACK • Department of Pediatrics, University of Tennessee School of Medicine, Crippled Children's Foundation Research Center at Le Bonheur Children's Medical Center, Memphis, Tennessee 38103.
Intestinal Lipid Metabolism, edited by Charles M. Mansbach II et al., Kluwer Academic/Plenum Publishers, 2001.

Table 1. Intestinal Apolipoproteins

Apolipoprotein	Major Site of Synthesis	Associated Lipoproteins	Function
B-100	Prenatal: small intestine, liver, yolk sac Postnatal: liver	Hepatic VLDL and LDL	Role in lipoprotein assembly and secretion Cholesterol delivery to cells by binding to LDL receptor
B-48	Small intestine Liver (rodents)	Intestinal chylomicrons and VLDL	Role in lipoprotein assembly and secretion
A-I	Prenatal: small intestine, liver, yolk sac Postnatal: small intestine and liver	Intestinal chylomicrons, VLDL, and HDL Hepatic HDL	Cofactor for lecithin: cholesterol acyltransferase (LCAT)
A-IV	Small intestine Liver (rodents)	Intestinal chylomicrons, VLDL, and HDL Hepatic HDL Significant fraction unassociated with lipoproteins in plasma	LCAT activator Reverse cholesterol transport Augmentation of apo C-II activation of lipoprotein lipase Post-prandial satiety factor Inhibition of gastric acid secretion Antioxidant
C-III	Prenatal: small intestine, liver, yolk sac Postnatal: small intestine and liver	Intestinal chylomicrons, VLDL, and HDL Hepatic VLDL and HDL	Inhibition of metabolism and hepatic uptake of triglyceride-rich lipoproteins and their remnants

1989; Black and Rohwer-Nutter, 1991; Chapman, 1986; Davidson and Glickman, 1985; Wu and Windmueller, 1979). Its major metabolic role is that of a cofactor for lecithin: cholesterol acyltransferase (LCAT), the enzyme responsible for the production of HDL cholesteryl ester by the transfer of a fatty acid from phosphatidylcholine to free cholesterol in HDL (Fielding *et al.*, 1972).

Apo A-IV is an abundant apolipoprotein synthesized by both the liver and the small intestine in rodents and by the intestine only in humans and in swine (Bisgaier *et al.*, 1985; Black *et al.*, 1990; Elshourbagy *et al.*, 1986, 1987; Fidge, 1980; Green *et al.*, 1980). Apo A-IV is a component of nascent intestinal lipoproteins, including chylomicrons and HDL, and becomes dissociated from chylomicrons soon after secretion. Several studies have suggested roles for apo A-IV in the activation of LCAT (Chen and Albers, 1985; Steinmetz and Utermann, 1985), in reverse cholesterol transport by promoting cellular cholesterol efflux (Stein *et al.*, 1986), serving as a ligand for HDL binding to hepatocytes (Dvorin *et al.*, 1986), and in activation of lipoprotein lipase by the augmentation of apo C-II transfer from HDL (Goldberg *et al.*, 1990). Recent work suggests a role for apo A-IV as a postprandial satiety factor and as an inhibitor of gastric acid secretion in the adult rat, with these effects mediated at the level of the central nervous system (Fujimoto *et al.*, 1992, 1993; Okumura *et al.*, 1994). Also, apo A-IV may function as an endogenous antioxidant (Qin *et al.*, 1998).

Apo C-III is a small peptide that is synthesized by both the liver and the intestine (Wu and Windmueller, 1979; Zannis *et al.,* 1985). This apolipoprotein is a component of triacylglycerol-rich lipoproteins and retards the hepatic uptake of these particles until their core has become depleted of triacylglycerol by lipoprotein lipase and hepatic lipase (Windler *et al.,* 1980). Apo C-III then dissociates and transfers to HDL. Apo C-III has also been shown to directly inhibit the lipolytic activities of lipoprotein lipase (Brown and Baginski, 1972) and hepatic lipase (Kinnunen and Ehnholm, 1976).

Regulation of apolipoprotein gene expression in the small intestine may occur at several levels and under the influence of many physiological factors. Tissue-specific basal expression of apolipoprotein genes in the small intestine, including expression during development, appear to be under transcriptional control. Regulation by factors such as dietary and biliary lipid and hormones may occur at the pre-, co-, or posttranslational levels, depending on the specific factor and apolipoprotein. Several models have been used to study the regulation of intestinal apolipoprotein gene expression, including intact animals, the mesenteric lymph-fistula rat, cultured intestinal epithelial cells, and, most recently, transgenic mice. Limited human data are also available. Although several consistent patterns of regulation have been described from studies using these models, in many cases conflicts have arisen when comparing results among species, developmental groups, and *in vivo* versus *in vitro* model systems. This chapter summarizes major findings from recent studies utilizing all of these experimental approaches.

2. Intestinal Lipoprotein Biogenesis

The biosynthesis of intestinal chylomicrons and VLDL is a complex process involving the assembly of core and surface lipids with apolipoproteins in a coordinated fashion in the endoplasmic reticulum (ER), followed by processing in the Golgi apparatus and secretion at the basolateral membrane (Hay *et al.,* 1986; Hussain *et al.,* 1996; Innerarity *et al.,* 1996; Sabesin and Frase, 1977). Information on the details of this assembly process in the small intestine is lacking. It has been generally thought that apolipoproteins are translated from mRNA in the rough ER, signal peptides are cleaved, glycosylation occurs, and the apolipoproteins are incorporated with lipid from the smooth ER in a transitional organelle. Apo B is the only apolipoprotein for which this assembly process has been reasonably well characterized, and these studies have been performed almost exclusively in cultured hepatocytes (Borchardt and Davis, 1987; Davis, 1999; Hamilton, 1994; Macri and Adeli, 1997; Wu *et al.,* 1996; Yao *et al.,* 1997). In the hepatocyte, apo B appears to be cotranslationally inserted into the inner leaflet of the ER (Pease *et al.,* 1991). From the inner leaflet, both apo B and associated membrane-derived lipid can bud into the lumen of the ER to form nascent lipoprotein particles and thereby rescue apo B from degradation. Apo B degradation appears to be mediated by the cytoplasmic ubiquitin-dependent proteasome (Davis, 1999).

One theory of triacylglycerol-rich lipoprotein assembly (the two-step theory), derived mainly from data on VLDL production in hepatocytes, holds that dense apo B-containing, relatively triacylglycerol-poor particles are formed first (first step), which then fuse with large non-apo B-containing triacylglycerol-rich particles to form a pre-chylomicron (second step; Hamilton, 1994; Innerarity *et al.,* 1996; Rustaeus *et al.,* 1999). Microsomal triacylglycerol transfer protein (MTP) plays an essential role in the lipidation of newly synthesized, ER membrane-bound apo B, allowing its release into the ER lumen (step one; Gordon *et al.*, 1995). Movement of the pre-chylomicron from the ER to the Golgi apparatus appears to be the rate-limiting step for cellular triacylglycerol flux in the adult rat small intestine (Kumar and Mansbach, 1997; Mansbach and Nevin, 1998). Apo B is essential for lipoprotein secretion and is the only apolipoprotein that forms an integral part of the lipoprotein particle and does not dissociate at any time during peripheral metabolism (Hamilton, 1994; Olofsson *et al.*, 1987). A recent study has confirmed that each chylomicron contains one molecule of apo B-48 (Phillips *et al.*, 1997). Details of the incorporation of other apolipoproteins into intestinal lipoproteins are not known.

3. Apolipoprotein B

3.1. Apolipoprotein B Gene Expression

The human apo B gene (*apob*) is localized in a 43 kb region of chromosome 2 and contains 29 exons and 28 introns (Blackhart *et al.*, 1986; Huang *et al.*, 1986; Knott *et al.*, 1985; Law *et al.*, 1985; Ludwig *et al.*, 1987). The majority of the introns are within the 5' terminal one third of the gene. The gene is flanked by nuclear matrix attachment sites, which may serve to isolate the gene in a looped chromatin domain, and which coincide with two DNAse I-sensitive sites (Levy-Wilson and Fortier, 1989). Several *cis*- and accompanying *trans*-acting regulatory elements for the apo B gene have been identified, primarily using *in vitro* transfection studies in Hep G2 (hepatocyte) and Caco-2 (enterocyte) cells using reporter gene constructs (Kardassis *et al.*, 1996), as shown in Fig. 1. A proximal promoter region between nucleotides -150 and $+124$ directs expression in cultured hepatocytes and enterocytes (Kardassis *et al.*, 1990). This proximal promoter region contains three regulatory elements, including a portion of the first exon of the apo B gene, that bind transcription factors CAAT/enhancer binding protein (C/EBP), members of the hepatic nuclear factor (HNF)-3 and -4 family, apolipoprotein regulatory protein-1 (ARP-1), and V-ERB-related receptor (EAR)-2 and -3 (Kardassis *et al.*, 1992; Ladias *et al.*, 1992; Metzger *et al.*, 1993). Further studies have shown that the second intron between nucleotides $+621$ and $+1064$ enhances the apo B promoter in transfected Hep G2 and Caco-2 cells and contains six regulatory elements that are recognized by HNF-1, C/EBP, and other unidentified proteins (Brooks *et al.*, 1991; Brooks and Levy-Wilson, 1992). The third intron between nucleotides $+1065$ and $+2977$ contains a 155 base pair (bp) fragment that also enhances the apo B promoter (Levy-Wilson *et al.*, 1992). However, the binding proteins for this region have not been well characterized. Finally, an upstream repressor element has been identified between nucleotides -3067 and -2734, which contains a ARP-1 binding site and may function by interfering with the function of HNF-1, which binds to the proximal promoter (Paulweber *et al.*, 1993).

After transcription, apo B mRNA undergoes splicing and polyadenylation. In the intestine, two stable transcripts of 7 and 14 kb are produced due to the presence of two different polyadenylation sites (Chen *et al.*, 1987; Cladaras *et al.*, 1986a, 1986b; Giannoni *et al.*, 1994; Huang *et al.*, 1985; Powell *et al.*, 1987; Teng *et al.* 1990). The physiological significance of this observation is not known. In the mature small intestine, both transcripts undergo a novel mRNA editing process in which the cytosine in position 6666 is deami-

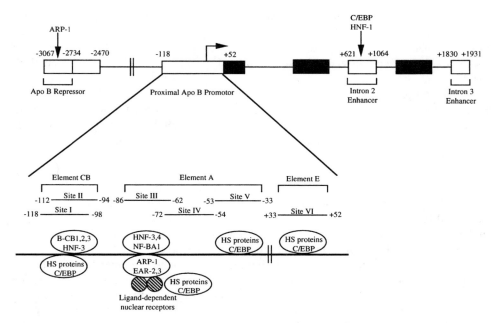

Figure 1. Regulatory elements and transcription factors responsible for transcriptional regulation of the apo B gene. This figure was adapted from Kardassis *et al.,* 1996. Abbreviations: ARP-1—apolipoprotein regulatory protein-1 (−); C/EBP—CAAT/enhancer binding protein (+); EAR-2,3—V-ERB-related receptor-2,3 (−); HNF-1,2,3—hepatic nuclear factor-1,2,3 (+); HS proteins—heat-stable proteins (+); NF-BA1—nuclear factor-BA1 (+).

nated to a uracil resulting in an in-frame stop codon (Chen *et al.,* 1987; Hospattankar *et al.,* 1987; Powell *et al.,* 1987; Teng *et al.,* 1993). This editing results in an mRNA species that translates apo B-48 protein. This editing is generally not regulated by dietary or hormonal manipulations in the small intestine (Chan *et al.,* 1997). This editing process and its regulation are discussed comprehensively in chapter 15.

3.1.1. *Apolipoprotein B Transgenic and Gene-Targeted Mouse Models*

The large size of the apo B gene hampered early efforts to produce transgenic mice expressing the human apo B gene (Kim and Young, 1998). Subsequent development of the P1 bacteriophage vector allowed development of a clone containing the complete 43 kb structural gene, as well as 19 kb of 5′ flanking sequences and 17.5 kb of 3′ flanking sequences (Callow et al., 1994; Linton et al., 1993; Young et al., 1994). Transgenic mice expressing this clone were found to have significant serum levels of human apo B. However, human apo B was found to be expressed only in the liver, not in the intestine. This finding suggested the requirement of a far distal promoter sequence for intestinal expression. Indeed, development of 145 and 207 kb bacterial artificial chromosomes (BACs) led to the resolution of this issue (Nielsen et al., 1997). The 207 kb BAC contained 120 kb of 5′ sequence and 35 kb of 3′ sequence. The 145 kb BAC contained 70 kb of 5′ sequence and 22 kb of 3′ sequence. Both BACs directed high levels of apo B expression in intestine of transgenic animals. Further studies established that the regulatory element controlling intestinal apo B expression is located between -33 and -70 kb 5′ to the structural gene (Nielsen et al., 1997).

Targeted gene disruption techniques have allowed production of apo B knockout mice. Homozygous apo B knockout mice ($apob^{-/-}$) demonstrated early gestational lethality, and those few surviving into late gestation exhibited severe malformations of the central nervous system and cranium (Farese et al., 1995; Huang et al., 1995). This finding may have been due to impaired transfer of lipid to the fetus because of defective lipoprotein assembly and secretion by the yolk sac, which normally expresses apo B in rodents (Farese et al., 1996a). Heterozygotes ($apob^{+/-}$) had reduced serum apo B and A-I and cholesterol levels (mainly HDL) and resistance to hypercholesterolemia on a high-fat diet, probably due to decreased hepatic lipoprotein synthesis and secretion. Heterozygotes exhibited no accumulation of fat in enterocytes on the high-fat diet, and cholesterol absorption was normal, suggesting no significant impairment of intestinal lipid absorption.

Apo B knockout animals carrying a single allele for the p158 human apo B transgene described previously ($HuBTg^{+/0}$, $apob^{-/-}$) were found to develop normally and were free of any developmental anomalies at birth (Young et al., 1995). However, because there was no intestinal apo B expression in these animals, in the suckling period they developed fat malabsorption and growth retardation. Histologic examination revealed enterocytes loaded with neutral fat, similar to the findings in intestine in humans with abetalipoproteinemia and homozygous hypobetalipoproteinemia. Approximately one third of the animals survived weaning and achieved normal adult size, although they did not absorb cholesterol and had very low vitamin E levels. However, they did not develop an essential fatty acid deficiency, suggesting some degree of triacylglycerol absorption, possibly via an alternate chylomicron-independent pathway. It is noteworthy that humans with abetalipoproteinemia experience spontaneous improvement in their coefficient of fat absorption with age, also suggesting induction of an alternative pathway. Although serum levels of liver-derived apo B and LDL cholesterol in these animals were the same as $HuBT^{+/0}$, $apob^{+/+}$ mice, serum HDL cholesterol levels were reduced by half. The investigators speculated that this may be related to reduced levels of intestinal apo A-I secretion, due to the block in chylomicron secretion.

Transgenic animals producing only apo B-48 ($apob^{48/48}$) or apo B-100 ($apob^{100/100}$) have also been produced and characterized (Farese et al., 1996b). With regard to intestinal apolipoprotein expression, a major question was whether lipid absorption with chylomicron formation and secretion could proceed normally with exclusive intestinal production of only apo B-100. The apo B-100 mice did not show any evidence of fat malabsorption and had retinyl palmitate absorption curves identical to those of the apo B-48 animals. Light microscopy did not show any abnormal accumulation of fat in enterocytes of apo B-100 animals either after a lipid bolus or chronic feeding of a high-fat diet. Additionally, animals with targeted disruption of a gene involved in the apo B mRNA editing complex, $apobec-1^{-/-}$, were found to develop normally with no evidence of fat malabsorption, despite exclusive production of apo B-100 (Hirano et al., 1996). Therefore, intestinal chylomicron assembly and secretion appear to proceed normally regardless of the size of the associated apo B protein.

3.1.2. Developmental Regulation of the Apo B Gene

In the developing rat, small intestinal apo B mRNA expression increases dramatically just prior to parturition, remains high during the first week of life, and declines to very low levels by the end of the second week of life through the suckling and weaning periods, followed by an increase to adult levels (Demmer *et al.,* 1986). It is also interesting that there is a 20-fold increase in placental apo B mRNA concentration during the last 48 hours of pregnancy, which suggests a role for this organ in maternal-to-fetal lipid transport immediately prior to parturition. Also, apo B mRNA expression is high in fetal membranes during late gestation. Hepatic apo B mRNA expression follows basically the same pattern as in the small intestine, except for high levels as early as day 18 of gestation when intestinal expression is still very low (Demmer *et al.,* 1986). Incorporation of radiolabeled methionine into apo B immunoprecipitates from tissue slices confirmed the synthesis of apo B protein in parallel with mRNA expression.

Apo B mRNA abundance has also been studied in early gestation human fetal tissues. In 6 to 12 week gestation fetuses, mRNA was detected by Northern blotting in predominantly the liver, the yolk sac, and the intestine with trace amounts detected in the stomach, the adrenal, and the kidney (Hopkins *et al.,* 1986). Synthesis and secretion of apo B from the intestine, the yolk sac, and the liver from fetuses in this gestational range were confirmed by *in vitro* radiolabeling and immunoprecipitation (Shi *et al.,* 1985). Other studies in fetal human small intestine and liver demonstrated no developmental changes in apo B mRNA abundance (Teng *et al.,* 1990). However, a study employing human fetal intestinal explants from 17 to 20 weeks gestation demonstrated that hydrocortisone suppressed apo B-100 synthesis and the secretion of triacylglycerol, phospholipid, and cholesteryl esters and stimulated synthesis of apo B-48 and A-I (Loirdighi *et al.,* 1997). This study suggests that corticosteroids may play a role in the regulation of intestinal apo B synthesis and secretion in the fetal small intestine.

Intestinal apo B expression in the developing piglet has been studied using a variety of techniques, including radiolabeling of intestinal explants *in vitro* and intestinal segments *in situ* (Black, 1992; Black and Davidson, 1989; Black and Ellinas, 1992; Black and Rohwer-Nutter, 1991). Jejunal and ileal apo B synthesis in fetal (40 day gestation), newborn (2-day-old), and suckling (1–2-week-old) pigs was studied in intact animals. Postnatal animals were given either low- or high-triacylglycerol isocaloric diets, and a bile-diverted group was also studied to test the effects of bile constituents on apolipoprotein synthesis (Black and Davidson, 1989; Black and Rohwer-Nutter, 1991). Jejunal and ileal segments in postnatal animals were then radiolabeled with [^{3}H]leucine *in situ,* and fetal intestine was radiolabeled by injection of [^{3}H]leucine into the umbilical vein. Apo B synthesis was very low in fetal intestine but increased dramatically in the newborn animals and then declined in the older suckling animals. There was a proximal to distal gradient of apo B synthesis in postnatal animals with lower synthesis in the ileum. Synthesis was not modulated by dietary or biliary lipid. However, when mucosal apo B mass was quantitated in intestinal mucosa in the newborn animals, mass was found to be regulated by dietary and biliary lipid (Black and Rohwer-Nutter, 1991). This lack of change in apo B synthesis despite changes in mass suggested a posttranslational regulatory mechanism.

The effects of epidermal growth factor (EGF), insulin, hydrocortisone, and thyroid hormone on apo B synthesis in jejunal explants from newborn piglets have also been investigated, and the only effect observed was a modest decrease in synthesis with treatment with the combination of EGF and hydrocortisone (Black and Ellinas, 1992). This study suggests that apo B synthesis in the newborn small intestine is not significantly regulated by growth factors or by hormones. In contrast, apo B expression at the posttranslational level in the neonatal period does appear to be regulated by lumenal lipid absorption.

3.2. Regulation of Intestinal Apo B Expression by Lumenal Fatty Acids

Studies in several model systems have provided data suggesting that, in most situations, the apo B gene is constitutively expressed in the small intestine with mRNA levels remaining constant, similar to the liver (Pullinger *et al.,* 1989). In most instances, absorbed lumenal lipid appears to regulate apo B secretion by co- or posttranslational mechanisms. The effects of dietary lipid on jejunal and ileal apo B synthesis in the rat were studied by Davidson *et al.* (1988b; 1986; 1987). Neither jejunal nor ileal apo B synthesis changed during acute triacylglycerol absorption, even after several days on a fat-free diet. Another recent study in the adult rat confirmed the lack of regulation of intestinal apo B mRNA by dietary lipid (Middleton and Schneeman, 1996). Previous studies in the mouse (Srivastava *et al.,* 1991) and the rabbit (Fisher *et al.,* 1988) demonstrating an increase in intestinal apo B mRNA levels with fat feeding may have been due to species differences or to the timing of the feeding. One study performed in normal adult human subjects demonstrated no change in jejunal apo B mRNA levels after an acute fatty meal (Lopez-Miranda *et al.,* 1994).

The lymph-fistula rat model has provided information on the regulation of intestinal apo B secretion by absorbed lipid. However, studies of mesenteric lymph apo B transport have yielded conflicting results. In studies by Renner *et al.* (1986) and by Nishimura *et al.* (1996), acute absorption of a bolus of an emulsion containing glycerol trioleate and taurocholate caused an increase in lymph apo B mass output, concomitant with an increase in triacylglycerol output. However, Hayashi *et al.* (1990) demonstrated that absorption of a continuous infusion of a lipid emulsion containing glycerol trioleate, phosphatidylcholine, and taurocholate did not change apo B jejunal synthesis or mass secretion in mesenteric lymph, despite the increase in triacylglycerol output. These results were interpreted as demonstrating that lipid absorption increases the size, but not the number, of mesenteric lymph chylomicrons in the rat. These conflicting results may be related to differences in method of lipid administration (bolus versus continuous), composition of the lipid emulsion, or possible contamination of lymph with plasma apo B.

Feeding studies have been performed in the newborn piglet comparing the effects of three dietary lipid formulations on intestinal apolipoprotein synthesis (H. Wang *et al.,* 1998a, 1996). Two-day-old piglets were fed acutely (24 hours) or chronically (7 days) with diets containing fat as medium-chain length triacylglycerols (MCT oil), intermediate-chain length triacylglycerols (coconut oil), and long-chain length unsaturated triacylglycerols (safflower oil). In the acute feeding group, jejunal apo B synthesis in the group receiving coconut oil was approximately half that of the other two groups. This same trend was also observed in the ileum. However, in the chronic feeding group, these differences were no longer present, suggesting a loss of intestinal apo B synthetic responsiveness in the older animal, possibly related to intestinal development.

The specific type of absorbed lumenal lipid may function to regulate intestinal apolipoprotein expression. The various exogenous (dietary) and endogenous (biliary) lumenal lipids include fatty acids of varying chain length and degree of saturation, cholesterol, phospholipid (primarily phosphatidylcholine), and bile acids. This issue is difficult to address in *in vivo* studies because of the technical problems of delivering a single lipid component to the microvillus membrane of the enterocyte. For example, fatty acids and cholesterol require bile acids for solubilization and absorption, yet bile acids may have an independent influence on enterocyte apolipoprotein expression. Therefore, *in vitro* studies using cultured intestinal epithelial cells have been most useful for investigating this area. The cell line used most often for such studies is the Caco-2 cell line, derived from a human colon carcinoma (Field and Mathur, 1995; Levy *et al.*, 1995). When plated on permeable filters in Transwell culture dishes, these cells differentiate to resemble enterocytes with an apical microvillus membrane and tight junctions. The Transwell culture dishes allow segregation of the apical and basolateral culture medium, allowing lipids to be added to the apical compartment and secreted lipoproteins to be collected from the lower basolateral medium compartment. Although convenient and reasonably physiological, this model system has its limitations. This cell line was derived from a colon carcinoma and therefore may not behave as a normal mature enterocyte. In fact, in many ways these cells behave as partially differentiated fetal enterocytes. For example, when maximally differentiated, these cells still produce a significant amount of apo B-100, reflecting reduced expression of the apo B mRNA editing mechanism, as in fetal small intestine (Jiao *et al.*, 1990; Wagner *et al.*, 1992). Although these cells express apo B and A-I, their expression of other apolipoproteins is limited, particularly that of apo A-IV and C-III (Levy *et al.*, 1995). Also, unlike normal enterocytes, these cells do not usually express intestinal fatty acid binding protein (I-FABP), and they esterify fatty acids predominantly via the glycerol-3-phosphate pathway (Levy *et al.*, 1995). Culture conditions appear to be critical to the patterns of gene expression by Caco-2 cells (Levy *et al.*, 1995). In fact, a recent study described culture conditions that induced expression of the I-FABP gene and enhanced the expression of the apo B mRNA editing gene (*apobec*), apo A-IV, and apo C-III genes (Le Beyec *et al.*, 1997). However, despite these limitations, several careful studies have provided insight into the regulation of apolipoprotein expression, particularly for apo B.

Most studies with Caco-2 cells have demonstrated an increase in basolateral secretion of apo B, triacylglycerol, cholesterol, and phospholipid when incubated with oleic acid (18:1) added to the apical medium (Dashti *et al.*, 1990; Field *et al.*, 1988; Traber *et al.*, 1987). However, apo B synthesis (Murthy *et al.*, 1992) and mRNA levels (Moberly *et al.*, 1990; Murthy *et al.*, 1992) did not change with the increase in apo B mass secretion after incubation with oleic acid. Study of the intracellular degradation of apo B with pulse-chase radiolabeling with [^{35}S]methionine demonstrated a decrease in apo B-100 degradation with oleic acid incubation (Murthy *et al.*, 1992), similar to that observed in hepatocytes (Yao *et al.*, 1997). However, the magnitude of this change would not appear to be sufficient to account for all of the increased apo B mass secretion. In another study, undifferentiated Caco-2 cells were transfected with a recombinant human apo B-48 cDNA (Luchoomum *et al.*, 1997). These cells also expressed MTP and were capable of triacylglycerol synthesis. After incubation with oleic acid, transfected cells secreted larger triacylglycerol-rich lipoproteins containing both recombinant and endogenous apo B. Although degradation of apo B-100 was decreased slightly, that of apo B-48 did not change. Under these experimental conditions, the availability of apo B appeared to determine the amount of triacylglycerol secreted.

Similar studies have been carried out in a newborn piglet intestinal epithelial cell line, IPEC-1, derived from a newborn, unsuckled piglet (Gonzalez-Vallina *et al.*, 1996). These cells differentiate when plated on permeable, collagen-coated Transwell filters in serum-free medium and become polarized with an apical microvillus membrane. These cells produce predominantly apo B-100 when undifferentiated but produce both apo B-100 and B-48 in the differentiated state due to an increase in apo B mRNA editing with over 50% of the apo B mRNA transcripts edited (Gonzalez-Vallina *et al.*, 1996). Basolateral secretion of apo B, triacylglycerol, phospholipid, and cholesteryl ester increases after apical incubation with oleic acid (Gonzalez-Vallina *et al.*, 1996; H. Wang *et al.*, 1997). When apo B synthesis and degradation were studied in the cells by [^{35}S]methionine radiolabeling after incubation with oleic acid, there was no significant difference in either synthesis or degradation. These findings may be explained by the presence of a large, slowly turning over intracellular pool of apolipoprotein, which is mobilized for lipoprotein assembly in response to fatty acid uptake. The most likely location for such an intracellular pool of apo B would be the ER, and this pool would require protection from degradation. Therefore, partial lipidation (possibly with endogenous triacylglycerol) of newly translated apo B may complete the first step of the two-step chylomicron assembly scheme. This partial lipidation may result in formation of a small apo B-containing particle in the ER lumen, which would be a reasonable location for an apo B storage pool. A small spherical apo B-containing particle has been isolated from rat intestinal epithelial cells (Magun *et al.*, 1985). In addition, intracellular pools of apo B, A-I, and A-IV not associated with lipoprotein particles have been identified in rat intestinal epithelial cells associated with smooth ER membranes (90%) in the fasting state, and the lipoprotein-bound fraction was shown to increase with fat feeding by 5% to 10% (Magun *et al.*, 1988).

In summary, these studies of apo B in Caco-2 and IPEC-1 cells suggest two mechanisms for regulation of apo B availability for lipoprotein production with oleic acid absorption. First, enterocytes may utilize a preformed pool of apo B. Second, to replenish this pool and possibly provide additional apo B during absorption, less newly synthesized apo B is degraded. This scheme would be particularly advantageous in the small intestine, where a large load of dietary triacylglycerol may appear at any time and would have to be efficiently absorbed.

The effects of other fatty acids on intestinal apo B expression have been studied. As expected, medium-chain length fatty acids do not affect apo B synthesis or secretion, because they bypass the lipoprotein synthetic pathway and are secreted directly into portal blood (Tso, 1994). Saturated fatty acids are taken up and esterified by Caco-2 and IPEC-1 cells (Field *et al.*, 1988; van Greevenbroek *et al.*, 1996; H. Wang *et al.*, 1997). However, relative to oleic acid, they are not as effectively secreted, particularly in the case of palmitic (16:0) and stearic (18:0) acids. Secretion of apo B and of newly synthesized phospholipid parallels these differences in triacylglycerol secretion (van Greevenbroek *et al.*, 1996; H. Wang *et al.*, 1997). The apparent relative inefficiency of the basolateral secretion of apo B and of newly synthesized triacylglycerol and phospholipid after incubation with long-chain length saturated fatty acids may be related to the incorporation of these fatty acids into cell membrane phospholipids, as well as triacylglycerol and phospholipid destined for secretion, in the ER. This may adversely affect membrane structure and impair lipoprotein assembly and secretion, including the incorporation of apo B into nascent triacylglycerol-rich lipoprotein particles. Incubation of IPEC-1 and Caco-2 cells with palmitic and stearic acids resulted in an intracellular accumulation of phospholipid (van Greevenbroek *et al.*, 1995;

H. Wang *et al.*, 1997). This accumulation resulted in the appearance of excessive intracellular membrane material by electron microscopy in the Caco-2 cells (van Greevenbroek *et al.*, 1995). A study in the lymph-fistula rat demonstrated decreased absorption of glycerol tristearate, as compared to glycerol trioleate (Bergstedt *et al.*, 1990). Although this was due in part to poor lipolysis, the absorbed glycerol tristearate was transported into mesenteric lymph less efficiently. Apo B secretion was not measured in this study. Nevertheless, these findings suggest that the *in vitro* observations in cultured cells may have relevance *in vivo*.

Long-chain length polyunsaturated fatty acids (LC-PUFAs) appear to have striking effects on lipoprotein metabolism. Linoleic (18:2n-6) and linolenic (18:3n-3) are considered essential fatty acids, and their dietary intake lowers serum cholesterol levels relative to saturated fatty acids (Kris-Etherton and Yu, 1997). Linoleic and linolenic acids are precursors of arachidonic (20:4n-6, AA) and docosahexaenoic (C22:6n-3, DHA) acids, respectively. Because the neonate cannot efficiently elongate and desaturate linoleic and linolenic acids to AA and DHA, AA and DHA may be essential fatty acids in the neonate (Innis, 1991). Both AA and DHA have been shown to be essential for normal nervous system and retinal development (Carlson *et al.*, 1994; Innis *et al.*, 1997; Innis *et al.*, 1994). The n-3 LC-PUFAs, abundant in marine oils, have been shown to reduce serum triacylglycerol levels in humans (Harris, 1989). Therefore, the effects of these fatty acids on intestinal lipid absorption and apolipoprotein expression have been of interest (Harris *et al.*, 1982).

Linoleic and linolenic acids are not as efficiently incorporated into cellular triacylglycerol or secreted as triacylglycerol-rich lipoproteins and do not stimulate apo B secretion to the same degree as compared to oleic acid in Caco-2 and IPEC-1 cells (Field *et al.*, 1988; van Greevenbroek *et al.*, 1996; H. Wang *et al.*, 1997). When linoleic acid was compared to its oxidized analogue, 13-hydroxy octadecadienoic (13-HODE), in Caco-2 cells, 13-HODE incubation resulted in decreased secretion of newly synthesized triacylglycerol and apo B and triacylglycerol and apo B mass without altering apo B mRNA abundance (Murthy *et al.*, 1998). This suggests that fatty acid oxidation may result in impaired triacylglycerol-rich lipoprotein and apo B secretion by the small intestine.

Studies of the effects of eicosapentaenoic acid (C20:5n-3, EPA) on apo B and lipid secretion in Caco-2 cells have yielded somewhat conflicting results. In experiments by Murthy *et al.* (1992; 1990), this fatty acid was shown to decrease synthesis and secretion of triacylglycerol relative to oleic acid. Oleic acid was predominantly incorporated into cellular triacylglycerols, whereas more EPA was incorporated into cellular phospholipids. Microsomal phospholipid from cells incubated with EPA was enriched in EPA, and microsomal triacylglycerol synthesis was reduced. Apo B synthesis and secretion were reduced after EPA treatment for 48 hours, and apo B mRNA levels were also reduced, suggesting pretranslational regulation by EPA. In contrast, Ranheim *et al.* (1992) found no differences in the effects of oleic acid and EPA on triacylglycerol synthesis and secretion, but they did observe reduced apo B secretion with EPA incubation relative to oleic acid after relatively shorter incubation times. However, when this group exposed cells to EPA for as little as 2 days before performing acute experiments, results were similar to those of Murthy *et al.* (1992; 1990). Results of similar studies from IPEC-1 cells generally agree with those of Murthy *et al.* (1992; 1990), demonstrating decreased apo B secretion and triacylglycerol synthesis and secretion after EPA incubation relative to oleic acid (Black *et al.*, 1998). Also, DHA resulted in a significant decrease in apo B secretion in IPEC-1 cells (Black *et al.*, 1998). However, DHA caused a modest decrease in triacylglycerol synthesis and secretion and actually resulted in higher phospholipid secretion relative to oleic acid and EPA in

IPEC-1 cells (Black *et al.,* 1998). A study in the lymph-fistula rat demonstrated impaired transport of DHA and EPA into mesenteric lymph relative to oleic acid (Ikeda *et al.,* 1995). Taken together, these results suggest that EPA and DHA can impair triacylglycerol synthesis and secretion and apo B expression, but exposure must occur for a long enough period of time for these fatty acids to be incorporated into microsomal membranes to exert an effect on lipoprotein assembly.

3.3. Regulation of Intestinal Apo B Expression by Lumenal Bile Acids, Cholesterol, and Phospholipid

Biliary lipids, composed of bile acids, cholesterol, and phosphatidylcholine, are a major source of lumenal lipid in the small intestine. *In vivo* studies in the adult rat have demonstrated that jejunal apo B synthesis, but not mRNA abundance, falls after bile diversion, and this decrease may be completely prevented by continuous replacement with taurocholate (Davidson *et al.,* 1986). In these studies, ileal apo B synthesis, but not mRNA levels, was down-regulated to an even greater degree by biliary diversion but was only partially restored by bile acid replacement (Davidson *et al.,* 1986). Further studies in a chronic bile-diverted rat model demonstrated that the reinfusion of either taurocholate or taurocholate plus lysophosphatidylcholine caused a re-expression of apo B synthesis (Davidson *et al.,* 1988b). It was also shown that cholesterol absorption was negatively correlated with jejunal apo B synthesis. These studies demonstrated important roles for biliary lipid components in the regulation of intestinal apo B synthesis in the adult rat. In contrast to findings in the rat, Field *et al.* (1994) found that taurocholate inhibited apo B secretion in Caco-2 cells by increasing its rate of degradation. Because treatment with a calcium ionophore produced the same effect, these investigators speculated that taurocholate might produce its effect on apo B secretion by the same mechanism by causing the release of intracellular calcium. The significance of this effect *in vivo* is unknown.

Phosphatidylcholine, independent of its hydrolysis to lysophosphatidylcholine and subsequent cellular uptake, has been shown to increase apo B secretion by increasing protein synthesis in Caco-2 cells (Field *et al.,* 1994). Mathur *et al.* (1996) also demonstrated an increase in basolateral secretion of triacylglycerol accompanied by a modest increase in triacylglycerol synthesis after incubation of Caco-2 cells with phosphatidylcholine. Although *in vivo* data from the rat suggest that changes in intestinal cholesterol absorption may regulate apo B expression (Davidson *et al.,* 1987), it is very difficult to sort out cholesterol as an independent variable, because this sterol requires solubilization with molecules such as bile acids, which may exert an independent effect on apo B. Caco-2 cells incubated with cholesterol or with 25-hydroxycholesterol did not demonstrate a change in apo B secretion compared to controls (Field *et al.,* 1994). Likewise, cholesterol did not induce any change in apo B secretion in IPEC-1 cells, whether solubilized with taurocholate or phosphatidylcholine (H. Wang *et al.,* 1999). Therefore, most likely, intestinal apo B expression is not significantly regulated by cholesterol.

3.4. Regulation of Intestinal Apo B Expression by Non-Lipid Factors

Phosphorylation of apo B-48, but not of apo B-100, occurs in Caco-2 cells, and the amount of phosphorylated apo B-48 is increased by incubation with okadaic acid, which inhibits protein serine/threonine phosphatase activity (Mathur *et al.*, 1993). However, treatment with okadaic acid decreased secretion of both apo B-48 and B-100 with an accompanying decrease in apo B mRNA levels. Although this is an interesting observation, the mechanisms and significance of the effects of phosphorylation and of okadaic acid on intestinal apo B expression are presently not clear.

Intracellular calcium ion concentrations may play a role in intestinal apo B and lipoprotein synthesis and secretion. Calcium antagonists inhibit lipid absorption by inhibiting triacylglycerol and chylomicron secretion (Levy *et al.*, 1992). Calcium ionophore treatment increased apo B synthesis and secretion in Caco-2 cells grown on plastic (Hughes *et al.*, 1988). However, in Caco-2 cells grown on filters in Transwell culture dishes, calcium ionophore treatment decreased apo B secretion by increasing the rate of apo B degradation with no effect on synthesis or on mRNA levels (Field *et al.*, 1994).

Cytokines may modulate apo B and lipid secretion in the small intestine. Treatment of Caco-2 cells with tumor necrosis factor-α (TNF-α) resulted in decreased secretion of triacylglycerol, phospholipids, and cholesteryl esters, as well as decreased apo B-48 and B-100 synthesis (Mehran *et al.*, 1995). Murthy *et al.* (1996) found that interleukin-1β (IL-1β), interleukin-6 (IL-6), and TNF-α all decreased the secretion of apo B in Caco-2 cells with IL-6 being the most potent. Interleukin-6 decreased synthesis and secretion of apo B-100 and B-48, as well as apo B mRNA abundance, without affecting protein degradation. In contrast, transforming growth factor-β (TGF-β) increased the secretion of apo B and triacylglycerol. TGF-β increased synthesis and secretion of apo B-100 and B-48, as well as apo B mRNA levels, without affecting protein degradation. These effects may be important in states characterized by small intestinal inflammation, such as Crohn's disease.

Vitamins and hormones may play a role in the regulation of intestinal apo B expression. Intestinal apo B-48 synthesis was decreased in hypothyroid rats and increased three- to fourfold after triiodothyronine administration (Davidson *et al.*, 1988a). Wagner *et al.* (1992) demonstrated an increase in apo B mRNA in Caco-2 cells after treatment with triiodothryronine and 1,25-dihydroxycholecalciferol with no accompanying change in apo B mass. Retinoic acid treatment resulted in increased apo B mRNA levels, but with an accompanying decrease in apo B secretion.

4. Apolipoproteins A-I, A-IV, and C-III

4.1. Apolipoprotein A-I/C-III/A-IV Gene Cluster

The genes for apo A-I, C-III, and A-IV form a cluster on chromosome 11 in humans (Karathanasis, 1985). The apo C-III gene is located 2.5 kb downstream from the apo A-I gene and 6.6 kb upstream from the apo A-IV gene. The apo C-III gene is transcribed in the opposite direction to that of the apo A-I and A-IV genes. Tissue-specific regulation of apo A-I, C-III, and A-IV gene transcription has been the focus of extensive study, and relevant *cis*- and *trans*-acting factors have been identified (Kardassis *et al.*, 1996). Figure 2 shows the organization of the human apo A-I, C-III, A-IV gene complex, as well as proximal and distal regulatory elements with associated transcription factors that have been identified to date. The transcriptional regulation of these genes has proven to be extremely complex. However, a few basic concepts have emerged from several studies (Kardassis *et al.*, 1996).

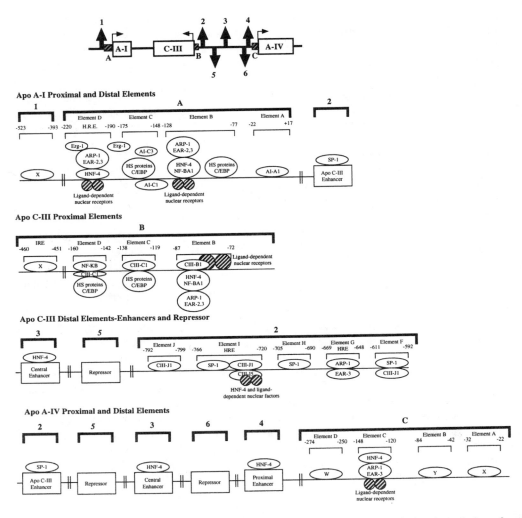

Figure 2. Regulatory elements and transcription factors responsible for transcriptional regulation of the apo A-I/C-III/A-IV gene cluster. Stippled boxes represent proximal promoters, upward arrows represent enhancers, and downward arrows represent repressors. This figure was adapted from Kardassis *et al.,* 1996 and from Vergnes *et al.,* 1997. Abbreviations: ARP-1—apolipoprotein regulatory protein-1 (−); C/EBP—CAAT/enhancer binding protein (+); EAR-2,3—V-ERB-related receptor-2,3 (−); Erg-1—early growth response element-1; HNF-4—hepatic nuclear factor-4 (+); HRE—hormone-response element; NF-BA1—nuclear factor-BA1 (+); NF-KB—nuclear factor kappa B (+); SP-1—stimulatory protein-1 (+); W, X, Y indicate nuclear activities of unknown identity.

An important observation has been that the distal apo C-III enhancer increases the activity of the promoters of all three genes in the cluster (Bisaha *et al.,* 1995; Ginsburg *et al.,* 1995; Ktistaki *et al.,* 1994; Lauer *et al.,* 1991; Ogami *et al.,* 1990; Talianidis *et al.,* 1995; Walsh *et al.,* 1993). The presence of this enhancer is necessary for maximum expression of apo A-IV in Hep G2 (liver) and Caco-2 (intestine) cells, regardless of its orientation. Other studies have shown that hepatic expression of the gene cluster requires only 5' regulatory elements of the apo A-I and C-III genes, whereas the intestinal expression of the apo A-I and A-IV genes requires elements located in the intergenic sequence between the apo C-III and A-IV genes (A. Walsh *et al.,* 1989; A. M. Walsh *et al.,* 1993). Recently, additional regulatory regions in the apo C-III/A-IV intergenic region have been identified (Vergnes *et al.,* 1997). An activating region in the center of the apo C-III/A-IV intergenic sequence modulates apo A-IV and C-III expression in Hep G2 and in Caco-2 cells (Vergnes *et al.,* 1997). Another positive element, located immediately upstream from the apo A-IV promoter, does not influence apo C-III expression. These regions are highly activated by HNF-4. Two negative elements have also been identified, one located near the apo A-IV gene and the other between the apo C-III distal enhancer and the central enhancer (Vergnes *et al.,* 1997). Studies in transgenic animals have provided information on the elements required for intestinal expression of apo A-I, C-III, and A-IV *in vivo.* Apo A-I expression in intestine requires 300 bp 5' to the apo A-I gene, as well as the apo C-III distal enhancer (Lauer *et al.,* 1991; A. Walsh *et al.,* 1989; A. M. Walsh *et al.,* 1993). Intestinal apo A-IV expression requires at least the apo C-III promoter 7.7 kb 5' of the apo A-IV promoter (Lauer *et al.,* 1991), and intestinal expression of the apo C-III gene requires a 5.9 kb intergenic sequence between the apo A-IV and C-III genes (de Silva *et al.,* 1994a). Recently, studies of transgenic mice demonstrated that the upstream $-700/-310$ fragment of the apo A-IV gene combined with the $-890/-500$ apo C-III enhancer is sufficient to direct a pattern of reporter gene expression similar to that of the endogenous apo A-1V gene (Le Beyee *et al.,* 1999).

Transcription factors have been identified that bind to specific regulatory elements of the apo A-I/C-III/A-IV gene complex to exert complex and sometimes antagonistic effects, as shown in Fig. 2. The three proximal regulatory elements of the apo A-I gene, A-IB, A-IC, and A-ID, are required for expression in the liver both *in vitro* and *in vivo* (Bisaha *et al.,* 1995; Papazafiri *et al.,* 1991; A. Walsh *et al.,* 1989). Elements A-IB and A-ID bind HNF-4, ARP-1, EAR-2 and -3, and members of the retinoic acid X receptor (RXR) family. Element A-IC binds several heat-stable factors, as well as C/EBP, which stimulate transcription (Ge *et al.,* 1994; Papazafiri *et al.,* 1991). Early growth response factor-1 (Erg-1) binds at sites -220 to -211 and -189 to -180 bps in the apo A-I promoter region and may play a role in liver regeneration (Kilbourne *et al.,* 1995). Several factors binding to the apo C-III distal enhancer and HNF-4 binding to the proximal apo A-I promoter are important for the expression of apo A-I in intestinal cells (Kardassis *et al.,* 1996).

Stimulatory protein-1 (SP1) is the major positive transcription factor that binds to the apo C-III distal enhancer to elements F, H, and I. Optimal activation of the apo C-III gene in hepatic and in intestinal cells requires positive regulatory factors that bind to the hormone response elements (HREs) in the distal enhancer, as well as all positive regulatory factors that bind to the distal regulatory elements F through J. Proximal element B binds HNF-4, ARP-1, EAR-2 and -3, and heterodimers of RXR with retinoic acid receptor (RAR), triiodothyronine receptor (T_3R) and peroxisome proliferation activator receptor (PPAR). HNF-4 positively activates the apo C-III promoter in liver and intestine, whereas ARP-1 and EAR-3 negate that action (Mietus-Snyder *et al.,* 1992; Ochoa *et al.,* 1993). Elements C and D bind C/EBP and two heat-stable proteins, and mutation of these elements to abolish binding greatly reduces hepatic and intestinal apo C-III gene transcription. These positive activators act synergistically with factors binding to distal enhancer elements. NF-κB binds to element D suggesting a potential role for this region in the acute phase response (Gruber *et al.,* 1994).

In summary, intestinal expression of the apo A-I gene is regulated by interactions between nuclear hormone receptors bound to apo A-I promoter elements D and B and factors bound to the apo C-III enhancer (Papazafiri *et al.*, 1991; Tzameli and Zannis, 1996; A. Walsh *et al.*, 1989; A. M. Walsh *et al.*, 1993). In a recent study in humans, apo A-I synthesis in intestinal endoscopic biopsy specimens correlated with steady-state apo A-I mRNA levels over a 10-fold range (Naganawa *et al.*, 1997). This variation appeared to be due to five distinct polymorphisms upstream of the apo C-III gene. Apo C-III intestinal gene expression is controlled by orphan and ligand-dependent nuclear hormone receptors bound to elements B, G, and I, and SP1 or other factors bound to the distal enhancer (de Silva *et al.*, 1994b; Ogami *et al.*, 1990; Talianidis *et al.*, 1995). Intestinal expression of the apo A-IV gene is controlled by synergistic interactions between nuclear hormone receptors bound to the apo A-IV promoter element C (HNF-4, ARP-1, and EAR-3), factors bound to the distal apo C-III enhancer (particularly SP1), and other factors such as HNF-4 bound to distal regulatory sites (Kardassis *et al.*, 1996; Ochoa *et al.*, 1993). None of the regulatory elements studied to date in this gene cluster have been characterized as lipid responsive in the small intestine. However, linkage studies in mouse strains expressing tremendous variations in hepatic apo A-IV gene expression in response to fat feeding have confirmed that control of hepatic apo A-IV mRNA levels involves both *cis*-acting elements linked to the apo A-IV gene and genetically distinct *trans*-acting factors (Reue *et al.*, 1993; Williams *et al.*, 1989).

4.1.1. Apolipoprotein A-I, C-III, and A-IV Transgenic and Gene-Targeted Mouse Models

Studies of transgenic mice expressing human apo A-I, C-III, and A-IV genes have contributed to understanding of the required gene regulatory elements for tissue-specific expression as discussed previously (Bisaha *et al.*, 1995; Lauer *et al.*, 1991; A. Walsh *et al.*, 1989; A. M. Walsh *et al.*, 1993). Mouse models have also enhanced our understanding of the physiological function of these apolipoproteins. Although there have been no specific studies addressing the effects of expression of human apo A-I and C-III in transgenic mice (Golder-Novoselsky *et al.*, 1992; Ito *et al.*, 1990; Mortimer *et al.*, 1997; Plump *et al.*, 1994; Tsukamoto *et al.*, 1997) or the targeted disruption of these genes in mice (Maeda *et al.*, 1994; Plump *et al.*, 1992; Voyiaziakis *et al.*, 1998; Williamson *et al.*, 1992; Zhang *et al.*, 1992) on intestinal lipid absorption, there was no obvious evidence of impaired or enhanced lipid absorption in these animal models. However, apo A-IV mouse models have been specifically studied to define a possible role for apo A-IV in fat absorption. The additional intestinal expression of human apo A-IV in transgenic mice did not enhance the absorption of lipid or of lipid-soluble vitamins (Aalto-Setala *et al.*, 1994). Also, apo A-IV knockout mice had normal lipid absorption as assessed by a retinyl palmitate tolerance test, as well as normal growth and weight gain (Weinstock *et al.*, 1997). Although these results suggest that apo A-IV does not play a major role in intestinal lipid absorption, it is possible that under certain circumstances, such as during the neonatal period when the diet is particularly lipid enriched, apo A-IV might serve to enhance fat absorption.

4.1.2. Developmental Regulation of the Apo A-I, A-IV, and C-III Genes

4.1.2a Apolipoprotein A-I. Apo A-I mRNA expression in the developing rat small intestine is very low in fetal intestine, but it increases immediately after parturition to approximately adult levels where it remains through adulthood (Elshourbagy *et al.*, 1985). As with apo

B, apo A-I mRNA is also expressed abundantly in rat fetal membranes. A proximal to distal gradient of apo A-I mRNA expression is established in the rat intestine in the perinatal period, which persists into adulthood (Rubin *et al.*, 1989). Hepatic apo A-I mRNA is relatively much more abundant in the rat fetus, remains at a constant level through parturition and suckling, and begins to decline during weaning to lower adult levels (Elshourbagy *et al.*, 1985). Cultured intestinal and hepatic explants from fetal and neonatal rats were used to study apo A-I synthesis by incorporation of [^{3}H]leucine (Perrin-Ansart *et al.*, 1988). Synthesis was barely detectable at day 18 of gestation and increased 2 days before the end of gestation. Apo A-I synthesis leveled off at birth in the intestine but continued to increase in the liver during suckling. Treatment of explants with dexamethasone enhanced apo A-I synthesis only at day 20 of gestation. Apo A-I synthesis by explants either at day 18 of gestation or at 0, 2, and 5 days of age were not affected by dexamethasone treatment. Thyroid hormone alone had no effect, but it enhanced the dexamethasone-induced synthetic stimulation. Administration of cortisone acetate to pregnant rats from day 14 of gestation resulted in stimulation of apo A-I synthesis only when it was prolonged after day 20 of gestation.

Apo A-I mRNA expression in the human fetus has been studied in abortuses between 6 and 12 weeks gestation by Northern blot hybridization and was high in the liver, the intestine, and the yolk sac (Hopkins *et al.*, 1986). Low levels of expression were also observed in fetal stomach and adrenal gland during this period of development. Similar studies in 20 to 22 week gestation fetuses demonstrated apo A-I mRNA expression predominantly in the liver and the intestine and low levels in the pancreas, the gonads, the kidney, and the adrenal (Zannis *et al.*, 1985). Human fetal intestine and liver from this gestational range also synthesized and secreted apo A-I in organ culture (Zannis *et al.*, 1982). Treatment of human fetal jejunal explants with hydrocortisone resulted in decreased synthesis and secretion of chylomicrons, VLDL, and apo B-100, but it increased secretion of HDL, apo A-I, and apo B-48 (Loirdighi *et al.*, 1997).

Intestinal apo A-I expression in the developing swine has been studied (Black and Davidson, 1989; Black and Ellinas, 1992; Black and Rohwer-Nutter, 1991; Trieu *et al.*, 1993). In the 40-day gestation fetal pig, apo A-I mRNA was expressed at a high level in the liver but was barely detectable in fetal intestine (Trieu *et al.*, 1993). Incorporation of [^{3}H]leucine into apo A-I in these organs in the fetus paralleled the mRNA abundance. In the 2-day-old newborn piglet, apo A-I mRNA expression was present in the small intestine as well as the liver. There was a proximal to distal gradient of expression in the intestine with robust expression in the jejunum and low levels in the ileum. This pattern of intestinal and hepatic apo A-I mRNA expression persists into adulthood (Trieu *et al.*, 1993). The effects of dietary and biliary lipid on intestinal apo A-I synthesis in neonatal swine have also been tested *in vivo* (Black and Rohwer-Nutter, 1991). In the 2-day-old newborn animal, dietary triacylglycerol induced a twofold increase in apo A-I synthesis in both the jejunum and the ileum, although ileal synthesis was lower than jejunal in all groups. Quantitation of mRNA levels in jejunal samples indicated regulation at the translational level (Trieu *et al.*, 1992). Although synthesis was reduced in the bile-diverted group, the difference did not reach statistical significance. Quantitation of mucosal apo A-I mass demonstrated changes in mass paralleling changes in lipid flux in the jejunum but not in the ileum (Black and Rohwer-Nutter, 1991). In older suckling animals, the responsiveness of apo A-I synthesis to lumenal lipid absorption was no longer present in either the jejunum or the ileum (Black and Davidson, 1989).

4.1.2b Apolipoprotein A-IV. In the developing rat, intestinal apo A-IV mRNA levels increase dramatically at parturition and gradually decline to near adult levels by the time of weaning (Elshourbagy *et al.*, 1985; Haddad *et al.*, 1986). Apo A-IV mRNA is also expressed in the fetal yolk sac in the rat. A proximal to distal gradient of intestinal apo A-IV mRNA expression is established in the perinatal period in the rat that persists into adulthood (Rubin *et al.*, 1989).

Data on the developmental expression of intestinal apo A-IV are not available for humans. However, plasma levels of apo A-IV increase to near adult levels during the first week of life (Steinmetz *et al.*, 1988). Because human liver does not significantly express apo A-IV, this finding suggests increased intestinal expression in the immediate neonatal period.

Intestinal apo A-IV expression in the developing swine has been characterized using both *in vivo* and *in vitro* techniques (Black and Ellinas, 1992; Black *et al.*, 1990). Apo A-IV synthesis was determined in 40-day gestation fetal, 2-day-old newborn, and 1- to 2-week-old suckling piglet small intestine (Black *et al.*, 1990). In the fetal animals, apo A-IV synthesis represented 2.1% of total protein synthesis, which was similar to jejunal basal synthesis rates in postpartum animals receiving a low-triacylglycerol duodenal infusion. A higher-to-lower proximal to distal synthetic gradient was present in the postpartum animals under basal conditions. With a high-triacylglycerol duodenal infusion for a 24-hour period, jejunal synthesis increased by sevenfold in the 2-day-old piglets. Analysis of jejunal apo A-IV mRNA levels demonstrated that this induction was mediated at the pretranslational level. In the older 1–2-week-old animals, a high-triacylglycerol diet increased jejunal apo A-IV synthesis by only twofold. Dietary lipid had no effect on ileal apo A-IV synthesis. Study of bile-diverted animals revealed no regulatory role for biliary lipid for apo A-IV synthesis in either proximal or distal small intestine. These studies clearly demonstrate that dietary lipid is responsible for the induction of proximal intestinal apo A-IV in the neonatal period. Furthermore, this responsiveness to dietary lipid decreases with maturation.

Studies of apo A-IV synthesis in newborn piglet intestinal explants were carried out with incubation with physiological concentrations of EGF, insulin, hydrocortisone, and thyroid hormone (Black and Ellinas, 1992). Insulin and EGF each alone caused a modest but significant increase in synthesis. The EGF-induced increase was abolished by cotreatment with hydrocortisone. The combination of insulin and hydrocortisone gave the greatest increase in synthesis. Thyroid hormone had no effect on explant apo A-IV synthesis. Therefore, glucocorticoids, EGF, and insulin, along with lumenal lipid absorption, may play regulatory roles in the developmental expression of apo A-IV in the neonatal small intestine.

4.1.2c Apolipoprotein C-III. Analysis of apo C-III mRNA abundance in the developing rat has demonstrated expression in the small intestine and the liver from the time of birth (Haddad *et al.*, 1986). During neonatal development, the apo C-III mRNA levels in liver showed a gradual decrease (days 2–8) and a subsequent slight increase (days 10–15), whereas in the intestine the levels remained initially unchanged (up to day 7) and subsequently showed a slight increase (days 8–15) (Haddad *et al.*, 1986).

Northern blot analysis of RNA from various tissues from 20 to 22-week human abortuses demonstrated apo C-III mRNA expression in the small intestine and the liver (Zannis *et al.*, 1985). Apo C-III mRNA was detected in the liver, the intestine, the yolk sac, and the adrenals of 6- to 12-week human embryos (Hopkins *et al.*, 1986). Apo C-III mRNA abundance was significantly higher in the yolk sac than in the liver.

In the developing swine, apo C-III mRNA was not detectable in 40-day gestation fetal small intestine, but was abundant in the 2-day-old piglet jejunum (Trieu *et al.,* 1993). However, by 2 to 3 weeks postnatally jejunal apo C-III mRNA expression had declined sharply and was undetectable in the adult jejunum. In newborn piglets fed low- and high-triacylglycerol diets, intestinal apo C-III synthesis and mRNA levels were shown to be inducible by fat feeding (Black *et al.,* 1996). Apo C-III mRNA was expressed abundantly in fetal, newborn, suckling, and adult liver.

A mechanism identified for developmental and tissue-specific expression of the apo A-I/C-III/A-IV gene cluster is DNA methylation (Shemer *et al.,* 1991; Weaver, 1991). Studies in the rat demonstrated that patterns of demethylation or methylation of these genes correlated well with their expression or lack of expression, respectively, in various tissues during development, including the liver and the small intestine (Shemer *et al.,* 1991; Weaver, 1991).

4.2. Regulation of Intestinal Apo A-I, A-IV, and C-III Expression by Lumenal Fatty Acids

4.2.1. Apolipoprotein A-I

Apo A-I is one of the most highly expressed apolipoproteins in the small intestine, and its mRNA encodes 1% to 2% of the total intestinal mRNA in the rat and in humans (J. I. Gordon *et al.,* 1982). The rat small intestine secretes apo A-I associated with chylomicrons, with VLDL, and with HDL particles (Green and Glickman, 1981; Green *et al.,* 1978; Magun *et al.,* 1985). In the adult rat, acute absorption of saturated or unsaturated dietary triacylglycerol or cholesterol did not regulate apo A-I expression (Davidson and Glickman, 1985; Davidson *et al.,* 1987). However, reintroduction of dietary triacylglycerol after several days on a fat-free diet was found to result in a small increment in jejunal apo A-I synthesis (Davidson and Glickman, 1985). Also in the rat, acute triacylglycerol absorption did not significantly stimulate apo A-I jejunal synthesis or secretion into mesenteric lymph (Hayashi *et al.,* 1990). Isolated rat jejunal enterocytes were studied after fat feeding and demonstrated a 50% fall in apo A-I content by 1 to 2 hours (Alpers *et al.,* 1982). By 4 hours, cellular levels had returned to prefeeding levels. Complementary studies in cultured intestinal explants revealed that over a 4-hour period, oleic acid stimulated an increase in both cell content and secretion of apo A-I (Alpers *et al.,* 1982). These findings suggest that oleic acid can regulate enterocyte apo A-I secretion and in the first few hours after fat feeding, secretion of apo A-I exceeds the synthetic capacity of the enterocyte. With chronic feeding of more complex lipid mixtures in the rat, regulation of apo A-I was found to be more complex. Go *et al.* (1988) found that chronic feeding of a high-saturated fat diet containing cholesterol and propylthiouracil resulted in decreased intestinal apo A-I mRNA abundance, but it increased apo A-I synthesis in intestinal explants and in polysome runoff studies, suggesting translational regulation. However, the effects of propylthiouracil on thyroid function was probably a confounding variable in this study. In Caco-2 cells, apo A-I production, as assessed by incorporation of [^{35}S]methionine into cellular and secreted apoprotein (Murthy *et al.,* 1992) and by secreted apo A-I mass (Dashti *et al.,* 1990), did not appear to be significantly regulated by fatty acids.

Feeding studies have been performed in the newborn piglet comparing the effects of three dietary lipid formulations on intestinal apolipoprotein synthesis (H. Wang *et al.,* 1998, 1996). Two-day-old piglets were fed acutely (24 hours) or chronically (7 days) with diets containing lipid as medium-chain length triacylglycerols (MCT oil), intermediate-chain length triacylglycerols (coconut oil), and long-chain length unsaturated triacylglycerols (safflower oil). In the acute feeding group, jejunal apo A-I synthesis in the group receiving safflower oil was approximately twice that of the other two groups. However, in the chronic feeding group, this difference was no longer present, suggesting a loss of jejunal apo A-I synthetic responsiveness in the older animal, possibly related to intestinal development.

In experiments with IPEC-1 cells, there was a progressive increase in basolateral apo A-I secretion with decreasing fatty acid saturation. Oleic acid caused the lowest, linoleic acid intermediate, and linolenic acid the highest apo A-I mass secretion (H. Wang *et al.,* 1997). In contrast to the case with apo B, apo A-I secretion overall did not correlate with triacylglycerol secretion, suggesting regulation independent of the secretion of triacylglycerol-rich lipoproteins. In addition, apo A-I secretion did not correlate with phospholipid secretion. It is likely that in the case of linoleic and linolenic acids, the observed increase in apo A-I secretion relative to that observed with oleic acid may represent increased secretion of apo A-I associated with smaller, relatively triacylglycerol-poor lipoproteins, such as HDL. In complementary newborn piglet studies, absorption of a lipid emulsion or safflower oil (both containing mainly linoleic and oleic acids) resulted in up-regulation of jejunal apo A-I synthesis (Black and Rohwer-Nutter, 1991; H. Wang *et al.,* 1996). Additionally, EPA increased apo A-I secretion in IPEC-1 cells (Black *et al.,* 1998). Therefore, in the neonate, intestinal apo A-I synthesis and secretion appear to be regulated by LC-PUFAs.

4.2.2. Apolipoprotein A-IV

Of all the apolipoproteins expressed in the small intestine, apo A-IV is the most responsive to lipid flux. In the adult rat, acute lipid absorption has been shown to up-regulate the synthesis and secretion of apo A-IV at the pretranslational level (Apfelbaum *et al.,* 1987; T. J. Kalogeris *et al.,* 1994; Nishimura *et al.,* 1996; Pessah *et al.,* 1985). The same observation has been made in the newborn piglet (Black *et al.,* 1990). However, as demonstrated for apo A-I, chronic feeding of complex lipid mixtures resulted in a more complicated pattern of regulation (Go *et al.,* 1988). Go *et al.* (1988) found that chronic feeding of a high-saturated fat atherogenic diet containing cholesterol and propylthiouracil resulted in decreased intestinal apo A-IV mRNA abundance, but it increased apo A-IV synthesis in intestinal explants and in polysome runoff studies, suggesting regulation at the translational level. However, the thyroid status of these animals probably was a confounding variable. Chronic feeding of cholestyramine and a fat-free diet to rats, maneuvers designed to reduce lumenal bile acids and dietary lipid, resulted in a reduction in apo A-IV mRNA levels in the jejunum (Sonoyama *et al.,* 1994).

The chain length of fatty acids appears to be important in the regulation of intestinal apo A-IV expression. However, data from studies of this issue have yielded conflicting results. It has been demonstrated in the rat that continuous intraduodenal infusion of long-chain length fatty acids (14:0, 18:0, 18:1, 18:2, and 20:4), but not short- or medium-chain length fatty acids (4:0, 8:0, and 12:0), increases mesenteric lymph lipid and apo A-IV output (T. J. Kalogeris *et al.,* 1996b). However, Nishimura *et al.* (1996) administered an intraduodenal bolus of glycerol trioleate (18:1) or glycerol tricaprylate (8:0) and demonstrated an increase in lymph apo A-IV output after both lipids. In addition, portal blood transport of apo A-IV and of free fatty acid increased after the glycerol tricaprylate bolus, and jejunal apo A-IV mRNA levels increased after administration of both lipids. Hepatic apo A-IV mRNA abundance increased only after glycerol tricaprylate was given. This study suggested that medium- and long-chain length fatty acids are both effective in up-regulating jejunal apo A-IV expression, and that medium-chain length fatty acids cause diversion of a portion of the newly synthesized apo A-IV into the portal circulation. In addition, the medium-chain length fatty acids transported in the portal blood apparently induced hepatic apo A-IV expression. Medium- and long-chain length fatty acids have been found to be equally effective in up-regulating jejunal apo A-IV, as well as apo C-III, expression at the pretranslational level in the newborn piglet (Black *et al.*, 1996). Unfortunately, the Caco-2 cell line has proven to be unsuitable for studying the regulation of apo A-IV expression by lipid factors. Although these cells do synthesize and secrete apo A-IV, gene expression is not regulated by lipid (Giannoni *et al.*, 1994; Traber *et al.*, 1987).

4.2.3. Apolipoprotein C-III

Studies of the regulation of intestinal apo C-III expression by dietary lipid have been rather limited. Rat jejunal enterocytes, isolated and cultured after fat feeding, did not demonstrate a change in cellular apo C-III levels (Alpers *et al.*, 1982). However, there was an increase in apo C-III mRNA levels following a lumenal triacylglycerol bolus in the adult rat small intestine (Blaufuss *et al.*, 1984). In the neonatal piglet, medium- and long-chain length fatty acids were equally effective in inducing apo C-III, as well as apo A-IV, synthesis and mRNA abundance (Black *et al.*, 1996). There was a strong linear correlation between apo C-III and A-IV synthesis and mRNA abundance, suggesting that expression of these genes might be coordinately regulated in the neonate. This coordinate intestinal regulation was not present in older suckling animals (H. Wang *et al.*, 1998).

4.3. Regulation of Intestinal Apo A-I, A-IV, and C-III Expression by Lumenal Bile Acids, Cholesterol, and Phospholipid

4.3.1. Apolipoprotein A-I

Biliary lipids play an important role in the regulation of apo A-I expression in the distal small intestine in the rat. Biliary diversion reduced ileal apo A-I synthesis in the rat, and reinfusion of taurocholate was not effective in fully restoring synthesis, suggesting the importance of other biliary constituents (Davidson and Glickman, 1985). Chronic feeding of cholestyramine, a bile-acid-binding resin, to rats resulted in a reduction in apo A-I mRNA levels in the ileum (Sonoyama *et al.*, 1994). Intestinal apo A-I synthesis was not influenced by either acute or chronic perturbations in intestinal cholesterol flux in the rat (Davidson *et al.*, 1987).

Studies in Caco-2 cells have demonstrated no regulation of apo A-I synthesis or secretion by bile acids, cholesterol, or phosphatidylcholine, all major components of bile (Field *et al.*, 1994). In contrast, in a newborn swine intestinal epithelial cell line, IPEC-1, taurocholate and phosphatidylcholine were found to significantly increase the basolateral secretion of apo A-I (H. Wang *et al.*, 1999). This regulation of apo A-I secretion occurred at the pretranslational level for taurocholate and at the posttranslational level for phosphatidylcholine. The regulation of apo A-I secretion by phosphatidylcholine did not involve changes in apo A-I synthesis or degradation. This effect of phosphatidylcholine on apo A-I secretion did not require significant hydrolysis or uptake of the phosphatidylcholine molecule. Cholesterol had no effect on the secretion of apo A-I, whether solubilized in taurocholate or in phosphatidylcholine. These effects of phosphatidylcholine and taurocholate, both major components of bile, may be important in the regulation of apo A-I synthesis and secretion in the newborn mammalian small intestine.

4.3.2. Apolipoprotein A-IV

Chronic feeding of a fat-free diet or cholestyramine, a bile-acid-binding resin, was found to reduce jejunal and ileal apo A-IV mRNA levels in the rat (Sonoyama, 1994). However, because chronic high-dose cholestyramine administration may cause bile acid deficiency and secondary fat malabsorption, a primary role for bile acids in the observed regulation of apo A-IV expression is not clear from this study. Studies in the lymph-fistulated adult rat have demonstrated a role for bile in the sustained lymphatic output of apo A-IV after cessation of duodenal lipid infusion (Kalogeris *et al.*, 1999). This regulation appears to occur at the posttranslational level. In the newborn and older suckling piglet, biliary diversion did not alter jejunal or ileal apo A-IV synthesis or mRNA abundance (Black *et al.*, 1990).

4.3.3. Apolipoprotein C-III

Intestinal apo C-III expression has not been shown to be significantly regulated by biliary lipids.

4.4. Regulation of Intestinal Apo A-I, A-IV, and C-III Expression by Non-Lipid Factors

4.4.1. Apolipoprotein A-I

Massive proximal small intestinal resection in the rat resulted in increased expression of apo A-I and cellular retinol binding protein II (CRBP II) in the residual ileum (J. L. Wang *et al.*, 1997). In this study, retinoic acid treatment stimulated early crypt cell proliferation in the intestinal remnant. The investigators proposed that retinoic acid might play an important role in intestinal adaptation, because the genes for both apo A-I and CRBP II have nuclear RXR response elements in their promoter regions (J. L. Wang *et al.*, 1997). Vitamin A administration has also been shown to increase apo A-I intestinal mRNA abundance in the normal adult rat (Nagasaki *et al.*, 1994).

Hormones and vitamins appear to regulate intestinal apo A-I expression. In the rat, intestinal apo A-I synthesis, but not mRNA levels, was decreased in hypothyroid rats and increased after triiodothyronine administration (Davidson *et al.*, 1988a). In Caco-2 cells, 1,25-dihydroxycholecaciferol and thyroid hormone increased apo A-I mRNA abundance (Wagner *et al.*, 1992). However, 1,25-dihydroxycholecaciferol treatment did not change apo A-I mass, whereas thyroid hormone decreased apo A-I mass, indicating a significant component of posttranscriptional regulation.

Cytokines appear to regulate apo A-I expression in the small intestine. In addition to reducing lipid and apo B synthesis, TNF-a treatment reduced apo A-I synthesis in Caco-2 cells (Mehran *et al.,* 1995).

4.4.2. Apolipoprotein A-IV

Although small intestinal lipid flux has been consistently found to increase apo A-IV expression, other circulating regulatory factors may play a role (Kalogeris *et al.,* 1997). Ileal, but not jejunal, infusion of triacylglycerol in the adult rat was shown to up-regulate apo A-IV expression in the ileum and in segments of the jejunum isolated from the nutrient stream in Thiry-Vella fistulas (Kalogeris *et al.,* 1996a). Peptide YY is a candidate mediator in this instance, because exogenous peptide YY administered to rats resulted in stimulation of jejunal apo A-IV synthesis (Kalogeris *et al.,* 1997). Interestingly, functional leptin receptors are expressed in the jejunum, and an intravenous infusion of leptin reduces apo A-IV mRNA levels in the jejunum in obese, leptin-deficient *ob/ob* mice (Morton *et al.,* 1998). Massive small bowel resection in rats has also been shown to rapidly up-regulate apo A-IV mRNA levels severalfold in remnant ileum within 48 hr (Rubin *et al.,* 1996; Sonoyama and Aoyama, 1997). One study suggests that a constituent of bile may be one factor responsible for this induction of apo A-IV expression (Sonoyama *et al.,* 1996).

Similar to observations for apo B and A-I, 1,25-dihydroxycholecalciferol and thyroid hormone treatment resulted in increased apo A-IV mRNA levels in Caco-2 cells without affecting mass secretion (Wagner *et al.,* 1992).

4.4.3. Apolipoprotein C-III

Vitamin A has been shown to increase intestinal apo C-III mRNA abundance in the adult rat (Nagasaki *et al.,* 1994).

Acknowledgment

Portions of this work were supported by NIH grant RO1 HD22551.

References

Aalto-Setala, K., Bisgaier, C. L., Ho, A., Kieft, K. A., Traber, M. G., Kayden, H. J., Ramakrishnan, R., Walsh, A., Essenburg, A. D., and Breslow, J. L., 1994, Intestinal expression of human apolipoprotein A-IV in transgenic mice fails to influence dietary lipid absorption or feeding behavior, *J. Clin. Invest.* **93**:1776–1786.

Alpers, D. H., Lancaster, N., and Schonfeld, G., 1982, The effects of fat feeding on apolipoprotein A-I secretion from rat small intestinal epithelium, *Metabolism* **31**:784–790.

Apfelbaum, T. F., Davidson, N. O., and Glickman, R. M., 1987, Apolipoprotein A-IV synthesis in the rat small intestine: Regulation by dietary triglyceride, *Am. J. Physiol.* **252**:G662–G666.

Bergstedt, S. E., Hayashi, H., Kritchevsky, D., and Tso, P., 1990, A comparison of absorption of glycerol tristearate and glycerol trioleate by rat small intestine, *Am. J. Physiol.* **259**:G386–393.

Bisaha, J. G., Simon, T. C., Gordon, J. I., and Breslow, J. L., 1995, Characterization of an enhancer element in the human apolipoprotein C-III gene that regulates human apolipoprotein A-I gene expression in the intestinal epithelium, *J. Biol. Chem.* **270**:19979–19988.

Bisgaier, C. L., Sachdev, O. P., Megna, L., and Glickman, R. M., 1985, Distribution of apolipoprotein A-IV in human plasma, *J. Lipid Res.* **26**:11–25.

Black, D. D., 1992, Effect of intestinal chylomicron secretory blockade on apolipoprotein synthesis in the newborn piglet, *Biochem. J.* **283**:81–85.

Black, D. D., 1995, Intestinal lipoprotein metabolism, *J. Pediatr. Gastroenterol. Nutr.* **20:**125–147.

Black, D. D., and Davidson, N. O., 1989, Intestinal apolipoprotein synthesis and secretion in the suckling pig, *J. Lipid Res.* **30:**207–218.

Black, D. D., and Ellinas, H., 1992, Apolipoprotein synthesis in newborn piglet intestinal explants, *Pediatr. Res.* **32:**553–558.

Black, D. D., and Rohwer-Nutter, P. L., 1991, Intestinal apolipoprotein synthesis in the newborn piglet, *Pediatr. Res.* **29:**32–38.

Black, D. D., Rohwer-Nutter, P. L., and Davidson, N. O., 1990, Intestinal apo A-IV gene expression in the piglet, *J. Lipid Res.* **31:**497–505.

Black, D. D., Wang, H., Hunter, F., and Zhan, R., 1996, Intestinal expression of apolipoprotein A-IV and C-III is coordinately regulated by dietary lipid in newborn swine, *Biochem. Biophys. Res. Commun.* **221:**619–624.

Black, D. D., Du, J., and Wang, H., 1998, Regulation of apolipoprotein secretion by long-chain polyunsaturated fatty acids in newborn swine intestinal epithelial cells, *Pediatr. Res.* **43:**98A.

Blackhart, B. D., Ludwig, E. M., Pierotti, V. R., Caiati, L., Onasch, M. A., Wallis, S. C., Powell, L., Pease, R., Knott, T. J., Chu, M. L., Mahley, R. W., Scott, J., McCarthy, B. J., and Levy-Wilson, B., 1986, Structure of the human apolipoprotein B gene, *J. Biol. Chem.* **261:**15364–15367.

Blaufuss, M. C., Gordon, J. I., Schonfeld, G., and Alpers, D. H., 1984, Biosynthesis of apolipoprotein C-III in rat liver and small intestinal mucosa, *J. Biol. Chem.* **259:**2452–2456.

Borchardt, R. A., and Davis, R. A., 1987, Intrahepatic assembly of very low density lipoproteins: Rate of transport out of the endoplasmic reticulum determines rate of secretion, *J. Biol. Chem.* **262:**16394–16402.

Brooks, A. R., and Levy-Wilson, B., 1992, Hepatocyte nuclear factor 1 and C/EBP are essential for the activity of the human apolipoprotein B gene second-intron enhancer, *Mol. Cell. Biol.* **12:**1134–1148.

Brooks, A. R., Blackhart, B. D., Haubold, K., and Levy-Wilson, B., 1991, Characterization of tissue-specific enhancer elements in the second intron of the human apolipoprotein B gene, *J. Biol. Chem.* **266:**7848–7859.

Brown, V. W., and Baginski, M. L., 1972, Inhibition of lipoprotein lipase by an apoprotein of human very low density lipoprotein, *Biochem. Biophys. Res. Commun.* **46:**375–381.

Callow, M. J., Stoltzfus, L. J., Lawn, R. M., and Rubin, E. M., 1994, Expression of human apolipoprotein B and assembly of lipoprotein (a) in transgenic mice, *Proc. Natl. Acad. Sci. USA* **91:**2130–2134.

Carlson, S. E., Werkman, S. H., Peeples, J. M., and Wilson, W. M., 1994, Long-chain fatty acids and early visual and cognitive development of preterm infants, *Eur. J. Clin. Nutr.* **48:**S27–S30.

Chan, L., Chang, B. H.-J., Nakamuta, M., Li, W.-H., and Smith, L. C., 1997, Apobec-1 and apolipoprotein B mRNA editing, *Biochim. Biophys. Acta* **1345:**11–26.

Chapman, M. J., 1986, Comparative analysis of mammalian plasma lipoproteins, *Methods Enzymol.* **128:**70–143.

Chen, C. H., and Albers, J. J., 1985, Activation of lecithin: cholesterol acyltransferase by apolipoproteins E-2, E-3, and A-IV isolated from human plasma, *Biochim. Biopys. Acta* **836:**279–285.

Chen, S.-H., Habib, G., Yang, C.-Y., Gu, Z.-W., Lee, B. R., Weng, S.-A., Silberman, S. R., Cai, S.-J., Deslypere, J. P., Rosseneu, M., Gotto, A. M., Lee, W.-H., and Chan, L., 1987, Apolipoprotein B-48 is the product of a messenger RNA with an organ-specific in-frame stop codon, *Science* **238:**363–366.

Cladaras, C., Hadzopoulou-Cladaras, M., Avila, R., Nussbaum, A. L., Nicolosi, R., and Zannis, V. I., 1986a, Complementary DNA derived structure of the amino-terminal domain of human apolipoprotein B and size of its messenger RNA transcript, *Biochem.* **25:**5351–5357.

Cladaras, C., Hadzopoulou-Cladaras, M., Nolte, R. T., Atkinson, D., and Zannis, V. I., 1986b, The complete sequence and structural analysis of human apolipoprotein B-100: Relationship between apo B-100 and apo B-48 forms, *EMBO J.* **5:**3495–3507.

Dashti, N., Smith, E. A., and Alaupovic, P., 1990, Increased production of apolipoprotein B and its lipoproteins by oleic acid in Caco-2 cells, *J. Lipid Res.* **31:**113–123.

Davidson, N. O., and Glickman, R. M., 1985, Apolipoprotein A-I synthesis in rat small intestine: Regulation by dietary triglyceride and biliary lipid, *J. Lipid Res.* **26:**368–379.

Davidson, N. O., Kollmer, M. E., and Glickman, R. M., 1986, Apolipoprotein B synthesis in rat small intestine: Regulation by dietary triglyceride and biliary lipid, *J. Lipid Res.* **27:**30–39.

Davidson, N. O., Magun, A. M., Brasitus, T. A., and Glickman, R. M., 1987, Intestinal apolipoprotein A-I and B-48 metabolism: Effects of sustained alterations in dietary triglyceride and muscosal cholesterol flux, *J. Lipid Res.* **28:**388–402.

Davidson, N. O., Carlos, R. C., Drewek, M. J., and Parmer, T. G., 1988a, Apolipoprotein gene expression in the rat is regulated in a tissue-specific manner by thyroid hormone, *J. Lipid Res.* **29:**1511–1522.

Davidson, N. O., Drewek, M. J., Gordon, J. I., and Elovson, J., 1988b, Rat intestinal apolipoprotein B gene expression: Evidence for integrated regulation by bile salt, fatty acid, and phospholipid flux, *J. Clin. Invest.* **82**:300–308.

Davis, R. A., 1999, Cell and molecular biology of the assembly and secretion of apolipoprotein B-containing lipoproteins by the liver, *Biochim. Biophys. Acta* **1440**:1–31.

de Silva, H. V., Lauer, S. J., Wang, J., Simonet, W. S., Weisgraber, K. H., Mahley, R. W., and Taylor, J. M., 1994, Overexpression of human apo C-III in transgenic mice results in an accumulation of apolipoprotein B-48 remnants that is corrected by excess apolipoprotein E, *J. Biol. Chem.* **269**:2324–2335.

Demmer, L. A., Levin, M. S., Elovson, J., Reuben, M. A., Lusis, A. J., and Gordon, J. I., 1986, Tissue-specific expression and developmental regulation of the rat apolipoprotein B gene, *Proc. Natl. Acad. Sci. USA* **83**:8102–8106.

Dvorin, E., Gorder, N. L., Benson, D. M., and A. M. Gotto, J., 1986, Apolipoprotein A-IV: A determinate for binding and uptake of high density lipoproteins by rat hepatocytes, *J. Biol. Chem.* **261**:15714–15718.

Elshourbagy, N. A., Boguski, M. S., Liao, W. S. L., Jefferson, L. S., Gordon, J. I., and Taylor, J. M., 1985, Expression of rat apolipoprotein A-IV and A-I genes: mRNA induction during development and in response to glucocorticoids and insulin, *Proc. Natl. Acad. Sci. USA* **82**:8242–8246.

Elshourbagy, N. A., Walker, D. W., Boguski, M. S., Gordon, J. I., and Taylor, J. M., 1986, The nucleotide and derived amino acid sequence of human apolipoprotein A-IV mRNA and the close linkage of its gene to the genes of apolipoproteins A-I and C-III, *J. Biol. Chem.* **261**:1998–2002.

Elshourbagy, N. A., Walker, D. W., Paik, Y. K., Bogusk, M. S., Freeman, M., Gordon, J. I., and Taylor, J. M., 1987, Structure and expression of the human apolipoprotein A-IV gene, *J. Biol. Chem.* **262**:7973–7981.

Farese, R. V., Ruland, S. L., Flynn, L. M., Stokowski, R. P., and Young, S. G., 1995, Knockout of the mouse apolipoprotein B gene results in embryonic lethality in homozygotes and neural tube defects, male infertility, and protection against diet-induced hypercholesterolemia in heterozygotes, *Proc. Natl. Acad. Sci. USA* **92**:1774–1778.

Farese, R. V., Cases, S., Ruland, S. L., Kayden, H. J., Wong, J. S., Young, S. G., and Hamilton, R. L., 1996a, A novel function for apolipoprotein B: Lipoprotein synthesis in the yolk sac is critical for maternal-fetal lipid transport in mice, *J. Lipid Res.* **37**:347–360.

Farese, R. V., Veniant, M. M., Cham, C. M., Flynn, L. M., Pierotti, V., Loring, J. F., Traber, M., Ruland, S., Stokowski, R. S., Huszar, D., and Young, S. G., 1996b, Phenotypic analysis of mice expressing exclusively apolipoprotein B-48 or apolipoprotein B-100, *Proc. Natl. Acad. Sci. USA* **93**:6393–6398.

Fidge, N. H., 1980, The redistribution and metabolism of iodinated apolipoprotein A-IV in rats, *Biochim. Biophys. Acta* **619**:129–141.

Field, F. J., and Mathur, S. N., 1995, Intestinal lipoprotein synthesis and secretion, *Prog. Lipid Res.* **34**:185–198.

Field, F. J., Albright, E., and Mathur, S. N., 1988, Regulation of triglyceride-rich lipoprotein secretion by fatty acids in Caco-2 cells, *J. Lipid Res.* **29**:1427–1437.

Field, F. J., Born, E., Chen, H., Murthy, S., and Mathur, S. N., 1994, Regulation of apolipoprotein B secretion by biliary lipids in Caco-2 cells, *J. Lipid Res.* **35**:749–762.

Fielding, C. J., Shore, V. G., and Fielding, P. E., 1972, A protein co-factor of lecithin: Cholesterol acyltransferase, *Biochem. Biophys. Res. Commun.* **46**:1493–1498.

Fisher, E. A., Anbari, A., Klurfeld, D. M., and Kritchevsky, D., 1988, Independent effects of diet and nutritional status on apolipoprotein B gene expression in the rabbit, *Arteriosclerosis* **8**:797–803.

Fujimoto, K., Cardelli, J. A., and Tso, P., 1992, Increased apolipoprotein A-IV in rat mesenteric lymph after lipid meal acts as a physiological signal for satiation, *Am. J. Physiol.* **262**:G1002–G1006.

Fujimoto, K., Fukagawa, K., Sakata, T., and Tso, P., 1993, Suppression of food intake by apolipoprotein A-IV is mediated through the central nervous system in rats, *J. Clin. Invest.* **91**:1830–1833.

Ge, R., Rhee, M., Malik, S., and Karathanasis, S. K., 1994, Transcriptional repression of apolipoprotein A-I gene expression by orphan receptor ARP-1, *J. Biol. Chem.* **269**:13185–13192.

Giannoni, F., Field, F. J., and Davidson, N. O., 1994, An improved reverse transcription-polymerase chain reaction method to study apolipoprotein gene expression in Caco-2 cells, *J. Lipid Res.* **35**:340–350.

Ginsburg, G. S., Ozer, J., and Karathanasis, S. K., 1995, Intestinal apolipoprotein A-I gene transcription is regulated by multiple distinct DNA elements and is synergistically activated by the orphan nuclear receptor, hepatocyte nuclear factor 4, *J. Clin. Invest.* **96**:528–538.

Go, M. F., Schonfeld, G., Pfleger, B., Cole, T. G., Sussman, N. L., and Alpers, D. H., 1988, Regulation of intestinal and hepatic apoprotein synthesis after chronic fat and cholesterol feeding, *J. Clin. Invest.* **81**:1615–1620.

Goldberg, I. J., Scheraldi, C. A., Yacoub, L. K., Saxena, U., and Bisgaier, C. L., 1990, Apolipoprotein C-II activation of lipoprotein lipase: Modulation by apolipoprotein A-IV, *J. Biol. Chem.* **265**:4266–4272.

Golder-Novoselsky, E., Forte, T. M., Nichols, A. V., and Rubin, E. M., 1992, Apolipoprotein A-I expression and high density lipoprotein distribution in transgenic mice during development, *J. Biol. Chem.* **267**:20787–20790.

Gonzalez-Vallina, R., Wang, H., Zhan, R., Berschneider, H. M., Lee, R. M., Davidson, N. O., and Black, D. D., 1996, Lipoprotein and apolipoprotein secretion by a newborn piglet intestinal cell line (IPEC-1), *Am. J. Physiol.* **271**:G249–G259.

Gordon, D. A., Wetterau, J. R., and Gregg, R. E., 1995, Microsomal triglyceride transfer protein: A protein complex required for the assembly of lipoprotein particles, *Trends Cell Biol.* **5**:317–321.

Gordon, J. I., Smith, D. P., Andy, E., Alpers, D. H., Schonfeld, G., and Strauss, A. W., 1982, The primary translation product of rat intestinal apolipoprotein A-I mRNA is an unusual preprotein, *J. Biol. Chem.* **257**:971–978.

Green, P. H. R., and Glickman, R. M., 1981, Intestinal lipoprotein metabolism, *J. Lipid Res.* **22**:1153–1173.

Green, P. H. R., Tall, A. R., and Glickman, R. M., 1978, Rat intestine secretes discoid high density lipoprotein, *J. Clin. Invest.* **61**:528–534.

Green, P. H. R., Glickman, R. M., Riley, J. W., and Quinet, E., 1980, Human apolipoprotein A-IV: Intestinal origin and distribution in plasma, *J. Clin. Invest.* **65**:911–919.

Gruber, P. J., Torres-Rosado, A., Wolak, M. L., and Leff, T., 1994, Apo C-III gene transcription is regulated by a cytokine inducible NF-kB element, *Nucleic Acids Res.* **22**:2417–2422.

Haddad, I. A., Ordovas, J. M., Fitzpatrick, T., and Karathanasis, S. K., 1986, Linkage, evolution, and expression of the rat apolipoprotein A-I, C-III, and A-IV genes, *J. Biol. Chem.* **261**:13268–13277.

Hamilton, R. L., 1994, Apolipoprotein B-containing plasma lipoproteins in health and disease, *Trends Cardiovasc. Med.* **4**:131–139.

Harris, W., 1989, Fish oils and plasma lipid and lipoprotein metabolism in humans: A critical review, *J. Lipid Res.* **30**:785–807.

Harris, W. S., Conner, W. E., Alam, N., and Illingworth, D. R., 1982, Reduction of postprandial triglyceridemia in humans by dietary *n*-3 fatty acids, *J. Lipid Res.* **29**:1451–1460.

Hay, R., Driscoll, D., and Getz, G., 1986, The biogenesis of lipoproteins, in: *Biochemistry and Biology of Plasma Lipoproteins* (A. M. Scanu & A. A. Spector, eds.), Marcel Dekker, Inc., New York, pp. 11–51.

Hayashi, H., Fujimoto, K., Cardelli, J. A., Nutting, D. F., Bergstedt, S., and Tso, P., 1990, Fat feeding increases size, but not number, of chylomicrons produced by small intestine, *Am. J. Physiol.* **259**:G709–G719.

Hirano, K.-I., Young, S. G., Farese, R. V., Ng, J., Sande, E., Warburton, C., Powell-Braxton, L. M., and Davidson, N. O., 1996, Targeted disruption of the mouse apobec-1 gene abolishes apolipoprotein B mRNA editing and eliminates apolipoprotein B-48, *J. Biol. Chem.* **271**:9887–9890.

Hopkins, B., Sharpe, C. R., Baralle, F. E., and Graham, C. F., 1986, Organ distribution of apolipoprotein gene transcripts in 6–12 week postfertilization human embryos, *J. Embryol. Exp. Morph.* **97**:177–187.

Hospattankar, A. V., Law, S. W., Lackner, K., and Brewer, H. B., Jr., 1986, Identification of low density lipoprotein receptor binding domains of human apolipoprotein B-100: A proposed consensus LDL receptor binding sequence of apo B-100, *Biochem. Biophys. Res. Commun.* **139**:1078–1085.

Hospattankar, A. V., Higuchi, K., Law, S. W., Meglin, N., and Brewer, H. B., Jr., 1987, Identification of a novel in-frame translational stop codon in human intestine apo B mRNA, *Biochem. Biophys. Res. Commun.* **148**:279–285.

Huang, L. S., Bock, S. C., Feinstein, S. I., and Breslow, J. L., 1985, Human apolipoprotein B cDNA clone isolation and demonstration that liver apolipoprotein B mRNA is 22 kilobases in length, *Proc. Natl. Acad. Sci. USA* **82**:6825–6829.

Huang, L. S., Miller, D. A., Bruns, G. A., and Breslow, J. L., 1986, Mapping of the human apo B gene to chromosome 2p and demonstration of a two-allele restriction fragment length polymorphism, *Proc. Natl. Acad. Sci. USA* **83**:644–648.

Huang, L.-S., Voyiaziakis, E., Markenson, D. F., Sokol, K. A., Hayek, T., and Breslow, J. L., 1995, Apo B gene knockout in mice results in embryonic lethality in homozygotes and neural tube defects, male infertility, and reduced HDL cholesterol ester and apo A-I transport rates in heterozygotes, *J. Clin. Invest.* **96**:2152–2161.

Hughes, T. E., Ordovas, J. M., and Schaefer, E. J., 1988, Regulation of intestinal apolipoprotein B synthesis and secretion by Caco-2 cells. Lack of fatty acid effects and control by intracellular calcium ion, *J. Biol. Chem.* **263**:3425–3431.

Hui, D. Y., Innerarity, T. L., Milne, R. W., Marcel, Y. L., and Mahley, R. W., 1984, Binding of chylomicron remnants and beta-VLDL to hepatic and extra-hepatic lipoprotein receptors: A process independent of apolipoprotein B-48, *J. Biol. Chem.* **259**:15060–15068.

Hussain, M. M., Kancha, R. K., Zhou, Z., Luchoomun, J., Zu, H., and Bakillah, A., 1996, Chylomicron assembly and catabolism: Role of apolipoproteins and receptors, *Biochim. Biophys. Acta* **1300**:151–170.

Ikeda, I., Sasaki, E., Yasunami, H., Nomiyama, S., Nakayama, M., Sugano, M., Imaizumi, K., and Yazawa, K., 1995, Digestion and lymphatic transport of eicosapentaenoic and docosahexaenoic acids given in the form of triacylglycerol, free acid and ethyl ester in rats, *Biochim. Biophys. Acta* **1259**:297–304.

Innerarity, T. L., Boren, J., Yamanaka, S., and Olofsson, S.-O., 1996, Biosynthesis of apolipoprotein B-48-containing lipoproteins, *J. Biol. Chem.* **271**:2353–2356.

Innis, S. M., 1991, Essential fatty acids in growth and development, *Prog. Lipid Res.* **30**:39–103.

Innis, S. M., Nelson, C. M., Rioux, M. F., and King, D. J., 1994, Development of visual acuity in relation to plasma and erythrocyte ω-6 and ω-3 fatty acids in healthy term gestation infants, *Am. J. Clin. Nutr.* **60**:347–352.

Innis, S. M., Akrabawi, S. S., Dierson-Schade, D. A., Dobson, M. V., and Guy, D. G., 1997, Visual acuity and blood lipids in term infants fed human milk or formulae, *Lipids* **32**:63–72.

Ito, Y., Azrolan, A., O'Connell, A., Walsh, A., and Breslow, J. L., 1990, Hypertriglyceridemia as a result of human apo C-III gene expression in transgenic mice, *Science* **249**:790–793.

Jiao, S., Moberly, J. B., and Schonfeld, G., 1990, Editing of apolipoprotein B messenger RNA in differentiated Caco-2 cells, *J. Lipid Res.* **31**:695–700.

Kalogeris, T. J., Fukagawa, K., and Tso, P., 1994, Synthesis and lymphatic transport of intestinal apolipoprotein A-IV in response to graded doses of triglyceride, *J. Lipid Res.* **35**:1141–1151.

Kalogeris, T. J., Tsuchiya, T., Fukagawa, K., Wolf, R., and Tso, P., 1996a, Apolipoprotein A-IV synthesis in proximal jejunum is stimulated by ileal lipid infusion, *Am. J. Physiol.* **270**:G277–G286.

Kalogeris, T. J., Monroe, F., Demichele, S. J., and Tso, P., 1996b, Intestinal synthesis and lymphatic secretion of apolipoprotein A-IV vary with chain length of intestinally infused fatty acids in rats, *J. Nutr.* **126**:2720–2729.

Kalogeris, T., Qin, X., Chey, W. Y., and Tso, P., 1997, Exogenous PYY stimulates jejunal apolipoprotein A-IV synthesis without affecting apo A-IV mRNA in rats, *Gastroenterol.* **112**:A884.

Kalogeris, T. J., Rodriguez, M. D., and Tso, P., 1997, Control of synthesis and secretion of intestinal apolipoprotein A-IV by lipid, *J. Nutrition* **127**:537S–543S.

Kalogeris, T. J., Fukagawa, K., Tsuchiya, T., Qin, X., and Tso, P., 1999, Intestinal synthesis and lymphatic secretion of apolipoprotein A-IV after cessation of duodenal fat infusion: Mediation by bile, *Biochim. Biophys. Acta* **1436**:451–466.

Kane, J. P., 1983, Apolipoprotein B: Structural and metabolic heterogeneity, *Ann. Rev. Physiol.* **45**:637–650.

Karathanasis, S. K., 1985, Apolipoprotein multigene family: Tandem organization of human apolipoprotein AI, CIII, and AIV genes, *Proc. Natl. Acad. Sci., U.S.A.* **82**:6374–6378.

Kardassis, D., Hadzopoulou-Cladaras, M., Ramji, D. P., Cortese, R., Zannis, V. I., and Cladaras, C., 1990, Characterization of the promoter elements required for hepatic and intestinal transcription of the human apo B gene: Definition of the DNA-binding site of a tissue-specific transcription factor, *Mol. Cell Biol.* **10**:2653–2659.

Kardassis, D., Zannis, V. I., and Cladaras, C., 1992, Organization of the regulatory elements and nuclear activities participating in the transcription of the human apolipoprotein B gene, *J. Biol. Chem.* **267**:2622–2632.

Kardassis, D., Laccotripe, M., Talianidis, I., and Zannis, V., 1996, Transcriptional regulation of the genes involved in lipoprotein transport: The role of proximal promoters and long-range regulatory elements and factors in apolipoprotein gene regulation, *Hypertension* **27**:980–1008.

Kilbourne, E. J., Widom, R., Harnish, D. C., Malik, S., and Karathanasis, S. K., 1995, Involvement of early growth response factor Egr-1 in apolipoprotein A-I gene transcription, *J. Biol. Chem.* **270**:7004–7010.

Kim, E., and Young, S. G., 1998, Genetically modified mice for the study of apolipoprotein B, *J. Lipid Res.* **39**:703–723.

Kinnunen, P. K. J., and Ehnholm, C., 1976, Effect of serum and C-apoproteins from very low density lipoproteins on human postheparin plasma hepatic lipase, *FEBS Lett.* **65**:354–357.

Knott, T. J., Rall, S. C., Innerarity, T. L., Jacobson, S. F., Urdea, M. S., Levy-Wilson, B., Powell, L. M., Pease, R. J., Eddy, R., Nakai, H., Byers, M., Priestley, L. M., Robertson, E., Rall, L. B., Betsholtz, C., Shows, T. B., Mahley, R. W., and Scott, J., 1985, Human apolipoprotein B: Structure of carboxyl-terminal domains, sites of gene expression, and chromosomal localization, *Science* **230**:37–43.

Kris-Etherton, P. M., and Yu, S., 1997, Individual fatty acid effects on plasma lipids and lipoproteins: Human studies, *Am. J. Clin. Nutr.* **65**:1628S–1644S.

Ktistaki, E., Lacorte, J.-M., Katrakili, N., Zannis, V. I., and Talianidis, I., 1994, Transcriptional regulation of the apolipoprotein A-IV gene involves synergism between a proximal orphan receptor response element and a distant enhancer located in the upstream promoter region of the apolipoprotein A-IV gene, *Nucleic Acids Res.* **22:**4689–4696.

Kumar, N. S., and Mansbach, C. M., 1997, Determinants of triacylglycerol transport from the endoplasmic reticulum to the Golgi in intestine, *Am. J. Physiol.* **273:**G18–G30.

Ladias, J. A., Hadzopoulou-Cladaras, M., Kardassis, D., Cardot, P., Cheng, J., Zannis, V., and Cladaras, C., 1992, Transcriptional regulation of human apolipoprotein genes Apo B, Apo C-III, and Apo A-II by members of the steroid hormone receptor superfamily HNF-4, ARP-1, EAR-2, and EAR-3, *J. Biol. Chem.* **267:**15849–15860.

Lauer, S. J., Somonet, W. S., Bucay, N., de Silva, H. V., and Taylor, J. M., 1991, Tissue-specific expression of the human apolipoprotein A-IV gene in transgenic mice, *Arteriosclerosis* **11:**1390a.

Law, S. W., Lackner, K. J., Hospattankar, A. V., Anchors, J. M., Sakaguchi, A. Y., Naylor, S. L., and Brewer, H. B., Jr., 1985, Human apolipoprotein B-100: Cloning, analysis of liver mRNA, and assignment of the gene to chromosome 2, *Proc. Natl. Acad. Sci. USA* **82:**8340–8344.

Le Beyec, J., Delers, F., Jourdant, F., Schreider, C., Chambaz, J., Cardot, P., and Pincon-Raymond, M., 1997, A complete epithelial organization of Caco-2 cells induces I-FABP and potentializes apolipoprotein gene expression, *Exper. Cell Res.* **236:**311–320.

Le Beyec, J., Chauffeton, V., Kan, H. Y., Janvier, P. L., Cywiner-Golenzer, C., Chatelet, F. P., Kalopissis, A. D., Zannis, V., Chambaz, J., Pincon-Raymond, P., and Cardot, P., 1999, The −700/−310 fragment of the apolipoprotein A-IV gene combined with the −890/−500 apolipoprotein C-III enhancer is sufficient to direct a pattern of gene expression similar to that for the endogenous apolipoprotein A-IV gene, *J. Biol. Chem.* **274:**4954–4961.

Levy, E., Smith, L., Dumont, L., Garceau, D., Garofalo, C., Thibault, L., and Seidman, E., 1992, The effect of a new calcium channel blocker (TA-3090) on lipoprotein profile and intestinal lipid handling, *Proc. Soc. Exp. Biol. Med.* **199:**128–135.

Levy, E., Mehran, M., and Seidman, E., 1995, Caco-2 cells as a model for intestinal lipoprotein synthesis and secretion, *FASEB J.* **9:**626–635.

Levy-Wilson, B., and Fortier, C., 1989, The limits of the DNAse I-sensitive domain of the human apolipoprotein B gene coincide with the location of chromosome anchorage loops and define 5′ and 3′ boundaries of the gene, *J. Biol. Chem.* **264:**21196–21204.

Levy-Wilson, B., Paulweber, B., Nagy, B. P., Ludwig, E. H., and Brooks, A. R., 1992, Nuclear-hypersensitive sites define a region with enhancer activity in the third intron of the human apolipoprotein B gene, *J. Biol. Chem.* **267:**18735–18743.

Linton, M. F., Farese, R. V., Chiesa, G., Grass, D. S., Chin, P., Hammer, R. E., Hobbs, H. H., and Young, S. G., 1993, Transgenic mice expressing high plasma concentrations of human apolipoprotein B-100 and lipoprotein (a), *J. Clin. Invest.* **92:**3029–3037.

Loirdighi, N., Menard, D., Delvin, D., and Levy, E., 1997, Selective effects of hydrocortisone on intestinal lipoprotein and apolipoprotein synthesis in the human fetus, *J. Cell. Biochem.* **66:**65–76.

Lopez-Miranda, J., Kam, N., Osada, J., Rodriguez, C., Fernandez, P., Contois, J., Schaefer, E. J., and Ordovas, J. M., 1994, Effect of fat feeding on human intestinal apolipoprotein B mRNA levels and editing, *Biochim. Biophy. Acta* **1214:**143–147.

Luchoomum, J., Zhou, Z., Bakillah, A., Jamil, H., and Hussain, M. M., 1997, Assembly and secretion of VLDL in nondifferentiated Caco-2 cells stably transfected with human recombinant apo B-48 cDNA, *Arterioscler. Thromb. Vasc. Biol.* **17:**2955–2963.

Ludwig, E. H., Blackhart, B. D., Pierotti, V. R., Caiati, L., Fortier, C., Knott, T., Scott, J., Mahley, R. W., Levy-Wilson, B., and McCarthy, B. J., 1987, DNA sequence of the human apolipoprotein B gene, *DNA* **6:**363–372.

Macri, J., and Adeli, K., 1997, Studies on intracellular translocation of apolipoprotein B in a permeabilized Hep G2 system, *J. Biol. Chem.* **272:**7328–7337.

Maeda, N., Li, H., Lee, D., Oliver, P., Quarfordt, S. H., and Osada, J., 1994, Targeted disruption of the apolipoprotein C-III gene in mice results in hypotriglyceridemia and protection from hypertriglyceridemia, *J. Biol. Chem.* **269:**23610–23616.

Magun, A. M., Brasitis, T. A., and Glickman, R. M., 1985, Isolation of high density lipoproteins from rat intestinal epithelial cells, *J. Clin. Invest.* **75:**209–218.

Magun, A. M., Mish, B., and Glickman, R. M., 1988, Intracellular apo A-I and apo B distribution in rat intestine is altered by lipid feeding, *J. Lipid Res.* 29:1107–1116.

Mansbach, C. M., and Nevin, P., 1998, Intracellular movement of triacylglycerols in the intestine, *J. Lipid Res.* 39:963–968.

Mathur, S. N., Born, E., Bishop, W. P., and Field, F. J., 1993, Effect of okadaic acid on apo B and apo A-I secretion by Caco-2 cells, *Biochim. Biophy. Acta* 1168:130–143.

Mathur, S. N., Born, E., Murthy, S., and Field, F. J., 1996, Phosphatidylcholine increases the secretion of triacylglycerol-rich lipoproteins by Caco-2 cells, *Biochem. J.* 314:569–575.

Mehran, M., Seidman, E., Marchand, R., Gurbindo, C., and Levy, E., 1995, Tumor necrosis factor-α inhibits lipid and lipoprotein transport by Caco-2 cells, *Am. J. Physiol.* 269:G953–G960.

Metzger, S., Hallaas, J. L., Breslow, J. L., and Sladek, F. M., 1993, Orphan receptor HNF-4 and bZIP protein C/EBP bind to overlapping regions of the apolipoprotein B gene promoter and synergistically activate transcription, *J. Biol. Chem.* 268:16831–16838.

Middleton, S., and Schneeman, B. O., 1996, Rat plasma triglycerides and hepatic fatty acid synthetase mRNA, but not apolipoprotein B and A-IV mRNA, respond to dietary fat content, *J. Nutr.* 126:1627–1634.

Mietus-Snyder, M., Sladek, F. M., Ginsburg, G. S., Kuo, C. F., Ladias, J. A. A., Darnell, J. E., and Karathanasis, S. K., 1992, Antagonism between apolipoprotein A-I regulatory protein 1, Ear 3/COUP-TF, and hepatocyte nuclear factor 4 modulates apolipoprotein C-III gene expression in liver and intestinal cells, *Mol. Cell. Biol.* 12:1708–1718.

Moberly, J. B., Cole, T. G., Alpers, D. H., and Schonfeld, G., 1990, Oleic acid stimulation of apolipoprotein B secretion from Hep G2 and Caco-2 cells occurs post-transcriptionally, *Biochim. Biophy. Acta* 1042:70–80.

Mortimer, B. C., Martins, I., Zeng, B. J., and Redgrave, T. G., 1997, Use of gene-manipulated models to study the physiology of lipid transport, *Clin. Exper. Pharm. Physiol.* 24:281–285.

Morton, N. M., Emilsson, V., Liu, Y. L., and Cawthorne, M. A., 1998, Lepin action in intestinal cells, *J. Biol. Chem.* 273:26194–26201.

Murthy, S., Albright, E., Mathur, S. N., and Field, F. J., 1990, Effect of eicosapentaenoic acid on triacylglycerol transport in Caco-2 cells, *Biochim. Biophys. Acta* 1045:147–155.

Murthy, S., Albright, E., Mathur, S. N., Davidson, N. O., and Field, F. J., 1992, Apolipoprotein B mRNA abundance is decreased by eicosapentaenoic acid in Caco-2 cells, *Arteriosclerosis and Thrombosis* 12:691–700.

Murthy, S., Mathur, S. N., Varilek, G., Bishop, W., and Field, F. J., 1996, Cytokines regulate apolipoprotein B secretion by Caco-2 cells: Differential effects of IL-6 and TGF-β1, *Am. J. Physiol.* 270:G94–G102.

Murthy, S., Born, E., Mathur, S., and Field, F. J., 1998, 13-hydroxy octadecadienoic acid (13-HODE) inhibits triacylglycerol-rich lipoprotein secretion by Caco-2 cells, *J. Lipid Res.* 39:1254–1262.

Naganawa, S., Ginsberg, H. N., Glickman, R. M., and Ginsberg, G. S., 1997, Intestinal transcription and synthesis of apolipoprotein A-I is regulated by five natural polymorphisms upstream of the apolipoprotein C-III gene, *J. Clin. Invest.* 99:1958–1965.

Nagasaki, A., Kikuchi, T., Kurata, K., Masushige, S., Hasegawa, T., and Kato, S., 1994, Vitamin A regulates the expression of apolipoprotein A-I and C-III genes in the rat, *Biochem. Biophys. Res. Commun.* 205:1510–1517.

Nielsen, L. B., McCormick, S. P., Pierotti, V., Tam, C., Gunn, M. D., Shizuya, H., and Young, S. G., 1997, Human apolipoprotein B transgenic mice generated with 207- and 145-kilobase pair bacterial artificial chromosomes. Evidence that a distant 5'-element confers appropriate transgene expression in the intestine, *J. Biol. Chem.* 272:29752–29758.

Nishimura, M., Seishima, M., Ohashi, H., and Noma, A., 1996, Effects of lipid administration on lymphatic apolipoprotein A-IV and B output and synthesis, *Am. J. Physiol.* 271:G322–G329.

Ochoa, A., Bovard-Houppermans, S., and Zakin, M. M., 1993, Human apolipoprotein A-IV gene expression is modulated by members of the nuclear hormone receptor superfamily, *Biochim. Biophys. Acta* 1210:41–47.

Ogami, K., Hadzopoulou-Cladaras, M., Cladaras, C., and Zannis, V. I., 1990, Promotor elements and factors required for hepatic and intestinal transcription of the human apo C-III gene, *J. Biol. Chem.* 265:9808–9815.

Okumura, T., Fukagawa, K., Tso, T., Taylor, I. L., and Pappas, T. N., 1994, Intracisternal injection of apolipoprotein A-IV inhibits gastric secretion in pylorus-ligated conscious rats, *Gastroenterol.* 107:1861–1864.

Olofsson, S.-O., Bjursell, G., Bostrom, K., Carlsson, P., Elovson, J., Protter, A. A., Reuben, M. A., and Bondjers, G., 1987, Apolipoprotein B: Structure, biosynthesis, and role in the lipoprotein assembly process, *Atherosclerosis* 68:1–17.

Papazafiri, P., Ogami, K., Ramji, D. P., Nicosia, A., Monaci, P., Cladaras, C., and Zannis, V. I., 1991, Promoter elements and factors involved in hepatic transcription of the human apo A-I gene: Positive and negative regulators bind to overlapping sites, *J. Biol. Chem.* **266:**5790–5797.

Paulweber, B., Sandhofer, F., and Levy-Wilson, B., 1993, The mechanism by which the human apolipoprotein B gene reducer operates involves blocking of transcriptional activation by hepatocyte nuclear factor 3, *Mol. Cell Biol.* **13:**1534–1546.

Pease, R. J., Harrison, G. B., and Scott, J., 1991, Cotranslational insertion of apolipoprotein B into the inner leaflet of the endoplasmic reticulum, *Nature* **353:**448–450.

Perrin-Ansart, M. C., Vacher, D., and Girard-Globa, A., 1988, Determination of apolipoprotein A-I synthesis in intestinal explants from fetal and neonatal rats, *Biochim. Biophys. Acta* **963:**541–548.

Pessah, M., Salvat, C., Amit, N., and Infante, R., 1985, Isolation and characterization of rat intestinal polyribosomes and RNA using absorption of fat. Increased translation *in vitro* of apo A-IV, *Biochem. Biophys. Res. Commun.* **126:**373–381.

Phillips, M. L., Pullinger, C., Kroes, I., Kroes, J., Hardman, D. A., Chen, G., Curtiss, L. K., Gutierrez, M. M., Kane, J. P., and Schumaker, V. N., 1997, A single copy of apolipoprotein B-48 is present on the human chylomicron remnant, *J. Lipid Res.* **38:**1170–1177.

Plump, A. S., Smith, J. D., Hayek, T., Aalto-Setala, K., Walsh, A., Verstuyft, J. G., Rubin, E. M., and Breslow, J. L., 1992, Severe hypercholesterolemia and atherosclerosis in apolipoprotein E-deficient mice created by homologous recombination in ES cells, *Cell* **71:**1–20.

Plump, A. S., Scott, C. J., and Breslow, J. L., 1994, Human apolipoprotein A-I gene expression increases high density lipoprotein and suppresses atherosclerosis in the apolipoprotein E-deficient mouse, *Proc. Natl. Acad. Sci. USA* **91:**9607–9611.

Powell, L. M., Wallis, S. C., Pease, R. J., Edwards, Y. H., Knott, T. J., and Scott, J., 1987, A novel form of tissue-specific RNA processing produces apolipoprotein B-48 in intestine, *Cell* **50:**831–840.

Pullinger, C. R., North, J. D., Teng, B.-B., Rifici, V. A., Ronhild de Brito, A. E., and Scott, J., 1989, The apolipoprotein B gene is constitutively expressed in Hep G2 cells: Regulation of secretion by oleic acid, albumin, and insulin, and measurement of the mRNA half-life, *J. Lipid Res.* **30:**1065–1077.

Qin, X., Swertfeger, D. K., Zheng, S., Hui, D. Y., and Tso, P., 1998, Apolipoprotein A-IV: A potent endogenous inhibitor of lipid oxidation, *Am. J. Physiol.* **274:**H1836–1840.

Ranheim, T., Gedde-Dahl, A., Rustan, A. C., and Drevon, C. A., 1992, Influence of eicosapentaenoic acid on secretion of lipoproteins in Caco-2 cells, *J. Lipid Res.* **33:**1281–1293.

Renner, F., Samuelson, A., Rogers, M., and Glickman, R. M., 1986, Effect of saturated and unsaturated lipid on the composition of mesenteric triglyceride-rich lipoproteins in the rat, *J. Lipid Res.* **27:**72–81.

Reue, K., Purcell-Huynh, D. A., Leete, T. H., Doolittle, M. H., Durstenfeld, A., and Lusis, A. J., 1993, Genetic variation in mouse apolipoprotein A-IV expression is determined pre- and post-transcriptionally, *J. Lipid Res.* **34:**893–903.

Rubin, D. C., Ong, D. E., and Gordon, J. I., 1989, Cellular differentiation in the emerging fetal rat small intestinal epithelium: Mosaic patterns of gene expression, *Proc. Natl. Acad. Sci. USA* **86:**1278–1282.

Rubin, D. C., Swietlicki, E. A., Wang, J. L., Dodson, B. D., and Levin, M. S., 1996, Enterocytic gene expression in intestinal adaptation: Evidence for a specific cellular response, *Am. J. Phyiol.* **270:**G143–G152.

Rustaeus, S., Lindberg, K., Stillemark, P., Claesson, C., Asp, L., Larsson, T., Boren, J., and Olofsson, S., 1999, Assembly of very low density lipoprotein: A two-step process of apolipoprotein B core lipidation, *J. Nutr.* **129:**463S–466S.

Sabesin, S. M., and Frase, S., 1977, Electron microscopic studies of the assembly, intracellular transport, and secretion of chylomicrons by rat intestine, *J. Lipid Res.* **18:**496–511.

Shemer, R., Eisenberg, S., Breslow, J. L., and Razin, A., 1991, Methylation patterns of the human apo A-I/C-III/A-IV gene cluster in adult and embryonic tissues suggest dynamic changes in methylation during development, *J. Biol. Chem.* **266:**23676–23681.

Shi, W.-K., Hopkins, B., Thompson, S., Heath, J. K., Luke, B. M., and Graham, C. F., 1985, Synthesis of apolipoproteins, alphafoetoprotein, albumin, and transferrin by the human foetal yolk sac and other foetal organs, *J. Embryol. Exp. Morph.* **85:**191–206.

Sonoyama, K., and Aoyama, Y., 1997, Different responses of apolipoprotein A-I, A-IV, and B gene expression during intestinal adaptation to a massive small bowel resection in rats, *Biosci. Biotech. Biochem.* **61:**1810–1913.

Sonoyama, K., Nishikawa, H., Kiriyama, S., and Niki, R., 1994, Cholestyramine and a fat-free diet lower apolipoprotein A-IV mRNA in jejunum and cholestyramine lowers apolipoprotein A-I mRNA in ileum of rats, *J. Nutr.* **124**:621–627.

Sonoyama, K., Nakamura, Y., Kiriyama, S., and Niki, R., 1996, Upregulation of apolipoprotein A-IV gene expression in residual ileum after massive small bowel resection requires the biliary secretion in rats, *P. S. E. B. M.* **211**:273–280.

Srivastava, R. A. K., Jiao, S., Tang, J. J., Pfleger, B. A., Kitchens, R. T., and Schonfeld, G., 1991, *In vivo* regulation of low-density lipoprotein receptor and apolipoprotein B gene expression by dietary fat and cholesterol in inbred strains of mice, *Biochim. Biophys. Acta* **1086**:29–43.

Stein, O., Stein, Y., Lefevre, M., and Roheim, P. S., 1986, The role of apolipoprotein A-IV in reverse cholesterol transport studied with cultured cells and liposomes derived from an ether analog of phosphotidylcholine, *Biochim. Biophys. Acta* **878**:7–13.

Steinmetz, A., and Utermann, G., 1985, Activation of lecithin: cholesterol acyltransferase by human apolipoprotein A-IV, *J. Biol. Chem.* **260**:2258–2264.

Steinmetz, A., Czekelius, P., Thiemann, E., Motzny, S., and Kaffarnik, H., 1988, Changes of apolipoprotein A-IV in the human neonate: Evidence for different inductions of apolipoproteins A-IV and A-I in the postpartum period, *Atherosclerosis* **69**:21–27.

Talianidis, I., Tambakaki, A., Toursounova, J., and Zannis, V. I., 1995, Conplex interactions between SP1 bound to multiple distal regulatory sites and HNF-4 bound to the proximal promoter lead to transcriptional activation of liver-specific human apo C-III gene, *Biochem.* **34**:10298–10309.

Teng, B.-B., Verp, M., Salomon, J., and Davidson, N. O., 1990, Apolipoprotein B messenger RNA editing is developmentally regulated and widely expressed in human tissues, *J. Biol. Chem.* **265**:20616–20620.

Teng, B.-B., Burant, C. F., and Davidson, N. O., 1993, Molecular cloning of an apolipoprotein B messenger RNA editing protein, *Science* **260**:1816–1819.

Traber, M. G., Kayden, H. J., and Rindler, M. J., 1987, Polarized secretion of newly synthesized lipoproteins by the Caco-2 human intestinal cell line, *J. Lipid Res.* **28**:1350–1363.

Trieu, V. N., Penn, D., Hay, R. V., and Black, D. D., 1992, Sequence and developmental expression of porcine intestinal and hepatic apolipoprotein A-I mRNA, *Gastroenterol.* **102**:A582.

Trieu, V. N., Hasler-Rapacz, J., Rapacz, J., and Black, D. D., 1993, Sequences and expression of the porcine apolipoprotein A-I and C-III mRNAs, *Gene* **123**:173–179.

Tso, P., 1994, Intestinal lipid absorption, in: *Physiology of the Gastroentestinal Tract,* Volume 2 (L. R. Johnson, ed.), Raven Press, New York, pp. 1867–1908.

Tsukamoto, K., Hiester, K. G., Smith, P., Usher, D. C., Glick, J. M., and Rader, D. J., 1997, Comparison of human apo A-I expression in mouse models of atherosclerosis after gene transfer using a second generation adenovirus, *J. Lipid Res.* **38**:1869–1876.

Tzameli, I., and Zannis, V. I., 1996, Binding specificity and modulation of the apo A-I promoter activity by homo- and heterodimers of nuclear receptors, *J. Biol. Chem.* **271**:8402–8415.

van Greevenbroek, M. M. J., Voorhout, W. F., Erkelens, D. W., van Meer, G., and de Bruin, T. W. A., 1995, Palmitic acid and linoleic acid metabolism in Caco-2 cells: Different triglyceride synthesis and lipoprotein secretion, *J. Lipid Res.* **36**:13–24.

van Greevenbroek, M. M. J., van Meer, G., Erkelens, D. W., and de Bruin, T. W. A., 1996, Effects of saturated, mono-, and polyunsaturated fatty acids on the secretion of apo B containing lipoproteins by Caco-2 cells, *Atherosclerosis* **121**:139–150.

Vergnes, L., Taniguchi, T., Omori, K., Zakin, M. M., and Ochoa, A., 1997, The apolipoprotein A-I/C-III/A-IV gene cluster: Apo C-III and apo A-IV expression is regulated by two common enhancers, *Biochim. Biophys. Acta* **1348**:299–310.

Voyiaziakis, E., Goldberg, I. J., Plump, A. S., Rubin, E. M., Breslow, J. L., and Huang, L. S., 1998, Apo A-I deficiency causes both hypertriglyceridemia and increased atherosclerosis in human apo B transgenic mice, *J. Lipid Res.* **39**:313–321.

Wagner, R. D., Krul, E. S., Moberly, J. B., Alpers, D. H., and Schonfeld, G., 1992, Apolipoprotein expression and cellular differentiation in Caco-2 intestinal cells, *Am. J. Physiol.* **263**:E374–E382.

Walsh, A., Ito, Y., and Breslow, J. L., 1989, High levels of the human apolipoprotein A-I in transgenic mice result in increased plasma levels of small high density lipoprotein (HDL) particles comparable to human HDL3, *J. Biol. Chem.* **264**:6488–6494.

Walsh, A. M., Azrolan, N., Wang, K., Marcigliano, A., O'Connell, A., and Breslow, J. L., 1993, Intestinal expression of the human apo A-I gene in transgenic mice is controlled by a DNA region 3' to the gene in the promoter of the adjacent convergently transcribed apo C-III gene, *J. Lipid Res.* **34:**617–623.

Wang, H., Zhan, R., Hunter, F., Du, J., and Black, D., 1996, Effect of acute feeding of diets of varying fatty acid composition on intestinal apolipoprotein expression in the newborn swine, *Pediatr. Res.* **39:**1078–1084.

Wang, H., Berschneider, H. M., Du, J., and Black, D. D., 1997, Apolipoprotein secretion and lipid synthesis: Regulation by fatty acids in newborn swine intestinal epithelial cells, *Am. J. Physiol.* **272:**G935–G942.

Wang, H., Hunter, F., and Black, D. D., 1998, Effect of feeding diets of varying fatty acid composition on apolipoprotein expression in newborn swine, *Am. J. Physiol.* **275:**G645–G651.

Wang, H., Roberson, R., Du, J., Eshun, J. K., Berschneider, H. M., and Black, D. D., 1999, Regulation of apolipoprotein secretion by biliary lipids in newborn swine intestinal epithelial cells, *Am. J. Physiol.* **276:**G353–G362.

Wang, J. L., Swartzbasile, D. A., Rubin, D. C., and Levin, M. S., 1997, Retinoic acid stimulates early cellular proliferation in the adapting remnant rat small intestine after partial resection, *J. Nutr.* **127:**1297–1303.

Weaver, L. T., 1991, Anatomy and embryology, in: *Pediatric Gastrointestinal Disease,* Volume 1 (W. A. Walker, P. R. Durie, J. R. Hamilton, J. A. Walker-Smith, and J. B. Watkins, eds.), B. C. Decker, Inc., Philadelphia, pp. 195–216.

Weinstock, P. H., Bisgaier, C. L., Hayek, T., Aalto-Setala, Sehayek, E., Wu, L., Sheiffele, P., Merkel, M., Essenburg, A. D., and Breslow, J. L., 1997, Decreased HDL cholesterol levels but normal lipid absorption, growth, and feeding behavior in apolipoprotein A-IV knockout mice, *J. Lipid Res.* **38:**1782–1794.

Weinstock, P. H., Bisgaier, C. L., Hayek, T., Aalto-Setala, Sehayek, E., Wu, L., Sheiffele, P., Merkel, M., Essenburg, A. D., and Breslow, J.L., 1997, Decreased HDL cholesterol levels but normal lipid absorption, growth, and feeding behavior in apolipoprotein A-IV knockout mice, J. Lipid Res. 38:1782–1794.

Williams, S. C., Grant, S. G., Reue, K., Carrasquillo, B., Lusis, A. J., and Kinniburgh, A. J., 1989, *Cis*-acting determinants of basal and lipid-regulated apolipoprotein A-IV expression in mice, *J. Biol. Chem.* **264:**19009–19016.

Williamson, R., Lee, D., Hagaman, J., and Maeda, N., 1992, Marked reduction of high density lipoprotein cholesterol in mice genetically modified to lack apolipoprotein A-I, *Proc. Natl. Acad. Sci. USA* **89:**7134–7138.

Windler, E., Chow, Y.-S., and Havel, R. J., 1980, Regulation of the hepatic uptake of triglyceride-rich lipoproteins in the rat, *J. Biol. Chem.* **255:**8303–8307.

Wu, A. L., and Windmueller, H. G., 1979, Relative contribution by liver and intestine to individual plasma apolipoproteins in the rat, *J. Biol. Chem.* **254:**7316–7322.

Wu, X., Zhou, M., Huang, L.-S., Wetterau, J., and Ginsberg, H. N., 1996, Demonstration of a physical interaction between microsomal triglyceride transfer protein and apolipoprotein B during the assembly of apo B-containing lipoproteins, *J. Biol. Chem.* **271:**10277–10281.

Yao, Z., Tran, K., and McLeod, R. S., 1997, Intracellular degradation of newly synthesized apolipoprotein B, *J. Lipid Res.* **38:**1937–1953.

Young, S. G., Farese, R. V., Pierotti, V. R., Taylor, S., Grass, D. S., and Linton, M. F., 1994, Transgenic mice expressing human apo B-100 and apo B-48, *Curr. Opin. Lipidol.* **5:**94–101.

Young, S. G., Cham, C. M., Pitas, R. E., Burri, B. J., Connolly, A., Flynn, L., Pappu, A. S., Wong, J. S., Hamilton, R. L., and Farese, R. V., 1995, A genetic model for absent chylomicron formation: Mice producing apolipoprotein B in the liver, but not in the intestine, *J. Clin. Invest.* **96:**2932–2946.

Zannis, V. I., Kurnit, D. M., and Breslow, J. L., 1982, Hepatic apo A-I and apo E and intestinal apo A-I are synthesized in precursor isoprotein forms by organ cultures of human fetal tissues, *J. Biol. Chem.* **257:**536–544.

Zannis, V. I., Cole, F. S., Jackson, C. L., Kurnit, D. M., and Karathanasis, S. K., 1985, Distribution of apo A-I, apo C-II, apo C-III, and apo E mRNA in human tissues. Time-dependent induction of apo E mRNA by cultures of human monocyte-macrophages, *Biochem.* **24:**4450–4455.

Zhang, S. H., Reddick, R. L., Piedrahita, J. A., and Maeda, N., 1992, Spontaneous hypercholesterolemia and arterial lesions in mice lacking apolipoprotein E, *Science* **258:**468–472.

CHAPTER 15

Recent Progress in the Study of Intestinal Apolipoprotein B Gene Expression

NICHOLAS O. DAVIDSON and SHRIKANT ANANT

1. Introduction

Apolipoprotein B (apo B) is a large (MW 550 kDa), hydrophobic protein that is difficult to work with and inaccessible to conventional protein biochemistry. The combination of its size and susceptibility to proteolysis, particularly in the absence of lipid, presents formidable challenges to scientists interested in understanding its biology and regulation. These difficulties notwithstanding, interest in the biology of apo B is fueled by its requisite importance in the process of cellular lipid export and in the ligand-directed uptake of lipoprotein particles from the systemic circulation. In particular, the role of apo B as a risk factor in atherosclerosis susceptibility has channeled attention toward understanding the functional domains in the protein responsible for interaction with the low-density lipoprotein receptor. A considerable amount has been learned in the dozen or so years since the first partial cDNA clones encoding apo B were isolated and characterized (Deeb *et al.*, 1985; Huang *et al.*, 1985; Knott *et al.*, 1985; Wei *et al.*, 1985). Indeed, the availability of these reagents was indispensable to the prediction of functional domains within the apo B protein. With the identification of the structural gene spanning the chromosomal *apob* locus and the recognition that a single genomic template has the capacity to encode two distinct molecular forms of the apo B protein, research into the molecular regulation of apo B gene expression has uncovered the existence of multiple levels of control. These include transcriptional mechanisms, posttranscriptional RNA editing, cotranslational translocation, and posttranslational processing. Each of these may potentially function as restriction points in the expression of apo B. This chapter addresses some of the recent developments in the area of apo B gene expression, with particular focus on the regulation of intestinal apo B and its significance in intestinal lipid metabolism.

NICHOLAS O. DAVIDSON • Department of Internal Medicine and Molecular Biology and Pharmacology, Washington University School of Medicine, St. Louis, Missouri 63110.   SHRIKANT ANANT • Department of Internal Medicine, Washington University School of Medicine, St. Louis, Missouri 63110.
Intestinal Lipid Metabolism, edited by Charles M. Mansbach II *et al.*, Kluwer Academic/Plenum Publishers, 2001.

2. Tissue-Specific Expression of Apo B

2.1. Transcriptional Regulation of Apo B Gene Expression in the Small Intestine

The structural gene encoding apo B spans over 43 kb on the short arm of human chromosome 2 (Knott *et al.*, 1985). The expression of apo B in adult humans is confined to the liver and the small intestine and a similar pattern of tissue-restricted expression is found in rats and in mice (reviewed in Young, 1990). Studies to examine the *cis*-acting regulatory elements of apo B were hindered by the large size of the structural gene, which made it difficult to clone into conventional lambda or cosmid vectors for transgene construction. Young and colleagues (1990), however, were able to take advantage of the P1 bacteriophage vector that allowed these workers to introduce a clone containing approximately 80 kb of chromosomal DNA spanning the entire human *apob* locus and that contained approximately 19 kb of 5' and approximately 17 kb of 3' flanking sequence (Linton *et al.*, 1993). Despite the presence of substantial flanking sequence molecules, expression of the human apo B transgene was restricted to the liver of these (transgenic) mice, with very low levels of human apo B mRNA detectable in the heart and no expression in the small intestine (Linton *et al.*, 1993). These findings were confirmed in a separate study by Rubin and colleagues who independently reported the use of a bacteriophage P1 clone containing approximately 19kb of 5' sequence and approximately 14 kb of 3' sequence flanking the human *apob* locus (Callow *et al.*, 1994). These findings indicated that the human apo B intestinal control element was located outside the region selected for the transgenic construction. The formal possibility that mouse intestine lacked the necessary *trans*-acting factors to interact with putative intestinal elements in these human apo B constructs was excluded by findings in transgenic mice expressing an 87 kb mouse P1 clone containing 33 kb of 5' flanking sequence and 11 kb of 3' flanking sequence, where hepatic but not intestinal expression of the murine apo B transgene was demonstrated (McCormick *et al.*, 1996). More recent studies by Young and colleagues, using transgenic mice generated with large bacterial artificial chromosomes (BACs) containing up to approximately 120 kb of 5' flanking sequence, have further refined the intestinal control element of apo B to a region located between -33 kb and -70 kb upstream of the gene (Nielsen *et al.*, 1997). These findings set the stage for further refinement of the intestinal control element, using smaller fragments of these BAC clones.

There are interesting parallels that emerge from the study of intestinal elements regulating the expression of apo B and the findings from earlier studies of the elements regulating intestinal expression of apo A-I. Similar to apo B, apo A-I is expressed in both the liver and the small intestine where its tissue-specific regulation is controlled by two physically remote elements. Studies by Gordon, Breslow and colleagues have demonstrated that intestinal expression of apo A-I in mice requires elements contained in a 260 base–pair region spanning nucleotides -780 to -520 of the convergently transcribed apo C-III gene, that is located approximately 9.2 kb 3' to the structural apo A-I gene (Bisaha *et al.*, 1995). Interestingly, however, despite the observation that human apo A-I is expressed in a regional gradient along the small intestine of these transgenic mice, nucleotides -780 to -520 of the apoC-III gene appear to lack the elements that suppress expression of apo A-I in crypt cells and in populations of enteroendocrine cells of the gut (Bisaha *et al.*, 1995). The interplay of these regulatory motifs in the regulation of apo B gene expression in the intestinal epithelium will be an interesting area for future investigation.

2.2. Posttranscriptional Regulation of Apo B Gene Expression in the Small Intestine

Two forms of apo B circulate in mammalian plasma, each representing the product of a single nuclear gene. The primary apo B transcript undergoes splicing and polyadenylation in both the liver and the small intestine. In the small intestine, the processed nuclear transcript undergoes site-specific deamination of a single cytidine nucleotide, producing uridine and thereby creating a UAA translational stop codon in the middle of the open reading frame (Fig. 1). This process, referred to as apo B RNA editing, is responsible for the production of apo B-48 (Chen *et al.,* 1987; Powell *et al.,* 1987). In the liver of most mammals, including humans, the nuclear apo B transcript is processed but not edited and encodes a full-length protein referred to as apo B-100. The small intestinal apo B transcript is extensively edited and generates apo B-48. Apo B RNA editing has many interesting features, and a considerable amount of information is available concerning the molecular regulation of this process. The essential components of this process are summarized in Fig. 1. Following transcription and processing of the nuclear transcript, the spliced and polyadeny-

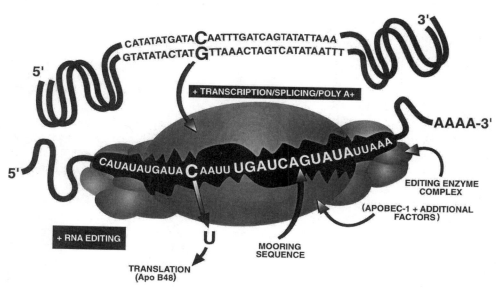

Figure 1. Intestinal apo B mRNA editing: An overview of the sequence requirements and enzymatic machinery. A common structural *apob* gene contains a genomically templated cytidine residue (outline) that is transcribed in the liver and the small intestine of all mammals. Following transcription and processing of the nuclear transcript by splicing and polyadenylation, the intestinal transcript is modified by C-to-U editing. The minimal *cis*-acting element is contained within approximately 30 nt spanning the edited base. This region, which is AU-rich, contains an 11 nt motif (shown as larger lettering) downstream of the edited base, referred to as a "mooring sequence." As shown in the figure, the apo B editing enzyme contains apobec-1, the catalytic subunit, as well as other essential protein factors, which coordinate the site-specific modification of nucleotide 6666 in apo B cDNA, changing a CAA to a UAA, translational stop codon. The edited apo B mRNA exits the nucleus into the cytoplasm where it is translated as a truncated protein, apo B-48.

lated apo B RNA undergoes C-to-U RNA editing within the nucleus (Lau *et al.*, 1991). Apo B RNA editing is mediated by a multicomponent enzyme complex that includes a single catalytic subunit, apobec-1, as well as other factors, yet to be identified, that are absolutely required for apo B RNA editing (Teng *et al.*, 1993). Following C-to-U editing of the intestinal apo B transcript, the edited RNA exits the nucleus and undergoes translational processing that results in the synthesis and secretion of apo B-48.

2.2.1. Cis-Acting Elements Directing Apo B RNA Editing

The minimal *cis*-acting elements required for apo B RNA editing are contained within a 30-nucleotide region spanning the edited base (Fig. 1). The entire region is rich in AU residues (approximately 70%) and includes a particularly crucial region located five nucleotides downstream of the edited C in which virtually all mutations decrease or eliminate apo B RNA editing (Shah *et al.*, 1991; Backus and Smith, 1992). This region is proposed to be involved in the site-specificity of C-to-U editing and has been referred to as a "mooring sequence" by Smith and colleagues (Smith and Sowden, 1996). In the context of this AU-rich RNA template and in the presence of a mooring sequence located at an optimal distance downstream of a cytidine residue, the multicomponent editing enzyme complex will bind to the template and deaminate the targeted cytidine residue. There appears to be a relatively lax specificity to the region immediately flanking the edited base because mutations within five nucleotides of either side are well tolerated (Chen *et al.*, 1990). Efficient RNA editing of this cassette requires an appropriate context that is composed of AU-rich bulk RNA and other elements located both 5' and 3' of the edited base that together direct high levels of editing activity as well as restricting this modification to a single base (Hersberger and Innerarity, 1998; Sowden *et al.*, 1998).

The precise delineation of the sequence requirements for apo B RNA editing is of more than theoretical interest. Although the site specificity of apo B RNA editing, under normal circumstances, involves deamination of a single cytidine in the middle of a template which spans over 14,000 nt, RNA editing may become unregulated and promiscuous under circumstances where apobec-1 is overexpressed (Sowden *et al.*, 1996; Yamanaka *et al.*, 1996). In addition, other RNA templates, such as NAT-1, may become substrates for apbec-1 under conditions of forced overexpression in transgenic animals (Yamanaka *et al.*, 1995, 1997). When apobec-1 is overexpressed, the requirements for a traditional mooring sequence are lost and C-to-U editing is likely mediated by elements distal to the canonical site at nt 6666 (Hersberger and Innerarity, 1998; Sowden *et al.*, 1998). The potential that other RNA templates may undergo C-to-U editing, coupled with the observation that transgenic mice and rabbits overexpressing apobec-1 at high levels in the liver develop hepatic dysplasia and carcinomas, raise the formal possibility that apobec-1 may potentially edit RNAs whose products are involved in growth regulation and transformation.

2.2.2. Apobec-1: An RNA-Specific Cytidine Deaminase

Apobec-1 is a 27 kDa protein, which functions as the catalytic subunit of the multicomponent apo B RNA editing enzyme. It has structural features in common with other cytidine deaminases, including the presence of a zinc-binding motif comprised of a histidine residue followed by another, usually valine residue and a catalytically active glutamic acid

(H-V-E) separated by 24 to 30 residues from a proline, a cysteine, two other residues, and another cysteine (P-C-X-X-C). This configuration is found in cytidine deaminases dating back to *E. coli* and is a so-called "signature" motif (Carter, 1995). These structural predictions were confirmed directly in experiments where apobec-1 was found to deaminate cytidine and deoxycytidine substrates that were presented as monomeric nucleosides (Navaratnam *et al.*, 1993b; MacGinnitie *et al.*, 1995). Nevertheless, for apobec-1 to perform deamination of a specific (cytidine) base in the context of an RNA template requires the presence of additional, complementation factors that are described in a later section. An additional feature of apobec-1, illustrated schematically in Fig. 1 and based on recent predictions from the crystal structure of the *E. coli* cytidine deaminase (Navaratnam *et al.*, 1998) is that the enzyme functions as a homodimer. This prediction was independently raised by Lau *et al.* in their analysis of the human apobec-1 protein, as evidenced by its ability to dimerize in solution (Lau *et al.*, 1994). In addition, apobec-1 binds to stretches of AU residues that are located both 5' and 3' of the edited base, and this binding occurs independent of interaction with the complementation factors (Anant *et al.*, 1995a, 1995b; Navaratnam *et al.*, 1995).

Apobec-1 is the sole catalytic subunit of the mammalian apo B RNA editing enzyme as evidenced by the results of gene targeting experiments in which disruption of both alleles eliminated apo B RNA editing (Hirano *et al.*, 1996; Morrison *et al.*, 1996; Nakamuta *et al.*, 1996). These experiments demonstrated conclusively that there is no genetic redundancy in apo B RNA editing, although the mice were otherwise completely unremarkable with the exception that their plasma contained exclusively apo B-100. Nevertheless, many important questions concerning the biology of apobec-1 were left unanswered by these knockout mice. Prominent among these was the observation, still unexplained, as to why apobec-1 is distributed widely in tissues of the rat and the mouse, including sites such as the spleen and the brain where there is virtually no apo B mRNA (Funahashi *et al.*, 1995; Nakamuta *et al.*, 1995). The implications raised by the presence of promiscuous RNA editing in the gain of function experiments in transgenic animals suggests that apobec-1 may actually target other RNA templates. The knockout experiments, although not formally disproving this possibility, would imply that this loss of function does not produce a lethal phenotype or impart any measurable effect on eating, on weight gain, or on reproductive health of the animals. However, a more subtle phenotype remains formally possible.

Apobec-1 gene expression is itself modulated in the rat and the mouse, particularly in the liver where it appears to be responsive to alterations in lipid loading as well as hormonal and nutritional events (reviewed in Davidson *et al.*, 1995). In addition to control at the level of apobec-1 mRNA abundance, the apobec-1 gene is regulated through the use of differential alternative promoter usage and alternative splicing, particularly in the rat and the mouse (Hirano *et al.*, 1997b; Nakamuta *et al.*, 1995; Qian *et al.*, 1997). In humans the *apobec-1* gene spans approximately 18 kb on chromosome 12, and it contains five exons, all of which are coding (Hirano *et al.*, 1997a). Although there was no evidence for alternative promoter usage or for the presence of untranslated exons as noted in the rat and the mouse apobec-1 genes, an interesting and unexpected finding in the human gene was the presence of a novel splice variant apobec-1 mRNA, which is expressed in the human small intestine (Hirano *et al.*, 1997a). This splice variant mRNA arises from the use of an alternative acceptor site within exon 3 of apobec-1 mRNA and results in exclusion of the 28 nt exon 2 from the mature transcript. The accompanying frame shift in the modified mRNA results in translation of the first six residues of apobec-1, followed by 30 novel residues. The 36-residue peptide generated as a result of this alternative splicing is expressed in cells of the small intestine, as evidenced by immunocytochemical localization (Hirano *et al.,* 1997a). The function of this peptide is unknown at present, although, based on structural homologies to other known growth factors, it is tempting to speculate that it may play a role in growth modulation.

2.2.3. *Other Trans-Acting Factors Involved in Apo B RNA Editing*

The mammalian apo B RNA editing enzyme includes apobec-1 as well as other protein factors, which together form an operational unit referred to as an *editosome* (Smith *et al.,* 1991). This structure has been characterized by Smith and coworkers as a multicomponent structure that forms as a protein–RNA complex with 11S mobility on glycerol gradients prior to the spontaneous assembly of a fully active 27S editosome (Backus and Smith, 1992; Smith *et al.,* 1991). Apobec-1, although required for apo B RNA editing, is insufficient when present alone. Identification and characterization of these additional factors has attracted considerable attention. Evidence for the existence of a 40 kDa protein that may be involved in apo B RNA editing was provided by Chan and colleagues who were able to cross-link a protein of this size to the apo B RNA flanking the edited base (Lau *et al.,* 1990). Other workers confirmed and extended these findings with the demonstration that proteins of approximately 66 and 44 kDa from rat liver could cross-link to apo B RNA templates spanning the edited base (Harris *et al.,* 1993; Navaratnam *et al.,* 1993a). A different approach was taken by Driscoll and coworkers, who used apobec-1 affinity chromatography to isolate proteins eluting with an apparent molecular mass of 65 kDa that, when added to apobec-1 and apo B RNA, resulted in C-to-U editing (Mehta *et al.,* 1996). Nevertheless, the nature of the proteins contained within this fraction is currently unknown. More recent studies have demonstrated that the 66 and 44 kDa proteins can be functionally distinguished and separated from one another using affinity chromatography (Yang *et al.,* 1997). Smith and colleagues in earlier work used yet another strategy to gain insight into the nature of the additional complementation factors required for apo B RNA editing, generating monoclonal antibodies against the 27S editosome and using these as reagents to deplete the complex of putative factors (Schock *et al.,* 1996). A protein factor of 240 kDa was tentatively identified, but its role in apo B RNA editing remains to be elucidated (Schock *et al.,* 1996). More recently, Lau *et al.* have used the yeast two-hybrid system to isolate an interacting gene whose product binds to apobec-1 and to apo B RNA (Lau *et al.,* 1997). This protein, referred to as ABBP-1, is a 36 kDa protein whose precise role in apo B RNA editing remains to be elucidated (Lau *et al.,* 1997). The identification and molecular characterization of interacting proteins involved in the assembly and function of the mammalian apo B RNA editosome is a major frontier and will undoubtedly remain the focus of investigation for some time. At the current time, there is no evidence that a single polypeptide, when added to recombinant apobec-1, will allow C-to-U editing of an apo B RNA.

2.2.4. *Apobec-1, Apo B RNA Editing, and Lipid Transport*

One of the cardinal features of intestinal lipoprotein secretion is the presence of apo B-48. This feature notwithstanding, it is unknown whether apo B-48 confers any functional advantage over apo B-100 in terms of the ability of the cell to mobilize triglyceride for the purposes of vectorial delivery of lipid destined for secretion. As noted previously, mice in which apobec-1 was eliminated through gene targeting demonstrated only apo B-100 in the intestine and thus present an opportunity to evaluate this question. Histological examination of the intestine of these mice demonstrated no residual fat droplets within enterocytes, suggesting that triglyceride absorption was not grossly impaired. Furthermore, intestinal chylomicrons isolated from mesenteric lymph of these mice revealed a normal size distribution (Morrison *et al.,* 1996). Finally, preliminary characterization of chylomicron clearance in apobec-1 gene-targeted mice demonstrated no differences from control, wild-type mice (Morrison *et al.,* 1996). Taken together, the evidence from apobec-1 knockout mice is that intestinal lipid absorption and chylomicron assembly and secretion are probably indistinguishable in animals synthesizing apo B-100 only versus apo B-48 only. These conclusions were independently validated by Young and colleagues, who used targeted mutagenesis of the murine apo B gene to generate mice that synthesize exclusively apo B-100 only or apo B-48 only (Farese *et al.,* 1996). Thus, using two quite distinct approaches, the conclusion from studies of mouse intestinal lipoprotein assembly is that apo B-100 and apo B-48 likely function interchangeably regarding lipid export.

Another aspect of the role of apobec-1 and of apo B RNA editing in lipoprotein metabolism concerns the catabolic fate of lipoprotein particles containing apo B-100 versus apo B-48. As alluded to in the introduction, lipoprotein uptake is a largely receptor-dependent process whose relationship to apo B involves two distinct pathways. Uptake of low-density lipoprotein (LDL) is mediated by an interaction between a region in the carboxyl terminus of apo B-100 and the LDL receptor. The requisite region for this interaction is missing from apo B-48 (Young, 1990). As a result, apo B-48-containing lipoprotein particles are directed to another receptor, referred to as the LDL-receptor-related protein, or LRP, in which apo E functions as the cognate ligand (reviewed in Herz and Willnow, 1995). Interestingly, LDL-receptor-gene-targeted mice have only mild hypercholesterolemia, provided they are fed a low-fat chow diet and accumulate only modest quantities of LDL (Ishibashi *et al.,* 1994). This situation is thus unlike the corresponding finding in humans with the disease familial hypercholesterolemia (FH), where elevated levels of total and of LDL cholesterol are found (Herz and Willnow, 1995). The absence of severe hypercholesterolemia in LDL receptor knockout mice is likely the result of the extensive editing of the hepatic apo B mRNA in this species, which favors the secretion of apo B-48 over apo B-100 and thus results in lower production rates of LDL. Apobec-1 knockout mice show no elevation in serum cholesterol or any evidence for accumulation of LDL, despite the fact that only apo B-100 circulates in their plasma (Hirano *et al.,* 1996). The likely explanation for this observation is that the augmentation in LDL production rates are adequately compensated for by the presence of sufficient numbers of LDL receptors. Consequently, it would be anticipated that intercrosses between LDL receptor knockout mice and apobec-1 knockout mice would be hypercholesterolemic and would demonstrate accumulation of LDL particles containing apo B-100. Indeed, this prediction has been directly confirmed in very recent studies from our laboratory.

2.3. Co- and Posttranslational Regulation of Apo B Gene Expression

Control of apo B gene expression may also be exerted at the level of co- and posttranslational processing of the nascent polypeptide, and much has been learned concerning this aspect of gene regulation through the study of hepatoma cells (reviewed in Pease and

Leiper, 1996; Yao *et al.*, 1997). It is beyond the scope of this chapter to present a discussion of this aspect of apo B gene regulation and the reader is referred to the two articles alluded to previously and the references therein for a balanced perspective. In principal, however, the current paradigm (see Yao *et al.*, 1997) generally holds that apo B undergoes cotranslational translocation of the nascent protein across the endoplasmic reticulum (ER) membrane (Ingram and Shelness, 1996), a process that may involve the interaction of apo B with chaperone proteins (Zhou *et al.*, 1995). During and following translocation into the lumen of the ER, apo B is "lipidated" through the actions of the microsomal triglyceride transfer protein (MTP), possibly by way of a physical interaction (Gordon *et al.*, 1996; Wang *et al.*, 1996). Lipidation of apo B is considered essential to its stability, because in the absence of MTP, as in abetalipoproteinemia, apo B is extensively degraded and virtually none is secreted (Benoist and Grand-Perret, 1997). Apo B degradation has been postulated to occur in the ER (Adeli *et al.*, 1997) and also in post-ER compartments (Wang *et al.*, 1995) and is thought to involve proteases, which are sensitive to the calpain inhibitor acetyl-leucyl-leucyl-norleucinal (Adeli *et al.*, 1997) as well as involving the proteasomal pathway (Fisher *et al.*, 1997). It is important to bear in mind, however, that much of the experimental paradigm has been generated through studies in human and in rat hepatoma cells and the possibility certainly exists that features of this paradigm may pertain exclusively to hepatic apo B gene expression.

Intestinal lipoprotein secretion and apo B gene expression has been examined *in vivo* using various animal species, including the rat and the mouse (Davidson *et al.*, 1986; Kim and Young, 1998). In addition, a considerable amount of work has been conducted using Caco-2 cells as a surrogate intestinal cell line in which to examine important features of apo B gene expression (Hughes *et al.*, 1987; Traber *et al.*, 1987). Lipoprotein secretion in these cells is highly dependent on the differentiation state of the cells, increasing exponentially as the cells manifest characteristics of a polarized epithelial monolayer (Hughes *et al.*, 1987; Traber *et al.*, 1987). A disappointing feature of this cell line, however, is that the synthesis of apo B at all stages of differentiation favors apo B-100 rather than apo B-48. This observation most plausibly reflects the low levels of apo B RNA editing activity, which in turn reflects limiting quantities of apobec-1 (Giannoni *et al.*, 1995). Hussain and coworkers have attempted to circumvent this problem in Caco-2 cells by creating clones of cells stably expressing high levels of apo B-48, driven by a heterologous promoter (Luchoomun *et al.*, 1997). These workers demonstrated that preconfluent (i.e., undifferentiated) cells were capable of lipoprotein secretion in the presence of apo B-48 and in the absence of other factors that are involved in coordinating spontaneous lipoprotein secretion as the cells differentiate. The conclusion that these authors reached was that the expression of apo B-48 was itself sufficient to allow the secretion of complex lipid from these cells (Luchoomun *et al.*, 1997). However, although the authors were able to exclude an important contribution to lipoprotein secretion from the expression of other apolipoprotein genes, there was detectable activity of the MTP even in preconfluent cells. Furthermore, the apo B-48-mediated secretion of lipoproteins from these cells was completely inhibited by a pharmacological MTP inhibitor suggesting that, although apo B-48 expression may be permissive for lipoprotein secretion from undifferentiated Caco-2 cells, MTP activity is absolutely required.

Murthy *et al.*, working with differentiated Caco-2 cells, examined the response of apo B gene expression to different types of fatty acid (oleic versus eicosapentaenoic acid;

Murthy *et al.*, 1992). These workers demonstrated that the secretion of apo B into the basolateral medium from cells incubated with the *n*-3 fatty acid was lower than that found with control cells or in cells incubated with oleic acid. Furthermore, and in keeping with findings in hepatoma cells (see Yao *et al.*, 1997 for a review), intracellular degradation of apo B was significantly reduced in cells exposed to oleate compared with either control or eicosapentaenoic-acid-treated cells (Murthy *et al.*, 1992). The major findings to emerge from this study are that apo B secretion from intestinal cells increases in response to lipid flux through mobilization of preformed pools of apo B, coupled with a variable effect on the intracellular synthesis and degradation of newly synthesized protein. These findings are consistent with earlier studies in rat intestine in which it was demonstrated that *de novo* synthesis of intestinal apo B-48 was not modulated in response to bolus triglyceride administration (Davidson *et al.*, 1986).

3. Concluding Remarks

The existence of multiple levels of regulation of apo B gene expression provides some measure of the importance of this protein in relation to lipid secretion. Further study of these various levels of control will undoubtedly provide new insight into the cell biology of this complex but vital process.

Acknowledgments

Work from the authors' laboratory, cited in this review, has been supported by grants HL-38180, DK 56260, Digestive Disease Research Core Center grant DK-52574, and Clinical Nutrition Research Unit grant DK 56341, all from the National Institutes of Health.

References

Adeli, K., Wettesten, M., Asp, L., Mohammadi, A., Macri, J., and Olofsson, S. O., 1997, Intracellular assembly and degradation of apolipoprotein B-100-containing lipoproteins in digitonin-permeabilized Hep G2 cells, *J. Biol. Chem.* **272**(8):5031–5039.

Anant, S., MacGinnitie, A. J., and Davidson, N. O., 1995a, Apobec-1, the catalytic subunit of the mammalian apolipoprotein B mRNA editing enzyme, is a novel RNA-binding protein, *J. Biol. Chem.* **270**(24):14762–14767.

Anant, S., MacGinnitie, A. J., and Davidson, N. O., 1995b, The binding of apobec-1 to mammalian apo B RNA is stabilized by the presence of complementation factors which are required for post-transcriptional editing, *Nucleic Acids Symp. Ser.* **33**:99–102.

Backus, J. W., and Smith, H. C., 1992, Three distinct RNA sequence elements are required for efficient apolipoprotein B (apo B) RNA editing *in vitro*, *Nucleic Acids Res.* **20**(22):6007–6014.

Benoist, F., and Grand-Perret, T., 1997, Co-translational degradation of apolipoprotein B100 by the proteasome is prevented by microsomal triglyceride transfer protein. Synchronized translation studies on Hep G2 cells treated with an inhibitor of microsomal triglyceride transfer protein, *J. Biol. Chem.* **272**(33):20435–20442.

Bisaha, J. G., Simon, T. C., Gordon, J. I., and Breslow, J. L., 1995, Characterization of an enhancer element in the human apolipoprotein C-III gene that regulates human apolipoprotein A-I gene expression in the intestinal epithelium, *J. Biol. Chem.* **270**(34):19979–19988.

Callow, M. J., Stoltzfus, L. J., Lawn, R. M., and Rubin, E. M., 1994, Expression of human apolipoprotein B and assembly of lipoprotein(a) in transgenic mice, *Proc. Natl. Acad. Sci. USA* **91**(6):2130–2134.

Carter, C. W., Jr., 1995, The nucleoside deaminases for cytidine and adenosine: Structure, transition state stabilization, mechanism, and evolution, *Biochimie* **77**(1–2):92–98.

Chen, S. H., Habib, G., Yang, C. Y., Gu, Z. W., Lee, B. R., Weng, S. A., Silberman, S. R., Cai, S. J., Deslypere, J. P., Rosseneu, M., Gotto, A. M., Jr., and Chan, L., 1987, Apolipoprotein B-48 is the product of a messenger RNA with an organ-specific in-frame stop codon, *Science* **238**(4825):363–366.

Chen, S. H., Li, X. X., Liao, W. S., Wu, J. H., and Chan, L., 1990, RNA editing of apolipoprotein B mRNA. Sequence specificity determined by *in vitro* coupled transcription editing, *J. Biol. Chem.* **265**(12):6811–6816.

Davidson, N. O., Kollmer, M. E., and Glickman, R. M., 1986, Apolipoprotein B synthesis in rat small intestine: Regulation by dietary triglyceride and biliary lipid, *J. Lipid Res.* **27**(1):30–39.

Davidson, N. O., Anant, S., and MacGinnitie, A. J., 1995, Apolipoprotein B messenger RNA editing: Insights into the molecular regulation of post-transcriptional cytidine deamination, *Curr. Opin. Lipidol.* **6**(2):70–74.

Deeb, S. S., Motulsky, A. G., and Albers, J. J., 1985, A partial cDNA clone for human apolipoprotein B, *Proc. Natl. Acad. Sci. USA* **82**(15):4983–4986.

Farese, R., Veniant, M. M., Cham, C. M., Flynn, L. M., Pierotti, V., Loring, J. F., Traber, M., Ruland, S., Stokowski, R. S., Huszar, D., and Young, S. G., 1996, Phenotypic analysis of mice expressing exclusively apolipoprotein B-48 or apolipoprotein B-100, *Proc. Natl. Acad. Sci. USA* **93**(13):6393–6398.

Fisher, E. A., Zhou, M., Mitchell, D. M., Wu, X., Omura, S., Wang, H., Goldberg, A. L., and Ginsberg, H. N., 1997, The degradation of apolipoprotein B-100 is mediated by the ubiquitin–proteasome pathway and involves heat shock protein 70, *J. Biol. Chem.* **272**(33):20427–20434.

Funahashi, T., Giannoni, F., DePaoli, A. M., Skarosi, S. F., and Davidson, N. O., 1995, Tissue-specific, developmental and nutritional regulation of the gene encoding the catalytic subunit of the rat apolipoprotein B mRNA editing enzyme: Functional role in the modulation of apo B mRNA editing, *J. Lipid Res.* **36**(3):414–428.

Giannoni, F., Chou, S. C., Skarosi, S. F., Verp, M. S., Field, F. J., Coleman, R. A., and Davidson, N. O., 1995, Developmental regulation of the catalytic subunit of the apolipoprotein B mRNA editing enzyme (apobec-1) in human small intestine, *J. Lipid. Res.* **36**(8):1664–1675.

Gordon, D. A., Jamil, H., Gregg, R. E., Olofsson, S. O., and Boren, J., 1996, Inhibition of the microsomal triglyceride transfer protein blocks the first step of apolipoprotein B lipoprotein assembly but not the addition of bulk core lipids in the second step, *J. Biol. Chem.* **271**(51):33047–33053.

Harris, S. G., Sabio, I., Mayer, E., Steinberg, M. F., Backus, J. W., Sparks, J. D., Sparks, C. E., and Smith, H. C., 1993, Extract-specific heterogeneity in high-order complexes containing apolipoprotein B mRNA editing activity and RNA-binding proteins, *J. Biol. Chem.* **268**(10):7382–7392.

Hersberger, M., and Innerarity, T. L., 1998, Two efficiency elements flanking the editing site of cytidine 6666 in the apolipoprotein B mRNA support mooring-dependent editing, *J. Biol. Chem.* **273**(16):9435–9442.

Herz, J., and Willnow, T. E., 1995, Lipoprotein and receptor interactions *in vivo, Curr. Opin. Lipidol.* **6**(2):97–103.

Hirano, K., Young, S. G., Farese, R. V., Jr., Ng, J., Sande, E., Warburton, C., Powell-Braxton, L. M., and Davidson, N. O., 1996, Targeted disruption of the mouse apobec-1 gene abolishes apolipoprotein B mRNA editing and eliminates apolipoprotein B-48, *J. Biol. Chem.* **271**(17):9887–9890.

Hirano, K., Min, J., Funahashi, T., Baunoch, D. A., and Davidson, N. O., 1997a, Characterization of the human apobec-1 gene: Expression in gastrointestinal tissues determined by alternative splicing with production of a novel truncated peptide, *J. Lipid. Res.* **38**(5):847–859.

Hirano, K., Min, J., Funahashi, T., and Davidson, N. O., 1997b, Cloning and characterization of the rat apobec-1 gene: A comparative analysis of gene structure and promoter usage in rat and mouse, *J. Lipid. Res.* **38**(6):1103–1119.

Huang, L. S., Bock, S. C., Feinstein, S. I., and Breslow, J. L., 1985, Human apolipoprotein B cDNA clone isolation and demonstration that liver apolipoprotein B mRNA is 22 kilobases in length, *Proc. Natl. Acad. Sci. USA* **82**(20):6825–6829.

Hughes, T. E., Sasak, W. V., Ordovas, J. M., Forte, T. M., Lamon-Fava, S., and Schaefer, E. J., 1987, A novel cell line (Caco-2) for the study of intestinal lipoprotein synthesis, *J. Biol. Chem.* **262**(8):3762–3767.

Ingram, M. F., and Shelness, G. S., 1996, Apolipoprotein B-100 destined for lipoprotein assembly and intracellular degradation undergoes efficient translocation across the endoplasmic reticulum membrane, *J. Lipid. Res.* **37**(10):2202–2214.

Ishibashi, S., Herz, J., Maeda, N., Goldstein, J. L., and Brown, M. S., 1994, The two-receptor model of lipoprotein clearance: Tests of the hypothesis in "knockout" mice lacking the low density lipoprotein receptor, apolipoprotein E, or both proteins, *Proc. Natl. Acad. Sci. USA* **91**(10):4431–4435.

Kim, E., and Young, S. G., 1998, Genetically modified mice for the study of apolipoprotein B., *J. Lipid. Res.* **39**:703–723.

Knott, T. J., Rall, S. C., Jr., Innerarity, T. L., Jacobson, S. F., Urdea, M. S., Levy-Wilson, B., Powell, L. M., Pease, R. J., Eddy, R., Nakai, H., Byers, M., Priestley, L. M., Robertson, E., Rall, L. B., Betsholtz, C., Shows, T. B., Mahley, R. W., and Scott, J., 1985, Human apolipoprotein B: Structure of carboxyl-terminal domains, sites of gene expression, and chromosomal localization, *Science* 230(4721):37–43.

Lau, P. P., Chen, S. H., Wang, J. C., and Chan, L., 1990, A 40 kilodalton rat liver nuclear protein binds specifically to apolipoprotein B mRNA around the RNA editing site, *Nucleic Acids Res.* 18(19):5817–5821.

Lau, P. P., Xiong, W. J., Zhu, H. J., Chen, S. H., and Chan, L., 1991, Apolipoprotein B mRNA editing is an intranuclear event that occurs posttranscriptionally coincident with splicing and polyadenylation, *J. Biol. Chem.* 266(30):20550–20554.

Lau, P. P., Zhu, H. J., Baldini, A., Charnsangavej, C., and Chan, L., 1994, Dimeric structure of a human apolipoprotein B mRNA editing protein and cloning and chromosomal localization of its gene, *Proc. Natl. Acad. Sci. USA* 91(18):8522–8526.

Lau, P. P., Zhu, H. J., Nakamuta, M., and Chan, L., 1997, Cloning of an Apobec-1-binding protein that also interacts with apolipoprotein B mRNA and evidence for its involvement in RNA editing, *J. Biol. Chem.* 272(3):1452–1455.

Linton, M. F., Farese, R. V., Jr., Chiesa, G., Grass, D. S., Chin, P., Hammer, R. E., Hobbs, H. H., and Young, S. G., 1993, Transgenic mice expressing high plasma concentrations of human apolipoprotein B-100 and lipoprotein(a), *J. Clin. Invest.* 92(6):3029–3037.

Luchoomun, J., Zhou, Z., Bakillah, A., Jamil, H., and Hussain, M. M., 1997, Assembly and secretion of VLDL in nondifferentiated Caco-2 cells stably transfected with human recombinant Apo B-48 cDNA, *Arterioscler. Thromb. Vasc. Biol.* 17(11):2955–2963.

MacGinnitie, A. J., Anant, S., and Davidson, N. O., 1995, Mutagenesis of apobec-1, the catalytic subunit of the mammalian apolipoprotein B mRNA editing enzyme, reveals distinct domains that mediate cytosine nucleoside deaminase, RNA binding, and RNA editing activity, *J. Biol. Chem.* 270(24):14768–14775.

McCormick, S. P., Ng, J. K., Veniant, M., Boren, J., Pierotti, V., Flynn, L. M., Grass, D. S., Connolly, A., and Young, S. G., 1996, Transgenic mice that overexpress mouse apolipoprotein B. Evidence that the DNA sequences controlling intestinal expression of the apolipoprotein B gene are distant from the structural gene, *J. Biol. Chem.* 271(20):11963–11970.

Mehta, A., Banerjee, S., and Driscoll, D. M., 1996, apobec-1 interacts with a 65-kDa complementing protein to edit apolipoprotein-B mRNA *in vitro*, *J. Biol. Chem.* 271(45):28294–28299.

Morrison, J. R., Paszty, C., Stevens, M. E., Hughes, S. D., Forte, T., Scott, J., and Rubin, E. M., 1996, Apolipoprotein B RNA editing enzyme-deficient mice are viable despite alterations in lipoprotein metabolism, *Proc. Natl. Acad. Sci. USA* 93(14):7154–7159.

Murthy, S., Albright, E., Mathur, S. N., Davidson, N. O., and Field, F. J., 1992, Apolipoprotein B mRNA abundance is decreased by eicosapentaenoic acid in Caco-2 cells. Effect on the synthesis and secretion of apolipoprotein B, *Arterioscl. Thromb.* 12(6):691–700.

Nakamuta, M., Oka, K., Krushkal, J., Kobayashi, K., Yamamoto, M., Li, W. H., and Chan, L., 1995, Alternative mRNA splicing and differential promoter utilization determine tissue-specific expression of the apolipoprotein B mRNA-editing protein (apo Bec1) gene in mice. Structure and evolution of apo Bec1 and related nucleoside/nucleotide deaminases, *J. Biol. Chem.* 270(22):13042–13056.

Nakamuta, M., Chang, B. H. J., Zsigmond, E., Kobayashi, K., Lei, H., Ishida, B. Y., Oka, K., Li, E., and Chan, L., 1996, Complete phenotypic characterization of apobec-1 knockout mice with a wild-type genetic background and a human apolipoprotein B transgenic background, and restoration of apolipoprotein B mRNA editing by somatic gene transfer of apobec-1, *J. Biol. Chem.* 271(42):25981–25988.

Navaratnam, N., Shah, R., Patel, D., Fay, V., and Scott, J., 1993a, Apolipoprotein B mRNA editing is associated with UV crosslinking of proteins to the editing site, *Proc. Natl. Acad. Sci. USA* 90(1):222–226.

Navaratnam, N., Morrison, J. R., Bhattacharya, S., Patel, D., Funahashi, T., Giannoni, F., Teng, B. B., Davidson, N. O., and Scott, J., 1993b, The p27 catalytic subunit of the apolipoprotein B mRNA editing enzyme is a cytidine deaminase, *J. Biol. Chem.* 268(28):20709–20712.

Navaratnam, N., Bhattacharya, S., Fujino, T., Patel, D., Jarmuz, A. L., and Scott, J., 1995, Evolutionary origins of apo B mRNA editing: Catalysis by a cytidine deaminase that has acquired a novel RNA-binding motif at its active site, *Cell* 81(2):187–195.

Navaratnam, N., Fujino, T., Bayliss, J., Jarmuz, A., How, A., Richardson, N., Somasekaram, A., Bhattacharya, S., Carter, C., and Scott, J., 1998, *Escherichia coli* cytidine deaminase provides a molecular model for Apo B RNA editing and a mechanism for RNA substrate recognition, *J. Mol. Biol.* 275(4):695–714.

Nielsen, L. B., McCormick, S. P., Pierotti, V., Tam, C., Gunn, M. D., Shizuya, H., and Young, S. G., 1997, Human apolipoprotein B transgenic mice generated with 207- and 145- kilobase pair bacterial artificial chromosomes. Evidence that a distant 5′-element confers appropriate transgene expression in the intestine, *J. Biol. Chem.* **272**(47):29752–29758.

Pease, R. J., and Leiper, J. M., 1996, Regulation of hepatic apolipoprotein-B-containing lipoprotein secretion, *Curr. Opin. Lipidol.* **7**(3):132–138.

Powell, D. R., Lee, P. D., Shively, J. E., Eckenhausen, M., and Hintz, R. L., 1987, Method for purification of an insulin-like growth factor-binding protein produced by human Hep G2 hepatoma cells, *J. Chromat.* **420**(1):163–170.

Qian, X., Balestra, M. E., and Innerarity, T. L., 1997, Two distinct TATA-less promoters direct tissue-specific expression of the rat apo B editing catalytic polypeptide 1 gene, *J. Biol. Chem.* **272**(29):18060–18070.

Schock, D., Kuo, S. R., Steinburg, M. F., Bolognino, M., Sparks, J. D., Sparks, C. E., and Smith, H. C., 1996, An auxiliary factor containing a 240-kDa protein complex is involved in apolipoprotein B RNA editing, *Proc. Natl. Acad. Sci. USA* **93**(3):1097–1102.

Shah, R. R., Knott, T. J., Legros, J. E., Navaratnam, N., Greeve, J. C., and Scott, J., 1991, Sequence requirements for the editing of apolipoprotein B mRNA, *J. Biol. Chem.* **266**(25):16301–16304.

Smith, H. C., Kuo, S. R., Backus, J. W., Harris, S. G., Sparks, C. E., and Sparks, J. D., 1991, *In vitro* apolipoprotein B mRNA editing: Identification of a 27S editing complex, *Proc. Natl. Acad. Sci. USA* **88**(4):1489–1493.

Smith, H. C., and Sowden, M. P., 1996, Base-modification mRNA editing through deamination—the good, the bad and the unregulated, *Trends Genet.* **12**(10):418–424.

Sowden, M., Hamm, J. K., and Smith, H. C., 1996, Overexpression of APOBEC-1 results in mooring sequence-dependent promiscuous RNA editing, *J. Biol. Chem.* **271**(6):3011–3017.

Sowden, M. P., Eagleton, M. J., and Smith, H. C., 1998, Apolipoprotein B RNA sequence 3′ of the mooring sequence and cellular sources of auxiliary factors determine the location and extent of promiscuous editing, *Nucleic Acids Res.* **26**(7):1644–1652.

Teng, B., Burant, C. F., and Davidson, N. O., 1993, Molecular cloning of an apolipoprotein B messenger RNA editing protein, *Science* **260**(5115):1816–1819.

Traber, M. G., Kayden, H. J., and Rindler, M. J., 1987, Polarized secretion of newly synthesized lipoproteins by the Caco-2 human intestinal cell line, *J. Lipid Res.* **28**(11):1350–1363.

Wang, C. N., Hobman, T. C., and Brindley, D. N., 1995, Degradation of apolipoprotein B in cultured rat hepatocytes occurs in a post-endoplasmic reticulum compartment, *J. Biol. Chem.* **270**(42):24924–24931.

Wang, S., McLeod, R. S., Gordon, D. A., and Yao, Z., 1996, The microsomal triglyceride transfer protein facilitates assembly and secretion of apolipoprotein B-containing lipoproteins and decreases cotranslational degradation of apolipoprotein B in transfected COS-7 cells, *J. Biol. Chem.* **271**(24):14124–14133.

Wei, C. F., Chen, S. H., Yang, C. Y., Marcel, Y. L., Milne, R. W., Li, W. H., Sparrow, J. T., Gotto, A. M., Jr., and Chan, L., 1985, Molecular cloning and expression of partial cDNAs and deduced amino acid sequence of a carboxyl-terminal fragment of human apolipoprotein B-100, *Proc. Natl. Acad. Sci. USA* **82**(21):7265–7269.

Yamanaka, S., Balestra, M. E., Ferrell, L. D., Fan, J., Arnold, K. S., Taylor, S., Taylor, J. M., and Innerarity, T. L., 1995, Apolipoprotein B mRNA-editing protein induces hepatocellular carcinoma and dysplasia in transgenic animals, *Proc. Natl. Acad. Sci. USA* **92**(18):8483–8487.

Yamanaka, S., Poksay, K. S., Driscoll, D. M., and Innerarity, T. L., 1996, Hyperediting of multiple cytidines of apolipoprotein B mRNA by apobec-1 requires auxiliary protein(s) but not a mooring sequence motif, *J. Biol. Chem.* **271**(19):11506–11510.

Yamanaka, S., Poksay, K. S., Arnold, K. S., and Innerarity, T. L., 1997, A novel translational repressor mRNA is edited extensively in livers containing tumors caused by the transgene expression of the apo B mRNA-editing enzyme, *Genes Dev.* **11**(3):321–333.

Yang, Y., Kovalski, K., and Smith, H. C., 1997, Partial characterization of the auxiliary factors involved in apolipoprotein B mRNA editing through APOBEC-1 affinity chromatography, *J. Biol. Chem.* **272**(44):27700–27706.

Yao, Z., Tran, K., and McLeod, R. S., 1997, Intracellular degradation of newly synthesized apolipoprotein B, *J. Lipid Res.* **38**(10):1937–1953.

Young, S. G., 1990, Recent progress in understanding apolipoprotein B, *Circulation* **82**(5):1574–1594.

Zhou, M., Wu, X., Huang, L. S., and Ginsberg, H. N., 1995, Apoprotein B-100, an inefficiently translocated secretory protein, is bound to the cytosolic chaperone, heat shock protein 70, *J. Biol. Chem.* **270**(42):25220–25224.

CHAPTER 16

The Role of Apolipoprotein A-IV as a Satiety Factor

PATRICK TSO and TAKASHI DOI

1. Sources of Lipid in the Lumen of the Gastrointestinal Tract

As much as 40% of the daily caloric intake in the Western diet is in the form of lipids ranging between 60 to 100 grams (Davenport, 1971). Triacylglycerol (TG) is the major dietary fat in humans. Major long-chain fatty acids present in the diet are palmitic (16:0), stearic (18:0), oleic (18:1), linoleic (18:2), and linolenic (18:3). In most infant diets, fat becomes an even more important source of calories. In human milk and in human formulas, as much as 50% of the total calories are present as fat (Hamosh, 1979). In human milk, there is also an abundance of medium-chain fatty acids. The human small intestine is presented daily with other lipids such as phospholipids (PL), cholesterol, and plant sterols. Both PL and cholesterol are major constituents of bile. In humans, the biliary PL is a major contributor of lumenal PL and as much as 11 to 20 grams of biliary PL enters the small intestinal lumen daily, whereas the dietary contribution is only between 1 to 2 grams (Northfield and Hofmann, 1975; Borgström, 1976). The small intestinal epithelium undergoes rapid turnover that also contributes to the lumenal PL and cholesterol. Although cholesterol is the predominant sterol of total dietary sterol in the Western diet, plant sterols account for 20% to 25% (Taylor and Gould, 1967; Gould et al., 1969).

2. Digestion of Lipids

Lipid digestion begins in the stomach. Lipase activity has been reported to be present in the human gastric juice (Schonheyder and Volquatz, 1946). In humans, the gastric lipase activity is contributed mainly by the stomach and the highest activity is detected in the fundus of the stomach. Human gastric lipase has a pH optimum of about 5.5 and, therefore, has

PATRICK TSO and TAKASHI DOI • Department of Pathology, University of Cincinnati College of Medicine, Cincinnati, Ohio 45267-0529.
Intestinal Lipid Metabolism, edited by Charles M. Mansbach II *et al.,* Kluwer Academic/Plenum Publishers, 2001.

been called acid lipase. It hydrolyzes medium-chain TG (predominantly 8–10 carbon chain length) better than the long-chain TG (Liao *et al.,* 1984). The main hydrolytic products of gastric lipase are diacylglycerols (DG) and free fatty acids (FA; Paltauf *et al.,* 1974; Roberts *et al.,* 1984). Gastric lipase does not hydrolyze PL or cholesteryl esters. In rodents, the main gastric lipase activity is derived from the lipase secreted by the salivary glands and is therefore called the lingual lipase. A survey of the animal kingdom clearly indicates that different animal species have either the lingual or the gastric lipase, but not both (DeNigris *et al.,* 1988).

Lipid is emulsified in the stomach (broken into small oil droplets). Lipid emulsion enters the small intestinal lumen as very fine lipid emulsion droplets less than 05 μm in diameter (Senior, 1964; Carey *et al.,* 1983). The combined action of the bile and the pancreatic juice brings about a marked change in the physical and chemical form of the lumenal lipid emulsion. Pancreatic lipase is secreted into the duodenum where it hydrolyzes TG to form 2-monoacylglycerol (2-MG) and FAs. The most potent gastrointestinal hormone that stimulates the release of enzymes by the pancreas is cholecystokinin (CCK; Solomon, 1994), and CCK-A receptor has been demonstrated to be present in the pancreas (Rosenzweig *et al.,* 1983). The pancreatic lipase works at the oil and water interphase. As a result, the rate of lipolysis is influenced by factors modifying the physicochemical properties of the interface as well as the surface area (Brockerhoff, 1968; Mattson and Beck, 1956; Simmonds, 1972).

In vitro studies using purified pancreatic lipase have demonstrated a potent inhibitory effect of bile salts on lipolysis of TG at concentration above the critical micellar concentration (Morgan *et al.,* 1969; Benzonana and Desnuelle, 1968). The inhibitory effect of bile salt is physiological because the concentration of bile salts in the duodenum is normally higher than the concentration of bile salt that is needed to observe the inhibitory effect. If this is the case, why, then, is pancreatic lipase so efficient in digesting TG? The explanation lies in the fact that the pancreas secretes another protein that counteracts this inhibition. The factor is called colipase. This factor was first isolated by Morgan *et al.* (1969) from rat pancreatic juice. The structure and mechanism of action of colipase have been elucidated by the elegant works from the laboratories of Dr. Desnuelle (Maylie *et al.,* 1971; Benzonana and Desnuelle, 1968) and Dr. Borgström (Borgström & Erlanson, 1973; Borgström *et al.,* 1979). Colipase acts by attaching to the ester bond region of the TG molecule. In turn, the lipase binds strongly to the colipase by electrostatic interactions, thereby allowing the hydrolysis of the TG by the lipase molecule (Erlanson-Albertsson, 1992). Colipase is secreted as a procolipase and is converted to the active form through the removal of a five-amino-acid fragment by trypsin. The five-amino-acid fragment released is called the enterostatin. Enterostatin has been demonstrated to inhibit fat intake—an action independent of the route of administration. For instance, it has been reported that both the intraduodenal (Mei and Erlanson-Albertsson, 1996), intraperitoneal (Okada *et al.,* 1996; Lin *et al.,* 1994), intravenous (Mei and Erlanson-Albertsson, 1992), and intracerebroventricular administration of enterostatin reduces fat intake, but both the dose and the time required for the administered enterostatin to exert its action are quite different. The physiological role of enterostatin in the inhibition of fat intake has often been questioned because intestinal digestion of peptides are extremely efficient and therefore only amino acids, di- and tripeptides, are taken up by the small intestinal mucosa. If so, one wonders how enterostatin is taken up by the small intestine and then transported into the circulation. This criticism should be reevaluated in view of recent reports that have documented the presence of both the mRNA as well as the enterostatin protein in the stomach and the small intestine of the rat (Sörhede *et al.,* 1996a, 1996b; Sörhede Winzell, 1998). Of course, it still remains to be demonstrated whether enterostatin is secreted by the stomach or by the intestine.

3. Apolipoprotein A-IV as a Satiety Factor

Apolipoprotein A-IV (apo A-IV) was discovered in the late 1970s, but its physiological role had not been firmly established until recently. Elegant *in vitro* experiments have suggested roles for apo A-IV in certain aspects of lipoprotein metabolism (Goldberg *et al.,* 1990; Fielding *et al.,* 1972; Bisgaier *et al.,* 1987; Dvorin *et al.,* 1988; Stein *et al.,* 1986), but there has been no direct evidence to date that apo A-IV plays such roles *in vivo*. The physiological relevance of the *in vitro* studies is still questioned because apo A-I can perform similar physiological roles that are prescribed to apo A-IV, and yet apo A-I is present at a higher concentration in blood than apo A-IV.

Recent *in vivo* studies (Fujimoto *et al.,* 1992, 1993a, 1993b) have provided evidence that apo A-IV may be involved in the inhibition of food intake following the ingestion of fat. Fujimoto *et al.* (1992) demonstrated that intravenous infusion of physiological saline in rats with indwelling right atrial catheter ate 3.90 ± 0.40 g during the first 30 min after refeeding (Table 1). Infusion of fasting intestinal lymph collected from donor lymph-fistu-

Table 1. Food Intake in 24-hr Fasted Rats after Infusion of 2 ml Test Solution Through Indwelling Artrial Catheter

	Food Consumption After Refeeding (g)	
Test Solutions	0–30 min	30–60 min
Control (physiological saline)	3.90 ± 0.40	1.31 ± 0.33
Mesenteric lymph samples		
Fasting Lymph	3.58 ± 0.33	1.75 ± 0.26
6–8hr after lipid infusion	0.60 ± 0.26^a	1.20 ± 0.11
6–8hr after lipid + L-81 infusion	3.91 ± 0.21^b	1.30 ± 0.25
5–7hr after cessation of L-81 infusion	0.40 ± 0.17^a	1.60 ± 0.22
Intralipid (nutrition control)		
2% Intralipid	4.00 ± 0.29	1.65 ± 0.38
Mesenteric lymph samples 6–8 hr after lipid infusion		
Immunoprecipitation with apo A-IV antiserum	3.41 ± 0.33^c	1.22 ± 0.46
Control (treated by normal goat serum)	0.87 ± 0.24^a	1.25 ± 0.48
Apolipoprotein A-IV (μg)		
60	3.35 ± 0.46	1.13 ± 0.31
135	2.14 ± 0.16^a	1.16 ± 0.26
200	$0.90 \pm 0.18^{a,d}$	1.10 ± 0.19
Apolipoprotein A-I (μg)		
200	3.90 ± 0.48	1.20 ± 0.25

Values are means $\pm$ SE. Five rats were treated in each group. $^a p < 0.01$ compared with values for saline control and fasting lymph. $^b p < 0.01$ compared with values for lymph sample from rats infused with lipid only. $^c p < 0.01$ compared with its control. $^d p < 0.01$ compared with value for 135 μg apo A-IV.

la rats had very little effect on food intake compared with saline infusion in the 24-hour fasted rats (see Table 1). In contrast, when compared to the fasting lymph, the mesenteric lymph samples collected during the sixth to eighth hour of dietary lipid infusion (active lipid absorption) markedly suppressed food intake during the first 30 min ($p < 0.01$). This suppression of food intake was not observed in the subsequent 30 min.

The fact that intestinal lymph collected from rats that are actively absorbing fat inhibits food intake is exciting because it seems to indicate that one or more factors are changing during active fat absorption that actually inhibit the absorption of fat. The lipid content of lymph increases as much as 10 to 15 times during fat absorption, making it possible that it is the lipid present in chylous lymph that is inhibiting food intake. To test whether the change in the lipid content of lymph is responsible for reducing food intake in 24-hr-fasted rats, rats were infused intravenously with a diluted Intralipid solution. When 2 ml of a 2% Intralipid in saline containing 42 μmol of triglyceride and 3.1 μmol of phospholipid (composition comparable to the lymph collected during active lipid absorption) was infused intravenously, food intake was not suppressed (Table 1). This result indicated that the anorectic effect of chylous lymph was probably not due to its lipid content.

If it is not the lipid component inhibiting food intake, we reasoned that it may be the apolipoprotein A-IV in the lymph that was responsible, because apo A-IV is the only apolipoprotein secreted by the small intestine that is markedly stimulated by lipid feeding (Hayashi *et al.*, 1990a, 1990b). Hayashi *et al.* (1990b) demonstrated that the stimulation of apo A-IV production by lipid feeding is associated with the formation and secretion of chylomicrons, and that the stimulation of apo A-IV production by fat absorption can be blocked by Pluronic L-81. Pluronic L-81 blocks the formation of chylomicrons. When the blockade of chylomicron formation by Pluronic L-81 is reversed by stopping Pluronic L-81 infusion, chylomicron formation and secretion resumes, and stimulation of apo A-IV secretion also appears. If the inhibition of food intake by intestinal lymph collected from animals actively absorbing lipid is caused by apo A-IV, what effect, then, will the lymph from an animal fed lipid plus Pluronic L-81 have on food intake? Table 1 shows that lymph from Pluronic L-81-treated animals did not inhibit food intake but that the lymph collected during the reversal of Pluronic L-81 inhibition was very potent in inhibiting the intake of food. This result strongly indicates that apo A-IV may be the factor in lymph that inhibits food intake, but the evidence is still not conclusive at this point.

To validate that apo A-IV does inhibit food intake, we studied the effect of apo A-IV-deficient chylous lymph on feeding. As Table 1 shows, the chylous lymph treated with normal goat serum suppressed food intake significantly in the first 30 min. In contrast, chylous lymph that was treated with apo A-IV antiserum had no effect on food intake—the animal consumed an amount of food similar to that consumed by the saline controls. In contrast, lymph treated with apo A-I antiserum was just as effective as the untreated lymph on inhibiting food intake. Next, 200 μg of either apo A-IV or apo A-I dissolved in 2 ml physiological saline was infused intravenously in 24-hr-fasted rats. As shown in Table 1, 200 μg of apo A-IV, an amount comparable to that present in 2 ml of lymph collected from a rat actively absorbing lipid, suppressed food intake significantly and to the same extent as the chylous lymph collected during 6 to 8 hr of lipid infusion ($p < 0.01$, apo A-IV vs. fasting lymph). Feeding was significantly suppressed by 135 μg apo A-IV ($p < 0.01$), but the effect was less than 200 μg of apo A-IV ($p < 0.01$). A small dose of 60 μg apo A-IV, equal to the amount in fasted lymph, did not suppress feeding. In contrast, 200 μg of apo A-I had no effect on food intake. No nonphysiological reaction including sedation, ataxia, or hyperthermia was observed after apo A-IV and chylous lymph infusion. These series of physiological studies finally led us to conclude that apo A-IV is a circulating signal released by the small intestine in response to fat feeding and is likely the mediator for the anorectic effect of a lipid meal. This function is unique to apo A-IV and is not shared by apo A-I. This is an important point because all of the functions that are attributed to apo A-IV in the *in vitro* studies can also be attributed to apo A-I even though plasma apo A-I concentration is higher than apo A-IV concentration.

4. Site of Apo A-IV Action

Feeding behavior is influenced by many circulating chemical factors, and chemosensitive monitoring systems for these factors exist both in the peripheral organs as well as the central nervous system (Novin *et al.*, 1981; Niijima, 1982). Fujimoto *et al.* (1993b) investigated whether the inhibition of food intake by apo A-IV was mediated centrally. Third cerebroventricular administration of apo A-IV decreased food intake in a dose-dependent manner (Table 2) and with a potency that was about 50-fold higher than intravenous administration (Fig. 1). In contrast, when apo A-I was infused into the third ventricle, it had no effect on food intake. The hypothesis that apo A-IV suppresses food intake via the central nervous system is further supported by the following experiment. When goat anti-rat apo A-IV serum was administered into the third ventricle in *ad libitum*-fed rats at 11:00 A.M. (during the light phase when rats usually do not eat), it resulted in feeding in all the animals tested (Table 3). A probable explanation for Fujimoto *et al.*'s (1993b) observation that administration of apo A-IV antiserum in the third ventricle elicits feeding is that apo A-IV antiserum removes any present endogenous apo A-IV.

Table 2. The Effect of Food Intake on the Infusion
of Apo A-IV into the Third Ventricle

	Food Consumption After Refeeding (g)	
	0–30 min	30–60 min
Apo A-IV		
0.5 µg	4.2 ± 0.3	1.0 ± 0.3
1.0 µg	2.8 ± 0.3^a	1.2 ± 0.4
2.0 µg	1.3 ± 0.2^a	1.0 ± 0.2
4.0 µg	0.6 ± 0.2^a	1.1 ± 0.2
Apo A-I		
4.0 µg	4.0 ± 0.9	1.1 ± 0.4
Saline control	4.4 ± 0.4	1.4 ± 0.2

Rats were fasted for 24 hr before the feeding study. Apo A-IV, apo A-I, or saline (total volume = 10 µl) was administered at 1 µl/min before refeeding. Values are expressed as mean ± SE. Fave rats were tested in each group. $^a p < 0.01$ compared with saline controls.

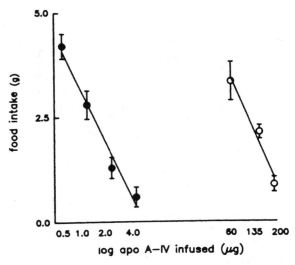

x-axis: log apo A–IV infused (μg)
y-axis: food intake (g)

Figure 1. The relationship between the suppression of food intake during the first 30 min of refeeding in 24-hr-fasted rats and the logarithm of the dose in μg of apo A-IV infused either into the third ventricle (closed circles) or intravenously (open circles, data derived from Fujimoto *et al.,* 1993b).

Five animals were studied at each dose and the values are expressed as mean $\pm$ SE. Food intake diminished linearly with increasing amounts of apo A-IV infused into the third ventricle ($Y = -1.78 \log X + 2.84$, $r = 0.98$, $p < 0.01$). (Used with permission from Fujimoto, K., Fukagawa, K., Sakata, T., and Tso, P., 1993, Suppression of food intake by Apolipoprotein A-IV is mediated through the central nervous system in rats, *J. Clin. Invest.* **91**:1830–1833).

Because available evidence suggests that *de novo* synthesis of apo A-IV in the brain is unlikely (Elsbourhagy *et al.,* 1987; Srivastava *et al.,* 1992), it has been proposed by Fujimoto *et al.* (1993a; 1993b) that apo A-IV released by the small intestine (or perhaps a fragment thereof) may traverse the blood–brain barrier and act in the central nervous system. This hypothesis is supported by a number of indirect evidences. First, Fujimoto *et al.*

Table 3. Feeding and Drinking Behavior During the Hour after Infusion of either Goat Anti Rat Apo A-IV and A-I Serum

Treatment	Incidence	Feeding Latency (min)	Duration (min)	Drinking Incidence
Goat anti rat				
apo A-IV	5/5[a,b]	25.9 $\pm$ 2.3	2.8 $\pm$ 0.5	3/5
apo A-I	0/5	n.d.	n.d.	0/5
Saline	0/5	n.d.	n.d.	0/5

The feeding and drinking response was measured at 11:00 h. Five animals were studied in each group. n.d. = not detected. [a]$p < 0.05$, compared with either apo A-I or saline; [b]Number of rats with elicited feeding response/Number of rats tested; both feeding latency and duration are expressed as mean $\pm$ SE.

(1993b) have demonstrated by rocket gel electrophoresis that apo A-IV, or a fragment of apo A-IV, is present in the third ventricular cerebrospinal fluid. Second, apo A-IV concentration in third ventricular cerebrospinal fluid increases as a result of lipid feeding. Third, using immunohistochemical technique, Fukagawa *et al.* (1995) demonstrated specific staining for apo A-IV in astrocytes and in tanycytes throughout both white and gray matter. The granular nature and perinuclear distribution of apo A-IV immunoreactivity suggests that apo A-IV may be contained in perinuclear organelles or vesicles. The demonstration of immunoreactive apo A-IV in tanycytes does not necessarily indicate a selective uptake mechanism for apo A-IV, because tanycytes are known to take up a variety of neurotransmitters and nonmetabolizable amino acids. However, the presence of apo A-IV immunostaining in astrocytes may indicate uptake of apo A-IV by astrocytes. Whether astrocytes are involved in satiety mechanisms associated with lipid feeding is unknown.

5. Mechanism of Inhibition of Food Intake by Apo A-IV

An important question regarding the inhibition of food intake by apo A-IV is how it works. Most of the evidence regarding this aspect of apo A-IV action is derived from the elegant work of Okumura *et al.* (1994, 1995, 1996). Okumura *et al.* demonstrated that intracisternal injections of purified apo A-IV inhibited gastric acid secretion (1994, 1995) and gastric motility (1996) in rats in a dose-dependent manner. As shown in Fig. 2, intracisternal infusion of physiological doses of apo A-IV markedly inhibited gastric motility as well as gastric acid secretion in a dose-dependent manner. The doses of apo A-IV used in these studies were thought to reproduce the levels of apo A-IV measured in cerebrospinal fluid after lipid feeding (Fujimoto *et al.*, 1993b). Intravenous infusion of similar doses of apo A-IV failed to elicit any gastric response. As proposed by Okumura *et al.* (1994), apo A-IV acts like an enterogastrone, that is, a humoral mediator released by the distal intestine that mediates the humoral inhibition of gastric acid secretion as well as motility by the inges-

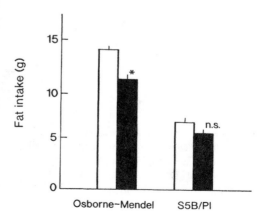

Figure 2. The effect of intracisternal injection of apo A-IV on gastric secretion in pylorus-ligated conscious rats. Under brief isoflurane anesthesia, rats received intracisternal injection of apo A-IV and the pylorus was ligated. Two hours after intracisternal injection, the animals were killed and the stomach was removed. (A) The volume was measured and (B) gastric acid output was determined. Each column represents the mean $\pm$ SE of 4 to 9 animals. * $p < 0.05$ when compared with saline control. (Used with permission from Okumura, T., Fukagawa, K., Tso, P., Taylor, I. L., and Pappas, T., 1994, Intracisternal injection of Apolipoprotein A-IV inhibits gastric secretion in pylorus-ligated conscious rats, *Gastroenterol.* **107**:1861–1864).

tion of fat. At present, it is not clear if there is a direct link between the effects of apo A-IV on food intake and its effects on gastric function. Apo A-IV could directly influence central feeding mechanisms; alternately, it could affect feeding through its effects on gastric function, especially via inhibition of gastric emptying (McHugh and Moran, 1985). Further work will be needed to clarify this important issue.

6. Is Apo A-IV a Short-Term Satiety Signal?

Although there is compelling evidence that apo A-IV is capable of inhibiting food intake acutely during the ingestion of a lipid meal, the temporal relationship between intestinal synthesis and secretion of apo A-IV and satiety has to be considered. The question is whether the increases in plasma levels of apo A-IV in response to lipid feeding are of sufficiently short latency and sufficiently large magnitude to elicit satiety. Rodriguez *et al.* (1997) demonstrated in a recent study that when a gastric bolus of 0.5 ml of a 20% intralipid solution (containing 100 mg of TG) was fed to rats, there was a significant increase of plasma apo A-IV within 15 minutes with the increment remaining statistically significant until 30 minutes after the meal (Fig. 3). Changes in plasma apo A-IV that were observed by Rodriguez *et al.* (1997) following a gastric lipid meal were similar to those observed by Fuji-

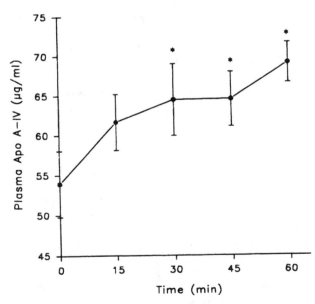

Figure 3. Plasma apo A-IV levels after a gastric bolus of 0.1 g of TG (0.5 ml of 20% Intralipid). Blood was collected every 15 min for 1 hour after delivery of lipid. Plotted values are means $\pm$ SE concentration for 6 rats. *Significant change from basal value ($p < 0.05$) (Used with permission from Rodriguez, M. D., Kalogeris, T., Wang, X. L., Wolf, R., and Tso, P., 1997, Rapid synthesis and secretion of intestinal Apolipoprotein A-IV after gastric fat loading in rats, *Am. J. Physiol.* **272**(4pt.2): R1170–R1177).

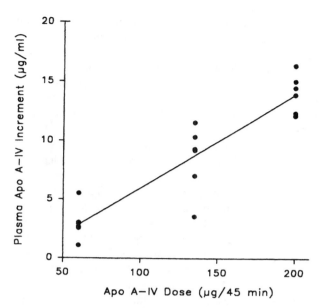

Figure 4. The effect of increasing dose of intravenously infused, purified apo A-IV on plasma apo A-IV levels. Values are increments in plasma concentration above levels measured in response to saline infusion; $n = 6$ rats. Linear regression analysis revealed a significant dose effect of apo A-IV on increment in plasma apo A-IV: $y = 0.079x - 1.889$; $r = 0.92$ (one-tailed), $p < 0.001$. (Used with permission from Rodriguez, M. D., Kalogeris, T., Wang, X. L., Wolf, R., and Tso, P., 1997, Rapid synthesis and secretion of intestinal Apolipoprotein A-IV after gastric fat loading in rats, *Am. J. Physiol.* **272**(4pt.2):R1170–R1177).

moto *et al.* (1992) in an experiment where intravenous apo A-IV produced a significant dose-dependent inhibition of food intake (Fig. 4). Rodriguez *et al.* (1997) therefore concluded that the increase in plasma levels of apo A-IV produced in response to lipid meals were sufficiently quick and large enough to produce satiety, thereby supporting a role for apo A-IV in the short-term control of food intake.

7. Role of Apo A-IV in the Long-Term Control of Food Intake

We currently have evidence that supports a role for apo A-IV in the long-term regulation of food consumption and body weight. First, those experiments that demonstrated the satiety effect of apo A-IV were performed in rats that had been deprived of food for relatively extended periods of time. Although this experimental paradigm ensured the absence of other food-intake-inhibitory influences, it may have produced a situation of unusually high sensitivity in rats to the food-intake-inhibitory effects of apo A-IV. This has been observed with other putative satiety factors. For example, severe food deprivation increases the sensitivity to the food-intake-inhibitory effects of a cholecystokinin analogue in dogs (Reidelberger *et al.*, 1986). On the other hand, it has been demonstrated that intravenous administration of apo A-IV decreases food intake in *ad libitum* fed rats (Fujimoto *et al.*, 1993a), thus strongly suggesting that exogenously administered apo A-IV controls food intake even under *ad libitum* conditions. However, what is still not clear is to what extent endogenous apo A-IV regulates food under *ad libitum* conditions, especially during the dark phase when most food intake occurs in rats. In rats with free access to food, central administration of apo A-IV antiserum stimulates feeding during the light cycle (Fujimoto *et al.*, 1993b). Similar studies done during the dark phase may help clarify the role of apo A-IV in the control of individual meals.

The second evidence indicating a potential role of apo A-IV in the long-term regulation of food intake is derived from the circadian rhythm of lymph and serum apo A-IV. Fukagawa *et al.* (1994) reported that in *ad libitum* fed rats, both the serum and intestinal lymph apo A-IV exhibited a circadian rhythm, with the level significantly higher during the dark period than the light period. When Fukagawa *et al.* (1994) examined the serum apo A-IV level in fasting rats, they found that the serum apo A-IV exhibited the same circadian rhythm as in the fed rats, but that the serum apo A-IV concentrations were significantly higher for all time points in the *ad libitum*-fed rats than in the fasting rats. This result indicated that although *ad libitum* feeding greatly increased the levels of serum apo-IV, it did not change the pattern of the inherent circadian rhythm of serum apo A-IV. The fact that serum apo A-IV increased during the dark phase, corresponding to the most active feeding period of the rat, is potentially of physiological significance and suggests that apo A-IV's physiological role may be a long-term regulator of food intake.

The third line of evidence supporting a potential long-term regulation of food intake and body weight by apo A-IV is derived from a recent paper reporting the down-regulation of intestinal apo A-IV mRNA levels by leptin (Morton *et al.*, 1998). This finding is extremely exciting. In addition to the effect of leptin on the transcriptional control of apo A-IV mRNA level, it has also been demonstrated that intestinal apo A-IV synthesis and secretion is up-regulated by insulin in both rodents and humans (Attia *et al.*, 1997; Black & Ellinas, 1992). As elegantly summarized in a recent review by Woods *et al.* (1998), energy homeostasis in the body is accomplished by a highly integrated and redundant neurohumoral system that prevents the impact of short-term fluctuations in energy balance on fat mass. Insulin and leptin are both hormones secreted in proportion to the body adiposity, and they play a critical role in this energy homeostasis. The fact that both leptin and insulin seem to regulate intestinal apo A-IV synthesis poses the possibility that apo A-IV may be involved in the long-term regulation of food intake and body weight.

8. Regulation of Intestinal Apo A-IV Synthesis and Secretion

We and others have demonstrated that the formation of chylomicrons stimulates the synthesis and secretion of chylomicrons. However, the mechanism of how this occurs is still unknown. This is a very interesting question for a molecular biologist to tackle. We have recently found that the formation and secretion of chylomicrons is not the only way to stimulate apo A-IV synthesis and secretion. Impetus for this finding came from studies in which we gave duodenal infusions containing graded doses of TG to rats and quantified both regional lipid distribution and mucosal synthesis of apo A-IV at various sites along the length of the intestine (Kalogeris *et al.*, 1994). We found that despite significant amounts of lipid present only in the proximal half of the small intestine, apo A-IV synthesis was stimulated in the proximal three fourths of the gut, even in segments where there was a negligible amount of lipid. This raises an interesting question of whether there may be factors other than lipid transport, but independent of the presence of lipid itself, that are capable of stimulating apo A-IV production by the gut. To address this question, a series of experiments were performed comparing the effects of proximal versus distal intestinal infusion of lipid on the synthesis of apo A-IV in both the proximal and distal intestine. Kalogeris *et al.* (1996a) found that after duodenal lipid infusion, both apo A-IV synthesis and mRNA levels were elevated two- to threefold compared with control infusions (glucose-saline) in the jejunum, but that ileal apo A-IV synthesis and mRNA levels were unaffected. Previous work from our laboratory demonstrated that under the conditions of our duodenal infusion, the amount of lipid reaching the ileum was negligible, suggesting that the lack of effect of duodenal lipid infusion on ileal apo A-IV expression was due to an insufficient exposure of the distal gut to lipid.

In contrast, delivery of lipid to the ileum stimulated both ileal as well as jejunal apo A-IV synthesis. Subsequent experiments in rats equipped with jejunal or ileal Thiry-Vella fistulas (segment of intestine isolated lumenally from the rest of the gastrointestinal tract) demonstrated the following interesting findings: ileally infused lipid elicits an increase in proximal jejunal apo A-IV synthesis independent of the presence of jejunal lipid, and both ileum and more distal sites may be involved in the stimulation. These results strongly suggest the existence of a signal arising from the distal gut that is capable of stimulating synthesis of apo A-IV in the proximal gut. These interesting findings have important physiological implications. The distal intestine is known to play an important role in the control of gastrointestinal function. Nutrient (especially lipid) delivered to the ileum results in the inhibition of gastric emptying (Lin *et al.*, 1990; MacFarlane *et al.*, 1983), decreased intestinal motility and transit (MacFarlane *et al.*, 1983; Spiller *et al.*, 1984), and decreased pancreatic secretion (Harper *et al.*, 1979). Ileal nutrient also inhibits food intake (Welch *et al.*, 1985; Meyer *et al.*, 1994). The mechanism for these effects have been collectively termed the "ileal brake" (Spiller *et al.*, 1984), and it appears to be related to the release of one or more peptide hormones from the distal intestine (Aponte *et al.*, 1985, 1989; Jin *et al.*, 1993; Pappas *et al.*, 1985, 1986a, 1986b; Savage, 1987). These effects have traditionally been considered operative only in the event of abnormal delivery of undigested nutrients to the distal gut, such as the malabsorptive state (Spiller *et al.*, 1984). However, growing evidence supports the notion that because of the rapid gastric emptying during the early phases of a meal, nutrient reaches the distal gut even under normal conditions (Rodriguez *et al.*, 1997; Lin *et al.*, 1990; Meyer *et al.*, 1994). We recently studied the intralumenal and mucosal distribution of a bolus of [^{3}H]triolein-labeled Intralipid (0.5 ml of a 20% emulsion) fed by gavage. By 15 to 30 min, radiolabeled lipid was spread evenly throughout the entire gut with 10% to 15% of the load recovered in the ileum and the cecum combined. Presence of substantial amounts of lipids in these distal sites persisted for at least 4 hr after the meal. When we examined apo A-IV synthesis by the small intestine, we found rapid stimulation (between 15–30 min) of apo A-IV synthesis throughout the intestine including the ileum. This was associated with significant stimulation of lymphatic output and plasma levels of apo A-IV by 30 min after the gastric lipid load (Rodriguez *et al.*, 1997). Consequently, it is becoming increasingly clear that even under normal conditions, a far greater length of the intestine could be involved both in the absorption of a lipid meal and the control of gastric and upper gut function than has been previously recognized. Thus, the "ileal brake" may play an important role in the normal control of gut function.

The most likely peptide to mediate the phenomenon of "ileal brake" is peptide tyrosine–tyrosine (PYY), which is a member of the peptide family including the pancreatic polypeptide (PP), neuropeptide Y (NPY), and fish pancreatic peptide Y (PY; Larhammar *et al.,* 1993). PYY is synthesized by the endocrine cells in the ileum and the large intestine (Aponte *et al.,* 1985; Adrian *et al.,* 1987; Hill *et al.,* 1991; Tatemoto, 1982; Taylor, 1985) and is released in response to intestinal nutrients, especially long-chain fatty acids (Hill *et al.,* 1991). However, PYY may not be the only mediator of the "ileal brake." For example, perfusion of the intestine with fat produces a greater suppression of pentagastrin-stimulated acid secretion than does PYY (Savage *et al.,* 1987), indicating that the enterogastrone effect of fat is mediated by more than one factor. We have now obtained evidence that PYY stimulates jejunal apo A-IV synthesis and secretion. Continuous intravenous infusion of physiological doses of PYY elicits significant increases in both synthesis and lymphatic transport of apo A-IV in rats (Kalogeris *et al.,* 1996b). We believe this is the first demonstration of the involvement of a gastrointestinal hormone in the control of expression and secretion of an intestinal apolipoprotein, thus bringing together two heretofore separate areas of research in gastrointestinal physiology.

9. Effect of Chronic High-Fat Feeding on Intestinal Apo A-IV Synthesis

Chronic feeding of a high-fat diet on intestinal apo A-IV synthesis is intriguing. Acute administration of a lipid meal results in marked stimulation of apo A-IV synthesis as well as mRNA levels both in the jejunum and the ileum (Apfelbaum *et al.,* 1987). In contrast, the chronic administration of a high-fat diet (30% by weight of diet as fat) results in stimulation of apo A-IV synthesis and mRNA levels only in the jejunum, but not in the ileum. This very interesting finding poses the question of whether the ileum becomes less responsive to lipid after chronic feeding of a high-fat diet or the adaptation of the digestive and absorptive processes results in fat no longer reaching the ileum. This question warrants further investigation.

Chronic feeding of a high-fat diet in humans results in a marked elevation of plasma apo A-IV levels during the first week following high-fat diet (Weinberg *et al.,* 1990). This elevation disappears, however, during the 2nd week of high-fat diet, thus leading the investigators to conclude that there is autoregulation of intestinal apo A-IV production in response to diets high in fat. Consequently, both rodent as well as human data seem to suggest that the intestinal apo A-IV synthesis and secretion become less responsive to fat after chronic feeding of high-fat diets. This issue certainly deserves further study because a good understanding of how intestinal apo A-IV synthesis is modified by chronic high-fat feeding may provide us with insightful clues as to why the high-fat diet predisposes both animals and humans to obesity.

10. Concluding Remarks

In summary, intestinal apo A-IV is a very interesting protein stimulated by dietary lipid with a potentially important physiological role in the integrated control of digestive func-

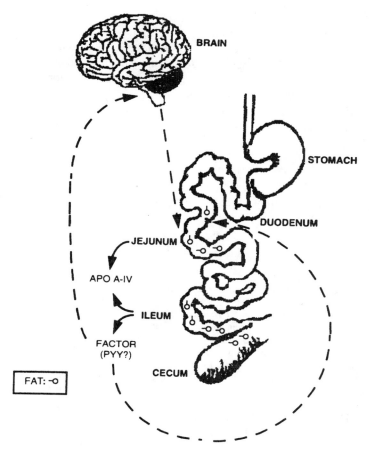

Figure 5. Proposed pathway for the control of apo A-IV synthesis by intestinal lipid. Lipid in the proximal intestine stimulates the synthesis and secretion of apo A-IV in a proximal–distal gradient in the intestine, depending on the total lipid load. This effect is dependent on presence of lipid in the regions where apo A-IV is expressed. Lipid in the distal gut (ileum, cecum) also stimulates apo A-IV in both the ileum and the proximal jejunum. This latter effect is independent of the presence of jejunal lipid and is probably mediated by a signal released in response to the presence of lipid in the distal intestine. This signal is probably peptide YY (tyrosine-tyrosine)(PYY), although other gut hormones have not been unequivocally ruled out. (Used with permission from Kalogeris, T. J., Rodriguez, M. D., and Tso, P., 1997, Control of synthesis and secretion of intestinal apolipoprotein A-IV by lipid, *J. Nutr.* **127**(3)**:**5375–5435).

tion and ingestive behavior as well as a presumed role in cholesterol and lipoprotein metabolism. In terms of its role in the regulation of upper gut function as well as satiety, many issues still remain to be addressed. For instance, we do not have any information about what molecular form of apo A-IV is involved, that is, free monomer, a homodimeric form (Weinberg and Spector, 1985), HDL-bound apo A-IV or, perhaps, apo A-IV-derived bioactive peptides. This important issue will have to be addressed before a comprehensive understanding of the physiology of apo A-IV can be achieved. Figure 5 integrates available evidence into a working model of the overall control of apo A-IV synthesis and secretion following the ingestion of lipid. Direct exposure of both the jejunum and ileum to fat results in the stimulation of apo A-IV synthesis and secretion by the respective segments of intestine. Exposure of the ileum to lipid also results in the secretion of a factor (PYY) or factors, which in turn stimulate further synthesis and release of apo A-IV by the jejunum. Whether or not the stimulation of apo A-IV biosynthesis by lipid absorption and by PYY is mediated by the same molecular mechanism inside the enterocyte is actively being investigated in our laboratory.

Acknowledgments

This work was supported by a grant from the National Institutes of Health DK 32288 and DK 53444.

References

Adrian, T. E., Bacarese, H. A., Smith, H. A., Chohan, P., Manolas, K. J., and Bloom, S. R., 1987, Distribution and postprandial release of porcine peptide YY, *J. Endocrinol.* **113:**11–14.

Apfelbaum, T. F., Davidson, N. O., and Glickman, R. M., 1987, Apolipoprotein A-IV synthesis in the rat intestine: Regulation by dietary triglyceride, *Am. J. Physiol.* **252:**G662–666.

Aponte, G. W., Fink, A. S., Meyer, J. H., Tatemoto, K., and Taylor, I. L., 1985, Regional distribution and release of peptide YY with fatty acids of different chain length, *Am. J. Physiol.* **249:**G745–G750.

Aponte, G. W., Park, K., Hess, R., Garcia, R., and Taylor, I. L., 1989, Meal-induced peptide tyrosine-tyrosine inhibition of pancreatic secretion in the rat, *FASEB J.* **3:**1949–1955.

Attia, N., Touzani, A., Lahrichi, M., Balafrej, A., Kabbaj, O., and Girard-Globa, A., 1997, Response of apolipoprotein A-IV and lipoproteins to glycaemic control in young people with insulin-dependent diabetes mellitus, *Diabet. Med.* **14**(3):242–247.

Benzonana, G., and Desnuelle, P., 1968, Action of some effectors on the hydrolysis of long-chain triglycerides by pancreatic lipase, *Biochim. Biophys. Acta* **164:**47–58.

Bisgaier, C. L., Sachdev, O. P., Lee, E. S., Williams, K. J., Blum, C. B., and Glickman, R. M., 1987, Effect of lecithin: Cholesterol acyl transferase on distribution of apolipoprotein A-IV among lipoproteins of human plasma, *J. Lipid Res.* **28:**693–703.

Black, D. D., and Ellinas, H., 1992, Apolipoprotein synthesis in newborn piglet intestinal explants, *Pediatr. Res.* **32**(5):553–558.

Borgström, B., 1976, Phospholipid absorption, in: *Lipid Absorption: Biochemical and Clinical Aspects* (K. Rommel, H. Goebell, and R. Böhmer, eds.), MTP Press Let., London, pp. 65–72.

Borgström, B., and Erlanson, C., 1973, Pancreatic lipase and colipase. Interactions and effects of bile salts and other detergents, *Eur. J. Biochem.* **37:**60–69.

Borgström, B., Erlanson-Albertson, C., and Wieloch, T., 1979, Pancreatic colipase—Chemistry and physiology, *J. Lipid Res.* **20:**805–816.

Brockerhoff, H., 1968, Substrate specificity of pancreatic lipase, *Biochim. Biophys. Acta* **159:**296–303.

Carey, M. C., Small, D. M., and Bliss, C. M., 1983, Lipid digestion and absorption, *Ann. Rev. Physiol.* **45:**651–677.

Davenport, H. W., 1971, *Physiology of the digestive tract,* 3rd ed., Yearbook Medical Publishers, Chicago.

DeNigris, S. J., Hamosh, M., Kasbekar, D. K., Lee, T. C., and Hamosh, P., 1988, Lingual and gastric lipases: Species differences in the origin of pancreatic digestive enzymes and species differences in localization of gastric lipase, Biochim. Biophys. Acta **959:**38–45.

Dvorin, E., Gorder, N. L., Benson, D. M., and Gotto, A. M., 1988, Apolipoprotein A-IV. A determinant for binding and uptake of high-density lipoproteins by rat hepatocytes, *J. Biol. Chem.* **261:**15714–15718.

Elsbourhagy, N. A., Walker, D. W., Paik, Y. K., Boguski, M. S., Freeman, M., Gordon, J. I., and Taylor, J. M., 1987, Structure and expression of the human apolipoprotein A-IV gene, *J. Biol. Chem.* **262:**7973–7981.

Erlanson-Albertsson, C., 1992, Pancreatic colipase. Structural and physiological aspects, *Biochim. Biophys. Acta* **1125:**1–7.

Fielding, C. J., Shore, V. G., and Fielding, P. E., 1972, A protein cofactor of lecithin: Cholesterol acyltransferase, *Biochem. Biophys. Res. Commun.* **46:**1493–1498.

Fujimoto, K., Cardelli, J. A., and Tso, P., 1992, Increased apolipoprotein A-IV in rat mesenteric lymph after lipid meal acts as a physiological signal for satiation, *Am. J. Physiol.* **262:**G1002–G1006.

Fujimoto, K., Machidori, H., Iwakiri, R., Yamamoto, K., Fujisaki, J., Sakata, T., and Tso, P., 1993a, Effect of intravenous administration of apolipoprotein A-IV on patterns of feeding, drinking and ambulatory activity of rats, *Brain Res.* **608:**233–237.

Fujimoto, K., Fukagawa, K., Sakata, T., and Tso, P., 1993b, Suppression of food intake by apolipoprotein A-IV is mediated through the central nervous system in rats, *J. Clin. Invest.* **91:**1830–1833.

Fukagawa, K., Gou, H. M., Wolf, R., and Tso, P., 1994, Circadian rhythm of serum and lymph apolipoprotein A-IV in *ad libitum*-fed and fasted rats. Am. J. Physiol. **267:**R1385–R1390.

Fukagawa, K., Knight, D. S., Hamilton, K. A., and Tso, P., 1995, Immunoreactivity for apolipoprotein A-IV in tanycytes and astrocytes of rat brain, *Neurosci. Lett.* **199:**17–20.

Goldberg, I. J., Scheraldi, C. A., Yacoub, L. K., Saxena, U., and Bisgaier, C. L., 1990, Lipoprotein C-II activation of lipoprotein lipase. Modulation by apolipoprotein A-IV, *J. Biol. Chem.* **265:**4266–4272.

Gould, R. G., Jones, R. J., LeRoy, G. V., Wissler, R. W., and Taylor, C. B., 1969, Absorbability of β-sitosterol in humans, *Metabolism* **18:**652–662.

Hamosh, M., 1979, The role of lingual lipase in neonatal fat digestion, in: *Development of Mammalian Absorptive Process, Ciba Foundation Symposium* 70 (K. Elliot and J. Whelan, eds.), Excerpta Medica, Amsterdam, pp. 69–98.

Harper, A. A., Hood, J. C. A., and Mushens, J., 1979, Inhibition of external pancreatic secretion by intracolonic and intraileal infusions in the cat, *J. Physiol. (London)* **292:**445–454.

Hayashi, H., Fujimoto, K., Cardelli, J. A., Nutting, D. F., Bergstedt, S., and Tso, P., 1990a, Fat feeding increases size, but not number, of chylomicrons produced by small intestine, *Am. J. Physiol.* **259:**G709–G719.

Hayashi, H., Nutting, D. F., Fujimoto, K., Cardelli, J. A., Black, D., and Tso, P., 1990b, Transport of lipid and apolipoproteins A-I and A-IV in intestinal lymph of the rat, *J. Lipid. Res.* **31:**1613–1625.

Hill, F. L. C., Zhang, T., Gomez, G., and Greeley, G. H., Jr., 1991, Peptide YY, a new gut hormone, *Steroids* **56:**77–82.

Jin, H., Gai, L., Lee, K., Chang, T. M., Li, P., Wagner, D., and Chey, W. Y., 1993, A physiological role of peptide YY on exocrine pancreatic secretion in rats, *Gastroenterol.* **105:**208–215.

Kalogeris, T. J., Fukagawa, K., and Tso, P., 1994, Synthesis and lymphatic transport of intestinal apolipoprotein A-IV in response to graded doses of triglyceride, *J. Lipid Res.* **35:**1141–1151.

Kalogeris, T. J., Tsuchiya, T., Fukagawa, K., Wolf, R., and Tso, P., 1996a, Apolipoprotein A-IV synthesis in proximal jejunum is stimulated by ileal lipid infusion, *Am. J. Physiol.* **270:**G277–G286.

Kalogeris, T. J., Sato, M., Wang, X., Monroe, F., and Tso, P., 1996b, Intravenous infusion of peptide YY stimulates jejunal synthesis and lymphatic secretion of apolipoprotein A-IV in rats, *Gastroenterol.* 110:A809 (Abstract).

Kalogeris, T. J., Rodriguez, M. D., and Tso, P., 1997, Control of synthesis and secretion of intestinal apolipoprotein A-IV by lipid, *J. Nutr.* **127**(3)**:**5375–5435.

Larhammar, D., Söderberg, C., and Blomgvist, A. G., 1993, Evolution of the neuropeptide Y family of peptides, in: *The Biology of Neuropeptide Y and Related Peptides* (W. F. Colmers and C. Wahlestadt, eds.), Humana Press, Totowa, NJ, pp. 1–41.

Liao, T. H., Hamosh, P., and Hamosh, M., 1984, Fat digestion by lingual lipase: Mechanism of lipolysis in the stomach and upper small intestine, *Pediatr. Res.* **18:**402–409.

Lin, H. C., Doty, J. E., Reedy, T. J., and Meyer, J. H., 1990, Inhibition of gastric emptying of sodium oleate depends upon length of intestine exposed to the nutrient, *Am. J. Physiol.* **259:**G1031–G1036.

Lin, L., Okada, S., York, D. A., and Bray, G. A., 1994, Structural requirements for the biological activity of enterostatin, *Peptides* **15:**849–854.

MacFarlane, A., Kinsman, R., and Read, N. W., 1983, The ileal brake: Ileal fat slows small bowel transit and gastric emptying in man, *Gut* **24:**471–472.

Mattson, F. H., and Beck, L. W., 1956, The specificity of pancreatic lipase or the primary hydroxyl groups of glycerides, *J. Biol. Chem.* **219:**735–740.

Maylie, M. F., Charles, M., Gache, C., and Desnuelle, P., 1971, Isolation and partial identification of a pancreatic colipase, *Biochim. Biophys. Acta* **229:**286–228.

McHugh, P. R., and Moran, T. H., 1985, The stomach: A conception of its dynamic role in satiety, *Prog. Psychobiol. Physiol. Psychol.* **11:**197–232.

Mei, J., and Erlanson-Albertsson, C., 1992, Effect of enterostatin given intravenously and intracerebroventrically on high-fat feeding in rats, *Regul. Pept.* **41:**209–218.

Mei, J., and Erlanson-Albertsson, C., 1996, Role of intraduodenally administered enterostatin in rat: Inhibition of food intake, *Obes. Res.* **4:**161–165.

Meyer, J. H., Elashoff, J. D., Doty, J. E., and Gu, Y. G., 1994, Disproportionate ileal digestion on canine food consumption. A possible model for satiety in pancreatic insufficiency, *Dig. Dis. Sci.* **39:**1014–1024.

Morgan, R. G. H., Barrowman, J., and Borgström, B., 1969, The effect of sodium taurodeoxycholate and pH on the gel filtration behaviour of rat pancreatic protein and lipases, *Biochim. Biophys. Acta* **175:** 65–75.

Morton, N. M., Emilsson, V., Liu, Y. L., and Cawthorne, M. A., 1998, Leptin action in intestinal cells, *J. Biol. Chem.* **273:**26194–26201.

Niijima, A., 1982, Glucose-sensitive afferent nerve fibers in the hepatic branch of the vagus nerve in the guinea pig, *J. Physiol. (London)* **332:**315–323.

Northfield, T. C., and Hofmann, A. F., 1975, Biliary lipid output during three meals and an overnight fat, *Gut* **16:** 1–11.

Novin, D., Rogers, R. C., and Hermann, G., 1981, Visceral afferent and efferent connections in the brain, *Diabetologia* **20:**331–336.

Okada, S., York, D. A., Bray, G. A., and Erlanson-Albertsson, C., 1996, Enterostatin (Val-Pro-Asp-Pro-Arg) the activation peptide of procolipase selectively reduces fat intake, *Physiol. Behav.* **49:**1185–1189.

Okumura, T., Fukagawa, K., Tso, P., Taylor, I. L., and Pappas, T. N., 1994, Intracisternal injection of apolipoprotein A-IV inhibits gastric secretion in pylorus-ligated conscious rats, *Gastroenterol.* **107:**1861–1864.

Okumura, T., Fukagawa, K., Tso, P., Taylor, I. L., and Pappas, T. N., 1995, Mechanisms of action of intracisternal apolipoprotein A-IV in inhibiting gastric acid secretion in rats, *Gastroenterol.* **109:**1583–1588.

Okumura, T., Fukagawa, K., Tso, P., Taylor, I. L., and Pappas, T. N., 1996, Apolipoprotein A-IV acts in the brain to inhibit gastric emptying in the rat, *Am. J. Physiol.* **270:**G49–G53.

Paltauf, F., Esfandi, F., and Holasek, A., 1974, Stereospecificity of lipases. Enzymatic hydrolysis of enantiomeric alkyl diacylglycerols by lipoprotein lipase, lingual lipase, and pancreatic lipase, *FEBS Lett.* **40:**119–123.

Pappas, T. N., Debas, H. T., and Taylor, I. L., 1985, Peptide YY: Metabolism and effect pancreatic secretion in dogs, *Gastroenterol.* **89:**1387–1392.

Pappas, T. N., Debas, H. T., Chang, A. M., and Taylor, I. L., 1986a, Peptide YY release by fatty acids is sufficient to inhibit gastric emptying in dogs, *Gastroenterol.* **91:**1386–1389.

Pappas, T. N., Debas, H. T., and Taylor, I. L., 1986b, The enterogastrone-like effect of peptide YY is vagally mediated in the dog, *J. Clin. Invest.* **77:**49–53.

Reidelberger, R. D., Kalogeris, T. J., and Solomon, T. E., 1986, Comparative effects of caerulein on food intake and pancreatic secretion in dogs, *Brain Res. Bull.* **17:**445–449.

Roberts, I. M., Montgomery, R. K., and Carey, M. C., 1984, Rat lingual lipase: Partial purification, hydrolytic properties and comparison with pancreatic lipase, *Am. J. Physiol.* **247:**G385–393.

Rodriguez, M. D., Kalogeris, T. J., Wang, X. L., Wolf, R., and Tso, P., 1997, Rapid synthesis and secretion of intestinal apolipoprotein A-IV after gastric fat loading in rats, *Am. J. Physiol.* **272:**R1170–R1177.

Rosenzweig, S. A., Miller, L. J., and Jamieson, J. D., 1983, Identification and localization of cholecystokinin-binding sites on rat pancreatic plasma membranes and acinar cells: A biochemical and autoradiographic study, *J. Cell Biol.* **96:**1288–1297.

Savage, A. P., Adrian, T. E., Carolan, G., Chattarjee, V. K., and Bloom, S. R., 1987, Effects of peptide YY (PYY) on mouth to caecum intestinal transit time and on the rate of gastric emptying in healthy volunteers, *Gut* **28:**166–170.

Schonheyder, G., and Volquatz, K., 1946, Gastric lipase in man, *Acta Physiol. Scand.* **11:**349–360.

Senior, J. R., 1964, Intestinal absorption of fats, *J. Lipid Res.* **5:**495–521.

Simmonds, W. J., 1972, Fat absorption and chylomicron formation, in: *Blood Lipids and Lipoproteins: Quantitation, Composition and Metabolism* (G. J. Nelson, ed.), Wiley Interscience, New York, pp. 705–743.

Solomon, T. E., 1994, Control of exocrine pancreatic secretion, in: *Physiology of the Gastrointestinal Tract,* 3rd ed. (L. R. Johnson, ed.), Raven Press, New York, pp. 1173–1207.

Sörhede, M., Erlanson-Albertsson, C., Mei, J., Nevalainen, T., Aho, A., and Sundler, F., 1996a, Enterostatin in gut endocrine cells—Immuocytochemical evidence, *Peptides* **17:** 609–614.

Sörhede, M., Mulder, H., Mei, J., Sundler, F., and Erlanson-Albertsson, C., 1996b, Procolipase is produced in rat stomach—A novel source of enterostatin, *Biochim. Biophys. Acta* **1301:**207–212.

Sörhede Winzell, M., 1998, Enterostatin in the gastrointestinal tract: Production and possible mechanism of action, Ph.D. diss., University of Lund.

Spiller, R. C., Trotman, I. F., Higgens, B. E., Ghatel, M. A., Grimble, G. K., Lee, Y. C., Bloom, S. R., Misiewics, J. J., and Silk, D. B. A., 1984, The ileal brake—Inhibition of jejunal motility after ileal fat perfusion in man, *Gut* **25:**365–374.

Srivastava, R. A. K., Srivastava, N., and Schonfeld, G., 1992, Expression of low-density lipoprotein receptor, apolipoprotein A-I, A-II and A-IV in various rat organs utilizing an efficient and rapid method for RNA isolation, *Biochem. Int.* **27:**85–95.

Stein, O., Stein, Y., and Roheim, P., 1986, The role of apolipoprotein A-IV in reverse cholesterol transport studied with cultured cells and liposomes derived from an ether analog of phosphatidylcholine, *Biochim. Biophys. Acta* **878:**7–13.

Tatemoto, K., 1982, Isolation and characterization of peptide YY (PYY), a candidate gut hormone that inhibits pancreatic exocrine secretion, *Proc. Natl. Acad. Sci. USA* **79:**2514–2518.

Taylor, C. B., and Gould, R. G., 1967, A review of human cholesterol metabolism, *Arch. Pathol.* **84:**2–14.

Taylor, I. L., 1985, Distribution and release of peptide YY in dog measured by specific radioimmunoassay, *Gastroenterol.* **88:**731–737.

Weinberg, R. B., and Spector, M. S., 1985, The self-association of human apolipoprotein A-IV. Evidence for an *in vivo* circulating dimeric form, *J. Biol. Chem.* **260:**14279–14286.

Weinberg, R. B., Dantzker, C., and Patton, C. S., 1990, Sensitivity of serum apolipoprotein A-IV levels to changes in dietary fat content, *Gastroenterol.* **98:**17–24.

Welch, I., Saunders, K., and Read, N. W., 1985, Effect of ileal and intravenous infusions of fat emulsions on feeding and satiety in human volunteers, *Gastroenterol.* **89:**1293–1297.

Woods, S. C., Seeley, R. J., Porte, D. J., and Schwartz, M. W., 1998, Signals that regulate food intake and energy homeostasis, *Science* **280:**1378–1383.

CHAPTER 17

The Possible Role of Intestinal Surfactantlike Particles in the Absorption of Triacylglycerols in the Rat

DAVID H. ALPERS, MICHAEL J. ENGLE, and RAMI ELIAKIM

1. Background

The process of triacylglycerol absorption has passed through at least three quantum leaps of progress. In the 1950s and 1960s there was a tremendous increase in knowledge concerning the intralumenal and intracellular fate of ingested triacylglycerols. By the end of this explosion of information the current concepts were defined: intralumenal hydrolysis, incorporation of fatty acids and monoglycerides into mixed bile salt micelles, uptake of lipid (not by pinocytosis) into the enterocyte with reformation of triacylglycerols, and finally, intracellular packaging and secretion into lymph via lipoproteins (Senior, 1964; Hofmann and Small, 1967; Johnston, 1968; Borgström, 1974; Friedman and Nylund, 1980). In addition, the intracellular pathways of fat absorption were described using transmission electron microscopy (Palay and Karlin, 1959; Strauss, 1966; Rubin, 1966; Sjostrand and Borgström, 1967). Although the details and mechanisms of the lumenal events were largely understood at that time, the precise intracellular events and their compartmentalization were unclear.

In the 1970s and 1980s a second major increase in understanding occurred, as the intracellular compartments of the cell were identified and characterized biochemically. Factors that regulated the production and secretion of intracellular triacylglycerols were identified and included phosphatidylcholine (Beil and Grundy, 1980; Parlier et al., 1989), regional differences in absorptive capacity (Wu et al., 1975), and the distinction between intracellular absorptive pathways for very low density lipoproteins (VLDL) and for chylomicrons (CM), using the detergent Pluronic L-81 as a discriminating agent (Tso et al., 1984). The nature of lipoproteins and their metabolism was increasingly understood (Hav-

DAVID H. ALPERS and MICHAEL J. ENGLE • Division of Gastroenterology, Department of Medicine, Washington University School of Medicine, St. Louis, Missouri 63110-1010.   RAMI ELIAKIM • Hadassah Medical School, Jerusalem, Israel.
Intestinal Lipid Metabolism, edited by Charles M. Mansbach II *et al.,* Kluwer Academic/Plenum Publishers, 2001.

el, 1980), and prechylomicrons were characterized in the intestinal cell (Mahley *et al.,* 1971; Redgrave, 1971).

In recent years, however, the process of fat absorption has seemed less clear than formerly anticipated. Inborn errors of metabolism have been found that involve either an unknown step unassociated with apolipoprotein production (Tangier disease) or an entirely new step mediated by the microsomal transfer protein (abetalipoproteinemia; Lin *et al.,* 1994). Although triacylglycerols in the native enterocyte are formed largely from the monoglyceride pathway, and the monoglyceride-acyltransferase has been associated with the smooth endoplasmic reticulum (SER), the location of the triacylglycerol synthetase has not been established. It is clear that the final production of lipid droplets is dependent on factors that are tissue specific (Kumar and Mansbach, 1997), but the precise nature of those factors is not yet known. In addition, despite much careful morphological analysis, and the demonstration of a vesicular membrane surrounding the fat droplet in the enterocyte (Sjostrand and Borgström, 1967), it is currently not understood how fat droplets get out of the cell. The relative roles of the Golgi and the SER have only begun to be investigated (Mansbach and Nevin, 1994). Reverse pinocytosis has not been found (Cardwell *et al.,* 1967), and the process of discharge has only rarely been observed microscopically. Moreover, the nature of the membrane surrounding the intracellular fat droplets has not been characterized.

2. Role of Intestinal Alkaline Phosphatase (IAP) in Fat Absorption

A positive correlation between IAP in serum or lymph and fat absorption has been reported many times previously in animals (Glickman *et al.,* 1970; Kleerekoper *et al.,* 1970; Miura *et al.,* 1979; Young *et al.,* 1981; Eliakim *et al.,* 1991a), in humans (Langman *et al.,* 1966; Day *et al.,* 1992; Domar *et al.,* 1993), and in models of experimental stress, such as hemorrhagic shock (W. Wang *et al.,* 1997). These confirmatory correlations have been found in both humans and in rats, even though IAP secretion is regulated by ABO secretor status in humans but not rats, and though circulating IAP accounts for only a small percent of the total serum AP in humans (Langman *et al.,* 1966). Still other studies in humans support the association of IAP and fat secretion. The increase of serum IAP secretion during diagnostic secretin–pancreozymin tests is blocked by somatostatin (Bayer and Pointner, 1980), raising the possibility that somatostatin may affect these processes *in vivo.* The levels of long-chain triacylglycerols and IAP were linearly related in chylous ascitic fluid (Malagelada *et al.,* 1977). On the other hand, there are some experimental models that do not appear at first glance to agree with the IAP/fat absorption association. In myo-inositol-deficient gerbils, fat absorption recovered 18 hr after repletion, but serum IAP did not rise, despite a partial recovery of brush border IAP activity (Chu *et al.,* 1987). Similarly, colchicine was found to decrease lipid, but not IAP, transport into the lymphatics (Miura *et al.,* 1982). However, in contrast with the lymph data, IAP was retained in the mucosa of fat-fed colchicine-treated animals. As IAP is not lipid associated in serum, some of it may be absorbed via the portal vein. Thus, some reports to the contrary, most studies support the relationship between secretion of IAP and lipid from the enterocyte.

IAP in both the rat and in humans rises in the serum after a meal containing triacylglycerol, but not after one containing carbohydrate or protein alone (Young *et al.,* 1981; Day *et al.,* 1992; Domar *et al.,* 1993). Moreover, the percent increase in serum far exceeds

Table 1. Comparison of Chymosomes and Surfactantlike Particles

Property	Chymosomes[a]	SLP[b]
Diameter size	100–150 nm	20–100 nm
Unilamellar	Yes	Yes
Enriched for IAP	Yes	Yes
PL content	PC and lyso PC	PC and lyso PC
Apical blebbing	Yes	No
Seen in/between cells	No	Yes
IAP in lumen	80% of mucosal total	1–3% of total
Surfactant proteins	Not tested	SP-A, SP-B
Microvillar enzymes	High content	Low content

[a]Data from Halbhuber *et al.*, 1994. [b]Data from Eliakim *et al.*, 1989, 1997.

that seen in the tissue itself. This discrepancy is consistent with an augmented secretion of the enzyme from the enterocyte. IAP in the tissue is a membrane-bound enzyme, attached to the apical membrane of the enterocyte by a glycan phosphatidylinositol linkage. IAP in lymph and serum, however, is not associated with membranes or with the lipoproteins (Glickman *et al.*, 1970; Miura *et al.*, 1979). Because there is abundant phospatidylinositol-specific phospholipase D in serum, IAP bound to any membrane in the blood might be released by the action of this enzyme (Eliakim *et al.*, 1990). The formation of blebs from microvilli of the apical membrane of enterocytes has been described, particularly in the calf (Halbhuber *et al.*, 1994). These blebs have been termed "chymosomes" and account for about 80% of the total IAP in the lumen and tissue combined (Table 1). Furthermore, they contain a full complement of microvillus membrane enzymes, most of which do not appear in the serum except in minute amounts. Although these chymosomes appear to be important in the calf, they cannot explain the appearance of IAP in the serum, because these structures are delivered into the intestinal lumen.

Some earlier studies had suggested that IAP might be associated with extracellular membranes in the intestinal mucosa, but the significance of these findings was not widely appreciated. IAP was found to correlate not with the transport of fatty acids across the brush border membrane, but with secretion into the lymph (Lam and Mistilis, 1973). This same study found that IAP increased after fat administration in the endoplasmic reticulum, and it suggested that IAP in lymph might originate from membranous structures that were produced inside the enterocyte. In support of this view, round AP-positive particles (50–100 nm diameter) have been observed both in the lumen and the lamina propria of the rat following fat feeding (Rufo *et al.*, 1973).

3. Identification of Surfactantlike Particles (SLPs)

In searching for a mechanism whereby IAP could be secreted basolaterally from the enterocyte, tissue was fixed with tannic acid to preserve membranes, as such mordants had been used previously to identify phopholipid-rich membranes in gastric epithelium (Kao

Table 2. Specific Activity of Digestive Enzymes in Rat and Human Intestinal Surfactantlike Particles (SLP) and Micorvillous Membranes (MVM)

	Enzyme Activity (ratio enzyme/maltase s.a.)			
Enzyme	SLP (rat)^a	MVM (rat)^a	SLP (hum)^b	MVM (hum)^b
Alkaline phosphatase	33.0	0.35	3.51	0.47
Sucrase-Isomaltase	0.245	0.213	0.52	0.55
Lactase	0.02	0.03	0.083	0.036
Trehalase	0.25	0.125	0.154	not tested

^{a}Data from Eliakim *et al.*, 1989. ^{b}Data from Mahmood *et al.*, 1993

and Lichtenberger, 1987; Hills *et al.*, 1983). Structures were found lying on the apical membrane of enterocytes in rat and human intestine that had a superficial appearance to pulmonary surfactant (DeSchryver-Kecskemeti *et al.*, 1990). These structures were purified from the surface of the cells in rat (Eliakim *et al.*, 1989) and in human intestine (Mahmood *et al.*, 1993), and were found to be enriched for IAP, but not for other microvillar hydrolytic enzymes (Table 2). The SLP isolated from both rat and human intestine was enriched for phospholipids, having a chol/PL ratio of 0.68 and 0.78, respectively, when compared with the chol/PL ratios of microvillus membranes isolated from those species (1.49 and 1.37; Mahmood *et al.*, 1993). The majority (75%) of the phospholipid was dipalmitoyl phosphatidylcholine (Eliakim *et al.*, 1989), even higher than the percent in lung surfactant (65%). In addition, the purified structure was a unilamellar membrane that had surface active properties and contained some pulmonary surfactant proteins. Table 1 demonstrates the comparison between this structure, termed "surfactantlike particle" or SLP, and the chymosomes reported from calf intestine. It would appear that SLP and chymosomes are different structures. Although both have not been isolated from the same species, there is no reason why an enterocyte might not produce both structures, presumably for different purposes. Subsequent studies have demonstrated the presence of SLP in the human and the mouse colon (Eliakim *et al.*, 1997) and in human and rat stomach (Goetz *et al.*, 1997), as well as in the small intestine. Surfactant protein A is much more abundant in the colon SLP than in the small intestinal SLP, although human small intestinal SLP contains much more SP-A than does the comparable rat SLP.

4. Production of Surfactantlike Particles (SLPs)

SLPs were found intracellularly by electron microscopy (Eliakim *et al.*, 1991b, De Schryver-Kecskemeti *et al.*, 1991) and did not contain tissue unspecific AP, as does pulmonary surfactant. Thus, it was very unlikely that SLP reached the apical membrane of enterocytes from the lung. Production of SLP was examined following triacylglycerol feeding, because that is the major stimulus for IAP increases in the serum (Eliakim *et al.*, 1991a). IAP production following fat feeding was elevated along with SLP in a dose-dependent fashion, but the specific activity (IAP units/mg SLP protein) rose with fat feeding (Yamagishi *et al.,* 1994a). Antibodies were raised against the rat SLP proteins (48, 68, 97, and 116 kDa), and were used in an ELISA after adsorption with microvillar membranes to remove IAP-reacting antibodies. Secretion was then studied in both apical (light mucosal scrapings) and basolateral (lamina propria and serum) directions for up to 7 hr after corn oil feeding (Yamagishi *et al.,* 1994a). SLP and IAP content in the enterocyte and in the lamina propria peaked 3 hr after fat feeding, and about 75% of the total mucosal content of IAP and SLP were found in the lamina propria. In addition, Golgi SLP content in the enterocyte fell to its nadir at 3 hr after fat feeding. In contrast, both IAP and SLP content peaked later at 5 hr in the lumen and in the serum. In addition, newly synthesized SLP proteins followed the same kinetics as pre-fat-fed SLP proteins, in that they peaked in the lamina propria at 3 hr after fat feeding, but not until 5 hr in the apical compartment (Alpers *et al.,* 1995). These data were consistent with the hypothesis that fat feeding induces both IAP and SLP production in the enterocyte and that these proteins are secreted basolaterally, moving largely to the lamina propria and later to the serum. Clearance from both compartments is quite rapid with a half-life of less than 10 min (Yamagishi *et al.,* 1994b). Because SLPs were never seen within the terminal web beneath the microvillar membrane, but were seen between the cells and within tight junctions, it was proposed that the SLP could move toward the lumen through tight junctions.

This possibility was examined further by studying SLP secretion in Caco-2 cells, a colonic tumor cell line that develops properties of fetal ileum after reaching confluence. Previous work had shown that the initial IAP secretion in these cells occurred basolaterally at 1 hr, followed by the appearance of newly synthesized IAP in the apical medium by 2 hr (Sussman *et al.,* 1989). As this timing supported the hypothesis that IAP (and by implication SLP) were secreted basolaterally and then moved through the tight junctions to the apical compartment, Caco-2 cells were treated with cytochalasin D to open up the tight junctions (Engle *et al.,* 1995). When this was done, the rate at which both IAP and SLP entered the apical compartment was increased. This finding was confirmed morphometrically by measuring increases in both the numbers and the surface area of apical SLP. Thus, SLP appear to be synthesized both in the enterocyte *in vivo* and in Caco-2 cells. *In vivo* synthesis and secretion is increased by fat feeding, and in the cell line direct evidence for basolateral secretion was obtained, whereas the kinetics of secretion *in vivo* also suggested initial basolateral secretion. These data provided a mechanism for the increase in IAP in the serum following fat feeding, one that did not involve release of microvillar membrane fragments, but did entail production of an entirely different membrane structure.

5. Role of Surfactantlike Particles (SLPs) in Fat Absorption

Although intracellular SLP was found in abundance by electron microscopy 3 to 5 hr after fat feeding in rat enterocytes, it was difficult to see in fasting animals. Because IAP was localized on the particle by immuno-electron microscopy, this marker was used to localize IAP within cells (Mahmood *et al.,* 1994). The intracellular membranous particles bearing IAP were commonly associated with fat droplets and were found near the basolateral membrane and in the intercellular spaces. To test the hypothesis that SLP was the intracellular membrane surrounding fat droplets, triacylglycerol absorption was inhibited by feeding rats Pluronic L-81, a nonionic detergent that does not alter uptake of fatty acids or

Table 3. Quantitation of SLP Seen by Electron Microscopy
in Enterocytes Following Corn Oil Feeding[a]

Time	Compartment	Relative score	
(hr)		Control	Pluronic L81
0	cytoplasm	0	0
1		0	3.3 ± 0.2
3		2.5 ± 0.3	3.6 ± 0.2
5		1.4 ± 0.1	3.8 ± 0.3
7		0	4.1 ± 0.2
0	intracellular	0	0
1		0	1.3 ± 0.2
3		2.4 ± 0.2	2.7 ± 0.1
5		2.4 ± 0.1	1.9 ± 0.1
7		0	2.1 ± 0.1
0	lumen	0	0
1		0.1 ± 0.1	0
3		1.8 ± 0.1	0.4 ± 0.3
5		1.8 ± 0.1	0.3 ± 0.2
7		0	0

[a]Data are derived from Mahmood *et al.* (1994) and represent the means of relative scores (scale of 0 to 4), derived from 42 samples (3 animals in each group).

monoglycerides from the lumen but does decrease the secretion of chylomicrons from the enterocyte. Pluronic L-81 produced a marked inhibition of fat-induced IAP and SLP secretion into the lumen and into the fraction of the mucosa remaining after removal of all epithelial cells, a fraction that would be expected to contain basolaterally secreted proteins. At the same time electron microscopy revealed large retained cytoplasmic lipid globules with tightly coiled membranes in apposition. A parallel absence of SLP overlying the apical membrane of the enterocyte was consistent with intracellular retention of the particle induced by Pluronic L-81 (Table 3).

Further evidence for association of SLP with fat droplets was obtained by using antibodies against IAP and SLP for immunocytochemical analysis of rat intestine after fat feeding (Zhang *et al.*, 1996). In fasting animals IAP was located largely in the microvillar membrane, whereas SLP was principally found in the lamina propria just beneath the basal membrane of the enterocyte. Triacylglycerol feeding produced a peak of staining for both IAP and SLP between 3 and 5 hr, noted intracellularly, in the lamina propria, and in apical compartments. Feeding Pluronic L-81 produced a decrease in reactivity of both antigens in the lamina propria, the lumen, and in intercellular space and redistributed both antigens from Golgi or cytosol to the circumference of intracellular fat droplets (Table 4). These data support the hypothesis that SLPs are, at least in part, the membranes that surround intracellular lipid droplets.

The rat is unique in having such a high content of IAP in the serum and in having two mRNAs encoding IAP (Engle and Alpers, 1992). It would be reasonable to assume that the development of the second mRNA would be in response to a special role for the product. The second IAP has a unique carboxyl terminus, with a sequence of 17 uninterrupted threonine residues. Once again the Caco-2 cell model was used to test the function of the two

Table 4. Distribution of Immunoreactive SLP in Rat Duodenal
Tissue Compartments Following Corn Oil Feeding^a

Time	Lumen	Cytosol	Paracellular	Sub-basal
		Intensity of SLP immunostaining		
(hr)		(Control/Pluronic L81)		
0	0.5/0.5	0.5/0	1/1	2.5/2
1	1/1	0/0.5	1/1	2.0/0.5
3	1.5/0.5	2.0/0.5	2.5/1	2.0/1.0
5	2.5/0.5	3.0/1.5	2.0/0.5	2.5/0
7	1.0/0.5	0.5/1.0	2.0/1	2.5/0

^{a}Data are derived from Zhang *et al.* (1996) and refer to the mean of scores (1 to 4 intensity scale), derived from six observations each from the outer 2/3 of the villus (3 separate animals in each group).

IAP genes (Tietze *et al.,* 1992). Transfection of both IAPs led to increased SLP production and secretion, but the effect of IAP-II with the oligothreonine sequence was much greater than that of the other IAP, whose structure is similar to that of IAPs in other species. By electron microscopy the SLPs were increased not only intracellularly but also apically, being particularly concentrated over the tight junctions. This localization would be expected if the SLP entered the apical compartment via the tight junctions.

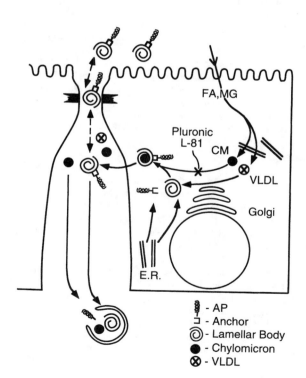

Figure 1. Cartoon depicting the proposed role of surfactantlike particles (lamellar body) in the transcytotic movement of triacylglycerols across the enterocyte. See text (section 6) for details.

6. Hypothesis

The current hypothesis regarding the regulation of SLP via IAP is illustrated in Fig. 1. After feeding the cell absorbs the products of triacylglycerol digestion—fatty acids and monoglycerides. These products in turn stimulate the formation of chylomicrons and VLDL. The fatty acids and monoglycerides (or some product of these lipids) stimulate IAP synthesis, which in turn increases SLP production and secretion. The lipid droplet is delivered to the basolateral membrane, and in the intercellular space the SLP and lipid droplet are dissociated. The lipid and most of the SLP flow into the vessels of the submucosa, and a small portion of the SLP migrates to the apical side of the enterocyte. The function of the apically located SLP and of the SLP in serum is not clear. However, the SLP does appear to play a role in fat absorption. Further studies will be needed to confirm this function and to discover others for this unique and interesting secreted membrane.

References

Alpers, D. H., Zhang, Y., and Ahnen, D. J., 1995, Synthesis and parallel secretion of rat intestinal alkaline phosphatase and a surfactant-like particle protein, *Am. J. Physiol.* **268** (Endocrinol. Metab. **31**):E1205–E1214.

Bayer, P. M., and Pointner, H., 1980, The action of somatostatin on intestinal alkaline phosphatase stimulated by secretin and cholecystokinin-pancreozymin, *Clin. Chim. Acta* **108**:129–134.

Beil, F. W., and Grundy, S. M., 1980, Studies on plasma lipoproteins during absorption of exogenous lecithin in man, *J. Lipid Res.* **21**:525–536.

Borgström B., 1974, Fat digestion and absorption, in: *Biomembranes* Volume 4B (D. H. Smyth, ed.), Plenum Press, New York, pp. 555–620.

Cardwell, R. R., Jr., Badenhausen, S., and Porter, K. R., 1967, Intestinal triglyceride absorption in the rat. An electron microscopic study, *J. Cell Biol.* **34**:123–155.

Chu, S. W., Geyer, R. P., and Walker, W. A., 1987, Myo-inositol action on gerbil intestine: Alterations in alkaline phosphatase activity upon phosphatidylinositol depletion and repletion *in vivo, Biochim. Biophys. Acta* **929**:220–225.

Day, A. P., Fheher, M. D., Chopra, R., and Mayne, P. D., 1991, Triglyceride fatty acid chain length influences the postprandial rise in serum intestinal alkaline phosphatase activity, *Ann. Clin. Biochem.* **29**:287–291.

DeSchryver-Kecskemeti, K., Eliakim, R., Carroll, S., Stenson, W. F., Moxley, M. A., and Alpers, D. H., 1990, Intestinal surfactant-like material: A novel secretory product of the rat enterocyte, *J. Clin. Invest* **84**:1355–1361.

DeSchryver-Kecskemeti, K., Eliakim, R., Green, K., and Alpers, D. H., 1991, A novel intracellular pathway for rat intestinal digestive enzymes (alkaline phosphatase and sucrase) via a lamellar particle, *Lab. Invest.* **65**:365–373.

Domar, U., Karpe, F., Hamsten, A., Stigbrand, T., and Olivecrona, T., 1993, Human intestinal alkaline phosphatase—Release to the blood is linked to lipid absorption, but removal from the blood is not linked to lipoprotein clearance, *Eur. J. Clin. Invest.* **23**:753–760.

Eliakim, R., DeSchryver-Kecskemeti, K., Nogee, L., Stenson, W. F., and Alpers, D. H., 1989, Isolation and characterization of a small intestinal surfactant-like particle containing alkaline phosphatase and other digestive enzymes, *J. Biol. Chem.* **264**:20614–20619.

Eliakim, R., Becich, M. J., Green, K., and Alpers, D. H., 1990, Both tissue and serum phospholipases release rat intestinal alkaline phosphatase, *Am. J. Physiol.* **259** (Gastrointest. Liver Physiol. 22):G618–G625.

Eliakim, R., Mahmood, A., and Alpers, D. H., 1991a. Rat intestinal alkaline phosphatase secretion into lumen and serum is coordinately regulated, *Biochim. Biophys. Acta* **1091**:1–8.

Eliakim, R., Becich, M. J., Green, K., and Alpers, D. H., 1991b, The developmental expression of intestinal surfactant-like particles in the rat, *Am. J. Physiol.* **261**(Gastrointest. Liver Physiol. 24):G269–G279.

Eliakim, R., Goetz, G. S., Rubio, S., Chaille-Heu, B., Shao, J.-S., Ducroc, R., and Alpers, D. H., 1997, Isolation and characterization of surfactant-like particles in rat and human colon, *Am. J. Physiol.* **272** (Gastrointest. Liver Physiol. 35):G425–G434.

Engle, M. J., and Alpers, D. H., 1992, The two mRNAs encoding rat intestinal alkaline phosphatase represent two unique nucleotide sequences, *Clin. Chem.* **38**:2506–2509.

Engle, M. J., Grove, M. L., Becich, M. J., and Alpers, D. H., 1995, Appearance of surfactant-like particles in apical medium of Caco-2 cells may occur via tight junctions, *Am. J. Physiol.* **268** (Cell Physiol. **37**):C1401–C1413.

Friedman, H. I., and Nylund, B., 1980. Intestinal fat digestion, absorption, and transport: A review, *Am. J. Clin. Nutr.* **33**:1108–1139.

Glickman, R. M., Alpers, D. H., Drummey, G. D., and Isselbacher, K. J., 1970, Increased lymph alkaline phosphatase after fat feeding; Effects of medium chain triglycerides and inhibition of protein synthesis, *Biochim. Biophys. Acta* **201**:226–235.

Goetz, G. S., Shao, J.-S., and Alpers, D. H., 1997. Purification and initial characterization of a surfactant-like particle (SLP) from human stomach, *Gastroenterol.* **112**:A128.

Halbhuber, K. J., Schulze, M., Rhode, H., Bublitz, R., Feuerstein, H., Walter, M., Linss, W., Meyer, H. W., and Horn, A., 1994. Is the brush border membrane of the intestinal mucosa a generator of "chymosomes"? *Cell Molec. Biol.* **40**:1077–1096.

Havel, R. J., 1980, Lipoprotein biosynthesis and metabolism, in: *Lipoprotein Structure,* (A. M. Scanu and F. R. Landsberger, eds.), New York Academy of Science, New York, pp. 16–29.

Hills, B. A., Butler, B. D., and Lichtenberger, L. M., 1983, Gastric mucosal barrier: Hydrophobic lining to the lumen of the stomach, *Am. J. Physiol.* **244** (Gastrointest. Liver Physiol. **7**):G561–G568.

Hofmann, A. F., and Small, D. S., 1967, Detergent properties of bile salts: Correlation with physiological function, *Ann. Rev. Med.* **18**:333–376.

Johnston, J. M., 1968, Mechanism of fat absorption, in: *Handbook of Physiology,* Volume 3 (J. Field, ed.), American Physiological Society, Washington, D.C., pp. 1353–1375.

Kao, Y. C., and Lichtenberger, L. M., 1987, Localization of phospholipid-rich zones in rat gastric mucosa: Possible origin of a protective hydrophobic lumenal lining, *J. Histochem. Cytochem.* **35**:1285–1298.

Kleerekoper, M., Horne, M., Cornish, C. J., and Posen, S., 1970, Serum alkaline phosphatase after fat ingestion: An immunological study, *Clin. Sci.* **38**:339–345.

Kumar, N. S., and Mansbach, C. M., 1997, Determinants of triacylglycerol transport from the endoplasmic reticulum to the Golgi in the intestine, *Am. J. Physiol.* **273** (Gastrointest. Liver Physiol. **36**):G18–G30.

Lam, K. C., and Mistilis, S. P., 1973, Role of intestinal alkaline phosphatase in fat transport, *Aust. J. Exp. Biol. Med. Sci.* **51**:411–416.

Langman, M. J. S., Leuthold, E., Robson, E. B., Luffman, J. E., and Harris H., 1966, Influence of diet on the "intestinal" component of serum alkaline phosphatase in people of differerent ABO blood groups and secretor status, *Nature* **212**:41–43.

Lin, M. C., Arbeeny, C., Bergquist, K., Kienzle, B., Gordon, D. A., and Wetterau, J. R., 1994, Cloning and regulation of hamster microsomal triglyceride transfer protein. The regulation is independent from that of other hepatic and intestinal proteins which participate in the transport of fatty acids and triglycerides, *J. Biol. Chem.* **269**:29138–29145.

Mahley, R. W., Bennett, B. D., Morre, D. J., Gray, M. E., Thistlethwaite, W., and LeQuire, V. S., 1971, Lipoproteins associated with Golgi apparatus isolated from epithelial cells of rat small intestine, *Lab. Invest.* **25**:435–444.

Mahmood, A., Mahmood, S., DeSchryver-Kecskemeti, K., and Alpers, D. H., 1993, Characterization of proteins in rat and human intestinal surfactant-like particles, *Arch. Biochem. Biophys.* **300**:280–286.

Mahmood, A., Yamagishi, F., Eliakim, R., DeSchryver-Kecskemeti, K., Gramlich, T. L., and Alpers, D. H., 1994, A possible role for rat intestinal surfactant-like particles in transepithelial triacylglycerol transport, *J. Clin. Invest.* **93**:70–80.

Malagelada, J. R., Stolbach, L. L., and Linscheer, W. G., 1977, Influence of carbon chain length of dietary fat on intestinal alkaline phosphatase in chylous ascites, *Am. J. Dig. Dis.* **22**:629–632.

Mansbach, C. M., and Nevin, P., 1994, Effect of brefeldin A on lymphatic triacylglycerol transport in the rat, *Am. J. Physiol.* **266** (Gastrointest. Liver Physiol. **29**):G292–G302.

Miura, S. M., Asakura, H., Miyairi, M., Morishita, T., Ishii, H., and Tsuchiya, M., 1979, Study on the fat absorption and transportation into intestinal lymph of rats. Differences in the absorption of saturated and unsaturated long chain fatty acids and the role of intestinal alkaline phosphatase, *JPN J. Gastroenterol.* **76**:871–880.

Miura, S., Asakura, H., Miyairi, M., Morishita, T., Nagata, H., and Tsuchiya, M., 1982, Effect of colchicine on intestinal alkaline phosphatase activity during linoleic acid absorption in rats, *Digestion* **23**:224–231.

Palay, S. L., and Karlin, L. J., 1959, An electron microscopic study of the intestinal villus. II. The pathway of fat absorption, *J. Biophys. Biochem. Cytol.* **5**:373–384.

Parlier, R. D., Frase, S., and Mansbach, C. M. II., 1989, Intraenterocyte distribution of absorbed lipid and effects of phosphatidylcholine, *Am. J. Physiol.* **256** (Gastrointest. Liver Physiol. **19**):G349–G355.

Redgrave, T. G., 1971, Association of Golgi membranes with lipid droplets (prechylomicrons) from within intestinal epithelial cells during absorption of fat, *Aust. J. Exp. Biol. Med. Sci.* **49**:209–224.

Rubin, C. E., 1966, Electron microscopic studies of triglyceride absorption in man, *Gastroenterol.* **50**:65–77.

Rufo, M. B., Malagelada, J. R., Linscheer, W. G., and Fishman, W. H., 1973, Metabolic variants of intestinal alkaline phosphatase in relation to fat absorption: *In situ* demonstration with the organ-specific inhibitors *l*-phenylalanine and *l*-homoarginine, *Histochemie* **33**:313–322.

Senior, J. R., 1964, Intestinal absorption of fat, *J. Lipid Res.* **5**:495–521.

Sjostrand, F. S., and Borgström, B., 1967, The lipid components of the smooth-surfaced membrane-bounded vesicles of the columnar cells of the rat intestinal epithelium during fat absorption, *J. Ultrastruct. Res.* **20**:140–160.

Strauss, E. W., 1966, Electron microscopic study of intestinal fat absorption *in vitro* from mixed micelles containing linolenic acid, monoolein and bile salt, *J. Lipid Res.* **7**:307–323.

Sussman, N. L., Eliakim, R., Rubin, D., Perlmutter, D. H., DeSchryver-Kecskemeti, K., and Alpers, D. H., 1989, Intestinal alkaline phosphatase is secreted bidirectionally from villous enterocytes, *Am. J. Physiol.* **257** (Gastrointest. Liver Physiol. **20**):G14–G23.

Tietze, C. C., Becich, M. J., Engle, M., Stenson, W. F., Eliakim, R., and Alpers, D. H., 1992, Caco-2 cell transfection by rat intestinal alkaline phosphatase cDNA increases surfactant-like particles, *Am. J. Physiol.* **263** (Gastrointest. Liver Physiol. **26**):G756–G766.

Tso, P., Drake, D. S., Black, D. D., and Sabesin, S. M., 1984, Evidence for separate pathways of chylomicron and very low-density lipoprotein assembly and transport by rat small intestine, *Am. J. Physiol.* **247** (Gastrointest. Liver Physiol. **10**):G599–G610.

Yamagishi, F., Komoda, T., and Alpers, D. H., 1994a. Secretion and distribution of rat intestinal surfactant-like particles following fat feeding, *Am. J. Physiol.* **266** (Gastrointest. Liver Physiol. **29**):G944–G952.

Yamagishi, F., Becich, M. J., Evans, B. A., Komoda, T., and Alpers, D. H., 1994b, The clearance of surfactant-like particle (SLP) proteins from the circulation in rats, *Am. J. Physiol.* **266** (Gastrointest. Liver Physiol. **29**):G596–G605.

Young, G. P., Friedman, S., Yedlin, S. T., and Alpers, D. H., 1981, Effect of fat-feeding on intestinal alkaline phosphatase activity in tissue and serum, *Am. J. Physiol.* **241** (Gastrointest. Liver Physiol. **4**):G461–G468.

Wang, W., Wang, P., and Chaudry, I. H., 1997. Intestinal alkaline phosphatase: Role in the depressed gut lipid transport after trauma-hemorrhagic shock, *Shock* **8**:40–44.

Wu, A. L., Clark, S. B., and Holt, P. R., 1975, Transmucosal triglyceride transport rates in proximal and distal rat intestine *in vivo*, *J. Lipid Res.* **16**:251–257.

Zhang, Y., Shao, J.-S., Xie, Q., and Alpers, D. H., 1996, Immunolocalization of intestinal alkaline phosphatase and surfactant-like particle proteins in rat duodenum, *Gastroenterol.* **110**:478–488.

CHAPTER 18

Inhibitors of Chylomicron Formation and Secretion

JOHN B. RODGERS

1. Synthesis of Triacylglycerols by the Enterocytes

Most of the lipid absorbed by the enterocytes leaves these cells as reesterified lipid in the form of chylomicrons (CM). CM formation and secretion is a complex process requiring proteins, phospholipids, and cholesterol in addition to triacylglycerols to form these lipoprotein particles. The first three compounds are also required to synthesize membranes and enzymes located in these absorbing cells that are involved in this process. The first step involved in CM formation is the reesterification of the products of triacylglycerol digestion, fatty acids and monoacylglycerols, that have been absorbed into the enterocytes (Johnson, 1978; Thomson and Dietschy, 1981). There are two pathways for reesterification in the enterocytes. The more important pathway for reesterification of the absorbed products of digested dietary lipid is the monoglyceride pathway, which utilizes 2-monoglyceride and free fatty acids as substrates (Mattson and Volpenhein, 1964). This involves reesterifying enzymes located on the cytoplasmic surface of the smooth endoplasmic reticulum (Bell *et al.*, 1981). As 2-monoacylglycerol is present in the enterocytes in significant amounts during the process of digestion and absorption of a lipid meal, this pathway is obviously efficient. The other pathway for triacylglycerol synthesis utilizes α-glycerol phosphate to donate the glycerol portion of the molecule and is therefore known as the α-glycerol phosphate pathway (Johnson, 1978). This pathway, which involves more steps than the monoglyceride pathway and is therefore less rapid, has to be utilized when triacylglycerol is being synthesized by the enterocytes under conditions when no 2-monoacylglycerol is being absorbed, which is the case while fasting. There are recent observations, however, to indicate that the α-glycerol phosphate pathway also involves the monoglyceride pathway with the phosphate group and the 1-acyl group hydrolyzed from phospatic acid formed by the glycerophospate pathway, thus producing 2-monoaclyglycerol that would then be available for the monoglyceride pathway (Yang and Kuksis, 1991). The triacylglycerol synthesized by

JOHN B. RODGERS • Department of Medicine, Albany Medical College, Albany, New York 12208.
Intestinal Lipid Metabolism, edited by Charles M. Mansbach II *et al.,* Kluwer Academic/Plenum Publishers, 2001.

either pathway is then translocated across the membranes of the smooth endoplasmic reticulum (SER) to the cisternae of this subcellular compartment where it associates with apolipoprotein B-48 (apo B-48) forming a nascent lipoprotein particle (Davis, 1993).

2. Synthesis of Apolipoprotein B-48 (Apo B-48)

According to studies done on hepatic samples, the apo B is synthesized by the rough endoplasmic reticulum (RER; Alexander *et al.*, 1976) and is secreted into the cytoplasm. It next translocates across the membranes of the SER at which time a considerable portion of the molecule remaining on the cytosolic side of the SER is cleaved off (Davis *et al.*, 1990). Most of the apo B synthesized by the enterocytes (apo B-48) is considerably smaller than that synthesized by the liver (apo B-100). Both are derived from a single gene but in the enterocytes there is a change at nucleotide 6666. A cytidine component in this location is deaminated, forming uridine, which converts this part of the structure to a termination codon (Garcia *et al.*, 1992), thus limiting the size of the apo B from the intestine. This alters the carboxy-terminal half that is the LDL-receptor-binding domain (Powell *et al.*, 1987; Chen *et al.*, 1987). This accounts for the fact that the apo B-48 on chylomicrons is not involved with the final uptake of the chylomicron remnant particles by the hepatocytes.

3. Microsomal Triglyceride Transfer Protein

The triacylglycerol that is synthesized by enzymes on the surface of the SER must then be translocated across the SER to the cysternae of the subcellular organelle, where nascent lipoproteins are formed by uniting the triacylglycerol droplet with lipoprotein membrane components including phospholipid, cholesterol, and apo B. This process of translocation of reesterified triacylglycerol across the SER is not well understood. Once the triacylglycerol is dissolved in, or is otherwise taken up into the outer membrane of the SER, it is postulated that the triacylglycerol droplet becomes surrounded by one leaflet of the endoplasmic reticulum, which subsequently buds off into the cisterna (Davis, 1993). Triacylglycerol solubility in the phospholipid membrane of the SER is quite limited, however. To concentrate triacylglycerol into a core associated with the lumenal leaflet of the SER, it is likely that the triacylglycerol droplet has to encounter a protein with hydrophobic components such as the microsomal transfer protein (MTP) described by Wetterau and Zilversmit (1984, 1985, 1986). MTP is found in relatively large amounts in the small intestine and the liver (Wetterau and Zilversmit, 1986) where triglyceride-rich lipoproteins are formed. The exact role of MTP in the formation and secretion of chylomicrons by the enterocytes is not actually known, but there is reason to believe that it is required for the lipid droplet to associate with the apo B in the cisternae of the SER.

Triglyceride-rich lipoproteins cannot be secreted by the intestine or the liver unless there is apo B in the lipoprotein membrane. Thus, in the disease of abetalipoproteinemia with little or no apo B found in the circulation, the serum levels of triglyceride and cholesterol are very low as triglyceride rich-lipoproteins are not present in the circulation. Originally this was thought to be due to a genetic defect resulting in a lack of apo B synthesis. This theory was disproven, however, by studies demonstrating that the enterocytes from these patients could synthesize apo B (Talmud *et al.*, 1988; Huang *et al.*, 1990). In subsequent investigations, Wetterau and his coworkers demonstrated that MTP was absent from intestinal biopsies from these patients (Wetterau *et al.*, 1992), indicating that a failure to synthesize MTP was the basis for the condition of abetalipoproteinemia and simultaneously demonstrating the essential role of this protein for the formation of nascent chylomicrons.

It has previously been observed that apo B is synthesized by the RER (Alexander *et al.*, 1976). Before apo B can associate with lipid particles, it must first dissociate from the RER membrane. An alternate explanation for the essential role of MTP for chylomicron formation is that it is necessary for this dissociation of apo B from the RER. MTP can transfer triacylglycerol and cholesteryl ester to the hydrophobic domains of apo B while it is still attached to RER. As triacylglycerol and cholesteryl esters are more hydrophobic than the phospholipid component of the RER membrane, these very hydrophobic lipids now associated with apo B may initiate the process of dissociation of apo B from the RER. The free apo B in the cytoplasm of the enterocytes could then migrate to the SER, where it subsequently translocates across the SER to finally associate with the triglyceride droplet in the cisternae of the SER.

4. Trafficking of Nascent Lipoproteins between Endoplasmic Reticulum and Golgi Apparatus

Nascent lipoprotein particles are next rapidly transferred to the Golgi cysternae (Sabesin and Frase, 1977). This process is apparently complex and not well understood. Recent studies on this question demonstrated that ATP, heat, cytosolic proteins, and possibly other components in the cytosol yet to be identified were required, indicating that there is an active transport process in the enterocytes involved with this transfer process (Kumar and Mansbach, 1997). Furthermore, these studies confirmed that the triacylglycerol droplets and apo B are united in the SER and migrate together to the Golgi. These investigators felt that their results indicated that the triacylglyceride droplets were enclosed in a proteophospholipid membrane, which facilitates this transport process (Kumar and Mansbach, 1997). Once the nascent lipoprotein particles are in the Golgi, enzymes of this subcellular organelle add terminal sugars to the glycoprotein on the surface of these particles (Goldberg and Kornfeld, 1983) to form the mature chylomicron particle. Golgi vesicles then migrate to the basolateral plasma membrane and the chylomicrons are secreted from the enterocytes by a process of reverse pinocytosis into the intercellular space (Sabesin and Frase, 1977).

As this process continues during the period of absorption of a lipid meal, lipid droplets enclosed in a membrane accumulate in the apical portion of the enterocytes (Cardell *et al.*, 1967). These investigators regarded these vesicles as a storage compartment for nascent chylomicrons that appear to separate from the SER prior to being transferred to the Golgi apparatus (Cardell *et al.*, 1967).

5. Secretion of Triglyceride-Rich Lipoproteins

The particle size of the triglyceride-rich lipoproteins being secreted by the enterocytes varies. During periods of fasting when triglyceride is being synthesized via the α-glycerol phosphate pathway, any reesterified lipid transported out of the enterocytes is secreted in particles the size of very low-density lipoproteins (VLDLs; Ockner *et al.,* 1964). When the intestine is actively absorbing dietary lipid, however, the monoglyceride pathway is mainly utilized for triglyceride synthesis. Under these conditions a small fraction of the reesterified lipid is still transported in VLDL-sized particles, but most of this absorbed lipid is transported in larger chylomicrons of varying sizes. Of interest is that the load of lipid presented for absorption does not affect the number of chylomicron particles secreted per unit time, but rather affects the size of the particles so that when maximal amounts of fat are being transported by the enterocytes, the percentage of chylomicron particles of relatively large size is increased (Hayashi *et al.,* 1990). There is reason to believe that the availability of apolipoproteins for chylomicron formation may affect chylomicron size with larger particles that are formed when the need for these membrane proteins is great (Glickman *et al.,* 1972). This might be anticipated, as less apolipoprotein is needed to transport a given mass of lipid when the lipid is transported in large as compared to small lipoprotein particles.

6. Protein Synthesis and Formation of Lipoproteins

During periods of lipid absorption, protein synthesis is required to synthesize enzymes and the subcellular membranes, including SER and Golgi involved with this process and the apolipoproteins, including apo B-48, that must be available before lipid can be secreted from the enterocytes in chylomicrons (Dobbins, 1970). Using puromycin, a known protein synthesis inhibitor, it was shown that lipid accumulated in the enterocytes in abnormal amounts when lipid was presented for absorption in rats treated with this agent (Sabesin and Isselbacher, 1965). Indeed, the morphology of the enterocytes from these rats resembled enterocytes from patients with abetalipoproteinemia, supporting the concept that synthesis of apolipoproteins, particularly apo B-48, by the enterocytes is essential for chylomicron secretion. It was subsequently shown that treatment of rats with acetoxycycloheximide, another protein synthesis inhibitor, decreased transport of absorbed lipid into lymph but did not completely abolish chylomicron secretion (Glickman and Kirsch, 1973). The lipid that was transported as chylomicrons under these conditions was transported in relatively large chylomicron particles, suggesting that the availability of apo B was limited and thus large particles were formed to maximize the utilization of what apo B-48 was available. Thus normal protein synthesis does appear to by necessary for normal chylomicron formation and secretion.

7. Chylomicron Retention Disorder

As described above, an "experiment in nature," abetalipoproteinemia, demonstrates how the failure to synthesize one specific protein, MTP, severely affects CM formation and transport. In another "experiment in nature," intestinal steatosis is also observed along with an inability of jejunal explants from patients with this disorder to secrete absorbed lipids (Levy *et al.,* 1987). In this condition of "chylomicron retention disease," the defect does not appear to be related to a failure to form nascent lipoproteins but to a problem with glycosylation of nascent chylomicrons, which normally takes place in the Golgi (Levy *et al.,* 1987). Thus, in addition to requiring specific proteins for chylomicron formation and secretion, the addition of specific carbohydrate components to the glycoproteins on the surface of nascent lipoproteins is also essential for the secretion of chylomicrons from the enterocytes.

8. Phospholipids and the Formation and Secretion of Lipoproteins

Another ingredient for sustaining normal chylomicron formation and secretion is phospholipid. Although no disease state has been identified where abnormal chylomicron secretion has been linked to a deficiency of phospholipid, defects of lipid transport by the enterocytes seen under experimental conditions in animals with essential fatty acid deficiency (Clark *et al.,* 1973; Levy *et al.,* 1992a) may be based on abnormal membrane structure and function related to the formation of phospholipids of abnormal composition.

Of interest in this regard is the observation that lumenal phospholipid, particularly phosphatidylcholine (PC), has been identified as a factor that increases the amount of absorbed dietary lipid that is secreted into lymph as chylomicrons by the enterocytes as determined by studies in both the normal and the bile duct fistula rat (O'Doherty *et al.,* 1973; Clark, 1978; Tso *et al.,* 1981b; Mansbach *et al.,* 1985; Mansbach and Arnold, 1986). This effect is not due to a delay in the digestion and uptake of lumenal lipid, but is rather due to an unexplained action of PC on the intracellular events that occur during chylomicron synthesis and secretion (Tso *et al.,* 1981b). Lumenal PC, either biliary or dietary, can be readily hydrolyzed to lyso-PC, quickly absorbed, and reesterified back to PC (Nilsson, 1968) to meet the needs of the enterocytes for this lipid. Additionally, PC can be synthesized *de novo* by the enterocytes utilizing α-glycerol phosphate, fatty acid, and choline (Kennedy, 1961). The PC present on chylomicrons can be derived from lumenal PC or from *de novo* synthesis by the enterocytes (Mansbach, 1977). There is also evidence to show that lumenal PC has a negative effect on enterocyte *de novo* synthesis of PC (Mansbach, 1977).

Lumenal PC in some way causes more efficient processing of absorbed digested lipids that are being reesterified by the monoglyceride pathway, resulting in greater transport of dietary lipid into lymph as chylomicrons (Mansbach *et al.,* 1985). Lumenal PC does not have this effect, however, on transport into lymph of endogenous lipids that are being reesterified by the α-glycerol phosphate pathway. In subsequent studies, it was determined that this effect of lumenal PC was the result of producing an expansion of the chylomicron precursor pool rather than changing the fractional turnover rate of this pool (Mansbach and Arnold, 1986). Thus, lumenal PC has an important effect on the intracellular trafficking of absorbed free fatty acids and 2-monoacylglycerol so that these dietary lipids are efficiently transported into the body as chylomicrons via intestinal lymph.

One final component of the CM particle that contributes to the formation of the lipoprotein membrane is cholesterol. No disease state or experimental model is known to show that abnormal CM transport is the result of a specific defect in cholesterol absorption or of its synthesis by the enterocytes.

9. Regional Differences in Formation and Secretion of Chylomicrons

There are regional differences in various segments of the small intestine for CM formation and secretion (Sabesin *et al.,* 1977). The jejunum is most efficient in this regard. In contrast, when large amounts of dietary fat are absorbed by the ileum, triglyceride accumulates progressively in the distal enterocytes, with more and larger lipid droplets observed to be retained than observed by the jejunum when it is actively absorbing lipid. Six hours after discontinuing the intralumenal fat infusion, ileal enterocytes were still filled with large lipid droplets whereas jejunal enterocytes contained very little lipid. It is not presently known what factors contribute to the inability of the ileum to transport CM out of the enterocytes at a rate comparable to that of the jejunum. However, it was felt that the delay in CM transport by the ileum wasn't the result of a deficiency of the availability of apo B. Neither relative lack of PC nor a regional difference in Golgi structure and function was felt to be the likely explanation for these observations.

10. Inhibition of Intestinal Chylomicron Secretion by Hydrophobic Surfactants

An experimental model of inhibition of lipid absorption is of interest in regard to the role of PC for CM transport by the enterocytes. Studies on the effects of surface-active compounds demonstrated that hydrophobic surfactants inhibit lipid absorption (Bochenek and Rodgers, 1977). The surfactants used in this study were block copolymers of polyethylene oxide (hydrophilic component) and polypropylene oxide (hydrophobic component). It was observed that those compounds with a molecular weight of 2000 to 3000 containing 80% to 90% hydrophobic component were the most effective inhibitors of lipid absorption. When a relatively high dose of one of these surfactants, Pluronic L-81, was given to rats, maldigestion of triglyceride with steatorrhea was produced, indicating that the surfactant interfered with pancreatic lipase. When a relatively small dose was given, however, there was no delay in digestion, in mucosal uptake, or in reesterification of the absorbed lipid by the enterocytes. Rather, there was a block in the exit of the reesterified lipid in the enterocytes, which was obvious by chemical and morphologic analyses (Brunelle *et al.*, 1979). Lipid and sterol balance studies performed on swine showed that the surfactant given chronically at a small dose produced increased fecal losses of neutral fat and neutral sterols but not acidic sterols (Brunelle *et al.*, 1979). Results were confirmed using rabbits treated with a similar surfactant containing 70% hydrophobic component (Rodgers *et al.*, 1983). Studies using rats with a mesenteric lymphatic fistula for collection of CM transported into intestinal lymph by the enterocytes absorbing a lipid meal infused into the duodenum demonstrated that these surfactants effectively blocked the process of transport of absorbed, reesterified lipid into lymph (Tso *et al.*, 1980, 1981a, 1982). Large lipid droplets were noted to accumulate in the cytoplasm of the enterocytes by electron microscopy (Fig. 1), and the only CM particles recovered in lymph were small particles of VLDL size (Fig. 2), similar to the particles secreted by the enterocytes during fasting (Tso *et al.*, 1981a). The surfactant, Pluronic L-81, was noted to have a rapid onset of action and, once such treatment was terminated, the inhibition of CM transport was also rapidly reversed (Tso *et al.*, 1981a, 1982).

The effect of hydrophobic surfactants on CM transport was not the result of inhibition of apo B synthesis or of its secretion into intestinal lymph (Hayashi *et al.*, 1990; Bochenek *et al.*, 1990), nor did this agent affect the addition of terminal sugars to the apo B of nascent CM, a process that takes place in the Golgi apparatus just prior to the release of CM into the lymph (Bochenek *et al.*, 1990).

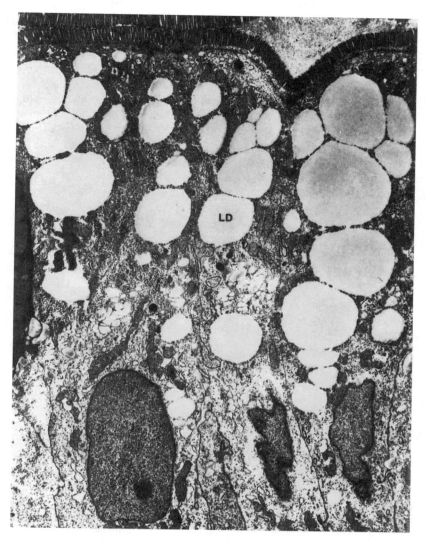

Figure 1. Electron micrograph of the intestinal epithelial cells following the infusion of lipid plus Pluronic L-81 (inhibitor of the formation of chylomicrons) for 4 hours. Please note the presence of large, non-membrane bound lipid droplets within the apical cytoplasm of intestinal epithelial cells (magnification: 8,000). (Reproduced with permission from Tso *et al.,* 1981a)

11. The Effect of Phospholipid on the Inhibitory Effects of Hydrophobic Surfactants on Chylomicron Secretion by Enterocytes

Other studies have indicated that there is a relationship between the hydrophobic surfactant and PC regarding chylomicron secretion. Thus, the inhibition of CM secretion into lymph by the surfactant can be reversed by supplying the lipid presented for absorption in the form of PC instead of the usual type of dietary lipid, triglyceride (Tso *et al.,* 1984). Sub-

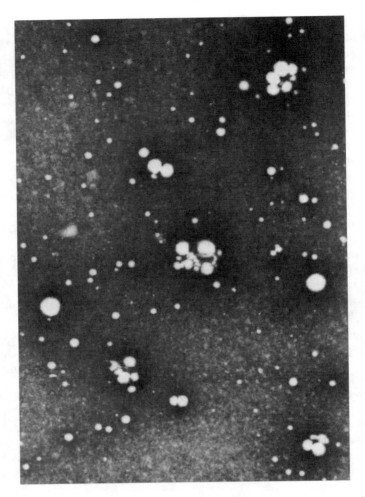

Figure 2. Electron micrograph of mesenteric lymph duct lymph obtained during a period when the rat was infused intraduodenally with triolein plus Pluronic L-81 (inhibitor of formation of chylomicrons). Lipoprotein particles visualized were small, measuring 200–800 angstroms. (magnification: 46,000). (Reproduced with permission from Tso *et al.*, 1981a)

sequent studies indicated that not all of the lipid presented for absorption had to be in the form of PC to reverse the inhibition of hydrophobic surfactant. A 50/50 mixture of triglyceride and PC infused into the duodenum was absorbed and was transported into lymph very efficiently despite the simultaneous treatment with Pluronic L-81 (Rodgers *et al.*, 1996). These data suggest that PC and Pluronic L-81 compete for positions on various membrane structures present in the enterocytes that are involved with CM formation and secretion. Both of these compounds are hydrophobic surfactants. It is well known that PC is present in many different membranes. That Pluronic L-81 and other similar hydrophobic surfactants might also accumulate at these locations appears likely.

The morphologic abnormalities observed in jejunal enterocytes exposed to hydrophobic surfactants while actively absorbing lipid are similar to that of the ileum presented with a large lipid load for absorption. Large lipid droplets accumulate in the cytoplasm in both cases. As discussed previously, as the reesterified lipid droplets with their apo B component leave the endoplasmic reticulum and enter the cytoplasm, small vesicles are formed. These nascent CM are later found in the Golgi apparatus for the final processing of CM prior to their secretion into lymph. One could postulate that PC has to be present in adequate amounts to form the proteophospholipid membranes felt to be required for transport of nascent chylomicrons from SER to Golgi (Kumar and Mansbach, 1997). A relative lack of PC in the ileal lumen for absorption into the ileal enterocytes may well explain the abnormality of ileal CM secretion. The presence of Pluronic L-81 in the jejunal enterocytes may displace PC from essential locations in the proteophospholipid membrane used to transport nascent chylomicrons to Golgi, causing large lipid droplets to accumulate in the cytoplasm.

12. Recovery from the Effects of Hydrophobic Surfactants

Exposure of the small bowel to hydrophobic surfactant does not cause permanent damage as CM transport resumes in a normal, if not an accelerated rate after treatment is terminated (Tso *et al.,* 1982). The triglyceride in the large lipid droplets that accumulates in the cytoplasm is not hydrolyzed and then reesterified prior to being incorporated into nascent lipoproteins for secretion as takes place in the liver, but instead is utilized without further processing to form nascent CM (Halpern *et al.,* 1988). Within 45 minutes after terminating treatment, small vesicles that are electron dense and apparently lipid filled appear in close proximity to the large lipid droplets (Fig. 3; Reger *et al.,* 1989). With more time the large lipid droplets become smaller as more vesicles are formed. Later these vesicles appear in Golgi. The Golgi membranes are difficult to detect during periods when the enterocytes are exposed to surfactant, but they become very prominent as the effect of the surfactant diminishes.

The reversal of the effect of hydrophobic surfactants in the enterocytes is likely related to a rapid decrease in the concentration of the agent in these cells due either to metabolism of the surfactant by drug-metabolizing enzymes in the enterocytes or by transport of the surfactant into the general circulation. Using radioactively labeled surfactant, it was observed that the agent was readily transported into the circulation of the rat and was promptly excreted in urine and in bile (Rodgers *et al.,* 1984). Metabolism of the surfactant by drug-metabolizing enzymes has not been studied.

13. Inhibition of Hepatic Secretion of Very Low Density Lipoproteins (VLDLs) by Hydrophobic Surfactants

Once hydrophobic surfactants are present in the body, it is no surprise to observe that they alter lipid metabolism. Using isolated perfused livers from rats pretreated with Pluronic L-81, Manowitz and colleagues observed decreased secretion of VLDLs into the perfusate compared to livers from rats fed a diet without the surfactant (Manowitz *et al.,* 1986). In another study, fasting rats were treated with Triton WR-1339 given intravenously. This

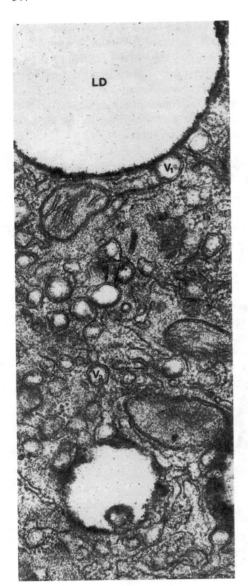

Figure 3. Four hours of intraduodenal infusion of lipid plus Pluronic L-81 followed by 45 min. of reversal. Osmium tetroxide fixation/tannic acid mordant. High-magnification view of cytoplasm to more adequately demonstrate the tannic acid mordanting at the surface of lipid droplets (LD). Membrane-limited vesicles (V_1) in juxtaposition to such droplets, containing electron-dense material, are sometimes seen opposed to the surface of lipid droplets (V_1); others (V_2) are scattered elsewhere in the apical cytosol. Such vesicles exhibit a typical unit membrane structure (arrows) not seen in osmium tetroxide-fixed cells alone (magnification: 40,000). (Reproduced with permission from Reger *et al.*, 1989)

caused a block of peripheral utilization of VLDLs secreted by the liver. In rats so treated, serum triglyceride, phospholipid, and cholesterol increased rapidly. In another group of fasting rats given Triton WR-1339 plus a small dose of Pluronic L-81 also administered intravenously, there was less of a rise observed for these serum lipids, supporting the conclusion that hydrophobic surfactants decrease hepatic VLDL secretion (Nutting and Tso, 1989).

14. The Effect of Hydrophobic Surfactant on the Development of Atherosclerosis

In rabbits given a high-fat, high-cholesterol diet—a model for inducing atherosclerosis—a relatively high dose of hydrophobic surfactant incorporated into the diet completely inhibited the development of hypercholesterolemia and the subsequent atherosclerosis, which was, indeed, observed in rabbits treated with this dietary program but without surfactant supplement (Rodgers *et al.*, 1983, 1987). Subsequent studies to determine whether the surfactant had an effect on the atherosclerotic process independent of its effects on circulating cholesterol were performed in rabbits treated with the atherogenic diet but with a dose of surfactant given as a dietary supplement that was too low to prevent the development of hypercholesterolemia. Under these conditions, not only was atherosclerosis not prevented, but, also, it appeared somewhat worse compared to that which was observed in rabbits given the atherogenic diet alone. Furthermore, VLDLs and LDLs were noted to be enriched in triglycerides, suggesting that small amounts of hydrophobic surfactants present in the circulation might alter the metabolism of triglyceride-rich lipoproteins, that is, CM and VLDL (Rodgers *et al.*, 1987). In a rabbit model of endogenous hypercholesterolemia produced by feeding a low-fat, cholesterol-free diet, a low dose of surfactant given as a dietary supplement did partially block the development of hypercholesterolemia. However, in this study, VLDLs, LDLs and HDLs from these animals were enriched with triglyceride, supporting the hypothesis that small amounts of these surfactants in the circulation cause abnormal metabolism of triglyceride-rich lipoproteins (Rodgers *et al.*, 1989).

15. Hydrophobic Surfactants Alter the Metabolism of Chylomicrons

A subsequent study in rats pretreated with Pluronic L-81 was performed in which CMs containing radioactively labeled triglyceride were injected into the jugular vein. Blood samples were obtained over the next 40 minutes to follow the disappearance of radioactively labeled triglycerides. The CMs used in this study were obtained from normal rats with a duodenal cannula for infusion of a lipid meal and a mesenteric lymph fistula for collection of the CMs formed and secreted into lymph by the normal rat intestine. Results showed a significant delay in clearance of the triglyceride of these CMs in rats pretreated with Pluronic L-81, indicating that the surfactant was interfering with the action of lipoprotein lipase (Rodgers *et al.*, 1993). Lipases such as lipoprotein lipase and pancreatic lipase are active at the lipid–aqueous interface and the quality of this interface affects the activity of these enzymes. Under certain conditions, hydrophobic surfactants have been shown to inhibit the activity of both of these enzymes, probably as a result of adversely affecting the quality of the interface.

Large doses of the surfactants probably inhibit the action of lipoprotein lipase. However, as so little triglyceride enters the circulation under these conditions as a result of the block of CM secretion by the intestine and VLDL secretion by the liver, the effect of the agent on lipoprotein lipase is not apparent.

16. Other Agents that Inhibit Lipid Absorption and Transport

Other agents have been studied for their effects on lipid absorption and transport into the body including ethanol, the calcium channel blocker TA-3090, and brefeldin. With acute ethanol treatment, Mansbach observed a delay in uptake of the products of triglyceride digestion, which was thought to be secondary to changes produced in the composition of the brush border membrane as a result of exposure to ethanol (Mansbach, 1983). Using human duodenal biopsy samples either incubated in the presence of a micellar solution of oleic acid plus ethanol or preincubated in this solution without ethanol and then organ cultured in a micellar-free medium containing ethanol, it was observed that ethanol decreased the synthesis of triglyceride from the absorbed oleate and also delayed secretion of triglyceride by the enterocytes (Zimmerman *et al.,* 1986). In studies utilizing the mesenteric lymph fistula rat infused intraduodenally with an emulsion containing oleate with and without ethanol, it was observed that rats exposed to ethanol transported less triglyceride into intestinal lymph (Hayashi *et al.,* 1992). In these studies, no attempt was made to determine whether the decreased transport of triglyceride by enterocytes exposed to ethanol was the result of a delay in absorption of oleate, a decrease in the rate of synthesis of triglyceride from absorbed oleate, a block in the secretion of CMs formed from the synthesized triglyceride, or a combination of all of these possible abnormalities. These acute studies on the effects of ethanol do not allow any conclusions to be made regarding whether or not chronic exposure to ethanol alters any of the steps involved with lipid absorption and transport from the intestine into circulation.

Levy *et al.* (1992b) conducted a study on the calcium channel blocker TA-3090 using mesenteric lymph fistula rats that were exposed to this agent either intraduodenally with a lipid meal or intraperitoneally prior to intraduodenal infusion of the lipid meal. In either case, TA-3090 caused a decrease in triglyceride secretion into intestinal lymph. Further studies using organ culture technique showed that the addition of this agent to the medium decreased triglyceride synthesis and secretion by enterocytes. It was the feeling of these investigators that intracellular calcium had a role in triglyceride synthesis and secretion by the enterocytes. These results supported earlier observations that calcium stimulates the secretion of triglyceride by hamster-everted jejunum (Strauss and Jacobs, 1981). Another effect of TA-3090 was a reduction of lymphatic flow, which has previously been shown to influence CM secretion into lymph (Tso *et al.,* 1985). Again, no studies were reported to indicate the effect of chronic treatment with this agent on the processes involved with lipid absorption and transport.

A study on brefeldin A, an agent that adversely effects Golgi formation and subsequent function, was performed in mesenteric lymph fistula rats given intraduodenal infusion of radioactively labeled triglyceride emulsion for 15 hours (Mansbach and Nevin, 1994). Between hours 4 to 7, brefeldin was included in the infusate. Output of radioactive triglyceride into lymph steadily increased with time until brefeldin was added. By the eighth hour (one hour after discontinuing brefeldin), triglyceride secretion into lymph was nearly completely blocked. The secretion of triglyceride into lymph was restored to normal 5 hours after treatment with brefeldin was terminated. Electron microscopy of treated enterocytes showed large lipid droplets plus numerous small, lipid-filled vesicles in the cytoplasm with essentially no Golgi observed. Golgi appeared to be collapsed into the endoplasmic reticulum. The morphology and function of the enterocytes quickly returned to normal within a short period after discontinuing brefeldin treatment. Although the hydrophobic surfactants discussed here only inhibit secretion of CM-sized lipoproteins and not VLDL-sized particles, brefeldin severely blocked secretion of all sized triglyceride-rich lipoproteins. This would be expected because without Golgi, essentially no triglyceride-rich lipoprotein particle can be secreted by the enterocyte no matter how small it is. No long-term studies were described using this agent.

17. Concluding Remarks

A number of experimental models are thus available to study various processes involved with lipid absorption and transport. Further studies using hydrophobic surfactants may help define the mechanism by which PC facilitates the synthesis and intracellular transport of nascent CM. Although these agents have subtle but definite adverse effects when given on a long-term basis, they are well tolerated when given to animals daily for months at a time in small to moderate doses. They can therefore be used for both acute and long-term studies on lipid absorption and metabolism.

References

Alexander, C. A., Hamilton, R. L., and Havel, R. L., 1976, Subcellular localization of B apoprotein of plasma lipoproteins in the rat liver, *J. Cell Biol.* **69:**241–263.

Bell, R. M., Ballas, L. M., and Coleman, R. A., 1981, Lipid topogenesis, *J. Lipid Res.* **22:**391–403.

Bochenek, W. J., and Rodgers, J. B., 1977, Effect of polyol detergents on cholesterol and triglyceride absorption: Hypolipidemic action of chronic administration of hydrophobic detergent, *Biochim. Biophys. Acta* **489:**503–506.

Bochenek, W. J., Weber, P., Slowinska, R., Tang, G., and Rodgers, J. B., 1990, Carbohydrate content of Apolipoprotein B-48 from rat chylomicrons of varying density, *Lipids* **25:**665–668.

Brunelle, C. W., Bochenek, W. J., Abraham, R., Kim, D. N., and Rodgers, J. B., 1979, Effect of hydrophobic detergent on lipid absorption in the rat and on lipid and sterol balance in the swine, *Dig. Dis. Sci.* **24:**718–725.

Cardell, R. R., Jr., Badenhausen, S., and Porter, K. R., 1967, Intestinal triglyceride absorption in the rat. An electron microscopic study, *J. Cell Biol.* **34:**123–155.

Chen, S. H., Habib, G., Yank, C. Y., Gu, G. W., Lee, B. R., Wang, S. A., Silberman, S. R., Cai, S. J., Deslypere, J. P., Rosseneu, M., Jr., Gotto, A. M., Li, W. H., and Chen, L., 1987, Apolipoprotein B-48 is the product of a messenger RNA with an organ-specific in-frame stop codon, *Science* **238:**363–366.

Clark, S. B., 1978, Chylomicron composition during duodenal triglyceride and lecithin infusion, *Am. J. Physiol.* **235** (Endocrinol. Metab. 2):E183–E190.

Clark, S. B., Ekkers, T. E., Singh, A., Balint, J. A., Holt, P. R., and Rodgers, J. B., Jr., 1973, Fat absorption in essential fatty acid deficiency: A model experimental approach to studies of the mechanism of fat malabsorption of unknown etiology, *J. Lipid Res.* **14:**581–588.

Davis, R. A., 1993, The endoplasmic reticulum is the site of lipoprotein assembly and regulation of secretion, in: *Subcellular Biochemistry, Endoplasmic Reticulum 21* (N. Borgese and J. R. Harris, eds.), Plenum Press, New York, pp.169–187.

Davis, R. A., Thrift, R. N., Wu, C. C., and Howell, K. E., 1990, Apolipoprotein B is both integrated into and translocated across the endoplasmic reticulum membrane, *J. Biol. Chem.* **265:**10005–10011.

Dobbins, W. O., 1970, An ultrastructural study of the intestinal mucosa in congenital B-lipoprotein deficiency with particular emphasis of intestinal absorptive cells, *Gastroenterol.* **50:**195–210.

Garcia, Z. C., Poksay, K. S., Bostrom, K., Johnson, D. F., Balestra, M. E., Shechter, I., and Innerarity, T. L., 1992, Characterization of apolipoprotein B mRNA editing from rabbit intestine, *Arterioscler. Thromb.* **12:**172–179.

Glickman, R. M., and Kirsch, K., 1973, Lymph chylomicron formation during the inhibition of protein synthesis: Studies of chylomicron apoproteins, *J. Clin. Invest.* **52**:2910–2920.

Glickman, R. M., Kirsch, K., and Isselbacher, K. J., 1972, Fat absorption during inhibition of protein synthesis: Studies of lymph chylomicrons, *J. Clin. Invest.* **51**:356–363.

Goldberg, D. E., and Kornfeld, S., 1983, Evidence for extensive subcellular organization of asparagine-linked oligosaccharide processing and lysosomal phosphorylation, *J. Biol. Chem.* **258**:3159–3165.

Halpern, J., Tso, P., and Mansbach, C. M., II, 1988, Mechanism of lipid mobilization by the small intestine after transport blockade, *J. Clin. Invest.* **82**:74–81.

Hayashi, H., Fujimoto, K., Cardell, J. A., Nutting, D. F., Bergstedt, S., and Tso, P., 1990, Fat feeding increases size, but not the number, of chylomicrons produced by small intestine, *Am. J. Physiol.* **259** (Gastroint. Liver Physiol. 22):G709–G719.

Hayashi, H., Nakata, K., Motohashi, Y., and Takano, T., 1992, Acute inhibition of lipid transport in rat intestinal lymph by ethanol administration, *Alcohol Alcoholism* **27**:627–632.

Huang, L. S., Inne, P. A., De, G. J., Cooper, M., Decklebaum, R. J., Kayden, H., and Breslow, J. L., 1990, Exclusion of linkage between the human apolipoprotein B gene and abetalipoproteinemia, *Am. J. Hum. Genet.* **46**:1141–1148.

Johnson, J. M., 1978, Esterification reactions in the intestinal mucosa and lipid absorption, in: *Disturbances in Lipid and Lipoprotein Metabolism* (J. M. Dietschy, A. M. Gotto, Jr., and J. A. Ontko, eds.), American Physiologic Society, Bethesda, MD, pp. 57–68.

Kennedy, E., 1961, Biosynthesis of complex lipids, *Federation Proc.* **20**:934–940.

Kumar, N. S., and Mansbach, C. M., 1997, Determinants of triacylglycerol transport from the endoplasmic reticulum to the Golgi in intestine, *Am. J. Physiol.* **273**(Gastrointest. Liver Physiol. 36):G18–G30.

Levy, E., Marcel, Y., Deckelbaum, R. J., Milne, R., Lepage, G., Seidman, E., Bendayan, M., and Roy, C. C., 1987, Intestinal apo B synthesis, lipids, and lipoproteins in chylomicron retention disease, *J. Lipid Res.* **28**:1263–1274.

Levy, E., Garofalo, C., Thibault, L., Dionne, S., Daoust, L., Lapage, G., and Roy, C. C., 1992a, Intralumenal and intracellular phases of fat absorption in essential fatty acid deficiency, *Am. J. Physiol.* **262** (Gastrointest. Liver Physiol. 25):G319–G326.

Levy, E., Smith, L., Dumont, L., Garceau, D., Garofalo, C., Thibault, L., and Seidman, E., 1992b, The effect of a new calcium channel blocker (TA-3090) on lipoprotein profile and intestinal lipid handling in rodents, *Proc. Soc. Expl. Biol. Med.* **199**:128–135.

Manowitz N. R., Tso, P., Drake, D. S., Frase, S., and Sabesin, S. M., 1986, Dietary supplementation with Pluronic L-81 modifies hepatic secretion of very low-density lipoproteins in the rat, *J. Lipid Res.* **26**:196–207.

Mansbach, C. M., II, 1977, The origin of chylomicron phosphatidylcholine in the rat, *J. Clin. Invest.* **60**:411–420.

Mansbach, C. M., II, 1983, Effect of ethanol on intestinal lipid absorption in the rat, *J. Lipid Res.* **24**:1310–1320.

Mansbach, C. M., II, and Arnold, A., 1986, Steady-state kinetic analysis of triacylglycerol delivery into mesenteric lymph, *Am. J. Physiol.* **251** (Gastrointest. Liver Physiol. 14):G263–G269.

Mansbach, C. M., II, and Nevin, P., 1994, Effect of brefeldin A on lymphatic triacylglycerol transport in the rat, *Am. J. Physiol.* **266** (Gastrointest. Liver Physiol. 29):G292–G302.

Mansbach, C. M., II, Arnold, A., and Cox, M. A., 1985, Factors influencing triacylglycerol delivery into mesenteric lymph, *Am. J. Physiol.* **249** (Gastrointest. Liver Physiol. 12):G642–G648.

Mattson, F. H., and Volpenhein, R. A., 1964, The digestion and absorption of triglycerides, *J. Biol. Chem.* **239**:2772–2777.

Nilsson, A., 1968, Intestinal absorption of lecithin and lysolecithin by lymph fistula rats, *Biochim. Biophys. Acta* **152**:379–390.

Nutting, D. F., and Tso, P., 1989, Hypolipidemic effect of intravenous Pluronic L-81 in fasted fats treated with Triton WR-1339: Possible inhibition of hepatic lipoprotein secretion, *Horm. Metabol. Res.* **21**:113–115.

Ockner, R. K., Hughes, F. B., and Isselbacher, K. J., 1964, Very low density lipoproteins in intestinal lymph: Role in triglyceride and cholesterol transport during fat absorption, *J. Clin. Invest.* **48**:2367–2373.

O'Doherty, P. J., Kakis, G., and Kuskis, A., 1973, Role of luminal lecithin in intestinal fat absorption, *Lipids* **8**:249–255.

Powell, L. M., Wallis, S. C., Pease, R. J., Edwards, Y. H., Knott, T. J., and Scott, J., 1987, A novel form of tissue-specific RNA producing produces apolipoprotein B in intestine, *Cell* **50**:831–840.

Reger, J. F., Frase, S., and Tso, P., 1989, Fine structure observations on rat jejunal epithelial cells during fat processing and resorption following L-81 exposure and reversal, *J. Submicrosc. Cytol. Pathol,* **21**:399–408.

Rodgers, J. B., Kyriakides, E. C., Kapuscinska, B., Peng, S. K., and Bochenek, W. J., 1983, Hydrophobic surfactant treatment prevents atherosclerosis in the rabbit, *J. Clin. Invest.* **71**:1490–1494.

Rodgers, J. B., Friday, S., and Bochenek, W. J., 1984, Absorption and excretion of the hydrophobic surfactant, 14C-Poloxalene 2930, in the rat, *Drug Metab. Disposition* **12**:631–634.

Rodgers, J. B., Slowinska, R., and Bochenek, W. J., 1987, Hydrophobic surfactant effects on aortic cholesterol accumulation and atherosclerosis in hypercholesterolemic rabbits, *Atherosclerosis* **64**:37–46.

Rodgers, J. B., Tang, G., and Bochenek, W. J., 1989, Hydrophobic surfactant inhibits hypercholesterolemia in pair-fed rabbits on a cholesterol-free, low fat diet, *Am. J. Med. Sci.* **296**:177–181.

Rodgers, J. B., Gray, L., and Tso, P., 1993, Treatment with hydrophobic surfactant inhibits chylomicron metabolism, *Clin. Res.* **41**:389A.

Rodgers, J. B., Beeler, D. A., and Tso, P., 1996, Relationship of phosphatidylcholine to hydrophobic surfactant on rat intestinal chylomicron secretion, *Experientia* **52**:671–676.

Sabesin, S. M., and Isselbacher, K. J., 1965, Protein synthesis inhibition: Mechanism for production of impaired fat absorption, *Science* **147**:1149–1151.

Sabesin, S. M., and Frase, S., 1977, Electron microscopic studies in the assembly, intracellular transport, and secretion of chylomicrons by rat intestine, *J. Lipid Res.* **18**:496–511.

Sabesin, S. M., Clark, S. B., and Holt, P. R., 1977, Ultrastructural features of regional differences in chylomicron secretion by rat intestine, *Expl. Molec. Pathol.* **26**:277–289.

Strauss, E. W., and Jacobs, J. S., 1981, Some factors affecting the lipoprotein secretory phase of fat absorption by intestine in vitro from golden hamster, *J. Lipid Res.* **22**:147–156.

Talmud, P. J., Lloyd, J. K., Muller, D. P. R., Collins, D. R., Scott, J., and Humphries, S., 1988, Genetic evidence that the apolipoprotein B gene is not involved in abetalipoproteinemia, *J. Clin. Invest.* **82**:1803–1806.

Thomson, A. B. R., and Dietschy, J. M., 1981, Intestinal lipid absorption: Major extracellular events, in: *Physiology of the Gastrointestinal Tract* (L. R. Johnson, ed), Raven Press, New York, pp. 1147–1220.

Tso, P., Balint, J. A., and Rodgers, J. B., 1980, Effect of hydrophobic surfactant (Pluronic L-81) on lymphatic lipid transport in the rat, *Am. J. Physiol.* **239** (Gastrointest. Liver Physiol. 2):G348–G353.

Tso, P., Balint, J. A., Bishop, M. B., and Rodgers, J. B., 1981a, Acute inhibition of intestinal lipid transport by Pluronic L-81 in the rat, *Am. J. Physiol.* **241** (Gastrointest. Liver Physiol. 4):G487–G497.

Tso, P., Kendrick, H., Balint, J. A., and Simmonds, W. J., 1981b, Role of biliary phosphatidylcholine in the absorption and transport of dietary triolein in the rat, *Gastroenterol.* **80**:60–65.

Tso, P., Buch, K. L., Balint, J. A., and Rodgers, J. B., 1982, Maximal lymphatic triglyceride transport rate from the rat small intestine, *Am. J. Physiol.* **242** (Gastrointest. Liver Physiol. 5):G408–G415.

Tso, P., Drake, D. S., Black, D. D., and Sabesin, S. M., 1984, Evidence for separate pathways of chylomicron and very low-density lipoprotein assembly and transport by rat small intestine, *Am. J. Physiol.* **247** (Gastrointest. Liver Physiol. 10):G599–G610.

Tso, P., Pitts, V., and Granger, D. N., 1985, Role of lymph flow in intestinal chylomicron transport, *Am. J. Physiol.* **249** (Gastrointest. Liver Physiol. 12):G21–G28.

Wetterau, J. R., and Zilversmit, D. B., 1984, A triglyceride and cholesteryl ester transfer protein associated with liver microsomes, *J. Biol. Chem.* **259**:10863–10866.

Wetterau, J. R., and Zilversmit, D. B., 1985, Purification and characterization of microsomal triglyceride and cholesteryl ester transfer protein from bovine liver microsomes, *Chem. Phys. Lipids* **38**:205–222.

Wetterau, J. R., and Zilversmit, D. B., 1986, Localization of intracellular triacylglycerol and cholesteryl ester transfer activity in rat tissue, *Biochim. Biophys. Acta* **875**:610–617.

Wetterau, J. R., Aggerbeck, L. A., Bouma, M. E., Eisenberg, C., Munck, A., Hermier, M., Schmitz, J., Gay, G., Rader, D. J., and Gregg, R. E., 1992, Absence of microsomal triglyceride transfer protein in individuals with abetalipoproteinemia, *Science* **258**:999–1001.

Yang, L. Y., and Kuksis, A., 1991, Apparent convergence (at 2-monoacylglycerol level) of phosphatidic acid and 2-monoacylglycerol pathways of synthesis of chylomicron triacylglycerols, *J. Lipid Res.* **32**:1173–1186.

Zimmerman, J., Gati, I., Eisenberg, S., and Rachmilewitz, D., 1986, Ethanol inhibits triglyceride synthesis and secretion by human small intestinal mucosa, *J. Lab. Clin. Med.* **107**:498–501.

CHAPTER 19

Intestinal Absorption and Metabolism of Peroxidized Lipids

TERRY S. LeGRAND and TAK YEE AW

1. Introduction

The intestine is unique among all fully differentiated organs in that it sits at the interface between the organism and its lumenal environment. In this regard, the intestine is a primary site of nutrient absorption and a critical defense barrier against dietary-derived mutagens, carcinogens, and oxidants. An important class of oxidants present in the human diet is lipid hydroperoxides, which are toxic products of oxidized polyunsaturated fats. Accumulation of peroxidized lipids in the gut lumen can contribute to impairment of mucosal metabolic pathways, enterocyte dysfunction independent of cell injury, and development of gut pathologies, such as cancer. Despite this recognition, and the implication of dietary peroxidized lipids in gut pathologies, we know little of the underlying mechanisms of the genesis of the disease processes or of the pathways of intestinal metabolism and lumenal disposition of dietary lipid hydroperoxides *in vivo*. This chapter summarizes our current understanding of the determinants of the intestinal absorption and metabolism of peroxidized lipids. In particular, we review the evidence supporting the pivotal role that GSH and NADPH play in the overall mucosal metabolism of toxic lipid hydroperoxides, and how reductant availability can be compromised under certain pathophysiological conditions, such as chronic hypoxia. The discussion is pertinent to understanding dietary lipid peroxides and GSH redox balance in intestinal physiology and pathophysiology and the significance of lumenal GSH in preserving the metabolic integrity of the intestinal epithelium.

TERRY S. LeGRAND • Department of Respiratory Care, The University of Texas Health Science Center, San Antonio, Texas 78245.   TAK YEE AW • Department of Molecular and Cellular Physiology, Louisiana State University Medical Center, Shreveport, Louisiana 71130-3932.
Intestinal Lipid Metabolism, edited by Charles M. Mansbach II *et al.,* Kluwer Academic/Plenum Publishers, 2001.

2. Peroxidized Lipids in Intestinal Pathology

2.1. Sources of Lumenal Lipid Hydroperoxides

A variety of natural mutagens and carcinogens, including peroxidized lipids, are found in the diet (Ames, 1983; Fink and Kritchevsky, 1981; Kinlen, 1983; Simic and Karel, 1980). These substances are capable of initiating degenerative processes via oxyradical production, and they promote digestive system disorders, such as intestinal inflammation and cancer (Grisham and Granger, 1988; Parks *et al.*, 1983). Lipid hydroperoxides are toxic products of peroxidized unsaturated fatty acids that are components of membrane lipids and of dietary fats (Girotti, 1985; Sevanian and Hochstein, 1985). Dietary intake of highly unsaturated fats represents a major source of lipid hydroperoxides in the intestinal lumen, and a daily intake of about 84 grams of dietary fat could be associated with an average hydroperoxide intake of 1.4 mmole/day (Wolff and Nourooz-Zadeh, 1996). Major dietary sources of hydroperoxides are lards, compound cooking fats, and fat intake associated with milk products, meat products, fish, and eggs (Wolff and Nourooz-Zadeh, 1996). Typical hydroperoxide contents have been measured in various tissues following experimental duodenal peroxidized lipid infusion, and totals of 0.4 µmole of hydroperoxide were found in intestinal mucosa, the liver, and the kidney, as estimated from the measured concentration of thiobarbituric acid reactive substances (TBARS) in the tissues (Aw *et al.*, 1992).

2.2. Lipid Hydroperoxides and Intestinal Malignant Transformation

Early studies provide evidence for cytotoxicity associated with excessive consumption of lipid hydroperoxides *in vivo*. In this regard, several investigators have demonstrated that the toxicity of dietary polyunsaturated oils correlates directly with their peroxide content (Andrews *et al.*, 1960; Kaneda *et al.*, 1955; Kimura *et al.*, 1984). It is notable that high intake of dietary polyunsaturated fatty acids can contribute to the lumenal accumulation of lipid hydroperoxides, even though such intake is unlikely to achieve cytotoxic levels of lipid hydroperoxides with human consumption of unsaturated fats. Nevertheless, ensuing peroxide-mediated lipid peroxidation and oxyradical generation can exert a significant impact on intestinal integrity by induction of tissue oxidative stress and redox imbalance. Reports in the literature link high intake of dietary fats to increased incidence of colon cancer (Carroll and Khor, 1975; Correa *et al.*, 1982; Reddy, 1983), consistent with the paradigm that dietary fat is a risk factor for the development of gut malignant transformation in humans. A crucial point that has not been appreciated heretofore is the fact that consumption of highly unsaturated fats can be associated with substantial intake of lipid hydroperoxides with potential deleterious effects on gastrointestinal homeostasis. In support of this suggestion, several laboratories have shown that local administration of oxygenated derivatives of unsaturated fatty acids promotes tumorigenesis in the colon (Bull *et al.*, 1984; Fischer *et al.*, 1980; Verma *et al.*, 1980). Significantly, hydroperoxy and hydroxy fatty acids elicited proliferative responses in colonic mucosa in rats that are correlated with stimulation of DNA synthesis and induction of ornithine decarboxylase (Bull *et al.*, 1984). In other studies, Hara *et al.* (1996) found that rats given peroxidized ethyl linoleate (peroxide value of 1400 mEq/kg lipid) developed mucosal hypertrophy of the colon. Taken together, these findings demonstrate the tumorigenic potential of oxidized lipids by their deleterious impact on normal intestinal cell turnover processes.

3. Intestinal Metabolism of Lipid Hydroperoxides

3.1. Intestinal GSH and Lipid Hydroperoxide Absorption and Disposition

Glutathione (GSH), together with several naturally occurring antioxidants, such as α-tocopherol (vitamin E), β-carotene (vitamin A), and ascorbic acid (vitamin C), as well as cellular antioxidant enzymes, such as superoxide dismutase, catalase, and GSH peroxidase, constitute a broad cellular mechanism that functions to detoxify reactive oxygen metabolites of endogenous or exogenous origin. The tripeptide, GSH (γ-glu-cys-gly) is the major free thiol in tissues (Kaplowitz *et al.,* 1985; Kauffman *et al.,* 1977) and possesses a reactive thiol (SH) group and a γ-glutamyl bond that is resistant to proteolytic cleavage (Kaplowitz *et al.,* 1985). As such, GSH is an excellent reductant in peroxide catabolism, and exogenously supplied GSH has been shown to afford protection against peroxide-induced cell damage in several cell types, such as enterocytes (Lash *et al.,* 1986), type II alveolar cells (Hagen *et al.,* 1986), renal proximal tubule cells (Hagen *et al.,* 1988), and endothelial cells (Andreoli *et al.,* 1986). Animal studies have shown that GSH is taken up intact from the lumenal surface of the small intestine *in vivo* to supplement the intracellular GSH pool (Hagen and Jones, 1987; Hagen *et al.,* 1990). These findings reinforce the idea that lumenal uptake of GSH may play a part in maintaining mucosal thiol balance that is crucial for normal intestinal function, including peroxide elimination. Oral administration of GSH was shown to prevent severe degeneration of jejunal and colonic epithelial cells induced by chronic GSH deficiency (Martensson *et al.,* 1990). This finding lends further support to a role for lumenal GSH in the maintenance of the functional integrity of the intestinal epithelium.

Because fats, many of which can be oxidized, comprise an appreciable percentage of the American diet (Kinlen, 1983; Fink and Kritchevsky, 1981; Simic and Karel, 1980), the GSH-dependent system in the gastrointestinal tract is anticipated to have a major role in removal of lumenal peroxidized lipids. The data to date implicate GSH as a key determinant of metabolism of toxic hydroperoxides by the intestine (Kowalski *et al.,* 1990; Aw and Williams, 1992; Aw *et al.,* 1992). Typically, GSH is supplied to the intestine by the diet and by biliary output from the liver. From 50% to 60% of GSH output from the liver goes to bile (Aw, 1994), is secreted into the duodenum, and is taken up intact by intestinal cells (Aw and Williams, 1992; Hagen and Jones, 1987; Hagen *et al.,* 1990; Lash *et al.,* 1986). Intestinal catabolism of lipid hydroperoxides has been shown to be an intracellular event (Kowalski *et al.,* 1990). The kinetics and extent of intracellular hydroperoxide metabolism largely depend on the efficiency of absorption of peroxides from the lumen (Aw *et al.,* 1992) and on a ready source of cellular reductant (GSH and NADPH) for peroxide reduction (Aw and Rhoads, 1994; Aw and Williams, 1992; Kowalski *et al.,* 1990).

Recent studies from our laboratory have provided important new insights into the driving force for lipid peroxide absorption and removal by the small intestine. In these studies, we have established a conscious lymph fistula rat model to delineate the role of GSH in intestinal disposition of lumenal lipid hydroperoxides *in vivo* (Aw *et al.,* 1992). The conceptual working hypothesis of the model is illustrated in Fig. 1. The model proposes that at a given dose of infused lipid hydroperoxide (represented by ROOH in Fig. 1), the amount of ROOH recovered from the intestinal lumen and lymph is dependent on the mucosal GSH status. In the GSH-sufficient state, increased intracellular peroxide metabolism drives lumenal ROOH absorption, which results in decreased lumenal retention. Moreover, en-

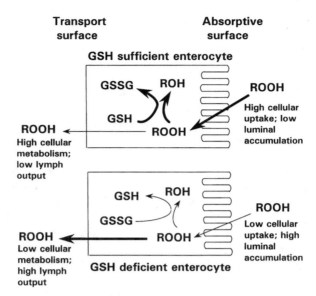

Figure 1. Conceptual working model for lipid hydroperoxide absorption and lymphatic transport in GSH-sufficient and GSH-deficient intestinal cells. Abbreviations are: ROOH, ROH: lipid hydroperoxide and hydroxide, respectively; GSH, GSSG: reduced and oxidized glutathione, respectively, Redrawn from Aw (1997).

hanced cellular metabolism decreases peroxide output into intestinal lymph (upper panel, Fig. 1). Conversely, in the case of intestinal GSH deficiency, decreased cellular peroxide catabolism is associated with reduced lumenal ROOH absorption and hence, increased lumenal ROOH retention. Decreased cellular metabolism promotes ROOH transport into lymph (lower panel, Fig. 1). Thus, it appears that the predominant driving force for elimination of lipid hydroperoxides from the intestinal lumen is largely derived from the metabolic capacity of the enterocytes to catabolize peroxides intracellularly. Conceptually, this mechanism for peroxide degradation represents a form of "metabolic trapping" that is analogous to regulation of biochemical reactions, whereby products are metabolically trapped and removed to ensure that the process continually favors substrate catabolism and product formation. It can be readily appreciated that the presence of GSH-dependent metabolic degradation of lipid peroxides offers an efficient model for rapid disposal of peroxidized lipids in the small intestine.

The utility of the lymph fistula rat model is illustrated experimentally by the finding that simultaneous quantification of hydroperoxide recovery in intestinal lumen and lymph provide reasonable estimates of the intestine's capacity for lipid hydroperoxide metabolism *in vivo* (Aw *et al.*, 1992). Figure 2 shows the profiles from experimentally determined data for hydroperoxide absorption and transport into lymph at varying levels of mucosal GSH. Using a variety of GSH inhibitors, for example, buthionine sulfoximine (BSO) to inhibit GSH synthesis, or GSH-depleting agents to lower GSH (Aw *et al.*, 1992), we consistently obtain higher lumenal and lymph recovery of infused lipid hydroperoxide (Fig. 2) that tracks with lower mucosal GSH concentrations. These peroxide recovery profiles are re-

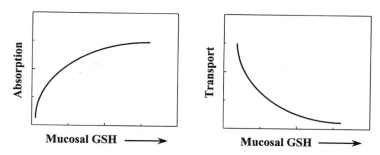

Figure 2. Intestinal lipid hydroperoxide absorption and lymphatic transport is determined by mucosal GSH. Increases in mucosal GSH promote intracellular metabolism of lipid hydroperoxides, which drives lumenal uptake of peroxide and decreases peroxide output into lymph. Profiles of lipid peroxide absorption and transport are based on experimentally determined data from Aw *et al.* (1992).

flective of an overall decreased absorption and metabolism of lumenal lipid hydroperoxides in association with reduced GSH levels (Fig. 2). Thus, the conclusion from these studies is that maintenance of mucosal GSH status is important for the quantitative removal of lumenal peroxidized lipids by the small intestine.

In additional studies, exogenous lumenal GSH supply in lymph fistula rats is shown to be an important source of reductant for mucosal metabolism of dietary peroxidized lipids during acute exposure of the intestine to a physiological lipid peroxide load (Aw and Williams, 1992). Supplementation of GSH-deficient lymph fistula rats (pretreated with BSO) with physiological concentrations of GSH ($100\mu M - 1$ mM) increases mucosal GSH while decreasing lumenal and lymph peroxide contents (Fig. 3). Such findings indicate that mucosal GSH levels can be effectively restored by exogenous GSH supplementation to support intracellular peroxide catabolism and attenuate peroxide transport into lymph. These results agree with previous studies in isolated intestinal everted sacs (Kowalski *et al.,* 1990).

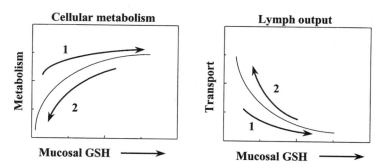

Figure 3. Exogenous GSH enhances intestinal metabolism of lipid hydroperoxides and attenuates peroxide output into lymph. 1) Exogenous GSH supplementation bolsters mucosal GSH concentration, enhances intracellular peroxide metabolism, and reduces lymphatic peroxide transport; 2) Blockade of intestinal uptake of GSH from gut lumen decreases mucosal GSH concentration, impairs intracellular peroxide metabolism, and enhances lymphatic peroxide output. Profiles are based on experimentally determined data from Aw and Williams (1992).

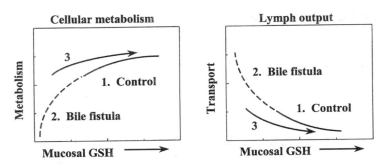

Figure 4. Diversion of biliary GSH decreases intestinal metabolism of lipid hydroperoxides and increases peroxide transport into lymph. 1) Lipid hydroperoxide metabolism in GSH-sufficient state is high and lymph peroxide output is low; 2) Diversion of biliary GSH in bile fistula animals decreases mucosal GSH and compromises cellular peroxide metabolism, which enhances lymphatic transport; 3) Exogenous GSH supply in bile fistula animals restores intracellular peroxide metabolism and decreases peroxide output into lymph. Profiles of metabolism and transport are based on experimentally determined data from Aw (1994).

Another significant finding is that biliary GSH is a major contributor to the endogenous lumenal GSH pool (Aw, 1994). Diversion of biliary GSH via a bile cannula leads to lipid hydroperoxide recovery from lumen and lymph that are substantially higher than controls (bile fistula, Fig. 4), consistent with decreased intestinal hydroperoxide metabolism in the bile fistula and lymph fistula animals. Exogenous GSH supplementation restores the intestine's capacity to metabolize lumenal peroxides in these animals (Fig. 4), indicating that GSH in bile is an important determinant of peroxide elimination by rat small intestine. It is not known to what extent biliary GSH contributes quantitatively to intestinal elimination of peroxidized lipids or the extent to which lumenal GSH preserves intestinal integrity with prolonged exposure of the gut to lipid peroxides. Given that lumenal GSH varies depending on biliary output, and that the human diet varies considerably in GSH levels (Jones *et al.*, 1992; Wierzbicka *et al.*, 1989) and in lipid composition and content, the previous considerations will have important physiological and pathophysiological implications.

3.2. GSH Redox Cycle Function and NADPH Supply in Intestinal Hydroperoxide Metabolism

The role of GSH in the metabolism of lipid hydroperoxides is inextricably tied to its function as a reductant in the GSH redox cycle, which is a key protective mechanism in the intestine against lipid peroxide challenge. In this system, GSH peroxidase reduces toxic peroxides at the expense of GSH, with concomitant production of GSSG (Fig. 5). GSH is regenerated from GSSG by GSSG reductase, utilizing reduced nicotinamide adenine dinucleotide phosphate (NADPH) as a reductant. The maintenance of GSH availability to support function of the GSH redox cycle ultimately depends on NADPH supply and is functionally coupled to the pentose phosphate pathway (Fig. 5). During peroxide challenge, the reduction of GSSG is critical to continued function of the GSH redox cycle. Typically,

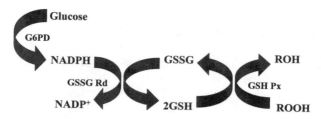

Figure 5. Glutathione redox cycle in hydroperoxide metabolism. G6PD: glucose 6-phosphate dehydrogenase; GSSG Rd: glutathione disulfide reductase; GSH Px: glutathione peroxidase. Redrawn from Aw and Rhoads (1994).

increased cellular GSSG results in activation of NADPH supply to maintain cell GSH homeostasis (Brigelius, 1985). NADPH is produced by a number of mechanisms, including mitochondrial NADH transhydrogenation (Weigl and Sies, 1977), NADP$^+$ specific dehydrogenases (Kauffman *et al.,* 1977), and the pentose phosphate pathway (Eggleston and Krebs, 1974). Although cellular NADPH homeostasis can be maintained by several intracellular sources, the major contributor to the total cellular NADPH pool is the pentose phosphate pathway (Eggleston and Krebs, 1974). This shunt is regulated by glucose 6-phosphate dehydrogenase activity and by the supply of glucose (Bousignone and Flora, 1972). Given the relatively low GSH synthetic rate in enterocytes (Williams and Aw, 1994) as compared to hepatocytes (Shan *et al.,* 1990), GSSG reduction may represent the major pathway for maintaining constant cell GSH in the intestine under high rates of lipid hydroperoxide metabolism. Consequently, an enhanced demand for GSSG reduction due to exaggerated GSH oxidation could exceed the availability of NADPH (Aw and Rhoads, 1994). The resultant depletion of the GSH pool would result in cessation of further peroxide degradation and would compromise intestinal cell integrity.

The endogenous rate of NADPH supply in enterocytes is low, but it is markedly stimulated by exogenous glucose administration (Aw and Rhoads, 1994). By contrast, the rate of NADPH supply in liver cells is five- to tenfold higher and appears to be minimally affected by exogenous glucose (Tribble and Jones, 1990). In enterocytes, glucose-stimulated NADPH production is blocked by 6-amino nicotinamide, an inhibitor of pentose phosphate shunt activity, indicating that the major source of enterocyte NADPH is derived from the function of this pathway. Furthermore, when NADPH supply and utilization are inhibited, glucose stimulation of peroxide (*tert*-butylhydroperoxide) detoxication does not occur (Aw and Rhoads, 1994), consistent with glucose enhancement of cellular NADPH availability to support GSH redox cycle function. Collectively, the findings show that intestinal metabolism of hydroperoxides is subject to regulation by glucose availability via enhanced reductant (NADPH) supply.

In liver cells, metabolism of organic hydroperoxides, such as *tert*-butylhydroperoxide (*t*BH), is efficient and appears to be independent of exogenous glucose supply. This is an interesting contrast to the results in intestinal cells wherein the endogenous rate of *t*BH metabolism in enterocytes is low in the absence of substrates (Aw and Rhoads, 1994) but increases markedly with exogenous glucose (Aw and Rhoads, 1994). Mechanistically, the glucose effect is mediated through stimulation of the pentose phosphate shunt to increase

NADPH supply for GSSG reduction (Aw and Rhoads, 1994). Metabolism of tBH in enterocytes from 24hr-fasted rats occurs at only one third the rate of cells from fed animals (Aw and Rhoads, 1994), thus providing further support for a role for glucose in intestinal peroxide metabolism. In other studies in the lymph fistula rat, the removal of glucose from the lipid peroxide infusate in GSH-sufficient animals results in elevated lumenal and lymph hydroperoxides, findings that are similar to those of the GSH-deficient state (Aw and Williams, 1992; Aw *et al.*, 1992). Despite an initial high mucosal GSH concentration, this reductant pool can readily be depleted, if it is not regenerated to keep pace with constant peroxide challenge. Thus, it appears that a constant glucose supply is critical to maintain steady state production of NADPH by the pentose phosphate shunt to support continued function of the GSH redox cycle to sustain intestinal peroxide elimination. The responsiveness of the gut to glucose suggests that nutrient availability is an important contributing factor in the catabolism of lumenal hydroperoxides by the small intestine. This raises the interesting possibility of regulation of intestinal lipid hydroperoxide disposition by lumenal and/or plasma glucose concentration. The organ-specific difference in quantitative NADPH supply and reliance on glucose for peroxide metabolism between the liver and small intestine underlie a greater sensitivity of the small intestine to lipid-hydroperoxide-induced injury.

Experimentally, the NADPH supply rate in cells can be quantified using the diamide-induced GSH redox cycle model. Diamide is a thiol oxidant that preferentially reacts with cellular GSH to produce GSSG at a constant rate (Aw and Rhoads, 1994; Tribble and Jones, 1990). Based on the redox cycle scheme illustrated in Fig. 5, the rate at which GSSG is reduced will be a direct measure of the cellular NADPH supply rate. Thus, in the presence of steady state infusion of diamide, constant GSH will only be maintained if the rate of GSH oxidation is matched by the rate of GSSG reduction by NADPH. Consequently, at the point where GSH oxidation exceeds GSSG reduction, maximal NADPH supply is achieved and GSH level falls. Experimentally, this "break point" at which cell GSH falls represents the critical infusion rate of diamide, which equals the rate of NADPH supply. Using this approach, our laboratory (Aw and Rhoads, 1994) and others (Tribble and Jones, 1990) have arrived at estimates of NADPH supply rates of 0.5 to 1.0 nmole·min^{-1} per 10^6 cells and 5 to 8 nmole · min^{-1} per 10^6 cells in enterocytes and in hepatocyes, respectively, in the presence of glucose.

Although it appears that GSH and NADPH supply are major players in intestinal handling of peroxidized lipids, the efficiency of the GSH redox cycle is likely to depend on the optimal function of the redox enzymes, GSH Px and GSSG Rd. Interestingly, studies from our laboratory and from others indicate that intestinal GSH Px and GSSG Rd functions are minimally modified under different physiological and pathophysiological states such as chronic O_2 deficiency (LeGrand and Aw, 1996), diabetes (Iwakiri *et al.*, 1995), and peroxide challenge (Reddy and Tappel, 1974; Vilas *et al.*, 1976). Hence, it appears that regulation of enzyme expression plays a relatively minor role in the overall catabolism of lipid hydroperoxides in the intestine. Notably, a unique isoform of GSH Px has been described in the small intestine, designated GSHPX-GI (Chu *et al.*, 1993). Although a physiological function has not been ascribed for this intestinal GSH peroxidase, the enzyme exhibits catalytic characteristics like the classical GSH Px (Chu *et al.*, 1993), which suggests a role in peroxide metabolism. However, the quantitative contribution of GSHPX-GI to intestinal peroxide elimination may be limited given the fact that GSHPX-GI has a catalytic rate one-tenth of that of GSH Px.

3.3. Compromised Lipid Hydroperoxide Catabolism and Intestinal Oxidative Stress

The responsiveness of the GSH redox cycle to substrate availability implies that lipid hydroperoxide elimination can be easily compromised under conditions of limited NADPH and GSH sources, with a resultant shift of cellular thiol redox status to a more oxidized state (Brigelius, 1985). In fact, a decrease in the GSH:GSSG ratio in tissues or the efflux of GSSG from cells are valid indicators of the presence of oxidative stress (Sies, 1985). The loss of thiol disulfide balance and the ensuing oxidation of protein thiols (Brigelius, 1985) will compromise functional integrity of the intestine. Importantly, ongoing oxidative stress can result in irreversible loss of tissue GSH, because GSSG is released from many cells during enhanced hydroperoxide catabolism (Jaeschke, 1990a, 1990b; Masuda *et al.*, 1993). During oxidative stress induced by Fe^{2+}/EDTA, Mizoguchi and associates (1994) observed a significant loss of GSH from isolated intestinal segments, which was reflected in a ten-fold increase in GSSG in the incubation medium. It was estimated that as much as 30% of the total GSH was lost from the cells. Because GSSG is not typically reduced extracellularly, the major organs do not take up GSSG from the blood (Uhlig and Wendel, 1992), the extracellular export of GSSG represents a significant mechanism for substantial quantitative loss of total cell GSH equivalents. Importantly, the inability of tissues to recover this GSSG pool translates to deleterious changes in cellular redox status. Because redox balance is pivotal for function of many cellular metabolic and transport pathways (Brigelius, 1985), the loss of thiol/disulfide integrity will severely impair tissue homeostasis.

4. Determinants of Intestinal Metabolism of Lipid Hydroperoxides during Chronic Hypoxia

4.1. Chronic Hypoxia and Effects on Intestinal Transport Functions

Chronic oxygen deficiency commonly occurs in disease states such as obstructive lung disease and cardiovascular dysfunction (Burrows *et al.*, 1972; Thurlbeck *et al.*, 1970). Adequate oxygen supply is required for the intestine and other tissues to function optimally (Aw *et al.*, 1987; Jones, 1981; Jones and Mason, 1978). Numerous studies have elucidated the effect of long-term hypoxia on organ function. Renal function is impaired by chronic oxygen deficiency (Jones *et al.*, 1995), and by nutrient (Lifshitz *et al.*, 1986) and drug absorption (Aw *et al.*, 1991) are altered. Hypoxia is also associated with decreased gastric emptying, which could impact pharmacodynamics in drug therapy regimes (Van Liere and Stickney, 1963). Other studies have shown that GSH transport by intestinal cells is inhibited by chronic hypoxia, and the rate of peroxidized methyl linoleate uptake is decreased in hypoxic compared to normoxic intestine (Bai and Jones, 1996). Our recent studies have shown that chronic hypoxia promotes intestinal oxidative stress, induces a state of mucosal metabolic instability, and compromises GSH-dependent detoxication in the intestine (LeGrand and Aw, 1996, 1997). Given the effect of hypoxia on gastrointestinal absorptive function and the important role that GSH plays in intestinal peroxide metabolism (Aw and Rhoads, 1994), disposition of lipid peroxides will be altered in the hypoxic intestine.

Recent studies from our laboratory have provided novel insights into the effects of long-term hypoxia on hydroperoxide absorption in the small intestine. For our studies, we have established a chronically hypoxic rat model wherein a normobaric hypoxic enclosure previously utilized by Aw and associates (1991) provides a relatively simple but effective mode of induction of hypoxia. A number of models have been employed to study intestinal hydroperoxide disposition under *in vitro* and *in vivo* conditions. Ideally, the lymph fistula rat model should provide quantitative measures of the absorption and metabolism of per-oxidized lipids in the hypoxic intestine. Unfortunately, the high mortality of hypoxic rat during surgery severely hampered the establishment of this conscious *in vivo* animal mod-el to directly monitor absorption and steady state lymphatic transport during constant in-traduodenal infusion of a given lipid hydroperoxide load. An alternate experimental mod-el of considerable utility for addressing the issue of intestinal disposition of lipid peroxides is that of the everted intestinal sac. This everted intestinal sac model was first described in 1954 by Wilson and Wiseman. It is a convenient *in vitro* method of quantifying transport of nutrients that we have successfully adapted for studying peroxide handling by the intesti-nal mucosa. Previous studies from different laboratories have carefully characterized the model, particularly in terms of adequate tissue oxygenation. Wilson and Wiseman (1954) have verified that lack of oxygenation of the intestinal segment can be overcome by dis-tention of the sac with an ample volume of buffer. Distention of the wall decreases wall thickness, which compensates for the additional serosal layers through which substances must be transported (Pritchard and Porteous, 1977; Wilson and Wiseman, 1954). Minimiz-ing excitement of animals prior to surgical removal of intestinal segments minimizes cate-cholamine release, which would impair mucosal blood flow and oxygen supply due to vil-lus vasoconstriction (Pritchard and Porteous, 1977). Results from these early studies have documented that active as well as passive transport processes can be quantified (Wilson and Wiseman, 1954).

4.2. Transport and Metabolism of Lipid Hydroperoxides During Chronic Hypoxia

Conflicting reports have been cited in the literature about the impact of chronic hy-poxia on absorption and transport of non-oxidized lipids, with some investigators showing no change in digestibility and utilization of dietary fat (Sridharan *et al.*, 1982), and others reporting decreases in fat absorption (Boyer and Blume, 1984). Under normal conditions, non-oxidized fatty acids are absorbed and then are partitioned between portal venous blood and lymph, dependent primarily on the chain length of the fatty acid. Oxidatively modified fatty acid hydroperoxides, which tended to be less lipophilic, could conceivably gain ac-cess to the systemic circulation directly via portal blood, in addition to transport as triglyc-erides via lymph. Indeed, our studies have shown that portal transport is an important path-way for oxidized ω3 fatty acids, particularly under GSH-deficient conditions (Aw *et al.*, 1993).

Speculation about absorption and lymphatic transport of fatty acid hydroperoxides have resulted in conflicting views as to whether lumenal peroxidized lipids are absorbed and processed by the intestine (Andrews *et al.*, 1960; Bergan and Draper, 1970; Bunyan *et al.*, 1968; Nagatsugawa and Kaneda, 1983). Several investigators were unable to measure lipid hydroperoxides in lymph (Andrews *et al.*, 1960; Bergan and Draper, 1970; Glavind, 1970), whereas others found appreciable amounts in both lymph (Nagatsugawa and Kane-da, 1983) and in tissues (Bunyan *et al.*, 1968; Kaneda *et al.*, 1955). The apparent discrep-ancies in these studies may be due to differences in mucosal GSH status under the various experimental conditions. From the discussion in previous sections, it is apparent that factors that contribute to alterations in intestinal GSH concentrations or to the function of the GSH redox cycle (Wendel, 1980; Fig. 1) are expected to significantly impact metabolic processing of dietary peroxidized lipids. One such pathophysiological factor could be chronic oxygen deficiency.

Chronic oxygen lack induces a variety of metabolic aberrations in the intestine (LeGrand and Aw, 1997). For example, the initial rate of hydroperoxide metabolism by chronically hypoxic enterocytes is greatly exaggerated compared to that of normoxic cells. In addition, exogenously supplied glucose significantly increases hydroperoxide metabolism in normoxic cells, but it has little effect on augmentation of peroxide elimination in hypoxic cells. Wide swings in mitochondrial oxygen consumption during substrate (glucose) or oxidant (*t*BH) challenge imply that hypoxic enterocytes exhibit loss of mitochondrial regulation and inherent metabolic instability, appearing to operate in a "hyperdynamic" state. In this state they continuously metabolize, not optimally, but maximally. It has been reported (Cerra, 1987) that patients with an unresolved focus of stress, such as burn, trauma, or sepsis, exhibit a prolonged hypermetabolic response, predisposing them to development of progressive metabolic dysregulation (Stoner, 1986). Similarly, prolonged oxygen deficiency appears to impose an ongoing stress on tissues, including the gastrointestinal tract. In this state, metabolism may be pushed to increasingly higher levels by a systemic insult. Continually elevated metabolism in the intestine could lead to failure of intestinal function in the face of a secondary challenge, such as exposure to lumenal lipid hydroperoxides.

In recent studies we found that even before exposure to peroxidized lipids, intestinal tissue from hypoxic rats exhibits significantly higher endogenous peroxide production than normoxic intestine, consistent with a generalized hypoxia-induced oxidative stress in the small intestine (LeGrand and Aw, 1996, 1997, 1998). Using the everted intestinal sac model, we have defined the profiles of hydroperoxide transport as illustrated in Fig. 6. Lipid peroxide transport into the lymphatic compartment by the normoxic gut segment is minimal (LeGrand and Aw, 1998) in accordance with our proposed hypothesis (Fig. 1), and it is consistent with the data obtained in lymph fistula animals (Fig. 2). In contrast, ablumenal peroxide recovery in the hypoxic intestine is significantly elevated (Fig. 6), which suggests

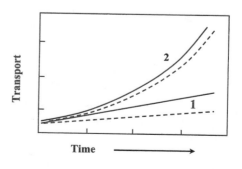

Figure 6. Chronic hypoxia increases lipid hydroperoxide transport over time. Chronic hypoxia attenuates intracellular metabolism of lipid hydroperoxides and increases peroxide transport into the ablumenal compartment of everted sacs (analogous to transport into lymph). 1) Lipid hydroperoxide transport in normoxic intestine is low, and is attenuated in the presence of exogenous glucose or GSH (dashed line); 2) Lipid hydroperoxide transport in hypoxic intestine is similar to that in normoxic intestine initially, then increases rapidly over time. Unlike normoxic intestine, however, transport is not attenuated in the presence of exogenous glucose or GSH (dashed line). Profiles of transport are based on experimentally determined data from LeGrand and Aw (1998).

an overall impairment of intracellular catabolism of lipid peroxides by the hypoxic gut. Moreover, whereas transport of peroxide is attenuated in the presence of exogenous glucose and GSH under normoxic conditions, these substrates have no effect on ablumenal peroxide recovery in hypoxic intestine (Fig. 6). These findings are consistent not only with a lack of an ability of the hypoxic intestine to utilize exogenous substrates to boost intracellular reductant supply for peroxide metabolism, but with an overall impairment of the metabolic function of the GSH redox cycle *per se* in peroxide detoxication. Thus, as with the normoxic intestine, GSH-dependent metabolism is the major determinant for peroxide elimination by the hypoxic intestine. Because the GSH status is compromised in the hypoxic intestine, development of intestinal oxidative stress can be exacerbated by lumenal peroxide challenge. The observations that the hypoxic intestine has an elevated endogenous level of lipid hydroperoxide in the absence of oxidant challenge (LeGrand and Aw, 1998) as well as a decrease in tissue GSH/GSSG status (LeGrand and Aw, 1996) are consistent with our suggestion.

In view of the fact that GSH uptake is impaired in the hypoxic intestine (Bai and Jones, 1996), it is not surprising that intestinal GSH-dependent metabolism of lipid peroxides is compromised in the chronically hypoxic state. Furthermore, our studies have shown that hypoxic enterocytes appear to be unable to utilize glucose for augmentation of hydroperoxide elimination (LeGrand and Aw, 1997). This inability of glucose to stimulate peroxide metabolism in the hypoxic intestine cannot be explained on the basis of decreased glucose uptake by enterocytes, because hypoxic intestine is just as capable of glucose uptake as normoxic intestine (LeGrand and Aw, 1998). Rather, there appears to be a reduction of glucose flux through the pentose phosphate shunt to support NADPH production for GSSG reduction. Our studies have shown that exogenous glucose significantly increases mitochondrial respiration (LeGrand and Aw, 1997), suggesting that both endogenous and exogenous glucose may preferentially be diverted to support primary metabolic demands, for example mitochondrial function, in the hypoxic intestine. Consequently, substrate availability is shunted away from secondary metabolic pathways, such as the pentose phosphate pathway, and from peroxide catabolism. This hypothesis is corroborated by direct measures of pentose-phosphate-pathway-dependent generation of NADPH in hypoxic enterocytes (LeGrand and Aw, 1998). By the diamide-induced GSH redox cycle approach, an estimated NADPH supply rate that approximates $0.4 \ nmol \cdot min^{-1}$ per 10^6 cells was obtained (LeGrand and Aw, 1998) that was 50% lower than the rate in normoxic intestine (Aw and Rhoads, 1994). Because the hyperdynamic metabolic state associated with chronic hypoxia diverts intracellular and exogenous glucose sources to enhance mitochondrial respiratory activity, this leaves the hypoxic intestine with little critical reserve to deal with additional challenges, such as the presence of lumenal lipid hydroperoxides.

It is significant that enhanced oxidative stress consequent to decreased peroxide removal in the hypoxic intestine causes oxidation of protein thiols. Because preservation of thiols is critical to the function of cellular enzymes and of other proteins, substantial protein thiol oxidation will have important consequences for cell integrity. Moreover, the added inability of the hypoxic intestine to restore cell GSH and redox homeostasis with exogenous glucose and GSH suggests that this organ is highly susceptible to oxidant-induced injury. It appears that the hypoxic intestine, already in a state of oxidative stress, is incapable of maintaining its functional integrity in the face of peroxide challenge (LeGrand and Aw, 1996, 1997). The compromised state of the hypoxic intestine could result in local damage to the mucosa, as well as enhanced transport of peroxidized lipid into the circulation via lymph. These oxidant-induced changes have important implications for the genesis of gut pathologies and atherosclerosis, respectively.

5. Concluding Remarks

The chemistry of lipid peroxidation and the deleterious role that lipid hydroperoxides play in cytotoxicity is well known. Although there is better appreciation of the contribution of lipid hydroperoxides to the pathogenesis of degenerative processes of the gut, our understanding of the physiological determinants of disposition of lumenal peroxidized lipids remains limited. Recent studies from our laboratory have provided important insights into the influence of lumenal GSH in intracellular hydroperoxide metabolism *in vivo*. Notably, use of the conscious, lymph fistula rat model has provided a quantitative and mechanistic definition of the role of mucosal and lumenal GSH in and the contribution of NADPH to intestinal intracellular peroxide elimination. Furthermore, the chronic hypoxia model allows for mechanistic study of hypoxia-induced compromise of intestinal hydroperoxide metabolism. This information is pertinent for understanding oxidant-mediated intestinal pathophysiology, given that consumption of highly unsaturated fats contributes to the accumulation of lipid peroxides in the intestinal lumen and to altered enterocyte redox homeostasis. The potential deleterious impact of redox imbalance on mucosal absorptive and transport function could have far-reaching implications regarding loss of intestinal integrity and the development of chronic disorders of the gut. Moreover, delineation of hypoxia-induced mucosal impairment of hydroperoxide detoxication could lead to understanding the intestinal responses in systemic disease states such as COPD and cystic fibrosis. Importantly, the fact that the intestinal epithelium can draw on a lumenal GSH pool to promote intracellular peroxide catabolism to preserve intestinal integrity could dictate future dietary treatment modalities in a variety of oxidative pathophysiological states, including chronic oxygen deficiency.

References

Ames, B. N., 1983, Dietary carcinogens and anticarcinogens: Oxygen radicals and degenerative diseases, *Science* **221**:1256–1264.

Andreoli, S. P., Mallett, C. P., and Bergstein, J. M., 1986, Role of glutathione in protecting endothelial cells against hydrogen peroxide oxidant injury, *J. Lab. Clin. Med.* **108**:190–198.

Andrews, J. S., Griffith, W. H., Mead, J. F., and Stein, R. A., 1960, Toxicity of air-oxidized soybean oil, *J. Nutr.* **70**:199–210.

Aw, T. Y., 1994, Biliary glutathione promotes the mucosal metabolism of lumenal peroxidized lipids by rat small intestine *in vivo*, *J. Clin. Invest.* **94**(September):1218–1225.

Aw, T. Y., 1997, Lumenal peroxides in intestinal thiol-disulfide balance and cell turnover, *Comp. Biochem. Physiol.* **118B**(3):479–485.

Aw, T. Y., and Rhoads, C. A., 1994, Glucose regulation of hydroperoxide metabolism in rat intestinal cells, *J. Clin. Invest.* **94**:2426–2434.

Aw, T. Y., and Williams, M. W., 1992, Intestinal absorption and lymphatic transport of peroxidized lipids in rats: effect of exogenous GSH, *Am. J. Physiol.* **263**:G665–G672.

Aw, T. Y., Andersson, B. S., and Jones, D. P., 1987, Suppression of mitochondrial respiratory function after short-term anoxia, *Am. J. Physiol.* **252**:C362–C368.

Aw, T. Y., Shan, X., Sillau, A. H., and Jones, D. P., 1991, Effect of chronic hypoxia on acetaminophen metabolism in the rat, *Biochem. Pharmacol.* **42**(5):1029–1038.

Aw, T. Y., Williams, M. W., and Gray, L., 1992, Absorption and lymphatic transport of peroxidized lipids by rat small intestine *in vivo:* Role of mucosal GSH, *Am. J. Physiol.* **262**:G99–G106.

Aw, T. Y., Smith, D. S., and Williams, A. D., 1993, Decreases in mucosal glutathione (GSH) promote portal transport of omega-3 fatty acid hydroperoxides *in vivo, FASEB J.* **7**(3):A385.

Bai, C., and Jones, D. P., 1996, GSH transport and GSH-dependent detoxication in small intestine of rats exposed *in vivo* to hypoxia, *Am. J. Physiol.* **271**:G701–G706.

Bergan, F. G., and Draper, H. H., 1970, Absorption and metabolism of 1-^{14}C-methyl linoleate hydroperoxide, *Lipids* **5**:976–982.

Bousignone, A., and Flora, A. D., 1972, Regulatory properties of glucose 6-phosphate dehydrogenase, *Current Topics in Cell Regulation* **6**:21–62.

Boyer, S. J., and Blume, F. D., 1984, Weight loss and changes in body composition at high altitude, *J. Appl. Physiol.* **57**:1580–1585.

Brigelius, R., 1985, Mixed disulfides: Biological functions and increase in oxidative stress, in: *Oxidative Stress.* 1st ed. (H. Sies, ed.), Academic Press, Inc., London, pp. 243–272.

Bull, A. W., Nigro, N. D., Golembieski, W. A., Crissman, J. D., and Marnett, L. J., 1984, *In vivo* stimulation of DNA synthesis and induction of ornithine decarboxylase in rat colon by fatty acid hydroperoxides autoxidation products of unsaturated fatty acids, *Canc. Res.* **44**:4924–4928.

Bunyan, J., Green, J., Murrell, E. A., Diplock, A. T., and Cawthorne, M. A., 1968, On the postulated peroxidation of unsaturated lipids in the tissues of vitamin E-deficient rats, *Br. J. Nutr.* **22**:97–110.

Burrows, B., Kettel, L. J., Niden, A. H., Rabinowitz, M., and Diener, C. R., 1972, Patterns of cardiovascular dysfunction in chronic obstructive pulmonary disease, *N. Engl. J. Med.* **286**:912–918.

Carroll, K. K., and Khor, H. T., 1975, Dietary fat in relation to tumorigenesis, *Prog. Biochem. Pharmacol.* **10**:308–353.

Cerra, F. B., 1987, Hypermetabolism, organ failure, and metabolic support, *Surgery* **101**:1–14.

Chu, F. F., Doroshow, J. H., and Esworthy, R. S., 1993, Expression, characterization, and tissue distribution of a new cellular selenium-dependent glutathione peroxidase, GSHPX-GI, *J. Biol. Chem.* **268**:2571–2576.

Correa, P., Strong, J. P., Johnson, W. D., Pizzolato, P., and Haenszel, W., 1982, Atherosclerosis and polyps of the colon: Quantification of precursors of coronary heart disease and colon cancer, *J. Chronic Dis.* **35**:313–320.

Eggleston, L. V., and Krebs, H. A., 1974, Regulation of the pentose phosphate cycle, *Biochem. J.* **138**:425–435.

Fink, D. J., and Kritchevsky, D., 1981, Introduction to the workshop on fat and cancer, *Cancer Res.* **41**:3684.

Fischer, S. M., Gleason, G. L., Mills, G. D., and Slaga, J. J., 1980, Indomethacin enhancement of TPA tumor promotion in mice, *Canc. Lett.* **10**:343–350.

Girotti, A. W., 1985, Mechanisms of lipid peroxidation, *J. Free Rad. Biol. Med.* **1**(2):87–95.

Glavind, J., 1970, Intestinal absorption and lymphatic transport of methyl linoleate hydroperoxide and hydroxyoctadecadienoate in the rat, *Acta Chem. Scand.* **24**:3723–3728.

Grisham, M. B., and Granger, D. N., 1988, Neutrophil-mediated mucosal injury: Role of reactive oxygen metabolites, *Dig. Dis. Sci.* **33**:6S–15S.

Hagen, T. M., and Jones, D. P., 1987, Transepithelial transport of glutathione in vascularly perfused small intestine of rat, *Am. J. Physiol.* **252**:G607–G613.

Hagen, T. M., Brown, L. A., and Jones, D. P., 1986, Protection against paraquat induced injury by exogenous GSH in pulmonary alveolar type II cells, *Biochem. Pharmacol.* **35**:4537–4542.

Hagen, T. M., Aw, T. Y., and Jones, D. P., 1988, Glutathione uptake and protection against oxidative injury in isolated kidney cells, *Kidney Int.* **34**:74–81.

Hagen, T. M., Wierzbicka, G. T., Bowman, B. B., Aw, T. Y., and Jones, D. P., 1990, Fate of dietary glutathione: Disposition in the gastrointestinal tract, *Am. J. Physiol.* **259**:G530–G535.

Hara, H., Miyashita, K., Ito, S., and Kasai, T., 1996, Oxidized ethyl linoleate induces mucosal hypertrophy of the large intestine and affects cecal fermentation of dietary fiber in rats, *J. Nutr.* **126**:800–806.

Iwakiri, T., Rhoads, C. A., and Aw, T. Y., 1995, Determinants of hydroperoxide detoxification in diabetic rat intestine: Effect of insulin and fasting on the glutathione redox cycle, *Metabolism* **44**(11):1462–1468.

Jaeschke, H., 1990a, Glutathione disulfide as index of oxidant stress in rat liver during hypoxia, *Am. J. Physiol.* **258**:G499–G505.

Jaeschke, H., 1990b, Glutathione disulfide formation and oxidant stress during acetaminophen-induced hepatotoxicity in mice *in vivo:* The protective effect of allopurinol, *J. Pharmacol. Exp. Ther.* **255**(3):935–941.

Jones, D. P., 1981, Hypoxia and drug metabolism, *Biochem. Pharmacol.* **30**(10):1019–1023.

Jones, D. P., and Mason, H. S., 1978, Gradients of O_2 concentration in hepatocytes, *J. Biol. Chem.* **253**(14):4874–4880.

Jones, D. P., Coates, R. J., Flagg, E. W., Eley, J. W., Block, G., Greenberg, R. S., Gunter, E. W., and Jackson, B., 1992, Glutathione in foods listed in the National Cancer Institute's Health Habits and History Food Frequency Questionnaire, *Nutr. Cancer* **17**:57–75.

Jones, D. P., LeGrand, T. S., Aw, T. Y., Dillehay, D., Cohen, S. D., Khairallah, E. A., and Manautou, J. E., 1995, Effect of oxygen deficiency on acetaminophen (APAP) toxicity in mice, *Toxicol.* **15**(1):152.

Kaneda, T., Sakai, H., and Ishii, S., 1955, Nutritive value or toxicity of highly unsaturated fatty acids, *J. Biochem.* **42**:561–573.

Kaplowitz, N., Aw, T. Y., and Ookhtens, M., 1985, The regulation of hepatic glutathione, *Ann. Rev. Pharmacol. Toxicol.* **25**:715–744.

Kauffman, F. C., Evans, R. K., and Thurman, R. G., 1977, Alterations in nicotinamide and adenine nucleotide systems during mixed-function oxidation of *p*-nitroanisole in perfused livers from normal and phenobarbital-treated rats, *Biochem. J.* **166**:583–592.

Kimura, T., Iida, K., and Takei, Y., 1984, Mechanisms of adverse effect of air-oxidized soybean oil-feeding in rats, *J. Nutr. Sci. Vitaminology* **30**:125–133.

Kinlen, L. J., 1983, Fat and cancer (editorial), *Br. Med. J.* **286**:1081–1082.

Kowalski, D. P., Feeley, R. M., and Jones, D. P., 1990, Use of exogenous glutathione for metabolism of peroxidized methyl linoleate in rat small intestine, *J. Nutr.* **120**:1115–1121.

Lash, L. H., Hagen, T. M., and Jones, D. P., 1986, Exogenous glutathione protects intestinal epithelial cells from oxidative injury, *Proc. Natl. Acad. Sci. USA* **83**:4641–4645.

LeGrand, T. S., and Aw, T. Y., 1996, Chronic hypoxia and glutathione-dependent detoxication in rat small intestine, *Am. J. Physiol.* **270**:G725–G729.

LeGrand, T. S., and Aw, T. Y., 1997, Chronic hypoxia, glutathione-dependent detoxication, and metabolic instability in rat small intestine, *Am. J. Physiol.* **272**:G328–G334.

LeGrand, T. S., and Aw, T. Y., 1998, Chronic hypoxia alters glucose utilization during GSH-dependent detoxication in rat small intestine, *Am. J. Physiol.* **274**:G376–G384.

Lifshitz, F., Wapnir, R. A., and Teichberg, S., 1986, Alterations in jejunal transport and (Na^+-K^+)-ATPase in an experimental model of hypoxia in rats, *Proc. Soc. Exp. Biol. Med.* **181**:87–97.

Martensson, J., Jain, A., and Meister, A., 1990, Glutathione is required for intestinal function, *Proc. Natl. Acad. Sci. USA* **87**:1715–1719.

Masuda, Y., Ozaki, M., and Aoki, S., 1993, K^+-driven sinusoidal efflux of glutathione disulfide under oxidative stress in the perfused rat liver, *FEBS Lett.* **334**(1):109–113.

Mizoguchi, T., Morita, Y., Nanjo, H., Tereada, T., and Nishihara, T., 1994, Responses of glutathione-related enzymes in isolated rat small intestine to Fe^{2+}-EDTA-mediated oxidative stress, *Biol. Pharm. Bull.* **17**(5):607–611.

Nagatsugawa, K., and Kaneda, T., 1983, Absorption and metabolism of methyl linoleate hydroperoxides in rats, *J. Jpn. Oil Chem. Soc.* **32**:362–366.

Parks, D. A., Bulkley, G. B., and Granger, D. N., 1983, Role of oxygen-derived free radicals in digestive tract diseases, *Surgery* **94**:415–422.

Pritchard, P. J., and Porteous, J. W., 1977, Steady-state metabolism and transport of *d*-glucose by rat small intestine *in vitro*, *Biochem. J.* **164**:1–14.

Reddy, B. S., 1983, Dietary fat and colon cancer, in: *Experimental Colon Carcinogenesis* (H. Antrup and G. M. Williams, eds.), CRC Press, Boca Raton, FL, pp. 225–239.

Reddy, K., and Tappel, A. L., 1974, Effect of dietary selenium and autoxidized lipids on the glutathione peroxidse system of gastrointestinal tract and other tissues in the rat, *J. Nutr.* **104**:1069–1078.

Sevanian, A., and Hochstein, P., 1985, Mechanisms and consequences of lipid peroxidation in biological systems, *Ann. Rev. Nutr.* **5**:365–390.

Shan, X., Aw, T. Y., and Jones, D. P., 1990, Glutathione-dependent protection against oxidative injury, *Pharmac. Ther.* **47**:61–71.

Sies, H., 1985, Hydroperoxides and thiol oxidants in the study of oxidative stress in intact cells and organs, in: *Oxidative Stress* (H. Sies, ed.), Academic Press, Inc., London, pp. 73–90.

Simic, M. G., and Karel, M., 1980, *Autoxidation in Food and Biological Systems,* Plenum Press, New York.

Sridharan, K., Malhotra, M. S., Upadhayay, T. N., Grover, S. K., and Dua, G. L., 1982, Changes in gastrointestinal function in humans at an altitude of 3,500 m, *Eur. J. Appl. Physiol. Occup. Physiol.* **50**:145–154.

Stoner, H. B., 1986, Metabolism after trauma and in sepsis, *Circ. Shock* **19**:75–87.

Thurlbeck, W. M., Henderson, J. A., Fraser, R., and Bates, D. V., 1970, Chronic obstructive lung disease: A comparison between clinical, roentgenologic, functional, and morphological criteria in chronic bronchitis, emphysema, asthma, and bronchiectasis, *Medicine* **49**:81–145.

Tribble, D. L., and Jones, D. P., 1990, Oxygen dependence of oxidative stress: Rate of NADPH supply for maintaining the GSH pool during hypoxia, *Biochem. Pharmacol.* **39**(4):729–736.

Uhlig, S., and Wendel, A., 1992, The physiological consequences of glutathione variations, *Life Sci.* **51**:1083–1094.

Van Liere, E. J., and Stickney, J. C., 1963, *Hypoxia,* The University of Chicago Press, Chicago.

Verma, A. K., Ashenden, C. L., and Boutwell, R. K., 1980, Inhibition of prostaglandin synthesis inhibitors of the induction of epidermal ornithine decarboxylase activity, the accumulation of prostaglandin, and tumor promotion caused by 12-*O*-tetradecanoylphorbol-13-acetate, *Canc. Res.* **140**:308–315.

Vilas, N. N., Bell, R. R., and Draper, H. H., 1976, Influence of dietary peroxides, selenium, and vitamin E on glutathine peroxidase of the gastrointestinal tract, *J. Nutr.* **106**:589–596.

Weigl, K., and Sies, H., 1977, Drug oxidations dependent on cytochrome P-450 in isolated hepatocytes: The role of the tricarboxylates and the aminotransferases in NADPH supply, *Eur. J. Biochem.* **77**:401–408.

Wendel, A., 1980, Glutathione peroxidase, in: *Enzymatic Basis of Detoxication* (W. B. Jakoby, ed.), Academic Press, New York, pp. 333–353.

Wierzbicka, G. T., Hagen, T. M., and Jones, D. P., 1989, Glutathione in food, *J. Food Compos. Anal.* **2**:327–337.

Williams, A. D., and Aw, T. Y., 1994, Glutathione (GSH) supply for lipid hydroperoxide metabolism in intestinal epithelial cells, *Gastroenterol.* **106**(4):1054.

Wilson, T. H., and Wiseman, G., 1954, The use of sacs of everted small intestine for the study of the transference of substances from the mucosal to the serosal surface, *J. Physiol.* **123**:116–125.

Wolff, S. P., and Nourooz-Zadeh, J., 1996, Hypothesis: UK consumption of dietary lipid hydroperoxides—A possible contributory factor to atherosclerosis, *Atheroslerosis* **119**:261–263.

CHAPTER 20

Intestinal Absorption of Fat-Soluble Vitamins

MOHSEN MEYDANI and KEITH R. MARTIN

1. Introduction and General Considerations

Vitamins are compounds that are required in small quantities for growth and maintenance of normal cell and organ functions; they are nutrients required primarily by vertebrates and not by lower organisms. Although vitamins may be grouped by their source or function, the most commonly used categorization relates to their chemical solubility, namely, hydrophobicity or hydrophilicity, which is directly associated with absorption processes.

Although essential for health, the detailed study of the influence of physiological and pathological factors on fat-soluble vitamin absorption has been limited. Fat-soluble vitamins A, D, E, and K are generally grouped and studied independently from water-soluble vitamins due to several key differences in structure and in function. First, fat-soluble vitamins differ metabolically because they do not form classical coenzyme structures or prosthetic groups that are easily dissociated from soluble apoenzymes (Moran and Greene, 1987). Second, depletion times are generally longer than for water-soluble vitamins due to the potential for storage in large fat depots. Third, the potential for toxicity of fat-soluble vitamins with excessive intake is problematic, although vitamins E and K may be tolerated at high intakes. Finally, the fat-soluble vitamins undergo distinct absorption processes and generally follow the requirements and pathways of lipid absorption and transport, as well as requiring carrier systems such as lipoproteins (Hollander, 1981). The disposition of fat-soluble vitamins after digestion is somewhat specific for each compound and is determined by their chemical nature. However, the early stages of intralumenal solubilization and intestinal mucosal uptake appear similar for all fat-soluble vitamins because of numerous overlapping physical properties (Mahan and Escott-Stump, 1996).

Although dietary fat-soluble vitamins are generally associated with proteins, the vita-

MOHSEN MEYDANI • Vascular Biology Program, Jean Mayer USDA Human Nutrition Research Center on Aging at Tufts University, Boston, Massachusetts 02111.   KEITH R. MARTIN • National Institute of Environmental Health Sciences, Research Triangle Park, North Carolina 27709-2233.
Intestinal Lipid Metabolism, edited by Charles M. Mansbach II *et al.,* Kluwer Academic/Plenum Publishers, 2001.

mins are released as a result of gastric acidity and/or via the action of proteolytic enzymes. The fat-soluble vitamins are solubilized and aggregated into small fat globules formed from lipids of ingested food through mechanical mixing of the chyme. These fat droplets are then emulsified within the lumen of the small bowel by secretion and interaction of bile acids and pancreatic juices. Pancreatic lipase, carboxylic ester hydrolase, and phospholipase A catalyze the cleavage of dietary triglycerides into long-chain fatty acids and monoglycerides, the hydrolysis of cholesterol esters, and the formation of lysophosphatides (Mayes, 1990).

Micelles formed from bile salts are an important medium for the digestion of esters of vitamins A and E. The amphiphilic bile salts form polymolecular multilamellar aggregates (mixed micelles) with monoglycerides, long-chain fatty acids, phospholipids, and cholesterol, creating a lipophilic core containing the fat-soluble vitamins, cholesterol, and other hydrophobic substances with an outer surface hydrated by the surrounding water phase. The formation of mixed micelles is necessary for the extraction of lipid-soluble molecules from the oily phases of food and subsequent solubilization of fat-soluble vitamins in an aqueous micellar phase from which they are taken up by the intestinal mucosa cells. Enzymes within pancreatic juice, and particularly the carboxyl ester hydrolases, act on solubilized micelles and hydrolyze the fat-soluble vitamin esters. Pancreatic carboxyl ester hydrolase is strongly dependent on the presence of bile salts for its activity. Thus, in addition to solubilizing the esterified vitamins, bile salts function as cofactors for enzymes that ultimately release the free vitamins (Mathias *et al.,* 1981).

The absorption of fat-soluble vitamins by the mucosal cells of the intestinal epithelium is associated with diffusion across the unstirred water layer adjacent to the microvillus surface and the plasma membrane lipid of the absorptive cells in the brush border. Presumably after collision of the micellar particles with the cell membrane of the enterocytes, the micellar components are partitioned into a monomeric phase, and the lipids and fat-soluble vitamins, but not bile salts, pass through the plasma membrane into the cytoplasm (Hollander, 1981). The absorption efficiency of fat-soluble vitamins has been tested in human beings with radiolabeled vitamins, and data indicate that the average absorption rate is 50% to 80% for vitamins A, D, and K but only 20% to 30% for vitamin E (Mahan and Escott-Stump, 1996).

Several dietary components, including pharmaceuticals and/or their delivery vehicles, other food constituents, and dietary lipids such as polyunsaturated fatty acids, may directly influence fat-soluble vitamin absorption due to similarities in lipid solubility and, as a result, may interact with fat-soluble vitamins during intestinal uptake. Elevated doses of vitamin A have been shown to decrease absorption of vitamins E and K through a mechanism involving competition for sites of intestinal absorption or transport, whereas high doses of vitamin E appear to augment vitamin A absorption (Hollander, 1981). Minimal data are available regarding the nature and function of the intestinal sites, mechanisms, and influencing factors associated with fat-soluble vitamin absorption.

At relatively high doses, absorption of vitamins A and E as water-miscible emulsions appears superior to oily solutions. The vitamins are taken up directly from water-miscible emulsions by the mucosal cells, circumventing the need for incorporation into lumenal mixed micelles, and are then transported both by lymphatic and portal pathways (Warwick *et al.,* 1976; Traber *et al.,* 1994). In the intestinal epithelium, vitamin A and negligible amounts of vitamin D are esterifed to long-chain fatty acids, whereas vitamins E and K remain unmodified. The fat-soluble vitamins are translocated from the absorptive cell membrane by simple diffusion or in association with binding proteins to lipid droplets formed on the smooth endoplasmic reticulum. The lipid droplets accumulate in the Golgi, where chylomicron and other triglyceride-rich lipoproteins are assembled. The fat-soluble vitamins and other lipids are transported in the aqueous medium of the extracellular space by these lipoproteins.

The chylomicrons and other triglyceride-rich lipoproteins are secreted from the Golgi into the lateral intercellular spaces. These particles enter the lacteals and intestinal lymphatics through the lamina propria and reach the systemic circulation via the thoracic duct. Movement of chylomicrons across the lymphatic endothelium occurs by passive diffusion and by active transport within pinocytotic vesicles. Chylomicrons in the lymph are the major route of transport for fat-soluble vitamins; however, the mode of lymphatic transport is determined in part by the concurrent absorption of other lipids. The fat-soluble vitamins are removed from the chylomicrons in the liver. Small amounts of the fat-soluble vitamins may be transported via the portal circulation, particularly in the case of the more polar derivatives of the vitamins such as retinoic acid (vitamin A) and menadione (vitamin K).

2. Vitamin A and Carotenoids

Vitamin A is found in significant quantities in liver, in kidney, in eggs, and in milk products and is primarily ingested in a form esterified with the long-chain fatty acid palmitic acid. The provitamin A carotenoids such as β-carotene are important sources of vitamin A, because the carotene molecule may be cleaved in the intestinal epithelium to initially form retinal through a putative NADP-dependent dioxygenase, with subsequent reduction to retinol (Wang, 1994).

The absorption of vitamin A involves a variety of dietary forms, and a small portion of retinol is transported directly to the portal blood as well as to the lymphatics. Vitamin A is normally absorbed as the free alcohol by an active transport process. As a result, detailed knowledge regarding the absorption of vitamin A in humans is limited. Intralumenal bile acid is necessary for maximum absorption. Both the pancreatic and the brush border retinal ester hydrolases convert retinal esters to retinol. The activity of these hydrolases is severely depressed by protein malnutrition and by pancreatic insufficiency. A bile salt-stimulated lipase appears to play a role in the digestion of retinal esters in human milk. In individuals aged 1 to 25 years with the malabsorption syndrome cystic fibrosis, administration of vitamin A alcohol in an oil–water emulsion was the best absorbed of vitamin A preparations (Warwick *et al.,* 1976; Johnson *et al.,* 1992).

Pancreatic and intestinal carboxylic acid hydrolases act on vitamin A in the intestinal lumen during solubilization of the esters and promote incorporation into mixed micelles (Mathias *et al.,* 1981). Vitamin A appears to be equally well absorbed in the proximal and distal portion of the small bowel. Within the range of physiological intralumenal concentrations of vitamin A, absorption is saturable and energy independent. Retinol, therefore, appears to penetrate the brush border membrane via a carrier-mediated, passive absorption process. At high intralumenal concentrations, vitamin A transport in the gut occurs via simple passive diffusion. Inside the intestinal epithelium, it appears that retinol is transported by a cellular retinol-binding protein (RBP) to the membranes of the smooth endoplasmic reticulum where it is esterified with long chain fatty acids, namely, palmitic acid, and is incorporated into lipid droplets. Because retinol is highly polar, small amounts of unesterified vitamin can be absorbed free of RBP directly into the portal vein, without prior incorporation into chylomicrons. Retinol is transported in human plasma bound to RBP (one molecule retinol to RBP) and to a pre-albumin acidic protein that also binds thyroxin. It is synthesized in the liver and is, therefore, decreased in plasma of individuals with acute viral hepatitis and protein calorie malnutrition due to impaired synthesis (Bender, 1995a).

Vitamin A deficiency as a result of inadequate dietary intake represents a major public health concern among children of Third World countries, particularly in the development of xerophthalmia (Gerster, 1997). Normal children given radiolabeled retinyl acetate in oil with nonlabeled vitamin (3,000 IU) excrete less than 5% of the label in their feces, indicating that intestinal absorption is almost complete; 17% of the absorbed label is excreted in the urine (Sivakumar and Reddy, 1978). Thus, about 80% of the administered dose is retained in the body. When the radiolabeled material is given along with massive doses (200,000 IU), only half of the administered dose appears to be retained. In children with kwashiorkor disease, 90% of the labeled vitamin administered is absorbed in the gut, indicating that vitamin A absorption is not significantly altered in protein energy malnutrition. As a result, the prevalence of signs of vitamin A deficiency in kwashiorkor disease appears to be due to dietary inadequacy and not faulty absorption (Sivakumar and Reddy, 1978).

Children with bronchopneumonia, acute gastroenteritis with diarrhea, or ascariasis (round worm infestation) were found to have significantly lower absorption efficiency than healthy children (Sivakumar and Reddy, 1978). The duration of the illness does not appear to be related to the extent of reduction in absorption, and recovery from the illness is associated with a return to normal levels of absorption. The defective absorption observed in these children was apparently not due to malabsorption of fat, as steatorrhea was not present, or to antibiotics, as the drugs did not alter absorption in healthy children (Reddy and Sivakumar, 1972). Thus, repeated attacks of respiratory tract infections, gastroenteritis, or diarrhea may enhance vitamin A requirements by interfering with intestinal absorption. In children whose dietary intakes are low and whose body stores of the vitamin are marginal, clinical manifestations of vitamin A deficiency may occur. The high incidence of infections in children of poor communities may contribute significantly to the widespread prevalence of vitamin A deficiency (Bates, 1995).

Because differences in absorption may arise between young and old, the relationship of supplemental and total vitamin A and supplemental vitamin E intake with fasting plasma biochemical indicators of vitamin A and vitamin E nutritional status was assessed among healthy elderly people and young adults. Although plasma retinol does not appear to be linked to vitamin A intake, plasma carotene levels are correlated with carotene intake. For elderly subjects, plasma carotene increased significantly with increasing total vitamin A intake, which includes provitamin A carotenoids. Fasting plasma retinyl esters were more sensitive to supplemental vitamin A in the elderly individuals than in the young adults, suggesting that elderly people who take vitamin A supplements may be at increased risk of overload (Krasinski *et al.,* 1989).

Insufficient production of bile acids in patients with various cholestatic syndromes and in particular individuals with biliary atresia suggest that vitamin A absorption is markedly impaired. Substantial reductions in the absorption of vitamin A in patients with cholestatic syndromes also suggest that intramuscular injections of vitamin A may be required to achieve and maintain normal levels of the vitamin (Moran and Greene, 1987).

One potentially significant source of vitamin A in the human diet derives from carotenoids. Carotenoids are a family of naturally occurring plant pigments comprised of more than 600 compounds that are divided into the hydrocarbon carotenes and the oxygenated xanthophylls. Initial interest in the carotenoids stemmed from their role as vitamin A precursors. Approximately 50 carotenoids possess pro-vitamin A activity and, in human plasma, consist of β-carotene, α-carotene, and cryptoxanthin with β-carotene having the highest potential vitamin A activity.

Although absorption, metabolism, and tissue distribution of carotenoids are not completely understood, carotenoids are absorbed in the intestinal mucosa of both animals and humans (Wang, 1994). The rate of absorption of carotenoids is affected by a number of conditions, including the amount of dietary lipids, proteins, and calories present in the body. Absorption of carotenoids in the intestine appears to be non-specific, which correlates with the process of passive diffusion. Once inside cells, β-carotene can either travel to the lateral surface of the cell and be incorporated into chylomicrons, or the molecule may be cleaved centrally or extrinsically by a putative dioxygenase. In humans, approximately 9% to 17% of orally administered labeled β-carotene is absorbed via the lymphatic system and 60% to 70% of the absorbed radioactivity is found as retinyl esters, whereas 15% remains as intact β-carotene. The efficiency of carotenoid absorption is 10% to 30% and decreases markedly with increasing intake. The plasma response is intrinsic to the carotenoid and appears to be influenced by the polarity of the molecule (van Vliet, 1996). Dietary fat is the most important factor because carotenoids are only absorbed in the presence of bile salts. In human serum, β-carotene is transported exclusively by lipoproteins and may be transferred or exchanged between lipoprotein groups in circulation. The initial increase in β-carotene in chylomicrons is followed by an increase first in very low density lipoproteins and in intermediate density lipoproteins and then, at later time points, in low density lipoproteins and in high density lipoproteins (Johnson and Russell, 1992). Carotenoids may be cleaved centrally or eccentrically to produce retinoids (Wang, 1994) with subsequent differential absorption of β-carotene and its metabolites into lymph or portal blood, which is dependent on the polarities of the metabolites produced.

In experimental animal models, several lumenal factors have been identified that affect β-carotene absorption. Absorption of β-carotene is increased by decreasing the unstirred water layer and by increasing the lumenal concentration of β-carotene, hydrogen ions, vitamin E, and fatty acids of various chain lengths. Normal gastric acidity (pH 1.3) has also been found to be an important factor in carotenoid absorption because an increase in pH to 6.4 was shown to significantly suppress the blood response in individuals given a 120 mg bolus dose of β-carotene (Tang *et al.,* 1996). Part of the absorbed β-carotene appears in the blood, and part is converted to retinol by an intracellular carotene oxygenase (Huang and Goodman, 1965; Olson, 1994). Several factors influencing the bioavailability of carotenoids have been identified and include the type of carotenoid, molecular linkage, the amount of the carotenoid consumed, the food matrix in which the carotenoid is present, absorption modifiers, nutrient status of the individual, genetic factors, other nongenetic, host-related factors, and interactions with other food components (de Pee and West, 1996).

Carotenoids are absorbed by duodenal mucosal cells by a mechanism involving passive diffusion, similar to that of cholesterol and the products of triglyceride lipolysis (Parker, 1996). One of the major concerns regarding the availability of carotenoids from food sources concerns their release from the physical matrix in which they are ingested and their dissolution in bulk lipid. In a study of Indonesian women, vitamin A status was substantially increased by intake of wafers containing β-carotene but not by a similar amount of β-carotene ingested as dark, leafy green vegetables, suggesting that the presence of fat, as well as the nature of the plant matrix, may influence the bioavailability of β-carotene. In addition, ingestion of raw versus cooked vegetables may have a marked impact on bioavailability (de Pee and West, 1996). The extent that carotenoids may compete with each other and/or with other lipid-soluble nutrients such as vitamin E during absorption is the subject of intense investigation. Evidence for the interaction between β-carotene and other carotenoids has been reported in the ferret in which canthaxanthin and lycopene reduced β-carotene absorption (van Vliet *et al.*, 1996). Interaction between β-carotene and lutein has also been reported. Evidence that β-carotene influences the bioavailability of vitamin E is incomplete and requires further study; however, it has been recently suggested in the literature that β-carotene does not interfere with the absorption of vitamin E (Meydani *et al.*, 1994; Fotouhi *et al.*, 1996; Parker, 1996).

3. Vitamin D

As a group, vitamin D is composed of vitamin D_1, which is a mixture of ergocalciferol and other sterols, vitamin D_2 (ergocalciferol), which is the active compound, and vitamin D_3 (cholecalciferol), which is formed in the skin and found in foods (Bender, 1995b). Vitamin D possesses many characteristics that make it unique among the fat-soluble vitamins (Moran and Greene, 1987). First, it is a steroid and in its main form acts in a manner similar to other steroid hormones. Second, there is no dietary requirement when children are exposed to sufficient amounts of sunlight because nonenzymatic photolysis of 7-dehydrocholesterol produces vitamin D in the skin. In general, sunlight is the primary source for maintaining vitamin D status. Third, it is the only vitamin known to be converted to a hormonal form. The active form of vitamin D, 1,25-dihydroxyvitamin D_3 ($1,25[OH]_2D_3$), which is formed by hydroxylation of vitamin D_3 in the liver and the kidney, fits the classic definition of a steroid hormone—it is produced in one tissue (kidney) and is carried through the blood in which it acts on other tissues via an intracellular (nuclear) receptor (vitamin D receptor, VDR) that has known response elements (VDR-response elements) in the promoter region of a variety of genes. This form of the vitamin modulates calcium homeostasis and bone mineralization (Haussler *et al.*, 1997). The fact that vitamin D exhibits such features has stimulated intense inquiry and, subsequently, advances in the elucidation of vitamin D metabolism and its mechanisms of action. The term vitamin D is now used to describe all steroids that qualitatively exhibit the biological activity of cholecalciferol.

Although limited in natural food sources, dietary sources with substantial levels of vitamin D include fatty fish, egg yolks, and fortified milk, which is absorbed in the jejunum or the ileum along with food fats (Linder, 1991). Thus, inhibition of normal fat absorption due to chronic pancreatitis, celiac disease, or biliary obstruction may result in vitamin D malabsorption (Thompson *et al.*, 1966). Only minor amounts of vitamin D_3 fatty acid esters are present naturally in the diet, and negligible amounts of vitamin D are esterified after passage through the intestinal epithelium. Interestingly, in some studies using human organ-culture systems, administration of $1,25[OH]_2 D_3$ has been shown to alter gastrointestinal physiology and/or morphology by significantly reducing cell proliferation in the duodenal epithelium (Thomas *et al.*, 1997). Vitamin D is taken up by the gut by means of a nonsaturable passive diffusion process. As with vitamin A, the more polar derivatives of vitamin D are absorbed most readily. Radiolabeled vitamin D_3 and 25-hydroxyvitamin D_3 (25[OH] D_3), a form present in foods of animal origin, given to humans indicate that absorption of the hydroxylated metabolite is less dependent on bile acids and predominantly occurs directly into the portal vein (Compston *et al.,* 1981).

Vitamin D and its metabolites are well accepted as key regulatory elements in bone mineral homeostasis (Bikle, 1994). The active metabolite of vitamin D_3, 1,25[OH]$_2$ D_3, controls calcium absorption in the human duodenum by mediating vitamin D receptor expression (Thomas *et al.,* 1997). As a result, vitamin D deficiency can lead to osteomalacia and may be important in the pathogenesis of age-related osteoporosis. Studies of vitamin D metabolism in young females, in elderly free-living females, and in elderly females living in nursing homes indicate that calcium absorption and 1,25[OH]$_2$ D_3 levels were significantly lower in individuals from nursing homes (Kinyamu *et al.,* 1997). Studies indicate that there is a decline in serum 25[OH] D_3, 1,25[OH]$_2$ D_3, and calcium absorption with advancing age, which may lead to secondary hyperthyroidism and bone loss (Francis, 1997). Vitamin D supplementation in older females improves calcium absorption, decreases parathyroid hormone levels, and reduces bone loss (Dawson-Hughes, 1996). Vitamin D may also reduce the incidence of hip and other nonvertebral fractures in the frail elderly who are likely to be vitamin D deficient (Francis, 1997). Moreover, reports also indicate that administration of both activated and nonactivated vitamin D reduce serum parathyroid levels and activated vitamin D is able to reduce bone resorption in postmenopausal, osteoporotic patients with vitamin D-sufficient status (Nakamura *et al.,* 1996).

4. Vitamin E

Vitamin E is a generic term for a number of tocopherol compounds each containing a chromanol ring and phytyl side chain, and it refers to any mixture of physiologically active tocopherols. Eight naturally occurring tocopherols have been identified, although only four have any physiological importance and include α, β, γ, and δ-tocopherol. Unlike other vitamins, vitamin E does not have specific enzymatic function but serves as the major lipid-soluble antioxidant *in vivo* (Meydani, 1995) and may modulate immunomodulatory factors including eicosanoid production (Meydani *et al.,* 1993).

Vegetable and seed oils and whole grains are the largest contributors of tocopherols to the diet, whereas animal products are generally poor sources of the vitamin. Esterified forms of vitamin E are almost completely deesterifed and hydroxylated prior to absorption. Studies have shown no apparent differences in the absorption of free α-tocopherol or forms esterified with acetate or succinate (Cheeseman *et al.,* 1995). Bile is essential for absorption of the vitamin, which proceeds by a nonsaturable passive diffusion process. The predominant route of absorption is via gastrointestinal lymphatics, with about 45% of a dose absorbed into the lymph as α-tocopherol and the rest appearing as metabolites in portal venous blood and feces (Bieri and Farrell, 1976). A marked decrease in absorption efficiency occurs as physiological amounts of the vitamin are increased to pharmacological doses.

Absorption of vitamin E, like vitamin A, requires the presence of pancreatic esterases as well as bile acids, although, unlike vitamin A, vitamin E is not reesterified during the absorption process. In conditions such as chronic pancreatitis in which the secretion of bile acid is markedly impaired due to inflammation and/or partial obstruction of the bile duct, absorption of vitamin E has been shown to be significantly reduced compared with other fat-soluble vitamins (Nakamura *et al.,* 1996). The appearance of vitamin E in the bloodstream following a bolus dose is biphasic with an initial peak at four hours and a second peak at eight hours. The middle portion of the small intestine is the site of maximum absorption in animals, and the efficiency of absorption varies from 20% to 40%, depending on the formation of mixed micelles and the availability of dietary lipids (Bender, 1995c). High concentrations of polyunsaturated fatty acid (PUFA) decrease absorption of vitamin E by suppressing intestinal uptake, but absorption is enhanced by the presence of medium-chain triglycerides. In cases of gastrointestinal pathology, vitamin E has been shown to be better absorbed from orally administered aqueous solutions such as Aquasol E than from oily solutions. In fact, administration of water-soluble RRR-α-tocopheryl glycol 1000 succinate (TPGS) is the treatment of choice in gastrointestinal malabsorption syndromes (Dimitrov *et al.,* 1996). Studies have also demonstrated the effectiveness of TPGS administration in increasing vitamin E levels in patients with short-bowel syndrome despite the presence of severe fat malabsorption (Traber *et al.,* 1994). However, in studies using three different single doses and multiple doses of TPGS or the lipid-soluble tocopheryl acetate given to 27 healthy volunteers for 4 weeks, it became clear that for normal adults and patients with normal lipid absorption, fat-soluble forms of vitamin E are more effective (Dimitrov *et al.,* 1996).

Absorption of vitamin E in the premature infant appears to be slightly less than absorption in the term infant. However, between 26 to 32 weeks gestation, absorption gradually improves, so that by 36 weeks postconception, it approaches that of a normal term infant. The absorption of vitamin E is related to the ability to digest and absorb dietary fats, an ability that is limited in all neonates (Avery and Fletcher, 1975). It has also been shown that human milk fat is better absorbed than cow's milk fat, and unsaturated fatty acids are better absorbed than saturated ones. The linear increase in vitamin E absorption parallels gestational maturity and appears directly related to the ability to absorb fats (Ehrenkranz, 1980). Total fat absorption has been found to improve significantly in premature infants fed formulas containing medium-chain triglycerides but restricted in long-chain fatty acids, although the relationship of this phenomenon to vitamin E absorption is not clear.

The tocopherol isomers (β, γ, and δ) and their corresponding tocotrienols occur in varying amounts in foods of plant origin, with γ-tocopherol often present in mixed diets in higher quantities than α-tocopherol. The isomers β, γ, and δ-tocopherol are more polar than α-tocopherol and, as a result, have lower rates of absorption (Bender, 1995c). It is not known whether these less methylated tocopherols compete with vitamin E for sites of absorption, alter the limited packing capacity of mixed micelles for vitamin E, affect partitioning of vitamin E in mucosal cell membranes, or antagonize incorporation of vitamin E into chylomicrons. In one study, humans were fed fat-rich, high-energy diets supplemented with different amounts of γ- and α-tocopherol, and plasma levels of the tocopherols were monitored at numerous time points (Meydani *et al.,* 1989). The findings indicated that there were differences in the distribution of the γ- and α-tocopherol isomers in postprandial lipoproteins. Although the tocopherol isomers are similarly absorbed, a cytosolic α-tocopherol-binding protein has been identified in the liver that specifically binds α-tocopherol with little affinity for the other isomers (Kayden and Traber, 1993; Sato *et al.,* 1993). Thus, this binding protein preferentially incorporates α-tocopherol into lipoproteins whereas other isomers are excreted in the bile (Traber and Kayden, 1989). In animals fed radiolabeled α-tocopherol, the extent of enterohepatic circulation of the tocopherol was assessed, and the results indicated that the magnitude of enterohepatic circulation was remarkably low (Lee-Kim *et al.*, 1988).

The potential interaction of other fat-soluble compounds may impair the absorption of vitamin E. In one study, male volunteers were fed vitamin E (800 mg/day) and β-carotene (30 mg/day) alone or in combination to determine the effects on plasma levels (Baker *et al.*, 1996). Intake of β-carotene, which may be cleaved to form vitamin A, did not affect plasma vitamin A levels. Feeding only vitamin E, β-carotene, or a combination of vitamin E with β-carotene significantly increased plasma vitamin E and carotene levels after 2 days.

5. Vitamin K

Vitamin K is a generic name for several structurally related compounds with the capacity to function as a cofactor for the mammalian microsomal enzyme γ-glutamylcarboxylase and includes vitamin K_1 (phylloquinone), vitamin K_2 (menaquinones), and vitamin K_3 (menadione; Lipsky, 1994; Gijsbers *et al.*, 1996). Naturally occurring vitamin K is fat soluble, but several synthetic water-soluble forms have been prepared. The main sources of vitamin K are green vegetables, which exclusively contain phylloquinone, and dairy products, in which a mixture of phylloquinone and the menaquinones are present. In humans, phylloquinone is acquired from dietary sources, and menaquinones are synthesized by intestinal bacteria. Studies have provided direct evidence for the absorption of menaquinones produced by intestinal microflora from the distal small bowel in humans (Conly *et al.*, 1994). Estimates indicate that approximately 20 μg menaqionones per gram dry weight is available for absorption in the colon. Although the menaquinones exhibit only 60% of the potency of phylloquinone, bacterial synthesis of menaquinones may provide a significant contribution to vitamin K status in humans (Moran and Greene, 1987).

The absorption of lipid-soluble vitamin K requires bile and pancreatic juices for maximum effectiveness. Absorption studies using everted gut sacs in rats have indicated that vitamin K absorption is energy-mediated and uses a saturable transport mechanism (Hollander *et al.*, 1977). The rate of lymphatic appearance of lumenally administered vitamin K is enhanced by increasing intralumenal concentrations of the bile salt sodium taurocholate, whereas variation of the intralumenal pH did not change the lymphatic appearance of the vitamin. Absorption of menaquinones in the colon and menadione in the distal intestine and colon occurs via passive diffusion and is facilitated by the presence of bile salts and fatty acids, whereas absorption of phylloquinone occurs primarily in the jejunum and is energy dependent (Hollander, 1973; Hollander *et al.*, 1976).

The absorption and transport of phylloquinone has been characterized in several human studies. In one study, 26 healthy individuals received 1.43 μg/kg body weight phylloquinone with a fat-rich meal; plasma samples were collected at numerous postprandial time intervals. Data indicate that although triacylglycerols were the major carrier of phylloquinone, low-density and high-density lipoproteins were found to contain small fractions of phylloquinone (Lamon-Fava *et al.*, 1998). Individuals consuming 1 mg phylloquinone in the form of spinach plus butter or as spinach without fat had substantially lower plasma levels of vitamin K than those taking a pharmaceutical preparation with the same amount of phylloquinone. As a result, the absorption of phylloquinone from the vegetable sources was significantly slower. This suggests that the bioavailability of dietary vitamin K (phylloquinones plus menaquinones) is lower than generally assumed and depends on the form in which the vitamin is ingested (Gijsbers *et al.*, 1996). Dihydro-vitamin K_1 is a dietary form of vitamin K_1 (phylloquinone) produced during the hydrogenation of vegetable oils. In a subsequent study, eight healthy individuals consumed hydrogenated fat as soybean oil or a partially hydrogenated soybean-oil-based stick margarine that contained dihydro-vitamin K_1. Dihydro-vitamin K_1 was detectable in plasma following dietary intake of a hydrogenated vitamin K_1-rich vegetable oil, but there were no significant changes in plasma vitamin K concentrations between the two diets (Booth *et al.*, 1996). To determine levels of vitamin K, the response of osteocalcin and other biochemical markers of vitamin K status to diets formulated to contain different amounts of phylloquinone was assessed in nine healthy individuals. Data indicate that undercarboxylated osteocalcin, plasma phylloquinone, and urinary γ-carboxyglutamic acid excretion are sensitive markers of vitamin K status because all of these markers respond to changes in dietary intake (Sokoll *et al.*, 1997).

Neonates are generally deficient in vitamin K, and the plasma concentrations of other factors associated with vitamin K are also lower. Prenatal maternal vitamin K supplementation has been suggested to improve the status of the fetus, but this may be dependent on route of administration. Supplementation of eight healthy, nonpregnant women of childbearing age with 5 mg of phylloquinone delivered by mouth or by intramuscular injection showed that the metabolic processing of supplemented vitamin K_1 differs depending on the route of administration (Hagstrom *et al.*, 1995). Infants beyond the neonatal period who have diarrhea and who are given antibiotics along with special formula containing low concentrations of vitamin K may develop hypoprothrombinemic bleeding within approximately 2 weeks (Shearer, 1995). Vitamin K deficiency in infants and in children with chronic cholestasis have been successfully treated with soluble vitamin K analogues that have been solubilized and administered in preparations of glycocholate and lecithin (Argao and Heubi, 1994). It appears that the intestinal flora contribute a significant amount of vitamin K for absorption during the infancy period, although no definitive evidence shows that intestinal bacteria are an important nutritional source of vitamin K (Lipsky, 1994).

Accumulating data suggest that the direct absorption of menaquinones produced by colonic bacterium is poor because bile salts, which are primarily produced in the small intestine, are not present to facilitate absorption (Gijsbers *et al.*, 1996). Competitive antagonism by structural analogues does not appear to occur, as no change in the absorption of phylloquinone occurs when menaquinones or menadione are administered concomitantly. Efficiency of absorption has been measured from 10% to 80%, depending on the vehicle in which the vitamin is administered and the extent of enterohepatic circulation (Linder, 1991). Isotopically labeled phylloquinone administered by mouth in physiological and in pharmacological doses appears in plasma within 20 minutes and peaks at 2 hours. Between 8% to 30% of administered radioactivity is recovered in the urine over a 3-day period, whereas total fecal radioactivity accounts for 45% to 60% of the administered dose over a 5-day period (Shearer *et al.*, 1970). Following oral administration of labeled vitamin K in an animal model, the vitamin is effectively transported to the liver, the kidneys, the heart, the skin, and the muscles.

6. Influence of Other Dietary Components on Absorption of Fat-Soluble Vitamins

Because absorption of fat-soluble vitamins is dependent on the absorption of fat from the gastrointestinal tract, a variety of drugs or other compounds interfering with fat absorption may have a detrimental impact on fat-soluble vitamin absorption. In recent years, pharmaceutical companies and the food industry have introduced a variety of low-fat or no-fat foods designed to modulate dietary lipid intake, which is believed to be associated with the incidence of degenerative diseases including coronary heart disease, obesity, and atherosclerosis (Gershoff, 1995). Introduction of fat substitutes into foods results in low-fat products with organoleptic properties that possess the taste and texture of foods with higher fat contents. Currently, variations include protein-, carbohydrate-, and fat-based fat substitutes as well as synthetic non-caloric fat substitutes (Gershoff, 1995). In the latter group, Olestra (sucrose polyester) is probably the most exhaustively studied fat substitute. Sucrose polyester contains a sucrose moiety esterified with 6 to 8 fatty acid molecules. It is neither hydrolyzed nor absorbed in the intestine and has not been shown to be toxic, carcinogenic, mutagenic, or teratogenic when fed to animals at doses of up to 10% of the total diet (Bergholtz, 1992). However, studies indicate that Olestra can impair the absorption of other dietary components, especially highly lipophilic compounds, when ingested at the same time. Olestra has been shown in some studies to decrease serum concentrations of carotenoids, of 25[OH]-ergocalciferol, and of phylloquinone, in a dose-responsive manner (Schlagheck *et al.*, 1997). Other studies have shown that Olestra markedly reduces the absorption of vitamins A, D, E, and K, and it may have a similar effect on other fat-soluble components of foods such as carotenoids and flavonoids. However, the impact on the absorption of fat-soluble vitamins in humans can be offset by adding these vitamins to food preparations in which fat is substituted by Olestra (Peters *et al.*, 1997).

Consumption of other non-nutritive compounds with either lipidlike physical qualities or the ability to alter lipid metabolism have been shown to decrease fat-soluble vitamin absorption. Orlistat is an inhibitor of gastric, carboxyl ester, and pancreatic lipases, which are necessary for lipid- and fat-soluble vitamin absorption. Orlistat specifically decreases the absorption of dietary fat through inhibition of enzymatic triglyceride hydrolysis. In one randomized, placebo-controlled study, administration of Orlistat to humans significantly decreased the absorption of vitamin E by 43% but had no impact on vitamin A absorption (Drent and van der Veen, 1993; Melia *et al.*, 1996). Ingestion of cholestyramine, a cholesterol-lowering drug that sequesters bile acids, has also been shown to reduce the absorption of vitamin E and carotenoids, namely, β-carotene and lycopene, up to 40% due to impairment of gastrointestinal absorption. Moreover, probucol, a cholesterol-lowering drug with antioxidant activity, has been reported to reduce vitamin E and carotenoid levels as much as 40% (Elinder *et al.*, 1995). Together, these results suggest that ingestion of synthetic, lipid-soluble compounds, including antioxidants and cholesterol-lowering drugs that may interfere with fat absorption, may impair fat-soluble vitamin status.

7. Conclusion

The absorption of fat-soluble vitamins A, D, E, and K follows closely the pathways and disposition of other dietary lipids and, as a result, functional or pathological alteration of gastrointestinal processes may interfere with fat-soluble vitamin absorption. In the intestinal epithelium, vitamins may remain unmodified (vitamins E and K) or may be esterified (vitamins A and D) with long-chain fatty acids and subsequently packaged into lipoproteins where they enter the circulation through the lacteals and the intestinal lymphatics systems. The absorption of vitamin A involves a variety of dietary forms, most of which are taken up in the proximal and distal portion of the small bowel, although a small portion of highly polar retinol is transported as free alcohol by active transport directly to the portal circulation. Fat malabsorption syndromes have been repeatedly shown to reduce fat-soluble vitamin absorption. Pro-vitamin A carotenoids are cleaved to form vitamin A and may substantially contribute to overall vitamin A status. In addition to dietary sources, exposure to sunlight represents a key source of vitamin D and the active form of vitamin D, 1,25-dihydroxy vitamin D_3, is formed in the liver and the kidney and is important in calcium absorption. Unlike other vitamins, vitamin E does not exhibit specific enzymatic function, but serves as the major lipid soluble antioxidant *in vivo*. Isomers of vitamin E all appear to be similarly absorbed, but the α-tocopherol isomer is the most biologically active form and is mainly retained. Administration of vitamin E in water-soluble form is effective in increasing levels in patients with short-bowel syndrome. Vitamin K includes several forms, but in humans, phylloquinone is acquired through the diet, whereas menaquinones are synthesized by intestinal bacteria, which represents a significant contribution to vitamin K status. Several effective water-soluble analogues of vitamin K are also available for individuals with malabsorption syndromes. Absorption of fat-soluble vitamins have been found to be affected by their interaction with each other and with other components of food. Several non-nutritive compounds that modulate lipid absorption may also interfere with the absorption of fat-soluble vitamins. Bile acid sequestrants and other cholesterol-lowering drugs have been shown to reduced fat-soluble vitamin status by interfering with their absorption. In general, absorption of fat-soluble vitamins is dependent on effective, optimal function of the gastrointestinal tract and is dependent on the presence of dietary lipid, bile acids, and pancreatic juices, as well as the activity of digestive enzymes. Collectively, there has been marked progress in understanding the absorption of fat-soluble vitamins; however, the interactions of these nutrients with each other and with other components of food, as well as compounds affecting gastrointestinal physiology, warrants further study in this area.

Acknowledgments

Some of the projects discussed in this chapter have been funded at least in part with Federal funds from the U.S. Department of Agriculture, Agricultural Research Service under contract number 53-K06–01. The contents of this publication do not necessarily reflect the views or policies of the U.S. Department of Agriculture, nor does mention of trade names, commercial products, or organizations imply endorsement by the U.S. Government.

The authors would like to thank Timothy S. McElreavy, M.A., for preparation of this manuscript.

References

Argao, E. A., and Heubi, J. E., 1994, Fat-soluble vitamin deficiency in infants and children, *Curr. Opin. Ped.* **5:** 562–566.

Avery, G. B., and Fletcher, A. B., 1975, Nutrition, in: *Neonatology Pathophysiology and Management in the Newborn* (G. B. Avery, ed.), Lippincott, Philadelphia.

Baker, H., DeAngelis, B., Baker, E., Khalil, M., and Frank, O., 1996, Human plasma patterns during 14 days ingestion of vitamin E, beta-carotene, ascorbic acid, and their various combinations, *J. Amer. Coll. Nutr.* **15:** 159–163.

Bates, C. J., 1995, Vitamin A, *Lancet* **345:** 31–35.

Bender, D. A., 1995a, Vitamin A: Retinol and beta-carotene, in: *Nutritional Biochemistry of the Vitamins* (D. A. Bender, ed.), Cambridge University Press, New York, pp. 19–49.

Bender, D. A., 1995b, Vitamin D, in: *Nutritional Biochemistry of the Vitamins* (D. A. Bender, ed.), Cambridge University Press, New York, pp. 51–85.

Bender, D. A., 1995c, Vitamin E: Tocopherols and tocotrienols, in: *Nutritional Biochemistry of the Vitamins* (D. A. Bender, ed.), Cambridge University Press, New York, pp. 87–105.

Bergholtz, C. M., 1992, Safety evaluation of Olestra, a non-absorbed fatlike fat replacement, *Crit. Rev. Food Sci. Nutr.* **32:** 141–146.

Bieri, J. G., and Farrell, P. M., 1976, Vitamin E, *Vit. Hormone* **34:** 31–75.

Bikle, D. D., 1994, Role of vitamin D, its metabolites, and analogs in the management of osteoporosis, *Rheum. Dis. Clin. N. Am.* **20:** 759–775.

Booth, S. L., Davidson, K. W., Lichtenstein, A. H., and Sadowski, J. A., 1996, Plasma concentrations of dihydrovitamin K1 following dietary intake of a hydrogenated vitamin K1-rich vegetable oil, *Lipids* **31:** 709–713.

Cheeseman, K. H., Holley, A. E., Kelly, F. J., Wasil, M., Hughes, L., and Burton, G., 1995, Biokinetics in humans of RRR-alpha-tocopherol: The free phenol, acetate ester, and succinate ester forms of vitamins E, *Free Rad. Biol. Med.* **19:** 591–598.

Compston, J. E., Merret, A. L., and Hammett, F. G., 1981, Comparison of the appearance of radiolabeled vitamin D_3 and 25-hydroxy-vitamin D_3 in the chylomicron fraction of plasma after oral administration in man, *Clin. Sci.* **60:** 241–243.

Conly, J. M., Stein, K., Worobetz, L., and Rutledge-Harding, S., 1994, The contribution of vitamin K_2 (menaquinones) produced by the intestinal microflora to human nutritional requirements for vitamin K, *Am. J. Gastroenterol.* **89:** 915–923.

Dawson-Hughes, B., 1996, Calcium and vitamin D nutritional needs of elderly women, *J. Nutr.* **126:** 1165S–1167S.

de Pee, S., and West, C. E., 1996, Dietary carotenoids and their role in combating vitamin A deficiency: A review of the literature, *Eur. J. Clin. Nutr.* **50:** S38–S53.

Dimitrov, N. V., Meyer-Leece, C., McMillan, J., Gilliland, D., Perloff, M., and Malone, W., 1996, Plasma α-tocopherol concentrations after supplementation with water- and fat-soluble vitamin E, *Am. J. Clin. Nutr.* **64:** 329–335.

Drent, M. L., and van der Veen, E. A., 1993, Lipase inhibition: A novel concept in the treatment of obesity, *Int. J. Obes. Related Metabol. Disord.* **17:** 241–244.

Ehrenkranz, R. A., 1980, Vitamin E and the neonate, *Am. J. Dis. Child.* **134:** 1157–1166.

Elinder, L. S., Hadell, K., Johansson, J., Molgaard, J., Holme, I., Olsson, A. G., and Walldius, G., 1995, Probucol treatment decreases serum concentrations of diet-derived antioxidants, *Arterioscler. Thromb. Vasc. Biol.* **15:** 1057–1063.

Fotouhi, N., Meydani, M., Santos, M. S., Meydani, S., Hennekens, C. H., and Gaziano, J. M., 1996, Carotenoid and tocopherol concentrations in plasma, peripheral blood mononuclear cells, and red blood cells after long-term β-carotene supplementation in men, *Am. J. Clin. Nutr.* **63:** 553–558.

Francis, R. M., 1997, Is there a differential response to alfacalcidol and vitamin D in the treatment of osteoporosis? *Calcified Tissue Int.* **60:** 111–114.

Gershoff, S. N., 1995, Nutrition evaluation of dietary fat substitutes, *Nutr. Reviews* **53:** 305–313.

Gerster, H., 1997, Vitamin A: Functions, dietary requirements and safety in humans, *Int. J. Vit. Nutr. Res.* **67:** 71–90.

Gijsbers, B. M. G., Jie, K.-S. G., and Vermeer, C., 1996, Effect of food composition on vitamin K absorption in human volunteers, *Br. J. Nutr.* **76:** 223–229.

Hagstrom, J. N., Bovill, E. G., Soll, R. F., Davidson, K. W., and Sadowski, J. A., 1995, The pharmacokinetics and lipoprotein fraction distribution of intramuscular vs. oral vitamin K1 supplementation in women of child-bearing age: Effects on hemostasis, *Thromb. Haem.* **74**: 1486–1490.

Haussler, M. R., Haussler, C. A., Jurutka, P. W., Thompson, P. D., Hsieh, J. C., Remus, L. S., Selznick, S. H., and Whitfield, G. K, 1997, The vitamin D hormone and its nuclear receptor: Molecular actions and disease states. *J. Endocrinol.* **154** (suppl): S57–S73.

Hollander, D., 1973, Vitamin K_1 absorption by everted intestinal sacs of the rat, *Am. J. Physiol.* **225**: 360–364.

Hollander, D., 1981, Intestinal absorption of vitamins A, E, D, and K, *J. Lab Clin Med* **97**: 449–462.

Hollander, D., Muralidhara, K. S., and Rim, E., 1976, Colonic absorption of bacterially synthesized vitamin K_2 in the rat, *Am. J. Physiol.* **230**: 251–255.

Hollander, D., Rim, E., and Muralidhara, K. S., 1977, Vitamin K_1 intestinal absorption *in vivo*. Influence of lumenal contents on transport, *Am. J. Physiol.* **232**: E69–E74.

Huang, H. S., and Goodman, D. S., 1965, Vitamin A and carotenoids. I. Intestinal absorption and metabolism of ^{14}C-labeled vitamin A alcohol and beta carotene in the rat, *J. Biol. Chem.* **240**: 2839–2844.

Johnson, E. J., and Russell, R. M., 1992, Distribution of orally administered beta carotene among lipoproteins in healthy men, *Am. J. Clin. Nutr.* **56**:128–135.

Johnson, E. J., Krasinski, S. D., Howard, L. J., Alger, S. A., Dutta, S. K., and Russell, R. M., 1992, Evaluation of vitamin A absorption by using oil-soluble and water-miscible vitamin A preparations in normal adults and in patients with gastrointestinal disease, *Am. J. Clin. Nutr.* **55**: 857–864.

Kayden, H. J., and Traber, M. G., 1993, Absorption, lipoprotein transport, and regulation of plasma concentrations of vitamin E in humans, *J. Lipid Res.* **34**: 343–358.

Kinyamu, H. K., Gallagher, J. C., Balhorn, K. E., Petranick, K. M., and Rafferty, K. A., 1997, Serum vitamin D metabolites and calcium absorption in normal young and elderly free-living women and in women living in nursing homes, *Am. J. Clin. Nutr.* **65**: 790–797.

Krasinski, S. D., Russell, R. M., Otradovec, C. L., Sadowski, J. A., Hartz, S. C., Jacob, R. A., and McGandy, R. B., 1989, Relationship of vitamin A and vitamin E intake to fasting plasma retinol, retinol-binding protein, retinyl esters, carotene, α-tocopherol, and cholesterol among elderly people and young adults: Increased plasma retinyl esters among vitamin A-supplement users, *Am. J. Clin. Nutr.* **49**: 112–120.

Lamon-Fava, S., Sadowski, J. A., Davidson, K. W., O'Brien, M. E., McNamara, J. R., and Schaefer, E. J., 1998, Plasma lipoproteins as carriers of phylloquinone (vitamin K1) in humans, *Am. J. Clin. Nutr.*, submitted for publication.

Lee-Kim, Y. C., Meydani, M., Kassarjian, Z., Blumberg, J. B., and Russell, R. M., 1988, Enterohepatic circulation of newly administered α-tocopherol in the rat, *Int. J. Vit. Nutr. Res.* **58**: 284–291.

Linder, M. C., 1991, Nutrition and metabolism of vitamins, in: *Nutritional Biochemistry and Metabolism with Clinical Applications* (M. C. Linder, ed.), Elsevier, New York, 153–189.

Lipsky, J. J., 1994, Nutritional sources of vitamin K, *Mayo Clin. Proc.* **69**: 462–466.

Mahan, L. K., and Escott-Stump, S., 1996, Digestion, absorption, transport, and excretion of nutrients. In: *Krause's Food, Nutrition, & Diet Therapy*, 9th ed. (L. K. Mahan, S. Escott-Stump, and P. Beyer, eds.), W. B. Saunders, Philadelphia, pp. 3–16.

Mathias, P. M., Harries, J. T., and Muller, D. P. R., 1981, The optimization and validation of assays to estimate pancreatic esterase activity using well-characterized micellar solutions of cholesterol oleate and tocopherol acetate, *J. Lipid Res.* **22**: 177–184.

Mayes, P. A., 1990, Digestion and absorption, in: *Harper's Biochemistry* (R. K. Murray, D. K. Granner, P. A. Mayes, and V. W. Rodwell, eds.), Appleton & Lange, Norwalk, CT, pp. 580–590.

Melia, A. T., Koss-Twardy, S. G., and Zhi, J., 1996, The effect of orlistat, an inhibitor of dietary fat absorption, on the absorption of vitamins A and E in healthy volunteers, *J. Clin. Pharm.* **36**: 647–653.

Meydani, M., 1995, Vitamin E, *Lancet* **345**: 170–175.

Meydani, M., Cohn, J. S., Macauley, J. B., McNamara, J. R., Blumberg, J. B., and Schaefer, E. J., 1989, Postpriandial changes in the plasma concentration of α- and γ-tocopherol in human subjects fed a fat-rich meal supplemented with fat-soluble vitamins, *J. Nutr.* **119**: 1252–1258.

Meydani, M., Meydani, S. N., and Blumberg, J. B., 1993, Modulation by dietary vitamin E and selenium of clotting whole blood thromboxane A_2 and aortic prostacyclin synthesis in rats, *J. Nutr. Biochem.* **4**: 322–326.

Meydani, M., Martin, A., Ribaya-Mercado, J. D., Gong, J., Blumberg, J. B., and Russell, R. M., 1994, β-Carotene supplementation increases antioxidant capacity of plasma in older women, *J. Nutr.* **124**:2397–2403.

Moran, J. R., and Greene, H. L., 1987, Nutritional biochemistry of fat-soluble vitamins, in: *Pediatric Nutrition: Theory and Practice* (R. J. Grand, J. L. Sutphen, and W. H. Dietz, eds.), Butterworths, Boston, pp. 51–67.

Nakamura, T., Takebe, K., Imamura, K., Tando, Y., Yamada, N., Arai, Y., Terada, A., Ishii, M., Kikuchi, H., and Suda, T., 1996, Fat-soluble vitamins in patients with chronic pancreatitis (pancreatitis insufficiency), *Acta Gastroenterologica Belgica* **59:** 10–14.

Olson, J. A., 1994, Vitamin A, retinoids, and carotenoids, in: *Modern Nutrition in Health and Disease* (M. E. Shils, J. A. Olson and M. Shike, eds.), Lea & Febiger, Philadelphia, pp. 287–307.

Parker, R. S., 1996, Absorption, metabolism, and transport of carotenoids, *FASEB J.* **10:** 542–551.

Peters, J. C., Lawson, K. D., Middleton, S. J., and Triebwasser, K. C., 1997, Assessment of the nutritional effects of olestra, a nonabsorbed fat replacement: Summary, *J. Nutr.* **127:** 1719S–1728S.

Reddy, V., and Sivakumar, B., 1972, Studies on vitamin A absorption in children, *Ind. Pediatr.* **9:** 307.

Sato, Y., Arai, H., Miyata, A., Tokita, S., Yamamoto, K., Tanabe, T., and Inoue, K., 1993, Primary structure of α-tocopherol transfer protein from rat liver, *J. Biol. Chem.* **268:** 17705–17710.

Schlagheck, T. G., Riccardi, K. A., Zorich, N. L., Torri, S. A., Dugan, L. D., and Peters, J. C., 1997, Olestra dose response on fat-soluble and water-soluble nutrients in humans, *J. Nutr.* **127:** 1646S–1665S.

Shearer, M. J., 1995, Vitamin D, *Lancet* **345:** 229–234.

Shearer, M. J., Barkhan, P., and Webster, G. R., 1970, Absorption and excretion of an oral dose of tritiated vitamin K_1 in man, *Br. J. Haemotol.* **18:** 297–308.

Sivakumar, B., and Reddy, V., 1978, Studies on vitamin A absorption in children, *Wld. Rev. Nutr. Diet.* **31:** 125–129.

Sokoll, L. J., Booth, S. L., O'Brien, M. E., Davidson, K. W., Tsaioun, K. I., and Sadowski, J. A., 1997, Changes in serum osteocalcin, plasma phylloquinone, and urinary g-carboxyglutamic acid in response to altered intakes of dietary phylloquinone in human subjects, *Am. J. Clin. Nutr.* **65:** 779–784.

Tang, G., Serfaty-Lacrosniere, C., Camilo, M. E., and Russell, R. M., 1996, Gastric acidity influences the blood response to a beta carotene dose in humans, *Am. J. Clin. Nutr.* **64:** 622–626.

Thomas, M. G., Sturgess, R. P., and Lombard, M., 1997, The steroid vitamin D_3 reduces cell proliferation in human duodenal epithelium, *Clin. Sci.* **94:** 375–377.

Thompson, G. R., Lewis, B., and Booth, C. C., 1966, Absorption of vitamin D_3-^{3}H in control subjects and patients with intestinal malabsorption, *J. Clin. Invest.* **45:** 94–102.

Traber, M. G., and Kayden, H. J., 1989, Preferential incorporation of α-tocopherol vs. γ-tocopherol in human lipoproteins, *Am. J. Clin. Nutr.* **49:** 517–526.

Traber, M. G., Schiano, T. D., Steephen, A. C., Kayden, H. J., and Shike, M., 1994, Efficacy of water-soluble vitamin E in the treatment of vitamin E malabsorption in short-bowel syndrome, *Am. J. Clin. Nutr.* **59:** 1270–1274.

van Vliet, T., 1996, Absorption of beta-carotene and other carotenoids in humans and animal models, *Eur. J. Clin. Nutr.* **50:** S32–S37.

van Vliet, T., van Vlissingen, M. F., van Schaik, F., and van der Berg, H., 1996, Beta-carotene absorption and cleavage in rats affected by the vitamin A concentration in the diet, *J. Nutr.* **126:** 499–508.

Wang, X.-D., 1994, Absorption and metabolism of beta-carotene, *J. Amer. Coll. Nutr.* **13:** 314–325.

Warwick, W. J., Hansen, L. G., and Sharp, H., 1976, Absorption of vitamin A in patients with cystic fibrosis, *Clin. Ped.* **15:** 807–810.

CHAPTER 21

Intestinal Metabolism of Interesterified Fats

CARL-ERIK HØY and HUILING MU

1. Stereochemistry of Triacylglycerols

In the human diet many different fats are relevant as sources of triacylglycerols. These include oils that originate from fruits such as palm oil and olive oil, or from seed such as corn, rapeseed, and soybean oil. Additionally, fats from adipose tissues and intramuscular fat droplets from pigs, cattle, poultry, and lamb as well as marine sources such as fish, seal, and whale oils are used in the human diet. These fats possess in general complicated fatty acid profiles involving a range of fatty acids, which could extend to 30 or more different fatty acids, with various chain lengths and unsaturation. In the triacylglycerols the fatty acids are distributed between the three positions of the glycerol moiety. The positions are numbered by the stereochemical numbering system (Fig. 1).

In most triacylglycerol molecules the fatty acids in the sn-1- position and in the sn3- position will be different and there will therefore exist two possible enantiomeric triglycerides with similar fatty acid profiles and with the same fatty acid located in the sn-2- position. In this chapter we use a shorthand notation, that is 16:0/18:2/18:1 for 1-palmitoyl-2-linoleyl-3-oleyl-sn-glycerol. This is also used for triacylglycerols analyzed by regiospecific methods that do not distinguish between sn-1- and sn-3- positions.

From the number of fatty acids present in most of these fats it is clear that the fats must consist of many individual triacylglycerol species. If no restriction on the location of fatty acids is assumed the number of possible triacylglycerol species that can be formed from n different fatty acids is n^3 (Bezard and Sempore, 1995). In general, the number is smaller due to preferential acylation of fatty acids to particular positions and in particular molecular species. An excellent example of this simplification of the species composition of natural fats is cocoa butter, which essentially contains three triacylglycerol species: 16:0/18:1/16:0, 18:0/18:1/18:0, and 16:0/18:1/18:0 (Bracco, 1994).

CARL-ERIK HØY and HUILING MU • Department of Biochemistry and Nutrition, Technical University of Denmark, DK-2800 Lyngby, Denmark.
Intestinal Lipid Metabolism, edited by Charles M. Mansbach II *et al.*, Kluwer Academic/Plenum Publishers, 2001.

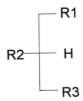

Figure 1. Stereochemistry of triacylglycerol structure. R1, R2 and R3 are the fatty acids acylated to the *sn1*-, *sn2*-, and *sn3*-positions, respectively, according to the stereochemical nomenclature.

2. Degradation of Fats

Fats are predigested by lingual and by gastric lipases as reviewed by Hamosh (1990), which are specific for the *sn*-3-position, but the major digestion results from hydrolysis with pancreatic lipase, which releases the fatty acids located in the *sn*-1,3- positions (Mattson and Beck, 1956). This results in absorption of the dietary fats as *sn*-2-monoacylglycerols and free fatty acids (Mattson and Volpenhein, 1964). These provide the starting material for the intestinal resynthesis of triacylglycerols for export as chylomicrons.

In general the polyunsaturated or monounsaturated fatty acids are located in the *sn*-2-position (Mattson and Volpenhein, 1961), but there are important exceptions such as human milk, which contains palmitic acid in the *sn*-2- position (Tomarelli *et al.*, 1968) and also bovine milk fat (Bracco, 1994), lard (Renaud *et al.*, 1995), and oils from whale or seal (Ackman, 1988) are enriched with saturated fatty acids in the *sn*-2- position. The location of the fatty acids within the triacylglycerol molecule affects the absorption as is evident from studies on fecal loss of fat. Aoyama *et al.* (1996) studied the absorption of 16:0/18:1/16:0 and found higher fecal losses of 16:0 than from 16:0/16:0/18:1. This agrees with earlier findings of Filer *et al.* (1969) indicating the importance of the location of 16:0 in the *sn*-2- position in human milk to reduce fecal loss of energy. In contrast with this de Schrijver *et al.* (1991) did not find any effects of triacylglycerol structure on absorption of native or randomized fats in the rat as determined from fecal lossess. The triacylglycerol structure also has important biological effects. For instance it influences the atherogenic potential of dietary fats such as peanut butter, which upon randomization and relocation of C20- and C22 fatty acid from the *sn*-3-position has a significantly lower effect in monkeys (Kritchevsky *et al.*, 1982).

3. Rationale for Production of Structured Triacylglycerols

In recent years it has become interesting to modify the fatty acid profile of naturally occurring triacylglycerols or to tailor make new triacylglycerols *de novo* to obtain particular functional properties, either physical or physiological. The purposes of these manipulations could be to manufacture fats with, for instance, reduced caloric content compared to normal fats aiming at lowering the energy intake of, for example, chocolate. These fats include Caprenin, which contains 8:0, 10:0, and 22:0 as well as Salatrim, which contains 2:0, 4:0, and 18:0. They are either based on short- or medium-chain fatty acids with decreased energy content per mole triacylglycerol compared to the normally occurring C16- and C18-fatty acids, or they contain long-chain fatty acids that are poorly absorbed in the intestine, due to low hydrolysis by pancreatic lipase or to formation of calcium soaps in the intestine, resulting in fecal losses of energy as described previously. Other fats such as Captex and emulsions such as FE 73403, which contain 8:0, 10:0, and 18:2, are intended for clinical nutrition. In clinical nutrition medium-chain triacylglycerols (MCT) were used for many years due to their easy absorption without formation of chylomicrons and to their direct utilization for energy production without deposition in adipose stores. This subject has been extensively reviewed by Bach and Babayan (1982) and by Babayan (1987). An obvious disadvantage of the MCT is the lack of polyunsaturated fatty acids. In situations of fat malabsorption, for instance following major intestinal surgery, a high incidence of essential fatty acid deficiency has recently been reported (Jeppesen *et al.,* 1997, 1998). In an attempt to overcome situations like this, fats can be made by interesterification of vegetable oils and medium-chain fatty acids to produce structured triacylglycerols, which combine the advantages of MCT of rapid digestion in the intestine and absorption of medium-chain fatty acids into the portal vein with enough polyunsaturated fatty acids to prevent essential fatty acid deficiency. The published work until now has focused on lipids with random triacylglycerol structure, partly due to the unavailability on a commercial scale of triacylglycerols with defined structures. The possible applications of structured lipids have been reviewed by Heird *et al.* (1986), and recently by Bell *et al.* (1997). The major effects have been improved absorption of linoleic acid in cystic fibrosis and the related fat malabsorption (McKenna *et al.,* 1985; Hubbard and McKenna, 1987) and a protein-sparing effect both with structured lipids based on safflower oil (Mok *et al.,* 1984) and on fish oil (Swenson *et al.,* 1991) without supporting tumor growth (Mendez *et al.,* 1992).

4. Production of Structured Triacylglycerols

4.1. Chemical and Enzymatic Randomization

Structured triacylglycerols can be manufactured by a chemical interesterification or by an enzymatic interesterification, with an unspecific lipase producing triacylglycerols with a random distribution of the fatty acids in the triacylglycerols. We refer to these fats as structured triacylglycerols or ST. This procedure in most cases results in the formation of a large number of different triacylglycerol species. Although an analysis of the fatty acid profile of an interesterified fat may indicate that a certain percentage of the fatty acids have been replaced by medium-chain ones, this does not ensure that a replacement has taken place in each triacylglycerol molecule (Fig. 2).

4.2. Regiospecific Interesterification

An alternative is to perform the interesterification using a regiospecific lipase that will result in replacement of the fatty acids in the sn-1- and sn-3- positions of the triacylglycerols. In this chapter we refer to these fats as specific structured triacylglycerols or SST (Fig. 3).

Also in this case only a part of the long-chain fatty acids in the sn-1- and in the sn-3- positions will be replaced, typically 80% (Mu *et al.,* 1998). It is important to realize that the interesterification process results in triacylglycerol formation according to the thermodynamic equilibrium of the reaction. A pure product will therefore not be formed by either the

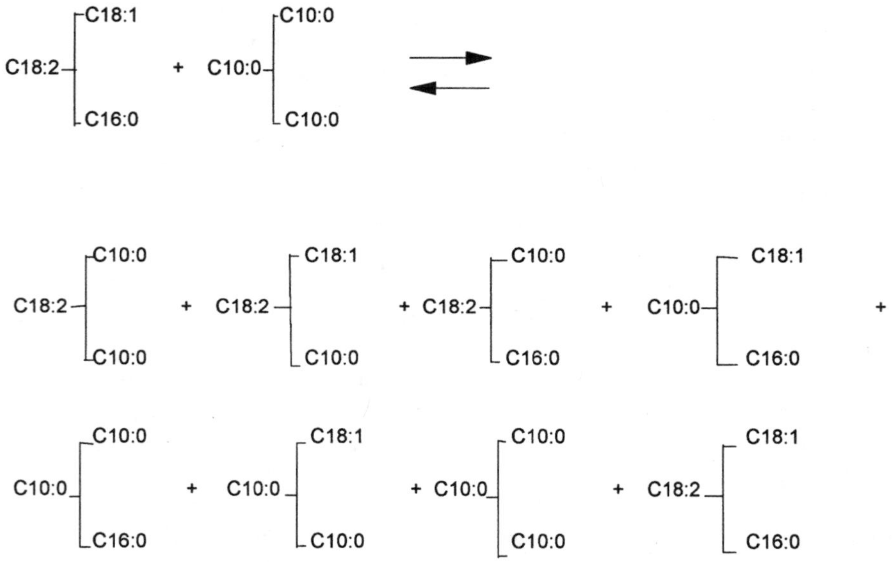

Figure 2. Chemical interesterification between a triacylglycerol representative of a vegetable oil and an MCT. The process is completely random and can be catalyzed, for example, with sodium methoxide.

chemical or the enzymatic process. The incorporation of fatty acids into the interesterified fat may be increased by additional interesterifications involving two or more steps with intermediate removal of liberated long-chain fatty acids. Furthermore, during the interesterification some rearrangement may occur, resulting in acylmigration between the *sn-2-* position and the *sn-1/3-* positions (Mu *et al.,* 1998), and a considerable formation of intermediate products such as diacylglycerols will also be present in the reaction mixture. The product mixture is more complex for randomly interesterified fats, but fats produced by regiospecific interesterification with lipases will also contain by-products. These problems in enzymatic interesterification have been thoroughly examined by Xu *et al.* (1998a). In order to make pure products, additional refining to remove starting material, both triacylglycerols and medium-chain free fatty acids, as well as liberated long-chain fatty acids, will be necessary (Xu *et al.,* 1998b).

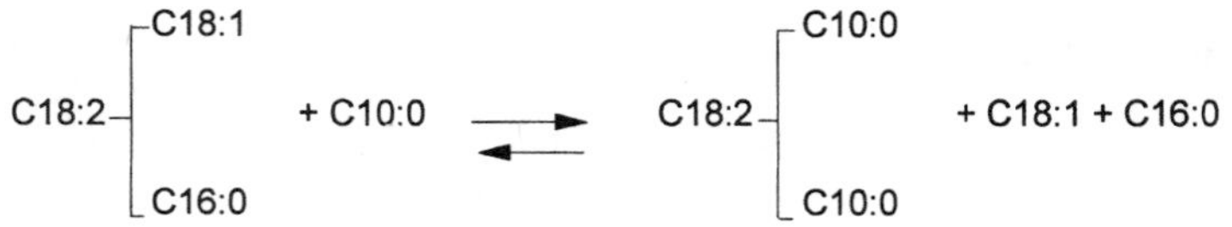

Figure 3. Enzyme-catalyzed interesterification between a triacylglycerol representative of a vegetable oil and a medium-chain fatty acid. The process can be performed using an immobilized regiospecific lipase, for example, from Mucor *miehi.* The process can be performed as a batch or as a continuous process.

5. Absorption of Synthesized Structured Triacylglycerols

5.1. Models for Intestinal Degradation and Absorption of Structured Triacylglycerols

Studies on the metabolism and intestinal absorption of ST as well as of SST have been based on *in vitro* as well as *in vivo* experiments, in most cases performed in rats. In this chapter we review papers based on the following models: *in vitro* digestion of triacylglycerols with lipases; lymphatic absorption following an intragastric or intraduodenal lipid bolus, or continuous duodenal infusion; bile and pancreatic juice diversion as a model for fat malabsorption; disappearance of fat from an irrigated intestinal loop; and determination of fecal fat loss.

5.2. *In Vitro* Degradation with Enzymes

Jandacek *et al.* (1987) prepared 2-oleyl-1,3-dioctanoyl-*sn*-glycerol, 8:0/18:1/8:0, in pure form and compared the degradation *in vitro* by pancreatic lipase including bile fluid from rats to the degradation of MCT (containing 67% 8:0, 23% 10:0, and 10% other fatty acids), 10:0/10:0/10:0, or 2:0/18:1/2:0. The rates of lipolysis, determined from the formation of free fatty acids, were 8:0/18:1/8:0 > MCT > 10:0/10:0/10:0 > 2:0/18:1/2:0, indicating that *in vitro* the specific structured triacylglycerol was well degraded by intestinal lipases. In a parallel experiment they examined the hydrolyses by a commercial porcine pancreatic lipase and found the following initial rates: MCT > 8:0/18:2/8:0 > sunflower oil. The total digestion at the end point was similar for MCT and for 8:0/18:2/8:0. They also examined the lipolysis of a series of structured triacylglycerols with random structure (Captex 810 series) ranging in medium-chain fatty acid (MCFA) contents from 32% to 80% and found that the hydrolysis rate was highest for the triacylglycerols with the highest contents of medium-chain fatty acids.

5.3. Intestinal Absorption of Structured Triacylglycerols Containing Linoleic Acid

Structured triacylglycerols and their metabolism have been studied in only a few papers. In the evaluation of the results reported in the literature it is important to keep in mind as outlined in sections 4.1 and 4.2 that in actual production, the triacylglycerols formed will be a mixture of many molecules. Some investigations, however, have been performed using pure triacylglycerols. Jandacek *et al.* (1987) used the irrigated, isolated small intestine loop model to demonstrate hydrolysis of the purified specific structured triacylglycerols 8:0/18:2/8:0 and 18:1/18:2/18:1, both with radiolabeled 18:2. They measured the label remaining in the loop after 45 min incubation and found that the remaining label from 18:1/18:2/18:1 was approximately 2½ times as large as was that from 8:0/18:2/8:0, which they interpreted as the result of a more rapid intestinal hydrolysis of the *sn*1 and *sn*3 fatty acids from a triacylglycerol containing solely medium-chain fatty acids in the *sn*-1/3- positions compared to a long-chain triacylglycerol, combined with a good absorption of the *sn*-2-monoacylglycerol. The absorptions of 18:2 as calculated were 45% from 8:0/18:2/8:0 and 17% from 18:1/18:2/18:1.

Ikeda *et al.* (1991) examined the absorption of fatty acids to the thoracic duct lymph using a wide array of purified SSTs containing 18:2, 8:0, and 10:0. The SSTs were either synthesized or were made by enzymatic interesterification. They used rats that were given an infusion of physiological saline / 5% glucose at a rate of 3 ml/hr until the end of the lymph collection period. The fats were given intragastrically as a 2 ml emulsion of 0.2 mmole fat, 47 mg sodium taurocholate, and 8.3 mg albumin. The lymph was collected for 24 hr. The authors found that the nature of the fats did not affect the lymph flow. They interpreted the appearance of 18:2 in the lymph as a measure of lymphatic absorption of 18:2, which, however, did not account for the contribution from endogenous fatty acids. They reported the following differences in absorption: 10:0/18:2/10:0 > 18:2/18:2/18:2 > 18:2/10:0/18:2 (Fig. 4) and 8:0/18:2/8:0 > 18:2/18:2/18:2 = 18:2/10:0/18:2 (Fig.5). In general the 24 hr recoveries of 18:2 were 60% to 80% of the amount of 18:2 administered.

They also demonstrated that 18:2 was better absorbed from 8:0/18:2/8:0 than from a physical mixture of 8:0/8:0/8:0 and 18:2/18:2/18:2, indicating that the degradation of triacylglycerols was facilitated by the location of medium-chain fatty acids in the *sn*-1/3-positions, which may imply that SST will be superior to a physical mixture of long-chain triacylglycerols and MCT as a source of 18:2, provided this is located in the *sn*-2- position (Fig 6).

In these experiments, low recoveries of 8:0 and 10:0 were observed compared to the amounts administered (Figs. 4 and 5). This was interpreted to reflect that the medium-chain fatty acids were preferentially transported by the portal vein system, an assumption that was supported by consistently higher recoveries in the lymph of 10:0 than of 8:0. When SST such as 18:2/8:0/18:2 and 18:2/10:0/18:2 are fed to rats, recoveries in the lymph of 8:0 and

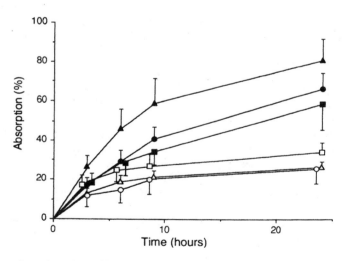

Figure 4. Absorption of capric and linoleic acids into thoracic duct lymph of rats upon intragastric administration of fat emulsions containing sodium taurocholate and albumin. ●, ■, or ▲ shows linoleic acid absorption in rats given 18:2/18:2/18:2, 18:2/10:0/18:2, or 10:0/18:2/10:0, respectively. ○ or △ shows capric acid absorption in rats given 10:0/10:0/10:0, 18:2/10:0/18:2, or 10:0/18:2/10:0, respectively. Data are means ± SE of 5 or 6 rats. Significant differences were not observed in this experiment. Reproduced from Ikeda *et al.* (1991) with the permission of AOCS Press.

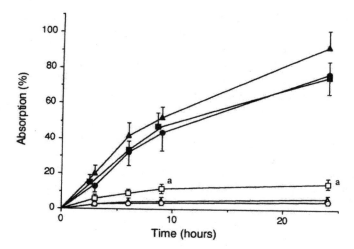

Figure 5. Absorption of caprylic and linoleic acids into thoracic duct lymph of rats. ●, ■, or ▲ shows linoleic acid absorption in rats given 18:2/18:2/18:2, 18:2/8:0/18:2, or 8:0/18:2/8:0, respectively. ○ or △ shows caprylic acid absorption in rats given 8:0/8:0/8:0, 18:2/8:0/18:2, or 8:0/18:2/8:0, respectively; a = caprylic acid absorption was significantly higher in 18:2/8:0/18:2 than in 8:0/8:0/8:0 and 8:0/18:2/8:0 groups at $p < 0.05$. Data are means $\pm$ SE of 5 to 7 rats. Reproduced from Ikeda *et al.* (1991) with the permission of AOCS Press.

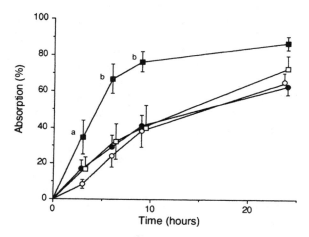

Figure 6. Absorption of linoleic acid into thoracic duct lymph of rats. ●, ■, or ○ shows linoleic acid absorption in rats given 18:2/8:0/18:2, 8:0/18:2/8:0, a 1 to 2 or a 2 to 1 mixture of 8:0/8:0/8:0 and 18:2/18:2/18:2, respectively; a = significantly higher than the other three groups at $p < 0.05$; b = significantly higher than the 1 to 2 mixture of 8:0/8:0/8:0 and 18:2/18:2/18:2 at $p > 0.05$. Data are means $\pm$ SE of 5 or 6 rats. Reproduced from Ikeda *et al.* (1991) with the permission of AOCS Press.

10:0 are higher than when SST such as 8:0/18:2/8:0 and 10:0/18:2/10:0 are given. This is due to the monoacylglycerols containing a MCFA in the $sn2$- position as a result of the intralumenal lipolysis in the former case, whereas in the latter case the MCFAs are absorbed as free fatty acids. 10:0 and 8:0 were better absorbed from 18:2/10:0/18:2 and from 18:2/8:0/18:2 than from fats containing MCT. Tso *et al.* (1995) also examined the lymphatic absorption following continuous infusion of 8:0/18:2/8:0, radiolabeled in 18:2, and compared this with 8:0/8:0/18:2, radiolabeled in 8:0, and 18:2/18:2/18:2, radiolabeled in 18:2. They found that the lymphatic output of linoleic acid from 8:0/18:2/8:0 resembled 18:2/18:2/18:2, but occurred at a lower rate, indicating that 18:2 was absorbed in the intestine, but presumably more distally. A recent paper has demonstrated a rapid uptake of linoleic acid in Caco-2 cells from 8:0/ C14–18:2/8:0, which was further increased by incubation with human gastric lipase, and which corrected the fatty acid profile charateristic of essential fatty acid deficiency in these cells (Spalinger *et al.*, 1998).

5.4. The Effects of Previous Dietary History

The previous dietary history is important for the absorption of fat. Intake of polyunsaturated fatty acids alters the intestinal membrane within few weeks (Brasitus *et al.*, 1985). The resulting increase in membrane phospholipid fluidity increases the absorption of C16- and C18- fatty acids, but not of medium-chain fatty acids (Thomson *et al.*, 1987). They later demonstrated that this affects the whole-body oxidation of fatty acids (Clandinin *et al.*, 1995). The effect of dietary history on the absorption of specific structured and randomized triacylglycerols was examined by M. M. Jensen *et al.* (1994). Prefeeding fish oil as compared to vegetable oil for 4 weeks thus increased the unsaturation of the intestinal phospholipids. Following ingestion of a specific structured triacylglycerol containing 8:0 and 10:0 in the $sn1/3$- positions and 18:2 in the sn-2- position or the same fat in a randomized form, an increased absorption into the thoracic duct lymph was found in the fish oil group of 18:2 from SST as well as from ST as compared to the vegetable oil group, whereas 10:0 absorption was similar. This agreed with the observations by Thomson *et al.* (1987). It was later demonstrated that dietary manipulation could alter the fatty acid profiles of the intestinal phospholipids within few hours (Poulsen *et al.*, 1997), indicating that the fatty acid profile of a single meal could change the intestinal membrane to a degree that could affect absorption of fat. A high recovery of 10:0 in the lymph following intake of ST compared to SST was reported (M. M. Jensen *et al.*, 1994), presumably reflecting the high content in the $sn2$- position as was also described by Ikeda *et al.* (1991). As also reported by G. L. Jensen *et al.* (1994), the recovery in the lymph was larger of 10:0 than of 8:0 in contrast to the dietary contents. Dietary medium-chain fatty acids in the form of MCT can also affect the intestinal membrane lipid composition and can increase the contents of 20:4n-6 within 40 hr (Wang *et al.*, 1996). The effect of this on the absorption of structured triacylglycerols has not yet been examined. Qu *et al.* (1996) examined the effects of protein deficiency on intestinal morphology and the effect of fat intake on restoration. Dietary protein deficiency induced atrophy of the colonic mucosa but not the small intestinal mucosa. Supplies of protein in combination with structured lipid made from soybean oil (5%), from MCT (50%), from menhaden oil (25%) and from tributyrin (20%) by random interesterification induced higher rate of colon cell proliferation than did a physical mixture of the same lipids, which may imply that colon absorption of short-chain fatty acids may be influenced by dietary fats.

Table 1. The Fatty Acid Compositions (wt%) and Structures of the Test Triglycerides[a]

Fatty acid	Defined MLM	Randomized MLM	MCT and soybean oil	Soybean oil
8:0	27.0	27.2	19.1	ND[b]
10:0	13.7	16.3	12.0	ND
16:0	2.2	2.0	7.3	10.6
18:0	0.8	0.7	2.5	3.6
18:1n-9	14.2	13.6	17.1	24.9
18:2n-6	37.8	35.6	36.3	52.7
18:3n-3	4.2	4.5	5.2	7.5
2-position				
8:0	ND	18.9	13.5	ND
10:0	ND	17.3	11.8	ND
16:0	2.5	2.6	0.4	0.6
18:0	ND	0.9	ND	ND
18:1n-9	24.9	14.6	16.8	23.4
18:2n-6	67.9	40.0	52.0	68.9
18:3n-3	6.6	4.7	5.4	7.1

[a]MLM defined triglycerides with medium-chain, long-chain, medium-chain fatty acids; MCT; medium-chain triglycerides. Reproduced from Christensen *et al.* (1995b) with the permission of AOCS Press.
[b]ND: not detected.

5.5. Intestinal Absorption of Structured Triacylglycerols in Pancreatic Deficiency

Christensen *et al.* (1995a) introduced a model of pancreatic deficiency leading to fat malabsorption typical of cystic fibrosis, for example. They inserted a cannula into the common bile and pancreatic duct and diverted the juice containing pancreatic lipase as well as bile acids. The fats examined were an SST made from soybean oil and medium-chain fatty acids made by enzymatic interesterification using an *sn*-1,3 specific lipase from Mucor *miehi,* the same fat in a randomized form, a third fat which was a physical mixture of soybean oil and MCT, and finally soybean oil (Table 1). The first three fats had similar contents of 18:2, but in the SST 18:2 was located in the *sn*2- position; in the ST 18:2 was distributed between *sn*-1-, *sn*-2- and *sn*-3- positions. The mixture of soybean oil and MCT contained the same level of 18:2, but 18:2 and 8:0 were not located in the same triacylglycerol molecules.

Fats were administered intragastrically as emulsions with taurocholate and choline after recovery from surgery. The lymph was collected from the thoracic duct during a period of 8 hours after the administration of fat and the contents of 18:2 and 18:3 was determined as a measure of lymphatic absorption. In the thoracic lymph significantly higher absorption of 18:2 and 18:3 was observed following intake of STT compared to the other fats (Figs. 7–8), indicating that location of medium-chain fatty acids in the *sn*-1,3- positions favored degradation of triacylglycerols and absorption of linoleic acid or linolenic acid as *sn*-2-monoacylglycerols compared to ST and a physical mixture, in agreement with the report of Ikeda *et al.* (1991). Soybean oil was poorly absorbed. As could be anticipated, because this was a model of malabsorption, the recoveries of linoleic acid in the lymph in this experiment were low, 4% to 10% depending on the fat administered (Table 2), compared to Ikeda *et al.* (1991), who reported recoveries of 50% to 70%.

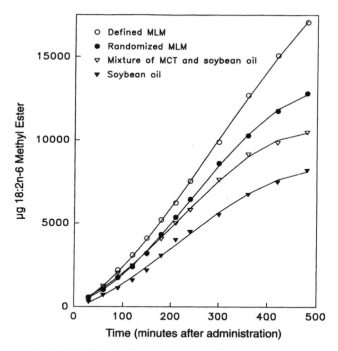

Figure 7. The amount (μg) of 18:2n-6 transported through the lymphatics accumulated over the experimental period after administration of defined MLM ($n = 6$), randomized MLM, mixture of MCT and soybean oil ($n = 5$), and soybean oil ($n = 4$). The curves indicate the best fitted cubic polynomials through the actual means (symbols). See Table 1 for abbreviations. Reproduced from Christensen *et al.* (1995a) with the permission of AOCS Press.

5.6. Intestinal Absorption of Structured Triacylglycerols Containing 20:5 and 22:6

The interest in effects of n-3 fatty acids on plasma lipids (Dyerberg *et al.*, 1975; Harris, 1989) and on brain and visual function (Neuringer *et al.*, 1988) has also initiated studies on the production of structured triacylglycerols containing 20:5n-3 (EPA) and 22:6n-3 (DHA) in order to supplement long-chain n-3 fatty acids in cases where the normal sources such as fish oils or fats from marine mammals cannot be applied. Christensen *et al.* (1995c) produced SST-containing 63 mole% 10:0 that was specifically located in the sn-1,3-positions as well as a ST that was manufactured by randomization of the SST. The two fats therefore had similar fatty acid profiles (Table 3), but different locations of the fatty acids.

Mesenteric duct cannulated rats were given 0.5 ml of the fats as an intragastric bolus and mesenteric lymph was collected for 24 hr (Christensen *et al.*, 1995c). Intragastric infusion of saline at 2 ml was given throughout the experiment. For these fats a higher lymphatic absorption of the polyunsaturated fatty acids EPA and DHA was observed for the first 8 hr of the experiment following intake of SST in which these fatty acids were located in the sn-2- position compared to the ST, in which a part of the EPA and DHA were randomly distributed (Fig. 9).

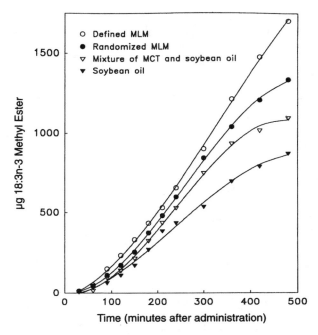

Figure 8. The amount (μg) of 18:3 *n*-3 transported through the lymphatics accumulated over the experimental period after administration of defined MLM ($n = 6$), randomized MLM ($n = 6$), mixture of MCT and soybean oil ($n = 5$), and soybean oil ($n = 4$). The curves indicate the best fitted cubic polynomials through the actual means (symbols). See Table 1 for abbreviations. Reproduced from Christensen *et al.* (1995a) with the permission of AOCS Press.

Table 2. Recovery (% of administered amount of fatty acid)

Administered fat	Fatty acid			
	8:0	10:0	18:2*n*-6[a]	18:3*n*-3[b]
Structured MLM	0.8 ± 0.2	1.6 ± 0.4	9.9 ± 1.7	8.9 ± 1.9
($n = 6$)				
Randomized MLM	0.7 ± 0.1	1.1 ± 0.2	5.7 ± 1.7	4.4 ± 1.0
($n = 6$)				
MCT and soybean oil	1.3 ± 0.5	4.6 ± 1.6	6.6 ± 1.6	5.1 ± 1.6
($n = 5$)				
Soybean oil	ND[c]	ND	3.7 ± 1.3	2.8 ± 1.3
($n = 4$)				

[a]Significance level: for comparisons of 18:2*n*-6, $p = 0.07$ for structured MLM vs. randomized LML; $p = 0.20$ for structured MLM vs. MCT and soybean oil; $p = 0.03$ for structured MLM vs. Soybean oil. [b]For comparisons of 18:3*n*-3, $p = 0.06$ for structured MLM vs. Randomized MLM; $p = 0.17$ for structured MLM vs. MCT and soybean oil; $p = 0.05$ for structured MLM vs. Soybean oil. [c]ND, not detected. Data are expressed as mean $\pm$ SEM. Student's *t*-test was used to test for significant differences between the dietary groups. See Table 1 for abbreviations. Reproduced from Christensen *et al.* (1995a) with the permission of AOCS Press.

Table 3. Fatty Acid Compositions in Triacylglycerols and in sn-2
Monoacylglycerols of Specific M-n3-M and random M-n3-M^a

Fatty acid	Triacylglycerols		Sn-2 Monoacylglycerols	
	Specific	Random	Specific	Random
	mol %			
C10:0	63.6	64.2	0.7	46.6
C14:0	0.6	0.2	1.2	0.3
C16.0	1.7	0.7	2.6	1.6
C16:1n-7	1.5	1.1	3.5	1.5
C18:0	0.3	0.2	0.7	0.5
C18:1	4.0	3.3	11.1	5.2
C18:2n-6	0.7	0.5	1.8	0.8
C18:3n-3	0.4	0.4	1.4	0.6
C18:4n-3	1.4	1.5	4.1	2.2
C20:1	1.3	0.9	2.8	1.4
C20:4n-6	0.9	1.0	2.8	1.6
C20:4n-3	0.6	0.6	1.7	0.9
C20:5n-3	10.2	11.7	30.7	17.6
C22:1	0.8	0.4	1.4	0.7
C22:5n-3	1.5	1.6	4.6	2.6
C22:6n-3	9.2	10.3	24.2	14.1

^{a}The results are averages of three determinations. Data for sn-2 monoacylglycerols were
obtained by Grignard degradation of the oil followed by isolation and analysis of the sn-2
monoacylglycerols. Reproduced from Christensen *et al.* (1995c) with the permission of
Am.Soc.Clin.Nutr.

This initial difference in absorption agreed with previous reports on the degradation *in
vitro* of fish oils (Bottino *et al.,* 1967) with pancreatic lipase, which was interpreted as evidence indicating that the presence of EPA and DHA in the sn-1,3- positions may provide
steric hindrance of the hydrolysis. In agreement also with the results of Bottino *et al.* (1967),
M. M. Jensen *et al.* (1994) found the highest rate of absorption of 10:0 for the SST compared to the randomized ST. This may be explained by the release of 10:0 as free fatty acid
occurring rapidly from the specific structured triacylglycerol (STT), whereas in the randomized fats a large proportion of the 10:0 would be absorbed as sn-2-monoacylglycerols
following slower degradation of triacylglycerols containing EPA and DHA in the sn1,3-po-
sitions. It should be noted, however, that the interpretation of the low hydrolysis noted *in
vitro* may not be extended to *in vivo* situation, which takes place over a period of several
hours with concurrent removal of fatty acids in the intestine. By examining the sn-1,2- and
sn-2,3-DAGs (-diacylglycerols) formed during pancreatic lipase digestion of fish oil, Yang
et al. (1990) concluded that there was no significant discrimination of n-3 fatty acids present in the sn-1,3-positions. In support of this, the accumulated absorptions of n-3 polyunsaturated fatty acids over a period of 24 hr were not statistically different (Christensen
et al., 1995c).

Lymph absorption experiments on larger animals involve complicated surgery and are
extremely expensive. Presumably due to experimental complications only one experiment
involving lymphatic absorption in larger animals has been published. G. L. Jensen *et al.*

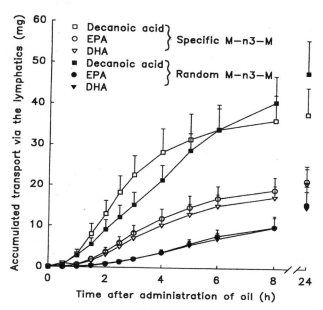

Figure 9. Accumulated transport (mg) through the lymphatics of decanoic acid, eicosapentaenoic acid (EPA), and docosahexaenoic acid (DHA) after intragastric administration of specific M-n3-M (n = 5), M = medium chain fatty acid, and random M-n3-M (n = 7). The profiles over the first 8 hr were significantly different for both EPA and DHA (both $p < 0.01$, repeated measures). However over 24 hr, total absorption of EPA and DHA were not significantly different (both p > 0.05), indicating slower and more prolonged absorption from the random M-n3-M. The overall transport of decanoic acid was not significantly different between the two oils (repeated measures). The amounts transported were determined by gas–liquid chromatography using internal standard. Average ± SEM. Reproduced from Christensen *et al.* (1995c) with the permission of Am.Soc.Clin.Nutr.

(1994) used a canine model to investigate the absorption of fatty acids from an interesterified fat manufactured from MCT (8:0 and 10:0) and fish oil by randomization as well as from a physical mixture of the same fats (Table 4).

In this experiment the fatty acid profiles were similar. From analyses of the equivalent carbon number of the fats it was clear that the interesterification had left a fraction of more than 20 wt% of MCT in the structured triglyceride fat. The authors observed similar lymphatic absorption of n-3 fatty acids from the two fats, but found 2.6-fold higher absorption of medium-chain fatty acids, mainly 10:0, from the structured triacylglycerol compared with the physical mixture, which was attributed to the presence of 10:0 in the *sn*-2-position. Although the content of 8:0 was three times the content of 10:0 in the diet, the recovery in the lymph of 10:0 was higher than of 8:0, which may reflect differences in the positioning of fatty acids during randomization, but it may also be caused by acylmigration of fatty acids from the *sn*-2 to *sn*-1,3- positions during degradation with pancreatic lipase, which is more rapid for 8:0 than for 10:0, combined with a higher absorption of 8:0 than of 10:0 via the portal vein. The high content of 8:0/8:0/8:0 (13% of the total fat) may also contribute

Table 4. Fatty Acid Composition of Diet[a]

Fatty acid	Structured Triacylglycerol	Physical mix (% by weight)[b]
8:0	40.16 ± 0.51	43.03 ± 0.62
10:0	12.74 ± 0.13	13.46 ± 0.52
12:0	0.51 ± 0.02	0.54 ± 0.15
14:0	2.34 ± 0.01	2.38 ± 0.08
16:0	6.44 ± 0.04	6.40 ± 0.15
16:1	5.72 ± 0.11	5.65 ± 0.02
16:3	1.27 ± 0.05	1.27 ± 0.03
18:0	1.52 ± 0.05	1.50 ± 0.07
18:1n-9	7.05 ± 0.03	6.86 ± 0.89
18:2n-6	4.56 ± 0.24	4.43 ± 0.13
18:3n-3	1.80 ± 0.08	1.45 ± 0.04
20:1n-9	4.08 ± 0.32	2.36 ± 0.23
20:4n-6	0.40 ± 0.14	0.17 ± 0.07
20:5n-3	5.75 ± 0.37	5.48 ± 0.40
22:4n-6	0.25 ± 0.06	0.23 ± 0.02
22:5n-3	0.82 ± 0.19	0.73 ± 0.04
22:6n-3	4.52 ± 0.28	4.05 ± 0.12
Sum MCFA[c]	53.41 ± 0.64	57.03 ± 0.98
Sum n-6	5.21 ± 0.11	4.83 ± 0.12
Sum n-3	12.89 ± 0.29	11.71 ± 0.46

[a]$\pm$ SE; n = 4 determinations. [b]Percent of total fatty acids. [c]Medium-chain fatty acids include 8:0–12:0. Reproduced from G.L. Jensen *et al.* (1994) with the permission of Am.Soc.Clin.Nutr.

to this effect because Ikeda *et al.* (1991) observed lower lymphatic recovery of 8:0 than of 10:0 from the respectives MCTs. The authors pointed out the possibility to direct medium-chain fatty acids towards extrahepatic tissues by interesterification, which introduces the medium-chain fatty acid into the *sn*-2-position.

The effects of dietary intervention with structured triacylglycerols as well as with specific structured triacylglycerols have been examined in rat pups from weaning (Christensen and Høy, 1997; Christensen *et al.*, 1998). In general, no effects on fatty acid profiles of phospholipids in the brain, the liver, or the retina were found. Similarly, no major functional effects, for example, visual function, auditory functions, or learning behavior were reported. This may result from the levels of *n*-3 applied, which were much higher than in maternal milk.

6. Intestinal Absorption of Naturally Occurring Structured Triacylglycerols

In a review on structured lipids it will be of interest to include the metabolism of triacylglycerols that occur in nature and are structured. Oils from marine mammals contain 20:5 and 22:6, which are located in the *sn*-1,3- positions contrary to fish oil, which contain the polyunsaturates located in the *sn*-2- position. Christensen *et al.* (1994) examined the lymphatic absorption of seal oil versus fish oil, using oils that had very similar fatty acid profiles, and found that the initial rate of absorption was higher following an intragastric bolus of fish oil compared with seal oil. However, after 24 hr the accumulated lymph recoveries of 20:5 and of 22:6 were similar for the two groups. In a recent study Ikeda *et al.* (1998) demonstrated that the distribution actually affected the arachidonic acid content of platelets.

7. Intestinal Resynthesis of Chylomicron Triacylglycerols

The intestinal resynthesis of triacylglycerols occurs mainly by the *sn*-2-monoacylglycerol pathway. This implies a conservation of the fatty acid located in the *sn*-2- position of the dietary triacylglycerols during digestion, absorption, and resynthesis of intestinal triacylglycerols for export as chylomicrons. This has been demonstrated with synthetic 12:0/ 12:0/18:1 by Åkesson *et al.* (1978), who found that 12:0 was conserved in the *sn*-2-position. Data by Yang and Kuksis (1991), who fed menhaden oil to rats and found retention of approximately 85% of the fatty acids in the *sn*-2-position, including 22:6*n*-3 and 20:5*n*-3, substantiated these findings. Further, Christensen and Høy (1996) observed that following intake of marine oils, either fish oil with *n*-3 fatty acid preferentially located in the *sn*-2-position or seal oil with the *n*-3 fatty acids preferentially located in the *sn*-1,3-positions, the fatty acids in the *sn*-2-position were conserved in the location of polyunsaturated fatty acids in the chylomicron triacylglycerols. The polyunsaturated fatty acids were found in the *sn*-2-position following fish oil feeding and in the *sn*-1,3-positions following seal oil feeding. The reacylation of the *sn*-2-monoacylglycerols by monoacylglycerol and diacylglycerol acyltransferases is stereospecific as shown by Lehner *et al.* (1993), favoring formation of the *sn*-1,2-diacylglycerol over *sn*-2,3-diacylglycerol by 9:1. This points to stereospecific production of the triacylglycerols during absorption, which may influence their degradation in the tissues. Triacylglycerol synthesis will reflect the absorbed fatty acids if longchain triacylglycerols are digested, but an endogenous supply of fatty acids circulated to the intestine will contribute to mucosal triacylglycerol if medium-chain fatty acids are absorbed and are preferentially transferred by the portal vein.

It is generally agreed that some medium-chain fatty acids will be transported by the lymphatics (M. M. Jensen *et al.*, 1994; G. L. Jensen *et al.*, 1994; Christensen *et al.*, 1995a, 1995c; Tso *et al.*,1995). The triacylglycerols formed will be more rapidly degraded by lipoprotein lipase (Hultin *et al.*, 1994) than long-chain triacylglycerols, and the long-chain fatty acids, which are combined with medium-chain fatty acids in plasma triacylglycerols, will be oxidized more rapidly.

Effects of the distribution of fatty acids in the triacylglycerols have been observed on the clearance of chylomicrons and tissue distribution of fatty acids (Christensen *et al.*, 1995b) as well as on the clearance of injected emulsions (Redgrave *et al.*, 1988; Mortimer *et al.*, 1988). The effects of intramolecular distribution of dietary fatty acids on the size of chylomicron have also been reported by Aoe *et al.* (1997), who found that chylomicrons were larger following gastric infusion of 18:1/16:0/18:1 than of 18:1/18:1/16:0 and that the triacylglycerol transport followed the same pattern, whereas apoprotein A-I transport was similar for the two fats, indicating that the dietary triacylglycerol structure affects lipoprotein metabolism. Renaud *et al.* (1995) demonstrated by comparing the effects of palm oil and lard both in either native or randomized form that the presence of a saturated fatty acid in the *sn*-2-position had significant negative effect on the conversion of linoleic acid into arachidonic acid that was abolished by randomization of the fat.

If medium-chain fatty acids are present in the dietary triacylglycerols, the location of the fatty acids will determine the absorption to the lymphatics relative to portal vein transport. This was demonstrated by Tso *et al.* (1995), who compared absorption of 8:0/18:2/8:0, C14-labeled in 18:2, with 8:0/8:0/18:2, C14-labeled in 8:0, and 18:2/18:2/18:2, C14-labeled in 18:2, in lymph fistulated rats given a continuous duodenal infusion of the fats. They found that despite degradation in the intestinal lumen there was a higher recovery of labeled 18:2 in the lower intestine following intake of 8:0/18:2/8:0 than from 8:0/8:0/18:2, indicating a delayed absorption of the monoacylglycerol. This was attributed to lack of endogenous fatty acids for resynthesis of chylomicron triacylglycerols for lymphatic export, and it resulted in a slow absorption of 2-linoleate contrary to the observations by Christensen *et al.* (1995a, 1995c). In both SSTs the lymphatic transport of 8:0 was low, indicating that 8:0 was transported in the portal vein whether it was absorbed as free fatty acid or as *sn*-2-monoacylglycerol. Tso *et al.* (1995) concluded that in order to absorb the linoleic acid from the *sn*-2- position efficiently it may be necessary to add additional long-chain triacylglycerols in order to provide fatty acids for resynthesis. In human chylomicrons Swift *et al.* (1990), in agreement, found 13% medium-chain fatty acids and a reduction of the chylomicron transport by 80% following intake of MCT for 6 days. Christophe *et al.* (1985) found that in malabsorption patients a diet containing MCT and linoleic acid as *sn*-2-monoacylglycerol the linoleic acid status improved, indicating that although long-chain fatty acids are limiting for chylomicron synthesis in absorption experiments this may be overcome in long-term feeding. This agrees with the findings of M. M. Jensen *et al.* (1994) and of G. L. Jensen *et al.* (1994) that randomization increased the lymph content of medium-chain fatty acids as compared with a physical mixture or an SST. You *et al.* (1998) pointed out that medium-chain fatty acids, when administered continuously at a low rate as MCT, could be elongated into palmitic acid and could introduce early onset of essential fatty acid deficiency. The situation was normalized by adding long-chain triacylglycerol and was not found for medium-chain fatty acids absorbed as *sn*-2-monoglyceride when long-chain fatty acids were available, as is the case during absorption of structured lipids.

It has recently been reported that dietary medium-chain fatty acids may affect the expression of calcium-binding proteins in the intestinal brush border cells (Devlin *et al.,* 1996). The importance of this observation in relation to enteral applications of structured lipids has not yet been examined.

The production of structured as well as of specific structured triacylglycerols has entered an interesting and promising phase in which the application of new enzyme technology will change our possibilities to tailor make fats for particular applications. In the first phase this new direction in fat technology will probably focus on increasing the uptake of energy and of polyunsaturated fatty acids in malabsorption and related disorders and to direct the fatty acids toward specific tissues for repair or for energy consumption. Another area will be the assembly of fats optimized for preterm infants for whom the fats will be tailored to contain long-chain *n*-3 fatty acids essential for development of the central nervous system and for minimizing the fecal loss. Finally, fats may be manufactured for slightly overweight persons to minimize their energy consumption while maintaining satiety.

References

Ackman, R. G., 1988, Some possible effects on lipid biochemistry of differences in the distribution on glycerol of long-chain *n*-3 fatty acids in the fats of marine fish and marine mammals, *Atherosclerosis* **70**:171–173.

Åkesson, B., Gronowitz, S., Herslof, B., and Ohlson, R., 1978, Absorption of synthetic, stereochemically defined acylglycerols in the rat, *Lipids* **13**:338–343.

Aoe, S., Yamamura, J., Matsuyama, H., Hase, M., Shiota, M., and Miura, S., 1997, The positional distribution of dioleyl-palmitoyl glycerol influences lymph chylomicron transport, composition and size in rats, *J. Nutr.* **127**: 1269–1273.

Aoyama, T., Fukui, K., Taniguchi, K., Nagaoka, S., Yamamoto, T., and Hashimoto, Y., 1996, Absorption and metabolism of lipids in rats depend on fatty acid isomeric position, *J. Nutr.* **126**:225–231.

Babayan, V. K., 1987, Medium-chain triglycerides and structured lipids, *Lipids* **22**:417–420.

Bach, A. C., and Babayan, V. K., 1982, Medium-chain triglycerides: An update, *Am. J. Clin. Nutr.* **36**:950–962.

Bell, S. J., Bradley, D., Forse, R. A., and Bistrian, B., 1997, The new dietary fats in health and disease, *J. Am. Diet. Assoc.* **97**:280–286.

Bezard, J.-A., and Sempore, B. G., 1995, Structural analysis of peanut oil triacylglycerols, in: *New Trends in Lipid and Lipoprotein Analyses* (J.-L. Sebedio and E. G. Perkins, eds.), AOCS Press, Champaign, IL, pp. 106–132.

Bottino, N. R., Vandenburg, G. A., and Reiser, R., 1967, Resistance of certain long-chain polyunsaturated fatty acids of marine oils to pancreatic lipase hydrolysis, *Lipids* **2**:489–493.

Bracco, U., 1994, Effect of triglyceride structure on fat absorption, *Am. J. Clin. Nutr.* **60**:1002S–1009S.

Brasitus, T. A., Davidson, N. O., and Schachter, D., 1985, Variations in dietary triacylglycerol saturation alter the lipid composition and fluidity of rat intestinal plasma membrane, *Biochim. Biophys. Acta* **812**:460–472.

Christensen, M. S., and Høy, C.-E., 1996, Effects of dietary triacylglycerol structure on triacylglycerols of resultant chylomicrons from fish oil- or seal oil-fed rats, *Lipids,* **31**:341–344.

Christensen, M. M., and Høy, C.-E., 1997, Early dietary intervention with structured triacylglycerols containing docosahexaenoic acid. Effect on brain, liver, and adipose tisssue, *Lipids* **32**:185–191.

Christensen, M. S., Høy, C.-E., and Redgrave, T. G., 1994, Lymphatic absorption of *n*-3 polyunsaturated fatty acids from marine oils with different intramolecular fatty acid distributions, *Biochim. Biophys. Acta* **1215**:198–204.

Christensen, M. S., Müllertz, A., and Høy, C.-E., 1995a, Absorption of triglycerides with defined or random structure by rats with biliary and pancreatic diversion, *Lipids* **30**:521–526.

Christensen, M. S., Mortimer, B.-C., Høy, C.-E., and Redgrave, T. G., 1995b, Clearance of chylomicrons following fish oil and seal oil feeding, *Nutr. Res.* **15**:359–368.

Christensen, M. S., Høy, C.-E., Becker, C. C., and Redgrave, T. G., 1995c, Intestinal absorption and lymphatic transport of eicosapentaenoic (EPA), docosahexaenoic (DHA), and decanoic acids: Dependence on intramolecular triacylglycerol structure, *Am. J. Clin. Nutr.* **61**:56–61.

Christensen, M. M., Lund, S. P., Simonsen, L., Hass, U., Simonsen, S. E., and Høy, C.-E., 1998, Dietary structured triacylglycerols containing docosahexaenoic acid given from birth affect visual and auditory performance and tissue fatty acid profile, *J. Nutr.* **128**:1011–1017.

Christophe, A., Verdonk, G., Robberecht, E., and Mahathanakhun, R., 1985, Effect of supplementing medium-chain triglycerides with linoleic acid-rich monoglycerides on severely disturbed serum lipid fatty acid patterns in patients with cystic fibrosis, *Ann. Nutr. Metab.* **29**:239–245.

Clandinin, M. T., Wang, L. C. H., Rajotte, R. VV., French, M. A., Goh, Y. K., and Kielo, E. S., 1995, Increasing the dietary polyunsaturated fat content alters whole-body utilization of 16:0 and 10:0, *Am. J. Clin. Nutr.* **61**:1052–1057.

de Schrijver, R., Vermeulen, D., and Viaene, E., 1991, Lipid metabolism responses in rats fed beef tallow, native or randomized fish oil and native or randomized peanut oil, *J. Nutr.* **121**:948–955.

Devlin, A., Innis, S. M., Wall, K., and Krisinger, J., 1996, Effect of medium-chain triglycerides on calbindin-D9k expression in the intestine, *Lipids* **31**:547–549.

Dyerberg, J., Bang, H. O., and Hjørne, N., 1975, Fatty acid composition of the plasma lipids in Greenland Eskimos, *Am. J. Clin. Nutr.* **28**:958–966.

Filer, L. J., Jr., Mattson, F. H., and Fomon, S. J., 1969, Triglyceride configuration and fat absorption by the human infant, *J. Nutr.* **99**:293–298.

Hamosh, M., 1990, *Lingual and Gastric Lipases: Their Role in Fat Digestion,* CRC Press, Boca Raton, FL.

Harris, W. S., 1989, Fish oils and plasma lipid and lipoprotein metabolism in humans: A critical review, *J. Lipid Res.* **30:**785–807.

Heird, W. C., Grundy, S. M., and Hubbard, V. S., 1986, Structured lipids and their use in clinical nutrition, *Am. J. Clin. Nutr.* **43:**320–324.

Hubbard, V. S., and McKenna, M. C., 1987, Absorption of safflower oil and structured lipid preparations in patients with cystic fibrosis, *Lipids* **22:**424–428.

Ikeda, I., Tomari, Y., Sugano, M., Watanabe, S., and Nagata, J., 1991, Lymphatic absorption of structured glycerolipids containing medium-chain fatty acids and linoleic acid, and their effect on cholesterol absorption in rats, *Lipids* **26:**369–373.

Ikeda, I., Yoshida, H., Tomooka, M., Yosef, A., Imaizumi, K., Tsuji, H., and Seto, A., 1998, Effects of long-term feeding of marine oils with different positional distribution of eicosapentaenoic and docosahexaenoic acids on lipid metabolism, eicosanoid production, and platelet aggregation in hypercholesterolemic rats, *Lipids* **33:**897–904.

Jandacek, R. J., Whiteside, J. A., Holcombe, B. N., Volpenhein, R. A., and Taulbee, J. D., 1987, The rapid hydrolysis and efficient absorption of triglycerides with octanoic acid in the 1 and 3 positions and long-chain fatty acid in the 2 position, *Am. J. Clin. Nutr.* **45:**940–945.

Jensen, G. L., McGarvey, N., Taraszewski, R., Wixson, S. K., Seidner, D. L., Pai, T., Yeh, Y.-Y., Lee, T. W., and DeMichele, S. J., 1994, Lymphatic absorption of enterally fed structured triacylglycerol vs. physical mix in canine model, *Am. J. Clin. Nutr.* **60:**518–524.

Jensen, M. M., Christensen, M. S., and Høy, C.-E., 1994, Intestinal absorption of octanoic, decanoic, and linoleic acids: Effect of triglyceride structure, *Ann. Nutr. Metabol.* **38:**104–116.

Jeppesen, P. B., Christensen, M. S., Høy, C.-E., and Mortensen, P. B., 1997, Essential fatty acid deficiency in patients with severe fat malabsorption, *Am. J. Clin. Nutr.* **65:**837–843.

Jeppesen, P. B., Høy, C.-E., and Mortensen, P. B., 1998, Essential fatty acid deficiency in patients receiving home parenteral nutrition, *Am. J. Clin. Nutr.* **68:**126–133.

Kritchevsky, D., Davidson, L. M., Weight, M., Kriek, N. P. J., and du Plessis, J. P., 1982, Influence of native and randomized peanut oil on lipid metabolism and aortic sudanophilia in the vervet monkey, *Atherosclerosis* **42:**53–58.

Lehner, R., Kuksis, A., and Itabashi, Y.,1993, Stereospecificity of monoacylglycerol and diacylglycerol acyltransferases from rat intestine as determined by chiral phase high-performance liquid chromatography, *Lipids* **28:**29–34.

Mattson, F. H., and Beck, L. W., 1956, The specificity of pancreatic lipase for the primary hydroxyl groups of glycerides, *J. Biol. Chem.* **219:**735–740.

Mattson, F. H., and Volpenhein, R. A., 1961, The specific distribution of fatty acids in the glycerides of vegetable fats, *J. Biol. Chem.* **236:**1891–1894.

Mattson, F. H., and Volpenhein, R. A., 1964, The digestion and absorption of triglycerides, *J. Biol. Chem.* **239:**2772–2777.

McKenna, M. C., Hubard, V. S., and Bieri, J. G., 1985, Linoleic acid absorption from lipid supplements in patients with cystic fibrosis with pancreatic insufficiency and in control subjects, *J. Ped. Gastroent. Nutr.* **4:**45–51.

Mendez, B., Ling, P. R., Istfan, N. W., Babayan, V., and Bistrian, B. R., 1992, Effects of different lipid sources in total and parenteral nutrition on whole body protein kinetics and tumor growth, *J. Paren. Enteral Nutr.* **16:**545–551.

Mok, K. T., Maiz, A., Yamazaki, K., Sobrado, J., Babayan, V. K., Moldawer, L. L., Bistrian, B. R., and Blackburn, G. L., 1984, Structured medium-chain and long-chain triglyceride emulsions are superior to physical mixtures in sparing body protein in the burned rat, *Metabolism* **33:**910–915.

Mortimer, B.-C., Simmonds, W. J., Joll, C., Stick, R. V., and Redgrave, T. G., 1988, Regulation of the metabolism of lipid emulsion model lipoproteins by a saturated acyl chain at the 2-position of triacylglycerol, *J. Lipid Res.* **29:**713–720.

Mu, H., Xu, X., and Høy, C.-E., 1998, Production of specific-structured triacylglycerols by lipase-catalyzed interesterification in a laboratory-scale continuous reactor, *JAOCS* **75:**1187–1193.

Neuringer, M., Anderson, G. J., and Connor, W., 1988, The essentiality of n-3 fatty acids for development and function of retina and brain, *Ann. Rev. Nutr.* **8:**517–541.

Poulsen, C., Christensen, M. S., and Høy, C.-E., 1997, Incorporation of n-3 polyunsaturated fatty acids of marine or vegetable origin into rat enterocyte phospholipids, *Nutr. Res.* **17:**149–162.

Qu, Z., Ling, P. R., Tahan, S., Sierra, P., Onderdonk, A. B., and Bistrian, B. R., 1996, Protein and lipid refeeding changes protein metabolism and colonic but not small intestnal morphology in protein-depleted rats, *J. Nutr.* **126:**906–912.

Redgrave, T. G., Kodali, D. R., and Small, D. M., 1988, The effect of triacyl-*sn*-glycerol structure on the metabolism of chylomicrons and triacylglycerol-rich emulsions in the rat, *J. Biol. Chem.* **263:**5118–5123.

Renaud, S., Ruf, J. C., and Petithory, D., 1995, The positional distribution of fatty acids in palm oil and lard influences their biologic effects in rats, *J. Nutr.* **125:**229–237.

Spalinger, J. H., Seidman, E. G., Lepage, G., Menard, D., Gavino, V., and Levy, E., 1998, Uptake and metabolism of structured triglyceride by Caco-2 cells: reversal of essential fatty acid deficiency, *Am. J. Physiol.* **275:** G652–G659.

Swenson, E. S., Selleck, K. M., Babayan, V. K., Blackburn, G. L., and Bistrian, B. R., 1991, Persistence of metabolic effects after long-term oral feeding of a structured triglyceride derived from medium-chain triglyceride and fish oil in burned and normal rats, *Metabolism* **40:**484–490.

Swift, L. L., Hill, J. O., Peters, J. C., and Greene, H. L., 1990, Medium-chain fatty acids: Evidence for incorporation into chylomicron triglycerides in humans, *Am. J. Clin. Nutr.* **52:**834–836.

Thomson, A. B. R., Keelan, M., Garg, M., and Clandinin, M. T., 1987, Spectrum of effects of dietary long-chain fatty acids on rat intestinal glucose and lipid uptake, *Can. J. Physiol. Pharmacol.* **65:**2459–2465.

Tomarelli, R. M., Meyer, B. J., Weaber, J. R., and Bernhart, F. W., 1968, Effect of positional distribution on the absorption of the fatty acids of human milk and infant formulas, *J. Nutr.* **95:**583–590.

Tso, P., Karlstad, M. D., Bistrian, B. R., and DeMichele, S. J., 1995, Intestinal digestion, absorption, and transport of structured triglycerides and cholesterol in rats, *Am. J. Physiol.* **268:**G568–577.

Wang, H., Dudley, A. W., Dupont, J., Reeds, P. J., Hachey, D. L., and Dudley, M. A., 1996, The duration of medium-chain triglyceride feeding determines brush border membrane lipid composition and hydrolase activity in newly weaned rats, *J. Nutr.* **126:**1455–1462.

Xu, X., Balchen, S., Høy, C.-E., and Adler-Nissen, J., 1998a, Pilot batch production of specific structured lipids by lipase-catalyzed interesterification: Preliminary study on incorporation and acyl migration, *JAOCS* **75:**301–308.

Xu, X., Skands, A., Adler-Nissen, J., and Høy, C.-E., 1998b, Production of specific structured lipids by enzymatic interesterification: Optimization of the reaction by response surface design, *Fett/Lipid* **100:**463–471.

Yang, L.-Y., and Kuksis, A., 1991, Apparent convergence (at 2-monoacylglycerol level) of phosphatidic acid and 2-monoacylglycerol pathways of synthesis of chylomicron triacylglycerols, *J. Lipid Res.* **32:**1173–1186.

Yang, L.-Y., Kuksis, A., and Myher, J. J., 1990, Lipolysis of menhaden oil triacylglycerols and the corresponding fatty acid alkyl esters by pancreatic lipase *in vitro:* A reexamination, *J. Lipid Res.* **31:**137–148.

You, Y. Q., Ling, P. R., Qu, Z., and Bistrian, B. R., 1998, Effect of continuous enteral medium-chain fatty acid infusion on lipid metabolism in rats, *Lipids* **33:**261–266.

CHAPTER 22

Structured Triacylglycerols in Clinical Nutrition

STEPHEN J. DeMICHELE and BRUCE R. BISTRIAN

1. Conventional Lipids in Clinical Nutrition

In recent years, intense interest and substantial investigation has been extended to dietary lipids because of their many nutritional, structural, and regulatory functions. A growing body of evidence is quite convincing that a forthcoming major advance in clinical nutrition will be the determination of the optimal types of lipid when formulating specialized diets. Among clinicians caring for the critically ill there is a consensus that dietary lipids are an important source of energy because of their caloric density and low osmolality. In addition they provide a source of essential fatty acids, linoleic acid (18:2n-6) and α-linolenic acid (18:3n-3), that are important for growth and for numerous physiological events related to their effects on eicosanoid metabolism and thus hemodynamics, inflammation, and immunocompetence in patients with various disease states. Dietary lipid also serves to facilitate the absorption of fat-soluble vitamins (A, D, E, K) and other important nutrients (i.e. carotenes, lycopene, flavonoids) that have not been established as essential.

The dietary management of patients with various diseases often involves enteral formulas containing lipid blends composed of long-chain triacylglycerols (LCT), principally from vegetable oils, and medium-chain triacylglycerols (MCT), derived from coconut, palm, and palm kernel oils. Some of the newer generation of disease-specific enteral formulas contain specialty lipids such as fish oil, borage oil, and structured triacylglycerols made from the interesterification of MCT and LCT. The physiological properties of specific n-6 and n-3 fatty acids can play a particularly important role in the nutritional support of stressed individuals undergoing a systemic inflammatory response (SIR). The incorporation of LCT into nutritional formulas has demonstrated both positive and negative physiological effects. Vegetable oils such as soybean, safflower, canola and corn are currently the most utilized form of LCT in conventional enteral feeding regimens (Gottschlich, 1992).

STEPHEN J. DeMICHELE • Strategic Discovery Research and Development, Ross Products Division, Abbott Laboratories, Columbus, Ohio 43215-1724.   BRUCE R. BISTRIAN • Department of Medicine, Beth Israel Deaconess Medical Center, Boston, Massachusetts 02215.
Intestinal Lipid Metabolism, edited by Charles M. Mansbach II *et al.,* Kluwer Academic/Plenum Publishers, 2001.

These lipids supply an isotonic source of nonprotein calories as well as a source of essential fatty acids. Further benefits of LCT include their oxidation, which generates significantly less CO_2 as compared with carbohydrate. This results in a lower respiratory quotient that can be beneficial for pulmonary patients with impaired ventilation who have trouble expiring CO_2 (Angelillo *et al.*, 1985). LCT improves nitrogen balance as effectively as carbohydrate, but in addition it reduces circulating triacylglycerols in critically ill animals and humans (Macfie *et al.*, 1981; Pomposelli *et al.*, 1986). Thus the replacement of some carbohydrate by LCT minimizes hyperglycemia and lipogenesis, which decreases the propensity for the development of hepatic steatosis and cholestasis, which can be observed with carbohydrate-only formulas, particularly in the setting of insulin resistance characteristic of the SIR (Sheldon *et al.*, 1978; Kaminski *et al.*, 1980; Allardyce, 1982).

Despite the near universal adoption of fat as a useful nutrient substrate, there are a variety of concerns regarding the clinical application of LCTs. The majority of the negative physiological effects, however, have been observed from the rapid intravenous administration of LCT emulsions at rates that exceed a patient's total energy expediture. Animal and clinical studies have shown that the physicochemical characteristics of chylomicrons in LCT emulsions may accumulate in the liver and the spleen when given in larger quantities (Belin *et al.*, 1976; du Toit *et al.*, 1978; Forbes, 1978) and may impair the reticuloendothelial system's ability to clear bacteria (du Toit *et al.*, 1978; Fischer *et al.*, 1980; Hamawy *et al.*, 1985; Sobrado *et al.*, 1985; Seidner *et al.*, 1989; G. L. Jensen *et al.*, 1990). Intravenous LCT emulsions administered at a rate in excess of the body's ability to metabolize the fat by oxidation or by normal storage has been reported to inhibit *in vitro* neutrophil chemotaxis and migration (Nordenstrom *et al.*, 1979), diminish bactericidal capacity (Fischer *et al.*, 1980), impair *in vitro* complement synthesis (Fischer *et al.*, 1980), and be associated with a deterioration in pulmonary function (Sundstrom *et al.*, 1973; Greene *et al.*, 1976; Freidman *et al.*, 1978; Levene *et al.*, 1980; Venus *et al.*, 1984, 1989). Clinical situations where parenteral and enteral LCT use might necessarily be limited include certain types of hypertriglyceridemias and pancreatitis, whereas enteral LCT often needs to be limited in pancreatic insufficiency and in other forms of severe fat malabsorption.

Medium chain triacylglycerols (MCT) have been developed as an alternate lipid source to LCT. MCT are semisynthetic triacylglycerols derived by the physical separation of tropical oils (palm kernel, coconut), which removes lauric acid (12:0) and other long chain fatty acids (LCFA) to leave predominantly caprylic (8:0) and capric acids (10:0). There are a number of unique aspects of MCT metabolism compared to LCT. These include lower melting points, smaller molecular size, no requirement for micelle formation before absorption, rapid absorption by gastric and intestinal mucosa with minimal requirement for pancreatic or biliary function, transport largely as free fatty acids bound to albumin in portal venous blood (Bach and Babayan, 1982; Bach *et al.*, 1988), and a metabolism as rapid as glucose (Johnson *et al.*, 1990; Metges and Wolfram, 1991). MCT are widely used in enteral formulas for the dietary management of patients with various medical conditions and are also available as intravenous lipid emulsions in Europe and in much of Asia. MCT alone may be less than ideal substrates because they may have toxic effects (intravenously) or may not be well tolerated (enteral) when given in large doses, and as well do not contain essential fatty acids. A mixture of both MCT and LCT can provide the benefits of both lipids while minimizing their potential side effects by limiting the dose of each. Animal and clinical studies have shown that the inclusion of MCT into intravenous solutions containing LCT have a thermogenic effect (Mok *et al.*, 1984; Mascioli *et al.*, 1991), are nitrogen sparing

(Mascioli *et al.*, 1991), and decrease TPN-associated intestinal atrophy (animals; Hinton *et al.*, 1998). Physical mixtures of LCT and MCT may maintain the bacterial clearance function of the reticuloendothelial system by reducing the amount of lipid that accumulates in the liver and the spleen with an isocaloric amount of LCT emulsion (Seidner *et al.*, 1989; Jensen *et al.*, 1990). Further clinical evidence supporting the improved tolerance of LCT and MCT emulsions comes from the observed benefits of reduced infectious morbidity and improved hepatic function in patients undergoing hepatic resection (Fan *et al.*, 1994).

2. Structured Triacylglycerols—An Alternate Lipid Source

In an effort to develop the optimal lipid source, structured triacylglycerols (STG) were developed containing chemically rearranged mixtures of medium-chain fatty acids (MCFA) and LCFA in order to retain the characteristics of both fats. These novel lipids could thus serve as ideal energy substrates under stressful conditions where frequent intolerance to LCT diets occur. The concept and role of tailor made STG was first advanced by Babayan (1987) as a new class of fats for nutritional and health care use in hospitalized patients. The initial clinical interest in STG was their proposed advantage in providing shorter chain length fatty acids as a source of immediate energy through oxidative metabolism while providing longer chain essential fatty acids for repletion and maintenance of tissue stores. The process of mixing specific proportions of MCT and LCT also provides an opportunity to preselect specific n-3, n-6, n-9, and medium/short-chain fatty acids that can be utilized for distinct nutritional and medical uses related to their unique metabolism.

STG containing MCFA are produced by mixing specific proportions of fractionated MCT or other tropical oils directly and various LCT oils (i.e., vegetable, fish), chemically allowing hydrolysis of the fatty acids to form free fatty acids and glycerol, followed by random reesterification into new triacylglycerol molecules (Mascioli *et al.*, 1988). This commercial process of interesterification leads to randomization so that all possible structures are found including LML, LMM, LLM, MLM, MMM, and LLL (where L and M represent LCFA and MCFA respectively at the sn-1, 2, and 3 position on the triacylglycerol molecule; Klemann *et al.*, 1994). Other interesterification processes in addition to transesterification can be employed including acidolysis (i.e., fatty acids are exchanged with fatty acids in triacylglycerols), glycerolysis (exchange of fatty acids between glycerol and triacylglycerols), or enzymatic (where STG of specific structures can be made). Numerous investigations have demonstrated that STG have different absorption and metabolic actions than identical physical mixtures of oils that have not been interesterified. The unique and important metabolic benefits of the enteral administration of STG in various disease states is outlined in the following sections.

3. STG in Critical Illness

3.1. Animal Models

Metabolic benefits of STG compared to equivalent physical mixtures have been demonstrated in various animal models of human disease. Early studies by Mok *et al.* (1984) and by Sobrado *et al.* (1985) assessed whether the intravenous administration of an

STG emulsion consisting of 60% MCT and 40% LCT had superior metabolic benefits as compared with emulsions containing either 100% LCT, 100% MCT, or a 50% LCT/50% LCT physical mixture in a rat model of burn injury (25% total body surface area). Rats were infused intravenously with isocaloric, isonitrogenous diets containing 36% of nonprotein calories as fat. Results showed that rats given STG had significantly higher weight gain, nitrogen balance, and serum albumin levels as compared with the other lipid groups. In addition, STG reduced lung sequestration of bacteria and normalized both hepatic and splenic uptake of bacteria (Sobrado *et al.*, 1985).

Further metabolic benefits with different STGs have been demonstrated in various animal models of trauma, burn injury, and endotoxic shock (Maiz *et al.*, 1984; Mok *et al.*, 1984; DeMichele *et al.*, 1988, 1989; Teo *et al.*, 1989, 1991; Swenson *et al.*, 1991; Gollaher *et al.*, 1993). Attenuation of the hypermetabolic and protein catabolic response to burn injury was observed in thermally injured rats receiving STG either parenterally or enterally. Rats receiving the STG containing both medium-chain and long-chain fatty acids showed the greatest gain in body weight, enhanced skeletal muscle and liver protein synthesis, and markedly increased nitrogen retention and serum albumin concentration over comparable groups of rats given LCT, MCT, or a physical mixture of LCT and MCT (Maiz *et al.*, 1984; Mok *et al.*, 1984; DeMichele *et al.*, 1988, 1989; Teo *et al.*, 1989, 1991; Swenson *et al.*, 1991; Gollaher *et al.*, 1993). Fatty infiltration of the liver and reticuloendothelial system overload were substantially reduced in burned guinea pigs given STG as compared to LCT (Sobrado *et al.*, 1985).

The collection of these studies suggests that diets containing STG as the primary fat source may reduce the catabolic response to injury compared to conventional fats or to physical mixtures of oils similar in fatty acid composition to the STG. These benefits appear to be related to a reduction in protein breakdown while improving skeletal muscle and liver protein synthetic rates. STG would appear to be useful in the hospitalized patient suffering from multiple burns, recovering from trauma, or following major surgery.

3.2. Clinical Testing

Recently the first clinical testing of a novel STG was performed by Kenler and associates (Kenler *et al.*, 1996). The authors compared the safety, gastrointestinal tolerance, and clinical efficacy of feeding an enteral diet containing a fish oil/MCT structured lipid (FOSL-HN) versus an isonitrogenous, isocaloric formula (O-HN) in patients undergoing major abdominal surgery for upper gastrointestinal malignancies. The STG consisted of an interesterified mixture of 60% MCT and 40% fish oil that served as the principal fat source of a nutritionally complete tube feeding formula. The control formula was a standard tube feeding diet containing a physical mixture of LCT (vegetable oil) and MCT. This prospective, blinded, randomized trial was conducted in 50 adult patients who were jejunally fed either FOSL-HN or O-HN for 7 days. Serum chemistries, hematology, urinalysis, gastrointestinal complications, liver and renal function, urinary prostaglandins, and outcome parameters were measured at baseline and on day 7 (Fig. 1). Comparisons were made in 18 O-HN and 17 FOSL-HN evaluable patients based on their ability to reach a tube feeding rate of 40 mL/hour.

It was clearly established in this trial that jejunal administration of FOSL-HN could be given safely in the early postoperative period with excellent tolerance. The fish oil/MCT STG was safe as evident by similar changes observed in serum chemistries, hematology,

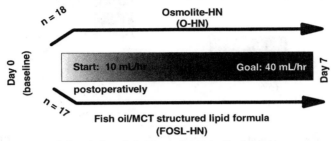

Design: Prospective, blinded, randomized trial
Patients: 50 adult patients (18 - 80 years of age)
Upper gastrointestinal malignancies, major abdominal surgery
Feeding jejunostomy placed during surgery

Daily caloric and protein needs were estimated to be 25 to 30 kcal/kg body weight and 1.2 to 1.5 g protein/kg.

Figure 1. Outline of the study design used to compare the safety, gastrointestinal tolerance, and clinical efficacy of feeding an enteral diet containing either a fish oil/MCT structured lipid (FOSL-HN) or an isonitrogenous, isocaloric formula, Osmolite HN (O-HN) in critically ill patients undergoing abdominal surgery for upper gastrointestinal malignancies.

and nitrogen balance parameters after 7 days of feeding as compared with O-HN (data not shown). Patients fed FOSL-HN experienced a 40% reduction in the total number of days with reported gastrointestinal complications (39 days vs. 23 days; $p = 0.036$) and a 50% decline in the total number of actual reported gastrointestinal complications (54 vs. 27; $p = 0.004$) compared with those fed O-HN. Seven-day administration of approximately 5.0 g EPA plus DHA as an STG produced no untoward side effects and did not raise blood glucose levels or prolong blood clotting parameters as compared with patients fed the control diet.

In addition, physiological data strongly suggest that the use of n-3 fatty acids in the form of STG may improve parameters of renal (Fig. 2) and liver (Fig. 3) function during the postoperative period through modulation of urinary prostaglandin levels (Fig. 4). Patient outcome data showed a 50% reduction in the total number of infections and positive cultures in patients receiving FOSL-HN compared with those fed O-HN. In addition there were significantly fewer infected patients with multiple infections (1/6 vs. 5/7; $p = 0.037$). Results from a follow-up study to this trial suggest that these benefits may be due in part to a reduction in eicosanoid production from peripheral blood mononuclear cells from patients given FOSL-HN as compared with control patients (Swails *et al.*, 1997).

4. STG in Cancer

Fish oil/MCT STG has also been used in animal models of cancer to assess its effect on tumor growth. Mendez et al. (1992) examined the short-term effects of three isocaloric/isonitrogenous Total Parenteral Nutrition (TPN) solutions, each containing a different lipid

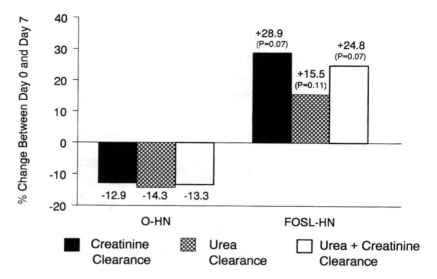

Figure 2. Renal function data for patients fed either Osmolite HN (O-HN; $n = 18$) or a fish oil/MCT structured lipid formula (FOSL-HN; $n = 17$). Values indicate the percent change of the mean day 7 value from the mean baseline (day 0) value for creatinine clearance, urea clearance, and the sum of creatinine and urea clearance.

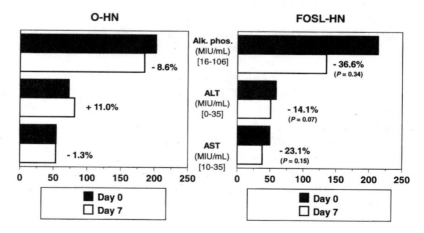

Figure 3. Liver function data for patients fed either Osmolite HN (O-HN; n=18) or a fish oil/MCT structured lipid formula (FOSL-HN; n=17). Percent values indicate the percent change of the mean day 7 value from the mean baseline (day 0) value for alkaline phosphatase (alk. phos.), alanine aminotransferase (ALT), and aspartate aminotransferase (AST). Bar values are means. Values in brackets indicate the normal range.

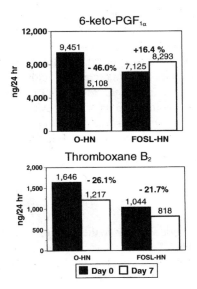

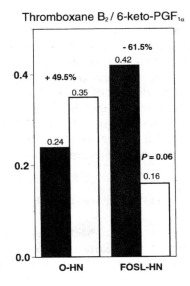

Figure 4. Urinary prostaglandin data for patients fed either Osmolite HN (O-HN; $n = 18$) or a fish oil/MCT structured lipid formula (FOSL-HN; $n = 17$). Percent values indicate the percent change of the mean day 7 value from the mean baseline (day 0) value for 6-keto-PGF$_{1\alpha}$, thromboxane B$_2$, and thromboxane B$_2$/6-keto-PGF$_{1\alpha}$. Bar values are means.

source, on host and tumor protein metabolism in rats with implantable Yoshida sarcoma tumors. Each diet contained 50% of nonprotein calories as either a structured lipid containing 60% MCT and 40% fish oil, a physical mix of MCT and fish oil, or LCT (50% safflower oil and 50% soybean oil; Liposyn II, Abbott Laboratories, North Chicago, IL). Results showed that tumor growth rate was reduced in rats given STG with significant increases in rates of tumor protein synthesis and tumor protein breakdown. STG also improved skeletal muscle protein synthesis and nitrogen balance as compared with the other groups. In a similar rat cancer model, fish oil/MCT STG led to an increase of nitrogen in the host, improved nitrogen balance, and an inhibition of tumor growth as compared with LCT (Crosby *et al.*, 1990; Ling *et al.*, 1991).

5. STG in Liver Disease

Fish oil/MCT STG could also have application in patients with sepsis-induced liver injury and dysfunction. Lanza-Jacoby and coworkers have shown that hypertriglyceridemia and fatty infiltration of the liver are two prominent clinical characteristics of gram-negative sepsis (Lanza-Jacoby *et al.*, 1982; Lanza-Jacoby and Tabares, 1990). This alteration in lipid metabolism can lead to an accumulation of liver lipids and a deterioration in liver function. Maintaining normal liver function is essential in preventing liver failure and in improving survival rates (Lanza-Jacoby *et al.*, 1995). Lanza-Jacoby and colleagues evaluated whether enteral feeding of a STG containing fish oil and MCT would prevent the hypertriglyceridemia and fatty infiltration of the liver during sepsis (Lanza-Jacoby *et al.*, 1995). Rats were intragastrically fed for 5 days nutritionally complete diets containing fish oil/MCT STG or LCT from soybean oil. On the fifth day, sepsis was induced by intravenous live *E. coli* and rats were terminated 24 hours later. STG reduced plasma free fatty acids of the control and septic rats by 50% as compared with control and septic rats given LCT. Both the rate of liver triacylglycerol esterification and the amount of liver triacylglycerol in septic rats fed STG was significantly lower as compared with livers from LCT rats. These findings suggest that feeding of fish oil/MCT STG may have an important role in the nutritional support of the patient with sepsis-associated fatty liver.

6. STG in Malabsorption Syndromes

Since 1990, research has broadened our knowledge base as to the absorption and transport benefits of interesterified blends of LCT and MCT versus their physical mixtures. The advantages of enterally fed STG may relate to differences in absorption, processing, and transport of the triacylglycerols. STGs that contain MCFA may provide a vehicle for rapid hydrolysis and absorption secondary to smaller molecular size and greater water solubility in comparison to LCT.

STG would have potential benefits in patients with cystic fibrosis who have a pancreatic obstruction with insufficient lipase. The STGs would be hydrolyzed as readily as MCT and could reduce the amount of pancreatic lipase required for hydrolysis and absorption of the fatty acids. Studies by Hubbard (Hubbard and McKenna, 1987), McKenna (McKenna *et al.*, 1985), and associates showed enhanced absorption of 18:2n-6 in cystic fibrosis patients fed STG containing long- and medium-chain fatty acids. Positional specificity of fatty acids on the glycerol moiety is likely to be a key factor. Because lipase-catalyzed hydrolysis of triacylglycerols splits the fatty acids from the sn-1 and sn-3 positions, an STG with MCFA in these positions with a LCFA in the sn-2 position may be appropriate for delivery of key essential fatty acids in patients with pancreatic insufficiency. Jandacek *et al.* (1987) confirmed this by performing *in vitro* lipase digestion and absorption studies using isolated intestinal loops. The results revealed rapid hydrolysis and absorption of a STG containing linoleic acid in the 2- position and MCFA in the 1- and 3- positions as compared with an LCT.

These observations raise the possibility of designer triacylglycerols with known fatty acids in each position on the glycerol moiety. Much interest has been generated regarding the preferential absorption of the 2-monoglyceride and which type of fatty acid occupies that position. Preduodenal lipases (lingual and gastric) and pancreatic lipases hydrolyze MCFA or LCFA in the sn-1 and sn-3 positions. The remaining fatty acid in the sn-2 position (2-monoglyceride) is quickly absorbed, reesterified with LCFA or with longer chain MCFA, formed into chylomicra lipid particles, and transported by the lymphatic system to the systemic circulation for eventual utilization by peripheral organs and tissues (Kayden *et al.*, 1967; Small, 1991). An LCFA in the sn-2 position such as linoleic, linolenic, or eicosapentaenoic acid, would be preferentially absorbed and retained in cell membrane phospholipids rather than be oxidized for energy. This improvement in absorption would have direct application in both children and adults with malabsorptive conditions. Triacylglycerols with MCFA in the sn-2 position (LML and LMM) would lead to absorption of the 2-

monoglyceride-MCFA, reesterification in the intestinal cell as LML, and transport via lymphatics to appear in systemic circulation as a STG (Bistrian, 1997). This pattern of absorption would be particularly likely if the MCFA were 10:0 or 12:0, because shorter chain fatty acids are more easily removed either in the intestinal lumen or intracellularly. Such triacylglycerols have been shown to have metabolic benefits in animal models of trauma as described earlier (Maiz *et al.*, 1984; Mok *et al.*, 1984; DeMichele *et al.*, 1988, 1989; Teo *et al.*, 1989, 1991; Swenson *et al.*, 1991; Gollaher *et al.*, 1993). Excellent studies by Christensen (Christensen *et al.*, 1994, 1995a,b), Jensen (M. M. Jensen *et al.*, 1994), and colleagues using a rat model of lipid absorption have shown that defined triacylglycerols with specific fatty acids in the 2 position on the glycerol moiety may provide a means to increase absorption of essential fatty acids in fat malabsorption. Details from these studies are covered in detail in another chapter. However, the benefits of different species of STG suggest that a randomized distribution rather than one species only is likely to be preferred in most clinical situations.

Tso and co-workers (Tso *et al.*, 1995) assessed the intestinal absorption of STG containing two MCFA (8:0) and one LCFA (18:2n-6) using a lymph fistula rat model. The digestion, absorption, and lymphatic transport of two structured triacylglycerols containing 1 LCFA and 2 MCFA either at the 1- and 3- or 1- and 2-position (Trilinolein vs. glyceryl 1,3-dioctanoate-2-[1-^{14}C]linoleate vs. rac-glycerol 1,2-di[1-^{14}C]octanoate-3-linoleate) was determined under steady state conditions. Results showed that octanoate, irrespective of its position on the triacylglycerol molecule, is poorly incorporated into lymph lipids and most likely was absorbed and transported via the portal circulation for the reasons previously discussed. The 2-monoglyceride with octanoate was absorbed significantly better than 2-monoglyceride with linoleate. In rats given 1,3-dioctanoate-2-[1-^{14}C] linoleate, significantly more ^{14}C lipid (mainly as 2-monoglyceride) was recovered in the lower small intestine, cecum, and colon. The poor uptake of 2-linoleate was not caused by limited lipolysis but by an inability of the small intestine to process the digested remnants (2-monoglyceride and free fatty acids) of the STG. The poor uptake of the 2-monoglyceride containing linoleate may also be due to the absence of other LCFA in the lipid diet. STG containing octanoic acid in the 1- and 3- positions and linoleic acid in the 2- position may not be advantageous to use as a sole source of dietary lipid but should be supplemented with LCT. The triacylglycerol/phospholipid output ratio was highest in the 1,3-dioctanoate-2-[1-^{14}C]linoleate group and it was significantly greater than the rac-glycerol 1,2-di[1-^{14}C]octanoate-3-linoleate) group. This may be an important finding that should be explored further, because this ratio affects both the formation, size, and metabolism of chylomicrons.

Significant advancements toward gaining further knowledge about the absorption and transport of STG have been made using STG consisting of random distributions of LCFA and MCFA on the glycerol moiety. A canine model was developed by Jensen and associates (G. L. Jensen *et al.*, 1994) to study the lymphatic absorption of these fats by cannulation of the cysterna chyli and placement of a transpyloric small bowel feeding tube. The diets provided identical fat contents at 36.8 g/l as fish oil/MCT STG or a physical mix containing 56.6 wt% of MCFA and 12.2 wt% of n-3 fatty acids from fish oil. Four dogs were enterally fed at 100 ml/hr in a crossover design with lymph samples collected at baseline and after 4 and 8 hours of feeding each diet. The MCFA accounted for 11% to 19% of total fatty acids found in lymph after 8 hours of feeding the STG diet, which was 58% to 74% higher than observed with the physical mix diet. The MCFA were found in the tri- and diglycerides with 10:0 in 1.5- to two-fold excess of 8:0, despite the 10:0/8:0 ratio of 0.3 in the original diets. There was a comparable increase in eicosapentaenoic acid and in docosahexaenoic acid and a decrease in linoleic acid and in arachidonic acid with either diet, but the MCFA/ n-3 ratio was 46% to 73% higher with STG compared to physical mix feeding. Molecular species analyses revealed that the MCFA in lymph were present on the same triacylglycerol as LCFA. In one dog that received a double crossover, a two- to threefold increase in the amount of lipid in lymph was detected with the STG diet as compared to the physical mix. These data confirm previous reports of significant MCFA absorption in lymph (G. L. Jensen *et al.,* 1989) and suggest that unique aspects of STG absorption and intestinal processing may account for their reported metabolic benefits.

It is felt that the previous observations are very important, especially the findings that one may enhance lipid absorption by forming STG with randomly mixed MCFA and LCFA on the same triacylglycerol backbone. These data were further explored by Tso and associates (Tso *et al.,* 1999) in a normal lymph fistula rat model and in a rat model of lipid malabsorption caused by ischemia/reperfusion (I/R)-induced injury (Fujimoto *et al.,* 1991). These models are set up to accurately measure and quantitate lipid absorption after a test meal. Tso *et al.* (1999) compared the digestion, absorption and lymphatic transport of a fish oil/MCT STG versus its physical mix equivalent in normal rat small intestine (sham-operated) and in intestine injured by I/R. Fasted rats were randomized into two treatments (control and I/R; Fig. 5). Under anesthesia, the superior mesenteric artery (SMA) was occluded for 20 min and then was reperfused in I/R rats. The SMA was isolated but not occluded in control rats. In both treatment groups, the mesenteric lymph duct was cannulated for lymph collection and a gastric tube was inserted. Each treatment group received 1 ml of a fish oil/MCT STG or its physical mix through the gastric tube followed by an infusion of phosphate-buffered saline for 8 hours. Lymph was collected hourly for 8 hours. Values in figs. 6 and 7 are means ± SEM for 7 rats/group. Results showed lymph triacylglycerol increased rapidly and maintained a 40% to 50% higher output with STG versus physical mix

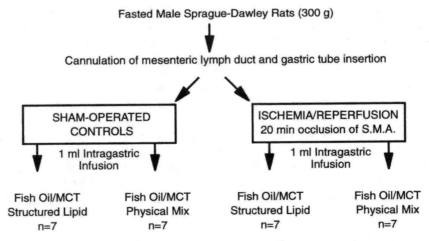

Figure 5. Outline of the study design used to compare the digestion, absorption, and lymphatic transport of a fish oil/MCT structured lipid versus its physical mix equivalent in normal rat small intestine (sham-operated controls) and in intestine injured by ischemia/reperfusion (20-minute occlusion of the superior mesenteric artery (S.M.A.) followed by reperfusion).

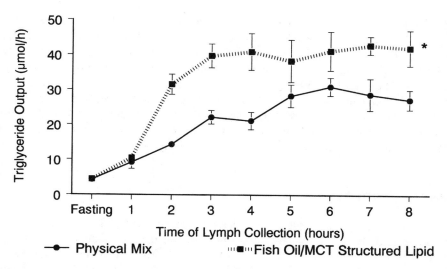

Figure 6. Lymphatic triacylglycerol output expressed as μmole per hour in sham-operated control rats. Lymph samples were collected 1 hour before (fasting) the intragastric infusion of either fish oil/MCT structured lipid or its physical mix and hourly for 8 hours during the infusion period. *$p < 0.01$ versus physical mix using repeated measures ANOVA. Values are means $\pm$ SEM; $n = 7$ rats per group.

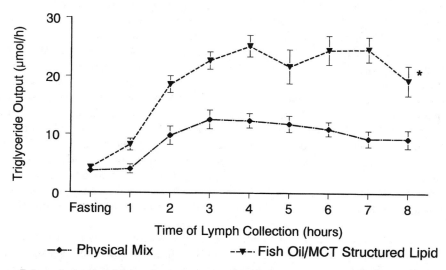

Figure 7. Lymphatic triacylglycerol output expressed as μmole per hour in rats injured by ischemia/reperfusion of the small intestine. Lymph samples were collected 1 hour before (fasting) the intragastric infusion of either the fish oil/MCT structured lipid or the physical mix and hourly for 8 hours during the infusion period. *$p < 0.01$ versus physical mix using repeated measures ANOVA. Values are means $\pm$ SEM; $n = 7$ rats per group.

in control rats (Fig. 6). Following I/R, lymphatic triacylglycerol output decreased 50% compared to controls (Figs. 6 and 7). Gastric infusion of STG significantly improved lipid transport by having a twofold higher triacylglycerol output to lymph as compared with its physical mix. This study provides important evidence supporting the use of randomly interesterified STG with MCFA to increase absorption of triacylglycerols and essential fatty acids under normal, healthy conditions and following induced intestinal dysfunction. Provision of oil blends in the form of STGs may have improved tolerance and lower gastrointestinal complications in malabsorption syndromes. Recent clinical studies from Kenler *et al.* (1996) as previously described have confirmed this hypothesis.

7. Lower Calorie Fats

For most of this chapter, we have discussed strategies using STG to facilitate the absorption of lipid and key essential fatty acids. However the food industry have designed STG to modify their energy content while maintaining the taste, mouth feel, flavor, and other organoleptic characteristics of fats. Nabisco (East Hanover, NJ) and Procter & Gamble (Cincinnati, OH) have produced triacylglycerols with mixtures of short-chain fatty acids with varying chain lengths and degrees of saturation that affect the digestion and absorption of these triacylglycerols, thereby reducing the energy content or its availability. The Nabisco Foods group developed a fat under the trademarked name of "Salatrim." Salatrim is a family of STGs that provides the physical properties of fat but with approximately half of the calories of a normal edible oil. This family of STG are produced by the interesterification of highly hydrogenated vegetable oils with triacylglycerols of acetic, propionic, and/ or butyric acids (Smith *et al.*, 1994). The resulting triacylglycerols can contain two short-chain fatty acids (acetic, propionic, and/or butyric) and one poorly absorbed (60%) long-chain saturated fatty acid, stearic acid. Thus the energy level of this family of STG is approximately 5.0 kcal/gm, significantly less than 9.0 kcal/gm for traditional fats. The Food and Drug Administration has recently accepted a Generally Recognized as Safe (GRAS) petition for Salatrim (Wheeler, 1995). Cultor Food Science Inc. has purchased the rights to manufacture and market Salatrim under the name Benefat. Benefat is currently used in a wide range of chocolate confectionery products, including coatings and centers, as well as in baked goods in the United States. Such products include Nabisco's SnackWell's candy products, Hershey's Reduced Fat Baking Chips, and Entenmann's reduced-fat cookies. It is also used in a range of baked goods in Japan and in a commercial ice cream in Taiwan (Haumann, 1997). Clinical studies have been conducted in normal individuals consuming elevated amounts of Salatrim as different snack foods, cookies, bonbons, ice cream, and dessert bars (Finley *et al.*, 1994). Salatrim was tolerated well with some individuals experiencing mild gastrointestinal discomfort and slight elevations in liver function enzymes.

The other new lower calorie fat called Caprenin was produced by Proctor & Gamble. The formula for this STG is 45% behenic acid (22:0) from hydrogenated rapeseed oil and 50% capric (8:0) and caprylic (10:0) acids in random distribution. This yields a fat that has a high melting point that is similar in texture and melting behavior to cocoa butter with an energy availability of approximately 5 kcal/gm (Finley *et al.*, 1994). Clinical study results, however, have been disappointing (Wardlaw *et al.*, 1995). Normal volunteers were fed diets containing either 60% of the fat calories as palm oil or palm kernel oil or 80% of the fat calories as butter for 3 weeks. For the next 6 weeks, the subjects received either the same diet or Caprenin. The group that received Caprenin displayed mild to moderate gastrointestinal complaints, a significant increase in the ratio of total cholesterol to HDL cholesterol, and a significant decrease in HDL cholesterol. These changes were not apparent in the volunteers given butter who later received Caprenin. The increase in LDL cholesterol in the caprenin individuals most likely was due to the behenic acid content of the fat. In contrast, clinical experience with Salatrim may have been more successful because of the use of a smaller chain saturated fatty acid, stearic acid, which is more neutral in its cholesterolemic effect (Bonanome and Grundy, 1988). M&M Mars (Hackettstown, NJ) first introduced Caprenin as part of the chocolate coating of the candy Milky Way II but has since removed this food component from the market.

8. Concluding Remarks

As discussed previously, the uses of STGs in clinical nutrition are growing as we learn more about their unique metabolism. Another such application is for the growing infant. Lipids are a key component of formulations for infant feeding. They fulfill a number of essential functions in growth and development and provide a well-tolerated energy source. Human milk is uniquely formulated to meet these needs by possessing unique triacylglycerols with 70% of the level of palmitic acid lies in the sn-2 position of the glycerol backbone (Filer *et al.*, 1970). Palmitic acid found in vegetable-oil-based formula mainly resides in the sn-1 and -3 positions, making the fatty acid more susceptible to form calcium soaps following digestion. These soaps can lead to constipation as well as contributing to the loss of dietary energy and calcium. STG made by reacting tripalmitin with unsaturated fatty acids using an sn-1,3-specific lipase was found to closely mimic the fatty acid distribution in human milk and is currently under commercial development under the trademarked name Betapol (Haumann, 1997). Clinical evidence suggests that infants fed Betapol had improved palmitic and oleic acid absorption, reduced formation of insoluble calcium soaps in stool, and improved calcium absorption as compared with infants fed a vegetable-oil-based formula (Carnielli *et al.*, 1995; Summers *et al.*, 1998). These results suggest that STG that mimic the stereoisometric structure of those from breast milk may have a valuable role in the design of preterm infant formula.

The potential for commercialization of the enzymatic process for large-scale synthesis of STGs needs to be explored (Akoh, 1995). Our complete understanding of the metabolism and metabolic benefits of STG in clinical nutrition will depend on the commercial availability of these unique lipids. Currently it is costly to manufacture quantities of specific and pure STG, so random transesterification of conventional oils will be the process of choice. The increased cost of MCT may be a barrier to their increased use in randomized STG by the food industry, unless a lower cost alternative is produced. One solution on the horizon may be through the genetic engineering of typical oil crop plants to yield novel oils with the desired fatty acid profiles and triacylglycerol structures (Merolli *et al.*, 1997).

In the near future it is anticipated that a broader range of starting fats will be used to create new STG with potential impacts on improving outcomes in acute clinical conditions or for serving as replacements for conventional lipids for disease prevention. The potential exists for designer triacylglycerols with custom fatty acid formulations for nutritional and pharmacological applications. MCFA could be targeted for lymphatic transport via interesterification with LCFA. Thus enteral feeding of such a STG would result in chylomicrons that contain mixed triacylglycerols of medium and long-chain fatty acids. This approach could provide a more efficient vehicle to deliver MCFA as a source of intracellular energy as well as to deliver specific LCFA to peripheral tissues. When transported by the portal route, MCFA are subject to first pass hepatic clearance and rapid oxidation. This unique metabolism of STG could have the potential to enhance the performance of athletes or of soldiers in military training by utilizing the rapid oxidation of MCFA for energy and by sparing the oxidation of glucose and muscle glycogen. The possible sparing of glycogen stores with moderate to heavy workouts or training could be an advantage to high-carbohydrate diets. Other fatty acids that have unique metabolic characteristics such as γ-linolenic acid to reduce inflammation, arachidonic and docosahexaenoic acids to restore their levels to normal in conditions such as prematurity, malabsorption, or end stage liver disease, or oleic acid for its cardiovascular benefits, could be provided as STG to enhance their utilization. Short-chain fatty acids could be used as a prominent component of a STG to further enhance fat absorption and potentially to provide specific intestinal mucosal benefits of short-chain fatty acids. The possible opportunities for novel STG are limited only by new knowledge about fatty acid metabolism as it becomes available and our own imagination.

Acknowledgments

The authors would like to thank Kathy Dailey for her expert assistance in the preparation of this chapter.

References

Akoh, C. C., 1995, Structured lipids—Enzymatic approach, *INFORM* **6:**1055–1061.

Allardyce, D. B., 1982, Cholestasis caused by lipid emulsions, *Surg. Gynecol. Obstet.* **154:**641–647.

Angelillo, V. A., Bedi, S., Durfee, D., Dahl, J., Patterson, A. J., and O'Donohue, W. J., 1985, Effects of low and high carbohydrate feedings in ambulatory patients with chronic obstructive pulmonary disease and chronic hypercapnia, *Ann. Intern. Med.* **103:**883–885.

Babayan, V. K., 1987, Medium chain triacylglycerols and structured lipids, *Lipids* **22:**417–420.

Bach, A., and Babayan, V. K., 1982, Medium-chain triacylglycerols: An update, *Am. J. Clin. Nutr.* **36:**950–962.

Bach, A., Storck, D., and Meraihi, Z., 1988, Medium chain triglyceride-based fat emulsions: An alternative energy supply in stress and sepsis, *J. P. E. N.* **12**(suppl)**:**82–88.

Belin, R. P., Bivins, B. A., Jona, J. Z., and Young, V. L., 1976, Fat overload with a 10% soybean oil emulsion, *Arch. Surg.* **111:**1391–1393.

Bistrian, B. R., 1997, Novel lipid sources in parenteral and enteral nutrition, *Pro. Nutr. Soc.* **56:**471–477.

Bonanome, A., and Grundy, S. M., 1988, Effect of dietary stearic acid on plasma cholesterol and lipoprotein levels, *N. Engl. J. Med.* **318**(19)**:**1244–1248.

Carnielli, V. P., Luijendijk, I. H. T., van Goudoever, J. B., Sulkers, E. J., Boerlage, A. A., Degenhart, H. J., and Sauer, P. J. J., 1995, Feeding premature newborn infants palmitic acid in amounts and stereoisomeric position similar to that of human milk: Effects on fat and mineral balance, *Am. J. Clin. Nutr.* **61:**1037–1042.

Christensen, M. S., Hoy, C.-E., and Redgrave, T. G., 1994, Lymphatic absorption of *n*-3 polyunsaturated fatty acids from marine oils with different intramolecular fatty acid distributions, *Biochim. Biophys. Acta* **1215:**198–204.

Christensen, M. S., Mullertz, A., and Hoy, C.-E., 1995a, Absorption of triacylglycerols with defined or random structure by rats with biliary and pancreatic diversion, *Lipids* **30**:521–526.

Christensen, M. S., Hoy, C.-E., Becker, C. C., and Redgrave, T. G., 1995b, Intestinal absorption and lymphatic transport of eicosapentaenoic (EPA), docosahexaenoic (DHA), and decanoic acids: Dependence on intramolecular triacylglycerol structure, *Am. J. Clin. Nutr.* **61**:56–61.

Crosby, L. E., Swenson, E. S., Babayan, V. K., Bistrian, B. R., and Blackburn, G. L., 1990, Effect of structured lipid-enriched total parenteral nutrition in rats bearing Yoshida Sarcoma, *J. Nutr. Biochem.* **1**:41–47.

DeMichele, S. J., Karlstad, M. D., Babayan, V. K., Istfan, N., Blackburn, G. L., and Bistrian, B. R., 1988, Enhanced skeletal muscle and liver protein synthesis with structured lipid in enterally fed burned rats, *Metabolism* **37**:787–795.

DeMichele, S. J., Karlstad, M. D., Bistrian, B. R., Istfan, N., Babayan, V. K., and Blackburn, G. L., 1989, Enteral nutrition with structured lipid: Effect on protein metabolism in thermal injury, *Am. J. Clin. Nutr.* **50**:1295–1302.

du Toit, D. F., Villet, W. T., and Heydenrych, J., 1978, Fat-emulsion deposition in mononuclear phagocytic system, *Lancet* **2**:898.

Fan, S. T., Lo, C. M., Lai, E. C., Chu, K. M., Liu, C. L., and Wong, J., 1994, Perioperative nutritional support in patients undergoing hepatectomy for hepatocellular carcinoma, *N. Engl. J. Med.* **331**:1547–1552.

Filer, L. J., Jr., Mattson, F. H., and Fomon, S. J., 1970, Triacylglycerol configuration and fat absorption by the human infant, *J. Nutr.* **99**:293–298.

Finley, J. W., Leveille, G. A., Dixon, R. M., Walchak, C. G., Sourby, J. C., Smith, R. E., Francis, K. D., and Otterburn, M. S., 1994, Clinical assessment of SALATRIM, a reduced-calorie triacylglycerol, *J. Agric. Food Chem.* **42**:581–596.

Fischer, G. W., Hunter, K. W., Wilson, S. R., and Mease, A. D., 1980, Diminished bacterial defense with Intralipid, *Lancet* **2**:819–820.

Forbes, G. B., 1978, Splenic lipidosis after administration of intravenous fat emulsions, *J. Clin. Path.* **31**:765–771.

Freidman, Z., Marks, K. H., Maisels, M. J., Thorson, R., and Naeye, R., 1978, Effect of parenteral fat emulsion on the pulmonary and reticuloendothelial systems in the newborn infant, *Pediatrics* **61**:694–698.

Fujimoto, K., Price, V. H., Granger, D. N., Specian, R., Bergstedt, S., and Tso, P., 1991, Effect of ischemia-reperfusion on lipid digestion and absorption in rat intestine, *Am J. Physiol.* **260**:G595–G602.

Gollaher, C. J., Fechner, K., Karlstad, M., Babayan, V. K., and Bistrian, B. R., 1993, The effects of increasing levels of fish oil containing structured triacylglycerols on protein metabolism in parenterally fed rats stress by burn plus endotoxin, *J. P. E. N.* **17**:247–253.

Gottschlich, M. M., 1992, Selection of optimal lipid sources in enteral and parenteral nutrition, *Nutr. Clin. Prac.* **7**:152–165.

Greene, H. L., Hazlett, D., and Demaree, R.,1976, Relationship between Intralipid-induced hyperlipemia and pulmonary function, *Am. J. Clin. Nutr.* **29**:127–135.

Hamawy, K. J., Moldawer, L. L., Georgieff, M., Valicenti, A. J., Babayan, V. K., Bistrian, B. R., and Blackburn, G. L., 1985, Effect of lipid emulsions on the reticuloendothelial system function in the injured animal, *J. P. E. N.* **9**:559–565.

Haumann, B. F., 1997, Structured lipids allow fat tailoring, *INFORM* **8**:1004–1011.

Hinton, P. S., Peterson, C. A., McCarthy, D. O., and Ney, D. M., 1998, Medium-chain compared with long-chain triacylglycerol emulsions enhance macrophage response and increase mucosal mass in parenterally fed rats, *Am. J. Clin. Nutr.* **67**:1265–1272.

Hubbard, V. S., and McKenna, M. C., 1987, Absorption of safflower oil and structured lipid preparations in patients with cystic fibrosis, *Lipids* **22**:424–428.

Jandacek, R. J., Whiteside, J. A., Holcombe, B. N., Volpenhein, R. A., and Tavlbee, J.D., 1987, The rapid hydrolysis and efficient absorption of triacylglycerols with octanoic acid in the 1 and 3 positions and long chain fatty acid in the 2 position, *Am. J. Clin. Nutr.* **45**:940–945.

Jensen, G. L., Mascioli, E. A., Meyer, L. P., Lopes, S. M., Bell, S. J., Babayan, V. K., Blackburn, G. L., and Bistrian, B. R., 1989, Dietary modification of chyle composition in chylothorax, *Gastroenterol.* **97**:761–765.

Jensen, G. L., Mascioli, E. A., Seidner, D. L., Istfan, N. W., Domnitch, A. M., Selleck, K., Babayan, V. K., Blackburn, G. L., and Bistrian, B. R., 1990, Parenteral infusion of long- and medium-chain triacylglycerols and reticuloendothelial function in man, *J. P. E. N.* **14**:467–471.

Jensen, G. L., McGarvey, N., Taraszewski, R., Wixson, S. K., Seidner, D. L., Pai, T., Yeh, Y. Y., Lee, T. W., and

DeMichele, S. J., 1994, Lymphatic absorption of enterally fed structured triacylglycerol vs. physical mix in a canine model, *Am. J. Clin Nutr.* 60:518–524.

Jensen, M. M., Christensen, M. S., and Hoy, C.-E., 1994, Intestinal absorption of octanoic, decanoic, and linoleic acids: Effect of triacylglycerol structure, *Ann. Nutr. Metab.* 38:104–116.

Johnson, R. C., Young, S. K., Cotter, R., Lin, L., and Rowe, W. B., 1990, Medium-chain triacylglycerol lipid emulsion: Metabolism and tissue distribution, *Am. J. Clin. Nutr.* 52:502–508.

Kaminski, D. L., Adams, A., and Jellinek, M., 1980, The effect of hyperalimentation on hepatic lipid content and lipogenic enzyme activity in rats and man, *Surg.* 88:93–100.

Kayden, H. J., Senior, J. R., and Mattson, F. H., 1967, The monoglyceride pathway of fat absorption in man, *J. Clin. Invest.* 46:1695–1703.

Kenler, A. S., Swails, W. S., Driscoll, D. S., DeMichele, S. J., Daley, B., Babineau, T. J., Peterson, M. B., and Bistrian, B. R., 1996, Early enteral feeding in postsurgical cancer patients: Fish oil structured lipid-based polymeric formula versus a standard polymeric formula, *Ann. Surg.* 223:316–333.

Klemann, L. P., Aji, K., Chrysam, M. M., D'Amelia, R. P., Henderson, J. M., Huang, A. S., Otterburn, M. S., and Yarger, R. G., 1994, Random nature of triacylglycerols produced by the catalyzed interesterification of short- and long-chain fatty acid triacylglycerols, *J. Agric. Food Chem.* 42:442–446.

Lanza-Jacoby, S., and Tabares, A., 1990, Triacylglycerol kinetics, tissue lipoprotein lipase, and liver lipogenesis in septic rats, *Am. J. Physiol.* 258:E676–E685.

Lanza-Jacoby, S., Lansey, S. C., Cleary, M. P., and Rosato, F. E., 1982, Alterations in lipogenic enzymes and lipoprotein lipase activity during gram-negative sepsis in the rat, *Arch. Surg.* 117:144–147.

Lanza-Jacoby, S., Phetteplace, H., and Tripp, R., 1995, Enteral feeding a structured lipid emulsion containing fish oil prevents the fatty liver of sepsis, *Lipids* 30:707–712.

Levene, M. J., Wigglesworth, J. S., and Desai, R., 1980, Pulmonary fat accumulation after Intralipid infusion in the preterm infant, *Lancet* 2:815–818.

Ling, P. R., Istfan, N. W., Lopes, S. M., Babayan, V. K., Blackburn, G. L., and Bistrian, B. R., 1991, Structured lipid made from fish oil and medium-chain triacylglycerols alters tumor and host metabolism in Yoshida-sarcoma-bearing rats, *Am. J. Clin. Nutr.* 53:1177–1184.

Macfie, J., Smith, R. C., and Hill, G. L., 1981, Glucose or fat as nonprotein energy source? A controlled clinical trial in gastroenterologic patients requiring intravenous nutrition, *Gastroenterol.* 80:103–107.

Maiz, A., Yamazaki, K., Sobrado, J., Babayan, V. K., Moldawer, L. L., Bistrian, B. R., and Blackburn, G. L., 1984, Protein metabolism during total parenteral nutrition (TPN) in injured rats using medium-chain triacylglycerols, *Metabolism* 33:901–909.

Mascioli, E. A., Babayan, V. K., Bistrian, B. R., and Blackburn, G. L., 1988, Novel triacylglycerols for special medical purposes, *J. P. E. N.* 2(suppl):127S–132S.

Mascioli, E. A., Randall, S., Porter, K. A., Kater, G., Lopes, S., Babayan, V. K., and Blackburn, B. R., 1991, Thermogenesis from intravenous medium-chain triacylglycerols, *J. P. E. N.* 15:27–31.

McKenna, M. C., Hubbard, V. S., and Pieri, J. G., 1985, Linoleic acid absorption from lipid supplements in patients with cystic fibrosis with pancreatic insufficiency and in control subjects, *J. Pediatr. Gastroenterol. Nutr.* 4:45–51.

Mendez, B., Ling, P. R., Istfan, N. W., Babayan, V. K., and Bistrian, B. R., 1992, Effects of different lipid sources in total parenteral nutrition on whole body protein kinetics and tumor growth, *J. P. E. N.* 16:545–551.

Merolli, A., Lindemann, J., and Del Vecchio, A. J., 1997, Medium-chain lipids: New sources, uses, *INFORM* 8:597–603.

Metges, C. C., and Wolfram, G., 1991, Medium- and long-chain triacylglycerols labeled with ^{13}C: A comparison of oxidation after oral or parenteral administration in humans, *J. Nutr.* 121:31–36.

Mok, K. T., Maiz, A., Yamazaki, K., Sobrado, J., Babayan, V. K., Moldawer, L. L., Bistrian, B. R., and Blackburn, G. L., 1984, Structured medium-chain triacylglycerols and long-chain triacylglycerols are superior to physical mixtures in sparing body protein in the burned rat, *Metabolism* 33:910–915.

Nordenstrom, J., Jarstrand, C., and Wiernik, A., 1979, Decreased chemotactic and random migration of leukocytes during Intralipid infusion, *Am. J. Clin. Nutr.* 32:2416–2422.

Pomposelli, J. J., Moldawer, L. L., Palombo, J. D., Babayan, V. K., Bistrian, B. R., and Blackburn, G. L., 1986, Short-term administration of parenteral glucose-lipid mixtures improves protein kinetics in porta caval shunted rats, *Gastroenterol.* 91:305–312.

Seidner, D. L., Mascioli, E. A., Istfan, N. W., Porter, K. A., Selleck, K., Blackburn, G. L., and Bistrian, B. R., 1989,

Effects of long-chain triacylglycerol emulsions on reticuloendothelial system function in humans, *J. P. E. N.* **13:**614–619.

Sheldon, D. F., Peterson, S. R., and Sanders, R., 1978, Hepatic dysfunction during hyperalimentation, *Ann. Surg.* **113:**504–509.

Small, D. M., 1991, The effects of glyceride structure on absorption and metabolism, *Annu. Rev. Nutr.* **11:**413–434.

Smith, R. E., Finley, J. W., and Leveille, G. A., 1994, Overview of SALATRIM, a family of low-calorie fats, *J. Agric. Food Chem.* **42:**432–434.

Sobrado, J., Moldawer, L. L., Pomposelli, J. J., Mascioli, E. A., Babayan, V. K., Bistrian, B. R., and Blackburn, G. L., 1985, Lipid emulsions and reticuloendothelial system function in healthy and burned guinea pigs, *Am. J. Clin. Nutr.* **42:**855–863.

Summers, L. K., Frayn, K. N., Quinlan, P. T., Ilic, V., and Fielding, B. A., 1998, The effect of triacylglycerol-fatty acid positional distribution on postprandial metabolism in subcutaneous adipose tissue, *Br. J. Nutr.* **79:**141–147.

Sundstrom, G., Zauner, C. W., and Arborelius, M., 1973, Decrease in pulmonary diffusing capacity during lipid infusion in healthy men, *J. Appl. Physiol.* **34:**816–820.

Swails, W. S., Kenler, A. S., Driscoll, D. F., DeMichele, S. J., Babineau, T. J., Utsunamiya, T., Chavali S., Forse, R. A., and Bistrian, B. R., 1997, Effect of a fish oil structured lipid-based diet on prostaglandin release from mononuclear cells in cancer patients after surgery, *J. P. E. N.* **21:**266–274.

Swenson, E. S., Selleck, K. M., Babayan, V. K., Blackburn, G. L., and Bistrian, B. R., 1991, Persistence of metabolic effects after long-term oral feeding of a structured triacylglycerol derived from medium-chain triacylglycerol and fish oil in burned and normal rats, *Metabolism* **40:**484–490.

Teo, T. C., DeMichele, S. J., Selleck, K. M., Babayan, V. K., Blackburn, G. L., and Bistrian, B. R., 1989, Administration of structured lipid composed of MCT and fish oil reduces net protein catabolism in enterally fed burned rats, *Ann. Surg.* **210:**100–107.

Teo, T. C., Selleck, K. M., Wan J. M. F., Pompeselli, J. J., Babayan, V. K., Blackburn, G. L., and Bistrian B. R., 1991, Long-term feeding with structured lipid composed of medium-chain and *n*-3 fatty acids ameliorates endotoxic shock in guinea pigs, *Metabolism* **40:**1152–1159.

Tso, P., Karlstad, M. D., Bistrian, B. R., and DeMichele, S. J., 1995, Intestinal digestion, absorption and transport of structured triacylglycerols and cholesterol in rats, *Am. J. Physiol.* **268:**G568–G577.

Tso, P., Lee, T., and DeMichele, S. J., 1999, Lymphatic absorption of structured triglycerides vs. physical mix in a rat model of fat malabsorption, *Am. J. Physiol.* **277**(Gastrointest. Liver Physiol. 40):G333–G340.

Venus, B., Patel, C. B., Mathru, M., and Sandoval, E. D., 1984, Pulmonary effects of lipid infusion in patients with acute respiratory failure, *Crit. Care Med.* **12:**293.

Venus, B., Smith, R. A., Patel, C., and Sandoval, E., 1989, Hemodynamic and gas exchange alterations during intralipid infusion in patients with adult respiratory distress syndrome, *Chest* **95:**1278–1281.

Wardlaw, G. M., Snook, J. T., Park, S., Patel, P. K., Pendley, F. C., Lee, M. S., and Jandacek, R. J., 1995, Relative effects on serum lipids and apolipoproteins of a caprenin-rich diet compared with diets rich in palm oil/palm kernel oil or butter, *Am. J. Clin Nutr.* **61:**535–542.

Wheeler, E., 1995, Commercial potential for Salatrim, *INFORM* **6:**1156–1159.

Index

DATE DUE

OCT 1 2 2002	
SEP 2 3 2002	
APR 2 8 2003	
OCT 0 7 2008	

DEMCO INC 38-2971